PESTICIDES IN THE ENVIRONMENT

Volume 2

Edited by

ROBERT WHITE-STEVENS

CHAIRMAN, BUREAU OF CONSERVATION AND
   ENVIRONMENTAL SCIENCE
COOK COLLEGE
RUTGERS UNIVERSITY–THE STATE UNIVERSITY OF NEW JERSEY
NEW BRUNSWICK, NEW JERSEY

MARCEL DEKKER, INC. New York and Basel

MARCEL DEKKER, INC.
270 Madison Avenue, New York, New York 10016

LIBRARY OF CONGRESS CATALOG CARD NUMBER: 77-138499
ISBN: 0-8247-1783-X
Current printing (last digit):
10 9 8 7 6 5 4 3 2 1

PRINTED IN THE UNITED STATES OF AMERICA

PREFACE

This text, the second volume of <u>Pesticides in the Environment</u>, considers certain specific aspects of the practical business of pest control.

In Chapter 1 Drs. Dwight Powell and Malcolm C. Shurtleff discuss the spectrum of major fungal and bacterial diseases which attack domestic crop plants. It is believed that there are in excess of one hundred thousand species and strains of fungi, bacteria, mycoplasmas, and viruses which have been identified as phytopathogens on food, fiber, and ornamental cultivars. Obviously it would not be feasible to present in any detail, all such pests. The authors have, however, discussed the major groups of fungi and other crop-plant parasites and diseases from the standpoint of morphology (physical characteristics), etiology (life cycles), phytopathology (disease symptoms), economic significance (induced losses), and control, by management, use of resistant cultivars, and application of appropriate and effective chemical pesticides (fungicides and bacteriacides). Typical and classical diseases are presented in detail.

One omnipresent feature of plant disease control is emphasized—prevention or prophylaxis. Where a crop plant pathogen cannot be effectively prevented by management (sanitation, soil amendments, rotation of crops, etc.) or by the use of resistant cultivars then it must be prevented by chemical methods. Characteristically phytopathogens cannot be effectively controlled by applying chemical pesticides <u>after</u> infection has become established or by therapy. Once the pathogen has penetrated the host tissue, it will generally seriously damage if not destroy the crop plant. It may be prevented from spreading or disseminating to other, as yet uninfected plants, by a chemical treatment, but this is, of course, a protective or prophylactic measure rather than a therapeutic treatment.

The recent invention and development of systemic fungicides and bacteriacides discussed in Chapter 1, has been a major advance in plant disease therapy, but they too are much more effective when applied prophylactically ahead of pathogen exposure. The vascular system in plants is substantially less effective and less rapid in the transport of chemical compounds than is that of the vertebrate or invertebrate organism, and the therapeutic effect of such pesticides is correspondingly reduced. Furthermore a number of

iii

plant pathogens (Fusarium, Ceratocystis, Verticillium, Xanthomonas spp.)
characteristically occlude or plug the vascular cells and preclude the transport of effective concentrations of the chemical. This is the case particularly with such diseases as the Dutch Elm Disease (Ceratocystis ulmi).

By far the most effective method for prevention of crop-plant disease
is to plant resistant cultivars. This is widely practiced wherever such resistant varieties and strains are available. However such resistant cultivars do not preclude the total use of chemical pesticides as the resistance
is generally limited to one, or at the most, three or four pathogens while
others will readily infect unprotected resistant plants. The practical grower,
therefore, has to employ chemical pesticides to prevent damage from those
pathogens to which his crop varieties are not resistant. Also, nature continually reshuffles the genetic material of the pathogens and fortuitously
generates new strains which can overcome the resistance of the resistant
cultivar. This has occurred several times with wheat varieties initially
bred for resistance to the fearsome black stem rust of wheat (Puccinia
graminis tritici); and quite recently it also occurred in the Southern Corn
Leaf Blight Disease (Helminthosporium maydis) of corn, where resistant
cultivars of field corn suddenly, in a single season, succumbed to an epiphytotic of blight induced by a new strain of the pathogenic organism. In
both cases the plant breeders met the situation by developing revised strains
of wheat and corn that were in turn resistant to the new virulent strain of
the pathogen.

It does serve to illustrate, however, that the battle for mankind to produce food and fiber is a continuing one and presumably always will be. By
the same process nature can and does evolve new strains of pathogens that
are resistant to chemical pesticides, and this, in turn, requires continuing
research into new chemical fungicides and bacteriacides with different modes
of action.

The use of so-called biological control methods other than the use of
resistant cultivars and management procedures discussed above, have
proven to be neither effective nor practicable in the control of fungal bacterial, mycoplasmal, or viral diseases of crop plants. The principle of
"setting a thief to catch a thief" has proven to be somewhat effective in the
control of invertebrate pests (insects, mites, nematodes, rodents), but for
plant disease control it must be presently regarded as virtually unsuccessful.
From the relatively vestigial investigations carried out to date on the biological control of plant pathogenic diseases, there does not appear to be any
reasonable possibility of such methods replacing chemical controls in the
foreseeable future with any degree of practicable economic validity.

From the standpoint of man's survival, with the burgeoning billions
now present on earth and predictively into the future, the control of phytopathogenic diseases on food and fiber crops will have to rely largely upon
chemical pest prevention.

Chapter 2 by Dr. Joseph N. Sasser deals with the neglected, and until recently relatively obscure, area of nematode (round worm) infestation of crop plants. It has been said, perhaps hyperbolically, that if all the plants and animals on the landscape could be instantly dissolved away, with the exception of the nematodes, every tree, plant, grass, or living creature would be outlined in a seething mass of nematodes. Certainly, as Dr. Sasser points out, they are ubiquitous and can cause untold and unmeasured destruction of useful crops and livestock. As they are largely invisible to the naked eye, the damage they do goes generally unrecognized and is usually misinterpreted by those unfamiliar with symptoms of nematode injury. Unproductive areas in a potato, tomato, peanut, or alfalfa field have often been interpreted as areas of soil that have been struck by lightning which has allegedly destroyed the fertility of the soil, when, in fact, the area may well harbor a substantial population of root nematodes.

Dr. Sasser has written a concise and explicit discussion of nematode morphology and the etiology of the major economic species which infest crop plants. He has also discussed the various control procedures developed to date. These include soil management (pH control, sanitation), rotations, resistant cultivars, physical treatments (heat, flooding, etc.), and chemical controls.

Again biological control is still in its infancy in the control of pathogenic nematodes. A few interesting discoveries in the parasitism of nematodes by other nematodes and by certain soil fungi have indiciated that some nematodes can be destroyed by predators as might be expected under natural conditions. Whether such predators and parasites could be marshalled in sufficient numbers and applied to nematode infested soils or waters in effective ways to control the economic infestation of domestic plants is, at present, speculative and exceedingly dubious.

As Dr. Sasser clearly points out, control of nematodes on infested crop plants will have to rely largely on chemical nematicides for some time into the foreseeable future.

Chapter 3 by Dr. Richard H. Gruenhagen deals with the practical aspects of pest control among home garden, public parkland, and highway ornamental plantings. As this area of pest control involves the maintenance of the aesthetic appearance as well as the economic protection of nursery stock, and home and public building grounds, it becomes rather difficult to make specific recommendations as to the legal and safe use of pesticide compounds. Recent requirements of the revised Federal Insecticide, Fungicide, and Rodenticide Act (FIFRA) in the form of the Federal Environment Pesticide Control Act (FEPCA) divide the available roster of registered pesticidal compounds into (a) unrestricted—which can be legally applied by anyone within the limits specified on the label of the container and (b) restricted—which can only be applied by registered, licensed pesticide applicators, and pest control operators, also within both the Federal and State established limits specified on the label.

At this time those compounds which are classified as (a) or (b) have not been completely and legally established, so general recommendations applicable throughout the United States cannot legally be made. The home gardener should, therefore, refer to his local county agent or his state land grant college or experiment station as to which pesticide he can legally apply for the particular pest he wishes to control.

Characteristically the home gardener needs only one or two insecticides, acaricides, fungicides, and herbicides each of which can be applied to control a broad spectrum of pests on his home plantings. He needs a broad spectrum of safe, preferably odorless, insecticides which can control a wide variety of insect pests over perhaps a dozen or more different plant species. It is obviously impracticable and exceedingly confusing for him to have to apply one chemical to control aphids on roses, another to control Japanese beetle on virginia creeper, and still another to control bagworms on his junipers. The same can be said for his use of fungicides to control black spot on roses, mildew on cucumbers, or scab on apple trees. Apart from the confusion of which to apply to what, how and when, he would have to invest in an array of products, and in some cases buy different equipment for various formulations, but most importantly he could readily jeopardize the safety of his children, his pets, and even the wildlife which visit his garden. He needs a herbicide which will remove broad leaved weeds (dandelions, plantains, chickweed, etc.) from his grass lawn, and another that will remove annual grasses (crab), and perhaps a third which will remove all plant growth from the driveway and sidewalks.

It is also important for the home gardener to be provided with pesticide compounds—i.e., insecticides, acaricides, fungicides—which are chemically compatible so he can mix them together and spray or dust them coincidently to protect most of the various species of plants in his garden. Individual treatment of specific plants with particular pesticides is frustrating, expensive, very laborious, and impracticable for the home gardener. Of course, herbicides are never applied in combination with insecticides or fungicides, and indeed should never be applied with the same equipment even sequentially.

With the rapid rise in suburban living, the opening of increasing areas of public parklands and forests, and the extensive use of ornamental plantings along highways and thoroughfares, the control of pests and the real and unsightly damage they create becomes more important both aesthetically and economically.

Chapter 4 by Dr. Arthur J. Hackett outlines the role of pesticides in the control of ectoparasites in domestic livestock and pets. Livestock disease in the U.S. is estimated to be in excess of $2 billion per year, half of which, at least, is probably attributable to ectoparasites (mostly insects and acarids) many of which vector bacterial, mycoplasmal, and viral diseases to their animal hosts. In many foreign countries such ectoparasites and the diseases they transmit severely curtail such animal protein as meat, eggs, and milk, desperately needed by the protein deficient populace.

Throughout South America, Africa, and Asia countless millions of children suffer from Kwashiorkor, a protein deficiency syndrome that results in anemia, retarded mental development, and early death. Precisely what a worldwide program in the control of ectoparasites among domestic livestock would contribute to these protein hungry peoples cannot be accurately estimated, but it could doubtlessly contribute to the reduction of Kwashiorkor by at least 25 percent, saving the lives of literally tens of millions of humans.

In Africa, alone, there is an area of some 4 million square miles, equivalent to the continental limits of the U.S., which is largely uninhabitable by man or cattle due to the presence of the tsetse fly (Glossinia morsitans) and its relative (Glossinia palpalis), the first of which vectors the dread Nagana parasite (Trypanosoma brucei) to cattle and the second which carries the related parasite (Trypanosoma gambiense) sleeping sickness to humans. Eradication of these flies would open up vast new lands for arable culture to supply the food needs of the burgeoning population of this continent.

Within the past 40 years, the elucidation of the life cycles of the major ectoparasites has substantially contributed to the methods of control, some by management, but most by chemical pesticides. Dr. Hackett delineates these developments in some detail, at least with respect to the major livestock and pet ectoparasites.

By far the most important discovery has resulted from the research which led to the systemic insecticides and acaricides. These are discussed in detail. They can be fed in the diet or applied to the skin in various ways whence they become absorbed into the animal's bloodstream and migrate to the peripheral circulatory system. During the active phase the therapeutic margin of safety is sufficiently wide to be innocuous to the host animal and, coincidently, toxic to the ectoparasite. There is the added advantage that systemic insecticides or acaricides can effectively control parasites after they have penetrated the host's body because the compound moves freely throughout all tissues where blood fluids circulate. This was not the case with the earlier dip or brush-on pesticides.

Most of these systemic compounds are organophosphates; they become readily metabolized into innocuous derivatives in the animal system, leaving no toxic residues in the edible products at harvest.

These four chapters consider four aspects of the use of pesticides from the point of view of practical field usage. Subsequent volumes will be devoted to such topics as the practical application of pesticides in forest management, stored food products, weed control, crop-insect and acarid control, public-health maintenance integrated control, and the impact of pesticides on wildlife.

Robert White-Stevens

CONTRIBUTORS TO VOLUME 2

Richard H. Gruenhagen, Department of Plant Pathology and Physiology, Virginia Polytechnic Institute and State University, Blacksburg, Virginia

Arthur J. Hackett, Department of Animal Sciences, Rutgers University, New Brunswick, New Jersey

Dwight Powell, Department of Plant Pathology, University of Illinois, Urbana, Illinois

J. N. Sasser, Department of Plant Pathology, North Carolina State University, Raleigh, North Carolina

Malcolm C. Shurtleff, Department of Plant Pathology, University of Illinois, Urbana, Illinois

CONTENTS

CONTENTS OF OTHER VOLUMES

PESTICIDES IN THE ENVIRONMENT

Chapter 1

THE ROLE OF FUNGICIDES IN CROP PRODUCTION

Dwight Powell and M. C. Shurtleff
Department of Plant Pathology
University of Illinois
Urbana, Illinois

The successful use of fungicides to control plant disease(s) requires a broad knowledge of many factors. These involve the fungicide, the disease, the causal agent, and the host. Thus, much of this chapter will be devoted to the general aspects of phytopathology, the science of plant diseases.

One should understand what causes a plant disease. What types of causal agents infect plants? Where is their source? How are they disseminated from plant to plant within a field or garden, from area to area, or from country to country? Where does a pathogen winter over and what environmental factors favor its development? What type of preventive measure is needed? If a fungicide is to be used, what type will be most effective? Such questions are innumerable. It is our hope to provide a clear, concise picture of the field of phytopathology in an abbreviated form so that the use of fungicides as an aid in crop production will be meaningful.

I. THE HISTORY OF PLANT PATHOLOGY

Walker (225) presents an excellent summary of the history of phytopathology. Much of the following has been taken from his discussion.

Before the 19th century, it was not firmly established that microorganisms were responsible for certain plant maladies.

Prevost (1755-1819), a Swiss Professor of Philosophy, studied the wheat bunt disease for about 10 years, beginning in 1797. He published this work, Memoirs on the Immediate Cause of Bunt or Smut of Wheat, and of Several Other Diseases of Plants and on Preventitives of Bunt. Thus, he became the first investigator to demonstrate clearly the pathogenic nature of any microorganism to plants. Prevost studied and described spore germination and found that spores applied to wheat seed resulted in infection to the plant. He was able to demonstrate the complete life cycle of the fungus. Prevost showed that solutions containing copper sulfate prevented spore germination. This explained the basis for seed treatment, which had been practiced for a century without any rational explanation. Needless to say, the theory of spontaneous generation was so entrenched in the minds of men at this time that Prevost's results and interpretations were not accepted. Approximately 40 years later, Louis Rene Tulasne (1815-1885) and Charles Tulasne (1817-1884), began extensive morphological studies of fungi. In 1847 they published their studies on the wheat bunt fungus, which confirmed Prevost's work with spore germination and relation of the fungus to the host plant.

Anton De Bary (1831-1888) studied the development of a number of fungi in a critical manner. He traced the development of the mycelium, sporiferous branches, and teliospores (chlamydospores) of the corn-smut fungus. De Bary also examined several of the common rust fungi, including the black stem rust of cereals and rust of bean. He did much to confirm, support, and establish the so-called germ theory of disease. Such noted scientists as Pasteur and Robert Koch added to this general concept. In 1882 Koch announced the poured-plate method of isolating bacteria using gelatin as a liquifiable solid medium. Koch, working with animal diseases caused by bacteria, developed Koch's Postulates for proving that a bacterium would cause a certain disease. He required that: (a) The organism must be associated in every case with the disease, and conversely the disease must not appear without it; (b) the organism must be isolated in pure culture and its specific morphological and physiological characteristics determined; (c) when the host is inoculated with the organism under favorable conditions, characteristic symptoms of the disease must develop; and (d) the microorganism must be reisolated and identified as that first isolated from the diseased host. While Koch worked with animal parasites, these postulates are one of the early lessons given a young student in phytopathology, since the same principles are applicable to plant diseases caused by bacteria and fungi.

Kuhn (1825-1910) was as important as De Bary in the early development of phytopathology. He published his epochal textbook in 1858, entitled, The Diseases of Cultivated Crops, Their Causes, and Their Control. This was the first text to be published in which fungi were regarded as causal factors in plant diseases. Kuhn had a great influence on the development and application of control measures. His work supplemented that of Prevost and De Bary and demonstrated entrance of the mycelium of the bunt fungus into wheat seedlings. He also followed its development in the host from that point to the formation of spores in the wheat "kernel."

While most botanists and phytopathologists were studying the fungi as causal agents of plant diseases, T. J. Burrill (1839-1916) (31), a Professor of Botany at the University of Illinois, in 1881 was the first to relate a bacterium as the causal agent of the fire blight disease of apple and pear. This work was confirmed by Arthur (8). Erwin F. Smith (1854-1927), soon after 1890, began to study bacterial diseases of plants and made a major contribution in 1895 (196).

Dmitrii Ivanowski (1864-1924), a Russian investigator who became interested in a mosaic disease of tobacco, noted that when he passed the juice from an infected plant through a Chamberland filter candle, the filtrate retained its infectivity. Since this filtering had removed bacteria and all other known microorganisms it was obvious (but not to him) that he had discovered a new type of infectious entity (Fig. 1). In 1897, M. W. Beijerinck (1851-1931), a Dutch bacteriologist, took up the study of tobacco mosaic. He found that the juice of infected plants, when filtered through porcelain, was sterile but still infectious. He demonstrated that the contagious entity was increased in some way in the infected plant. This separates it from the classification of a toxin or toxic substance susceptible to dilution and which would not

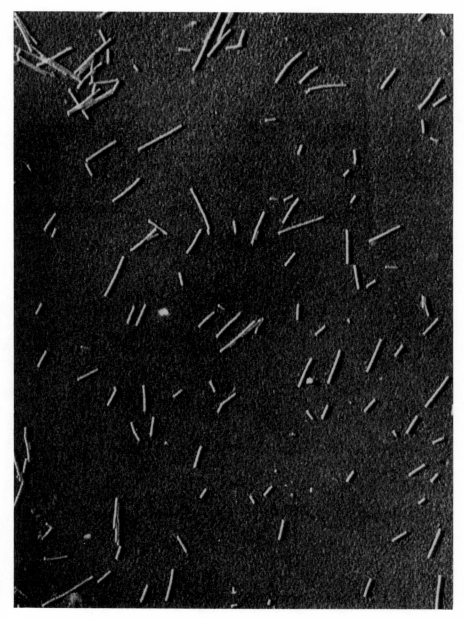

FIG. 1. Particles of the tobacco mosaic virus (15 × 300 nm). (Courtesy of H. H. Thornberry, University of Illinois.)

increase in the plant.   He subsequently referred to this entity as a virus
(225).

It is unnecessary to give a detailed account of the history of fungicides
in this chapter.   Horsfall (103) has given an excellent chronological annota-
tion of the landmarks in the history of fungicides.   Most fungicide usage
prior to 1850 was strictly on a trial and error basis.   Botanists had found
that three inorganic materials, copper sulfate, mercuric chloride, and
sulfur were effective in preventing certain plant maladies.   In some cases
the association between the use of fungicides and the control of a disease was
very definite and significant even though the nature of the disease was unknown.   For example, Robertson, in 1824, reported that sulfur was specific
for powdery mildew of peach (103).   Yet, the fungus causing this disease
was not studied until almost 50 years later by De Bary in 1870 (16).   Sulfur
is one of the two principal fungicides used today to control powdery mildew
diseases.

Sulfur is one of the earliest chemicals used on plants, probably because
it is a natural product and easily obtainable.   Horsfall (103) tells of Homer,
about 1000 B.C., referring to "the pest-averting sulfur."   In later years,
sulfur was used as flowers of sulfur, which are small crystals produced by
sublimation.   Because sulfur is neither soluble nor suspendible in water, it
was used only in dust form.   By supplying adjuvants having various properties such as wetting, spreading, and dispersion, sulfur powders were made
wettable.   In other words, they could be made to suspend in water.   Wilcoxon
and McCallan (230, 231) demonstrated an inverse relationship between fungitoxicity and particle size of sulfur.   They also showed that the smaller the
particle the more tenacious the residue.   Thus, where most sulfur formulations had particle sizes of approximately 25 to 50 μm sulfurs with particles
below 10 μm were now developed.

A variation of the microfine wettable powders was the development of
flotation paste sulfur.   This was a byproduct of fuel gas from coal.   When
bituminous coal is carbonized to produce coke and gas, the sulfur in the
coal is driven off as hydrogen sulfide.   This is removed by absorption in an
alkaline solution that, when oxidized with a catalyst, yields elemental sulfur in a finely divided state and is recovered by a flotation process (71).
Sauchelli (190) showed that 95% of the sulfur particles in flotation paste were
not over 3 μm in diameter.   Flotation paste sulfur was sold to fruit growers
during the 1930s and 1940s in 50-gal drums weighing about 500 lb.   The
paste contained 50% sulfur and 50% water.   It would suspend in water as long
as it remained wet.   When allowed to become dry, it caked badly and would
not rewet.   Because of its small particle size it was a very effective and
superior sulfur fungicide but was difficult to handle physically and thus had
a short period of popularity.   A dehydrated flotation sulfur was never comparable to the paste in fungicidal effectiveness.

Lime-sulfur is the supernatant liquid obtained from the chemical reaction of calcium hydroxide, sulfur, and hot water. It is not just a combination of lime (calcium hydroxide) and sulfur in water, as the name signifies. While the active ingredients used for the preparation of lime-sulfur are relatively simple, the chemistry of this mixture is extremely complex. It is thought to be fungicidally active as calcium polysulfide (71). It was first developed by Kenrick, in 1833, who proposed a self-boiled lime-sulfur for grape powdery mildew (104). It was not until 1851 that Grison, a gardener in France, showed that the solution resulting from such a mixture was fungicidal. The mixture was then exploited and called Eau Grison (French for "water of Grison"). Lime-sulfur still did not become popular, probably because it is extremely phytotoxic to the foliage of grapes and many other plants unless it is used in very low doses. Hale, in 1888 (103), showed that lime-sulfur would control peach leaf curl when applied during the dormant period. From that period, until the advent of organic fungicides starting about 1940, lime-sulfur was one of the most widely used fungicides.

Copper sulfate was also found to be a good fungicide during the mid-18th century. Due to phytotoxicity, its use was confined to seed treatment. Proust, in 1800, first studied the chemistry of copper sulfate on wheat seed by the addition of lime. Millardet, in Bordeaux, France, in 1882 discovered that Proust's mixture was effective in protecting grape foliage from the downy mildew disease (103). This mixture was called bordeaux mixture. From a dosage standpoint it is always designated by a three-numbered formula such as 2:4:100. This means 2 lb of copper sulfate, and 4 lb of hydrated lime in 100 gal of water. Many such copper sulfate-hydrated lime-water ratios such as 1/2:1:100, 8:8:100, and so on are used depending on the crop and disease involved. This mixture was by far the most popular fungicide from its discovery until the advent of the organic fungicides in the 1940s.

As with lime-sulfur, bordeaux mixture was found to be extremely complex chemically. Pickering (165), however, did an excellent study of this mixture and concluded that tribasic copper sulfate was eventually formed. Other workers (28, 141, 232) have supported this view.

Copper sulfate is water-soluble but when hydrated lime is added to the copper sulfate solution an insoluble precipitate is formed. Thus, many studies have been made through the years to produce an insoluble copper, which would be as effective as bordeaux mixture yet would not need a protective agent such as lime. Insoluble or neutral coppers as tribasic copper sulfate, COCS (copper oxychloride sulfate), copper phosphate, and copper oxides were developed. All had special uses and found a successful market, but none were considered as effective fungicidally as bordeaux mixture. A point of historical value is the work of Heuberger and Horsfall (98) on cuprous oxide. They found that the colors ranged from red through yellow and orange to green to brown, as controlled by the presence or absence of cupric oxide. Cuprous oxide, in the yellow form, had the smallest particles and was the most effective fungicide. Both the direct fungicidal and protective value of the cuprous oxide varied with particle size; the smaller the particle the more effective.

Attempts to utilize copper as inorganic complexes or in organic structures have met with considerable success (92, 171).

Mercuric chloride (corrosive sublimate) was first used as a wood preservative by Homberg in 1705 (103). In 1890 (119) it was used as a wheat seed treatment. It is regarded as both an effective disinfectant and disinfestant. The very toxic nature of this material to humans and animals has limited its use.

Mercurous chloride (calomel) is fairly insoluble in water, thus it does not have the fungicidal potential of mercuric chloride. It is used more as a dust or suspension in water. It found widespread use as a turf fungicide, especially in combination with mercuric chloride (152).

Riehm, in 1913, used an organic mercury as a seed treatment for wheat bunt (103). Many such mercury compounds were developed and found very useful for seed treatments, turf disease control, and so forth. Howard (105) found that certain organic mercuries were suitable for foliar fungicides. These were cationic, water-soluble compounds and very effective as contact fungicides. One of the first compounds was phenylmercuric triethanol ammonium lactate marketed as Puratized Agricultural Spray. Later it was shown that phenyl mercury acetate (PMA) was very effective and easy to formulate as a wettable powder. It soon became a very popular fungicide. Winter (234) showed that phenyl mercury compounds caused fruit thinning of apples while the methyl mercurys did not. However, the latter were never as effective as fungicides. With the introduction of the Miller Amendment in 1953 to the Federal Food, Drug, and Cosmetic Act (Public Law 518), mercury was given a zero tolerance and market acceptance of organic mercury fungicides for food crops dwindled to practically nothing. In the United States they have now virtually banned all uses of fungicides in the environment.

In 1934, Tisdale and Williams (211) reported the fungicidal activity of the dithiocarbamates. Martin (103), independently, made the same discovery. This started a new era that introduced the organic fungicides to crop production. Thiram came first followed by ziram and ferbam (103). Goldsworthy et al. (79) published on the use of the dialkyl dithiocarbamates as fungicides. Dimond et al. (59) introduced a water-soluble sodium ethylenebis dithiocarbamate (nabam) which produced an insoluble residue. In 1947, Heuberger et al. (97) added zinc sulfate to a nabam solution and showed that zineb (zinc ethylene bisdithiocarbamate) was formed and was an effective fungicide. The importance of this group of fungicides is emphasized by the fact that they controlled fungi that heretofore had not been effectively controlled by sulfur or copper compounds. Powell et al. (181) showed that ferbam was exceptionally good for the control of both cedar-apple rust and apple blotch. Zineb has proven to be as effective and is now widely used for this purpose. Maneb (manganese ethylene bisdithiocarbamate) also has proved to be an excellent fungicide and is used extensively on vegetables. Maneb, zineb, and ferbam are used to control many diseases of trees, shrubs, and flowers.

In 1940, Cunningham and Sharvelle (51) introduced chloranil as a seed treatment. In the next few years important fungicides such as dichlone (1943); glyodin (1946); and captan (1952) entered the agricultural field (103).

Dodine (Cyprex) was studied and developed by Hamilton starting in 1954 (87). It was introduced in 1956 by the American Cyanamic Company for experimental testing. Powell et al. (182) showed that it was effective in controlling apple scab under field conditions. Cation (38) in 1959 showed that it was extremely effective for control of cherry leaf spot. Dodine has developed into one of our most popular apple and cherry fungicides. It is also used to control numerous diseases of trees and shrubs.

Dinocap (Karathane) was shown to be specific for the control of powdery mildew by Sprague in 1949 (200). With sulfur, it is one of the most widely used fungicides for powdery mildew diseases.

Dicloran (Botran) was first described by Clark et al. (43) in 1960 as a new fungicide active against Botrytis spp. Ogawa et al. (161) in 1961 showed this compound to be effective against Rhizopus rot of sweet cherries. Powell (172) found that field application of dicloran (Botran) would control postharvest Rhizopus rot of peaches but would not prevent apple scab or Botryosphaeria fruit rot of apple.

The antibiotics have not developed as rapidly as expected when cycloheximide (Actidione) and streptomycin were shown to have promise. Cycloheximide was the first antifungal antibiotic to be reported (66). Previously all antibiotics were shown to be bactericidal. Ford et al. (69) have presented an excellent summary and bibliography of cycloheximide through 1958. While it was highly fungicidal and controlled cherry leaf spot at dosages as low as 1 to 2 parts per million (ppm) (37, 39), it was also highly phytotoxic. This has probably limited its commercial use as much as any other factor. Cycloheximide is now formulated with other fungicides, such as captan and thiram, to increase its protective qualities as well as to "safen" it from being phytotoxic.

Streptomycin, a bactericide, was first reported to be effective in reducing the fire blight disease of apple in 1952 by Murneek (154). The following year many papers were published which supported Murneek's studies (7, 80, 96, 235). Gray (83) showed that streptomycin was more effective when used with a humectant such as glycerol. Streptomycin is still the only effective bactericide for fire blight. Its effectiveness is markedly improved when applied at night (173), a finding thought to be due primarily to its increased absorption by the host.

Oxytetracycline, first studied by Goodman (81), was mixed with streptomycin formulations to forestall the development of strains of bacteria resistant to streptomycin. It was shown to have promise in controlling bacterial spot of peach by Dunegan et al. (65). This work was revived by Keil (117) who attempted to develop a more effective formulation. Oxytetracycline-streptomycin formulations are also used as bare-root dips by nurserymen to control crown gall, a serious bacterial disease of woody plants.

II. CAUSES OF PLANT DISEASES

In the study of plant pathology it is important to distinguish between the "disease-causing organism" and the disease. The disease-causing organism is generally referred to as the pathogen. When a pathogen infects the host and symptoms appear, the host is considered diseased, or disease has been produced. The expression, "the disease is spreading to other plants," is heard often. Actually, the pathogen is spreading from plant to plant and causing the disease. When fungicides are applied, it is for the purpose of destroying the pathogen and preventing the disease. We do not, however, adhere too closely to this concept and commonly speak of controlling the disease.

A diseased plant is one that is not normal considering the environmental conditions of its growth and symptoms, or one in which signs of either a morphological or a physiological nature are visible. Under this definition the cause of plant diseases can be extremely wide and varied. Phytopathologists consider that plant diseases may be either noninfectious or infectious. Noninfectious diseases may be caused by low or high temperatures, an excess or deficiency of water or light, chemical or mechanical injuries, abnormal atmospheric conditions, unfavorable soil-moisture relations, mineral excesses and deficiencies, etc. (Figs. 2 through 4). Here the disease or causal agent cannot be carried from one plant to another nor can it be transmitted from an unhealthy to a healthy plant.

Infectious diseases are those in which the causal agent or pathogen can be disseminated from plant to plant to cause new infections. These diseases are caused by fungi, bacteria, viruses, mycoplasm, and nematodes. This chapter is concerned only with plant diseases caused by fungi and bacteria.

A fungus is one of a large group of primitive plants that have no leaves, flowers, or chlorophyll (Fig. 5). They generally reproduce by the formation

FIG. 2. Lightning damage in a soybean field.

FIG. 3. Potash deficiency of soybean.

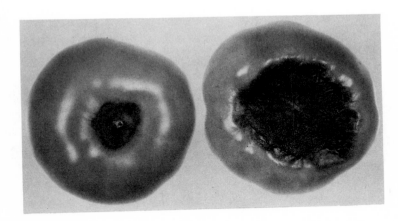

FIG. 4. Blossom-end rot of tomato caused by calcium deficiency.

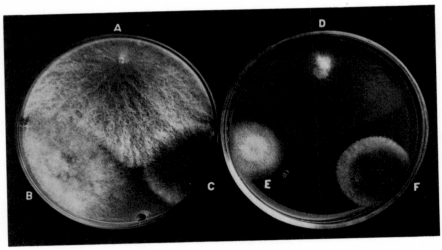

FIG. 5. Several fungi growing on agar in Petri plates from bits of corn tissue. (A) Diplodia zeae; (B) Nigrospora oryzae; (C) Gibberella zeae; (D) Sclerotium or Rhizoctonia bataticola and Fusarium moniliforme; (F) Helminthosporium sp.

of spores. Since they do not contain chlorophyll, they must grow on either living or dead plant or animal tissue. If a fungus can survive on dead plant debris it is called a saprophyte. If it infects living plant tissue it is called a parasite. Phytopathologists prefer pathogen to parasite since the latter has a common usage with organisms that infect animals. Some fungi invade both living and dead plant tissue. These are referred to as facultative pathogens. Many fungi will grow only on living plants and they are called obligate pathogens.

   Basically a fungus has five main morphological structures. The sporophore, or spore-bearing structure, produces spore(s) homologous to the seed of higher plants. When a spore germinates (Fig. 6) it sends out a germ tube or root-like structure called a hypha. This is a fungus filament that grows and branches to form a mass of hyphal threads referred to as mycelium. As the mycelium grows and ages it becomes a thallus; "a general term for the vegetative portion of a nonvascular plant" (198).

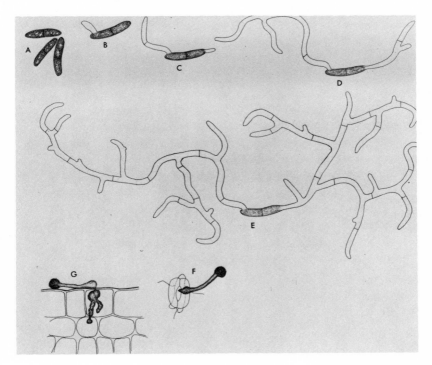

FIG. 6.   (A-E) Stages in the germination of a spore of Diplodia zeae; (G) germination and/or formation of an appressorium above the point of direct penetration, and mycelium within two host cells; (F) germination followed by stomatal penetration.

There are many variations of these basic structures upon which these fungi are classified. Alexopoulos (5) points out "that within a morphological series, parasites (pathogens) are considered more advanced than saprobes, (saprophytes), obligate parasites (pathogens) more advanced than facultative ones, and highly specialized obligate parasites (pathogens) more advanced than less specialized species."

Fungi are taxonomically grouped into four main classes. The Phyco-mycetes (Fig. 7) are characterized by nonseptate hyphae; the asexual spores are produced in a structure called a sporangium. Sporangia are borne at the tips of specialized hyphae called sporangiophores. Sporangiospores, arising from them are either motile or nonmotile. If motile, they possess

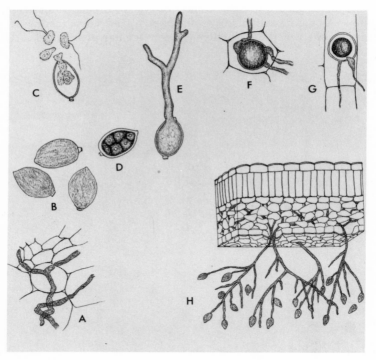

FIG. 7. Structures of Phycomycetes. (A) somatic mycelium within host cells; (B) typical, lemon-shaped sporangia; (C) release of zoospores from a sporangium; (D) formation of spores within a sporangium; (E) my-celium (germ tube) germination of a sporangium; (F) oogonium with three antheridia; (G) dormant, thick-walled oospore and remnants of antheridium and hypha; (H) sporophores with sporangia of Phytophthora infestans (the late blight fungus) emerging from the undersurface of a potato leaf.

two flagella that help them to move through a water medium. The motile spores are called zoospores. While they may be more commonly produced by aquatic Phycomycetes they are also formed by terrestrial species. Zoospores are dependent upon free water for motility. If nonmotile, the spores are called aplanospores. These are borne mostly on terrestrial Phycomycetes and are largely disseminated by wind. Sexual spores develop as a part of the life cycle of these fungi and are usually considered resting spores that carry the fungus through the winter or other dormant periods. While the Phycomycetes are either aquatic or soil-inhabiting fungi they may also infect certain terrestrial angiosperms. The soil-inhabiting fungi include some very important genera, such as Pythium, Phytophthora, and Aphanomyces, which are responsible for many root diseases. The terrestrial forms include notably the downy mildews.

The second class of fungi are the Ascomycetes. These fungi have septate hyphae and specialized sexual and asexual spore-bearing structures. The sexual fruiting structures are called ascocarps, which produce asci and ascospores. There are three main types of ascocarps (Fig. 8): (a) Cleistothecia are closed, round, often black structures usually found on the leaf surface late in the season. Upon maturing the wall disintegrates and releases one or more asci and ascospores (e.g., powdery mildew) (Figs. 8 and 9). (b) Perithecia are ostiolate, oval or round bodies imbedded in plant tissue (Fig. 10). Usually they are slightly pyriform (pear-shaped) with the beaked portion of the perithecium extruding, ever so slightly, above the surface of the epidermis. In the immature stage one can find numerous sac-like structures which are called asci (Figs. 8 and 11). As each ascus matures it generally contains eight ascospores. Ascospores are discharged through the ostiole. One perithecium may contain as many as 400 ascospores. Usually the ascospores are discharged when the mature perithecium absorbs water during spring rains. (c) The third type of ascocarp is called an apothecium, a disc or saucer-shaped structure in which the asci are produced in the open (Figs. 8 and 12). Here, also, each ascus contains eight ascospores. Apothecia may or may not be stalked. In case they are not stalked they are imbedded superficially in the leaf tissue.

Since the powdery mildews are obligate pathogens, the cleistothecia are always formed on the surface of the living leaf or other plant tissue. Perithecia and apothecia, however, may develop only saprophytically. The asexual stage of these fungi are usually passed upon the living plant.

There are many types of macroscopic, asexual fruiting structures (Fig. 13): (a) The pycnidium is a pear shaped, hollow, chitinous-like structure lined with conidiophores and, except for the ostiolate portion, is usually imbedded in plant tissue. It produces spores called conidia. These ooze in a tendril-like fashion from the ostiole during very humid or wet periods. (b) Acervuli are typically flat and open, containing a bed of short conidiophores growing side by side and arising from a thallus (Figs. 13 and 14).

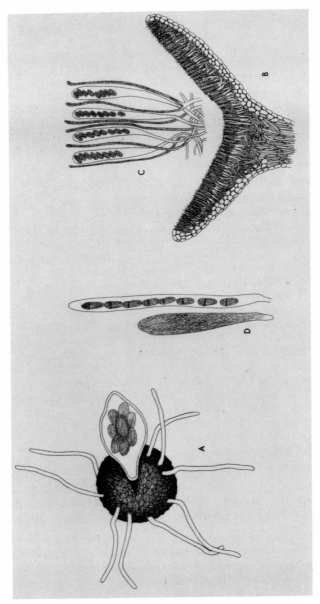

FIG. 8. Types of ascocarps. (A) cleistothecium of a powdery mildew fungus (Sphaerotheca) showing the flexous appendages and an ascus with eight ascospores, (B) cross-section of an apothecium of the brown rot fungus (Monilinia) with a row of asci lining the "cup"; (C) four asci (containing eight ascospores) and sterile hyphae of Monilinia; (D) an immature ascus of Venturia inaequalis and one ready to discharge its eight ascospores.

FIG. 9. Close-up of powdery mildew (<u>Erysyphe graminis</u>) on a wheat leaf. The black specks are cleistothecia.

Conidia are borne at the tips of conidiophores. Variations exist, such as acervuli remaining fairly well closed until maturity when opening occurs. (c) <u>Sporodochia</u> can be characterized as a cluster of conidiophores arising from a single narrow base with conidia borne on the tips. These appear to be analogous to a bouquet of flowers in a vase. They are characteristic of the fungus that causes brown rot of fruit. In some fungi, conidia are produced vertically in chains on the conidiophores without specialized structures, such as in the powdery mildews (Fig. 15). In another group conidia are budded from the ascospores. Examples of the Ascomycetes include many of our most important pathogenic fungi such as those that cause Dutch elm disease, oak wilt, chestnut canker, apple scab, brown rot of stone fruits, cherry leaf spot, bitter rot, and numerous leaf spot diseases.

The <u>Fungi Imperfecti</u> is a third class of fungus that logically follows the Ascomycetes (with some exceptions). As far as anyone can determine they lack a sexual or perfect stage and reproduce only by means of asexual structures. They have essentially the same asexual fruiting structures that are found in the Ascomycetes. It is presumed by mycologists that, with relatively few exceptions, they represent "the conidial stages of the Ascomycetes whose ascigerous stages are either rarely formed, or have been dropped from the life cycle in the evolution of these organisms" (5).

As mentioned above, the sexual (perithecia and apothecia) stages of the Ascomycetes are commonly found in vegetative debris while the asexual conidia develop on living host tissue. For this reason many fungi have been

FIG. 10. Dead wheat head covered with perithecia of <u>Gibberella zeae</u>, the scab fungus.

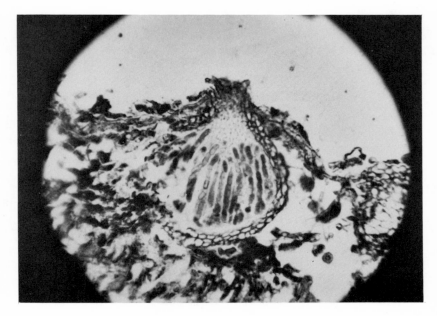

FIG. 11. Cross-section through a perithecium of the apple scab fungus (Venturia inaequalis) showing mature asci and ascospores.

FIG. 12. Brown rot apothecia ready to discharge ascospores of Monilinia fructicola.

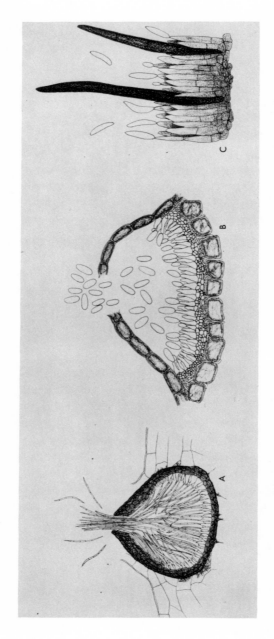

FIG. 13. Asexual fungal fruiting structures. (A) pycnidium of Septoria; (B) acervulus of Gloeosporium; (C) sporodochium of Volutella.

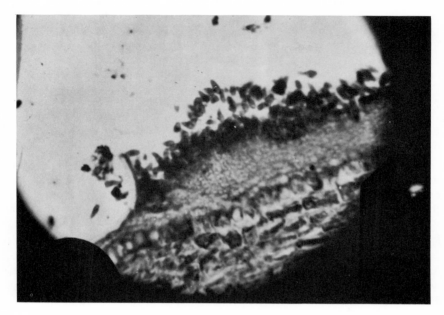

FIG. 14. The acervulus stage of the apple scab fungus (Venturia inae-qualis) on an apple fruit. Note the conidia or summer spores that have been released.

FIG. 15. Powdery mildew on a cucumber leaf caused by Erysiphe cichoracearum.

assigned two different scientific names. The fungus causing apple scab was found first on the living host and the conidia were described and named Fusicladium dendriticum (Wal.) Fcl. Years later the perithecial stage was found developing in dead apple leaves. This phenomenon was shown to occur during the winter months. After a thorough study, the ascospores from these perithecia were shown to infect apple leaves. These infections soon produced the Fusicladium spores. A new name, Venturia inaequalis (Cke.) Winter, was assigned to this stage of the fungus. Earlier the apple scab fungus belonged to the Fungi Imperfecti but now, with the discovery of the ascigerous stage, it is a member of the Ascomycetes.

The International Rules of Botanical Nomenclature states that a fungus may have only one name and Venturia inaequalis was chosen as the valid name. The use of the conidial name has some advantages, however, and the International Botanical Congress in 1950 decided to legalize the use of form-names for conidial stages, still recognizing the name of the sexual stage as the official name of the fungus (5).

The Fungi Imperfecti have a form order Mycelia Sterila that contains fungi that do not produce spores or, at least, spores have never been observed. These fungi are identified by their hyphal characteristics, and sclerotia. A sclerotium is normally considered a resting body.

The genus Rhizoctonia was created in 1815 by DeCandolle (16). Through the years many species were assigned to this genus, and one was described by Kuhn as R. solani (16), the cause of a common potato disease. Walker (225) states, "the relation of R. solani to the Basidiomycetes (the fourth class of fungi to be discussed later) was not established until early in the twentieth century. The basidial stage is, in fact, an inconspicuous saprophytic stage of the pathogen." While the basidial stage had been described, none of the investigators associated it with R. solani. Rolfs, in 1903, was the first to germinate the basidiospores and secure a fungus identical with R. solani (Fig. 16) (225).

Alexopoulos (5) states, "The Fungi Imperfecti are therefore, conidial stages of Ascomycetes, or more rarely, Basidiomycetes, whose sexual stages have not been discovered or no longer exist."

There are many important diseases caused by these fungi such as apple blotch, vascular wilts caused by Verticillium and the complex genus Fusarium, the many Alternaria diseases, etc. (Figs. 17 through 20). We will mention many more such diseases as we discuss control programs.

The Basidiomycetes are the most advanced and complex of all the fungi and are thought by many to have originated from the Ascomycetes. These fungi include such popularly known forms and diseases as the mushrooms (toadstools), puffballs, stinkhorns, most wood-rotting fungi, rusts, and smuts. All types of host relationships exist. The rusts are obligate

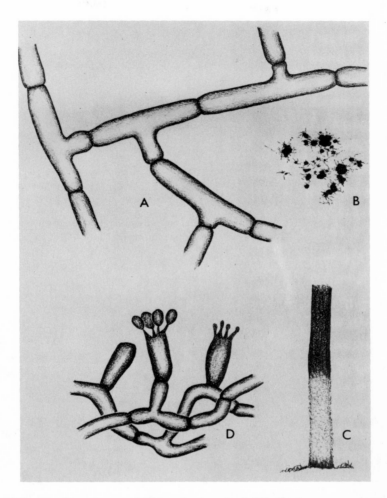

FIG. 16. Structures of Rhizoctonia solani (<u>Thallephorus cucumis</u>).
(A) typical thick-walled, barrel-shaped cells of <u>Rhizoctonia</u> mycelium showing the typical right-angled branching; (B) sclerotia on the surface of a
potato tuber; (C) the sexual stage (<u>Thallephorus</u>) fruiting on a tomato stem;
(D) three basidia, immature to mature with four basidiospores.

FIG. 17. Anthracnose of muskmelon caused by the fungus <u>Colletotri-chum lagenarium</u>. (Courtesy of the Department of Plant Pathology, University of Minnesota.)

FIG. 18. Verticillium wilt of tomato caused by the fungus <u>Verticillium albo-atrum</u>.

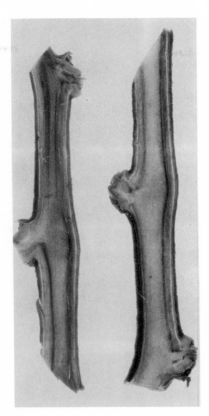

FIG. 19. Tomato wilts. Discolorations typical of (left) Fusarium wilt;
(right) Verticillium wilt.

pathogens, the smuts are facultative, while mushrooms (toadstools), and
most wood-rotters, etc., may be facultative, or secondary pathogens, or
live only saprophytically.

The order Uredinales contains the rust fungi. These have a promycelium which arises from the germination of a teliospore, a diploid cell.
During its development, reduction-division occurs to form haploid nuclei.
The promycelium normally becomes three-septate so that four uninucleate
cells result, two with <u>plus</u> nuclei and two with <u>negative</u> nuclei. These
soon migrate into lateral buds that enlarge to become <u>basidiospores</u>. In the
case of <u>Puccinia graminis</u> Pers., which causes black stem rust of cereals
and grasses, the basidiospores are carried by air to barberry (Fig. 21).
The barberry leaf is infected on the upper surface and <u>pycnia</u> develop. Each
pycnium contains <u>spermatia</u> (pycniospores), which are the male gametangia,

FIG. 20. Early blight of tomato (<u>Alternaria solani</u>). Infection causes
yellowing and early drop of leaves.

and <u>receptive hyphae,</u> which are the female organs. Spermatia are incompatible with the receptive hyphae of the same pycnium. For fertilization,
a pycniospore from a plus pycnium must unite with a receptive hypha from
a minus pycnium or <u>vice versa</u>. The spermatia are exuded from the ostiole
of the pycnium in a nectar-like droplet, which is thick, sticky, and sweet.
Insects are believed to play an important role in disseminating the spermatia.

When a compatible union of a spermatium and receptive hypha occurs,
a mycelial mat of binucleate cells develops in the leaf tissue. This results
in the development of the <u>aecium</u> (Fig. 22).

The aecium is always borne on the lower surface of the leaf. It is
described as a group of binucleate hyphal cells within an infected host.
These cells give rise to chains of aeciospores by successive and conjugate
divisions of the nuclei. The aecium appears as a small inverted cup (aecial
cup) after the aeciospores start to be liberated. When aeciospores are released they are carried by wind back to the cereal crops. Infection occurs
and binucleate mycelium develops in the host tissue eventually to form the
<u>uredium</u> (Fig. 23).

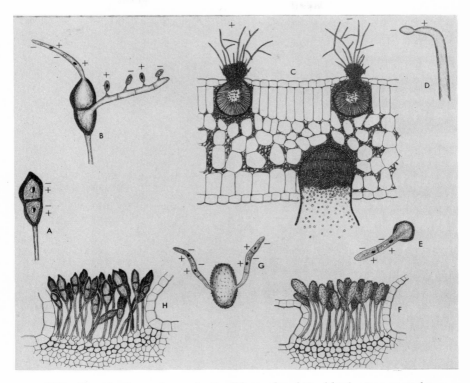

FIG. 21. Various stages in the life cycle of the black stem rust fungus, Puccinia graminis. (A) reduction-division of the nuclei in the teliospore; (B) the germinating teliospore with promycelium or basidium and basidiospores; (C) cross-section of a barberry leaf showing two pycnia with their receptive hyphae on the upper surface, and an aecium (cluster cup) on the undersurface showing chains of aeciospores; (D) union of a basidiospore and tip of a receptive hypha; (E) germination of an aeciospore with its dicaryotic (N + N) mycelium; (F) longitudinal section of the red rust stage—growth of the uredospores has ruptured the epidermis; (G) germination of a uredospore; (H) section through a pustule of the black rust stage—the teliospores form in the same pustules as the uredospores as the cereal or grass host matures.

FIG. 22. Black stem rust (aecial stage), <u>Puccinia graminis</u>, on under-surface of barberry leaves. (USDA Photograph.)

The uredium is a fruiting structure very similar to the acervulus of the Ascomycetes. It produces the uredospores, which are also binucleate. The uredospores are the most effective means of dissemination. These spores form in early spring and continue forming until the grain or grass is mature, spreading the fungus from plant to plant. They are comparable to conidia since they are the only repeating stage.

As the grain matures, the uredia develop into another fruiting struc-ture called the telium which begins to produce two-celled, binucleate telio-spores. Occasionally telia are produced directly from mycelium resulting in late uredospore infections (5). Thus, the life cycle has been completed since the teliospores give rise to the promycelium, etc. We have described a rust fungus which has five distinct stages or spores; teliospores, basidio-spores, pycniospores, aeciospores, and uredospores (Fig. 21).

As one might suspect, there are variations from this so-called long cycle and some rust fungi have only three or four spore types. Uredospores

FIG. 23. Black stem rust (uredia) on wheat culms and spikelets, caused by the fungus Puccinia graminis tritici.

are not found in some of the rusts. For example, the fungus causing the
cedar-apple rust, Gymnosporangium juniperi-virginianae Schw. does not
have uredospores. Here the cedar gall (Fig. 24) (found on Juniperus spp.
or on red cedar) produces the teliospores (Fig. 25) which give rise to the
basidiospores (Fig. 26). These are disseminated by wind to nearby apple
trees where they infect the upper surface of the leaves (Fig. 27). Fruit infections (Fig. 28) may also develop, but these are generally terminal-type
infections that mar the fruit but do not function to continue the life cycle.
When leaves become infected, pycnia form, ooze pycniospores, and aecia
develop on the undersurface of the leaf giving rise to aeciospores. These
are transported by wind to infect Juniperus spp. where a gall is initiated.
After 2 years, the gall is ready to produce teliospores. This rust does not
produce uredospores (Fig. 29).

Since the rusts are obligate pathogens many of them are highly specialized and infect only one host. This is an autoecious or monoecious rust.
An example is the fungus Gymnoconia interstitialis (Schl.) Lag., which
causes the orange rust disease of brambles. Some rusts require two distinct hosts, which are usually of a completely nature botanically, to complete
their life cycle. These are called heteroecious. In addition to the wheat
stem rust and cedar-apple rust described above another good example of a

FIG. 24. Cedar-apple rust galls on red cedar twigs caused by the fungus Gymnosporangium juniperi-virginianae.

heterecious rust is the white pine blister rust (<u>Cronartium ribicola</u>), which has the alternate hosts of white pine, a gymnosperm, and currants or gooseberries that are angiosperms.

The second order of the Basidiomycetes, which contains many plant pathogens, is known as the <u>Ustilaginales</u> or smuts. The smuts are not obligate pathogens. Many have been induced to complete their entire life cycles on artificial media. The smut fungi have some spore forms similar to the rust fungi. Asexual reproduction is often by means of conidia, which may arise from either uninucleate or binucleate mycelium. Some conidia resemble the basidiospores of other Basidiomycetes. Both basidiospores and conidia may bud repeatedly in a yeast-like fashion. This is a common means of reproduction in the smut fungi. Alexopoulos (5) claims that the teliospores of the smuts are the characteristic structures of the Ustilaginales. Because of the manner in which they are formed, many authors refer to them as chlamydospores.

Alexopoulos (5) gives the life cycle of <u>Ustilago maydis</u> (DC). Corda as an example of how smuts may develop (Fig. 30). The binucleate teliospores undergo karyogamy and become uninucleate diploid spores (Fig. 31). When germination occurs a promycelium forms a short germ tube into which

FIG. 25. Photomicrograph showing teliospores of the cedar-apple rust fungus and a section of the spore horn.

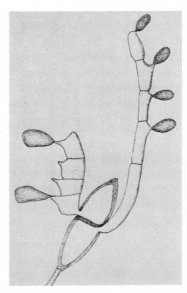

FIG. 26. Germination of a teliospore of the cedar-apple rust fungus
(Gymnosporangium juniperi-virginianae) showing the "dwarf" and "elongate"
types of basidia, each with four basidiospores.

FIG. 27. Cedar-apple rust on upper surface of apple leaves.

the diploid nucleus migrates. Meiosis occurs and the resulting four haploid nuclei distribute themselves more or less uniformly in the promycelium.
Septa form to separate the nuclei and form four uninucleate cells. Mitosis
now occurs and one daughter nucleus migrates into a bud that develops at
the side of each promycelium cell; the other remains behind. The uninucleate buds are the basidiospores and these become the primary source of
inoculum. As the basidiospores germinate and the resulting hyphae enter
the host tissue, they fuse with other hyphae and form diploid nuclei. Gall
formation (in which the teliospores are borne) and basidiospore dissemination are continuous throughout the active growing period.

Bacteria are classified here with plant life, although they could just as
well fit into animal life. They are considered most primitive in structure
and development. They are unicellular microorganisms, widely distributed
in air, soil, water, living plants and animals, and dead organic matter.
They multiply by either binary or transverse fission. The rate of productivity varies within species but they multiply very rapidly. They exhibit
three fundamental shapes: (a) spherical, (b) spiral or curved rods, and
(c) rods (Fig. 32). All plant pathogenic species are rod formers. Bacteria
are extremely small and can be seen individually only with the laboratory
or light microscope at about 400 to 600 times magnification. They are

FIG. 28. Aecial stage of cedar-apple rust on fruit. (Courtesy of
Duane Moore, University of Wisconsin.)

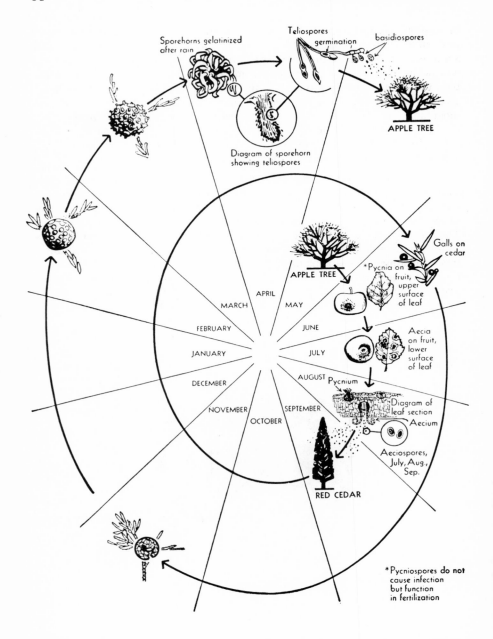

FIG. 29. Disease cycle of cedar-apple rust caused by <u>Gymnosporan-gium juniperi-virginianae</u>. (Courtesy of USDA.)

measured in microns, a unit of measure equal to 1/1,000,000 of a meter (39.34 in.).

Bacteria are classified botanically in a division of the plant kingdom known as the Protophyta and Class Schizomycetes (189). There are several different orders of bacteria, but only two contain plant pathogens. The order Pseudonadales is characterized by elongate cells, straight rods, occasionally coccoid. They are Gram-negative and are motile by means of flagella. The genus Pseudomonas contains species that cause blossom blast of pears, blister spot of apple, halo blight of beans, southern bacterial wilt of tobacco and potato, etc. A second genus, Xanthomonas, contains species of bacteria responsible for bacterial wilt of corn (Fig. 33), bacterial spot of stone fruits (Figs. 34 through 36), black rot of crucifers, common blight of bean, bacterial spot of pepper and tomato (Fig. 37), angular leaf spot of cotton, angular leaf spot of strawberry, and so forth.

The order Eubacteriales have simple undifferentiated cells either spherical or straight rods. The bacteria are motile with flagella arranged peritrichously. They are also Gram-negative. The genus Agrobacterium

FIG. 30. Corn ear converted to common smut (Ustilago maydis).

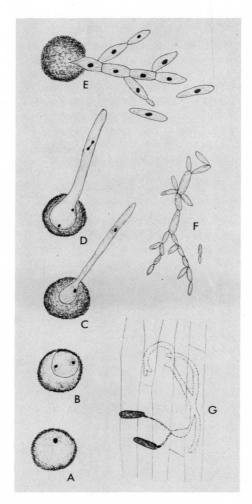

FIG. 31. Various stages in the life cycle of the corn smut fungus, Ustilago maydis. (A) karyogamy (2N condition) of a teliospore; (B) the nucleus has divided into two parts (N + N); (C) germination of the teliospore showing the promycelium (basidium) and migration of one nucleus; (D) four-nucleate stage; (E) basidium with a number of basidiospores or sporidia; (F) a sporidium budding in a yeast-like manner; (G) germination of two haploid basidiospores (+ and -) with penetration of the epidermal cells of a corn leaf, followed by fusion of the haploid hyphae to form a dicaryotic (N + N) mycelium.

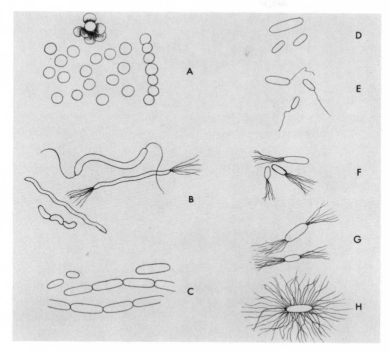

FIG. 32. Bacteria. At the left are the three fundamental types: (A) coccus; (B) spirillum; (C) bacillus. To the right showing various types of flagellation: (D) nonmotile; (E) a single polar flagellum; (F) polar flagella; (G) bipolar flagella; (H) peritrichous flagella.

is soilborne and attacks roots causing the common crown gall disease. Many species of the genus Erwinia are serious pathogens (Figs. 38 through 40). Most notable are E. caratovora, which causes bacterial soft rots of many different plants, and E. amylovora, the incitant of the fire blight disease of apple, pear, quince and many ornamentals. Another member of this family, Corynebacterium michiganense, causes tomato canker (Fig. 41).

There are probably fewer than 50 serious bacterial diseases of plants. They are extremely destructive because effective methods of prevention have not been developed.

An interesting phenomenon with fungi and bacteria is the tendency toward biological or physiological specialization. This denotes the presence of more than one race of a particular species, which are identical morphologically and vary only in their infectivity on different varieties or species

FIG. 33. Bacterial wilt of sweet corn caused by Xanthomonas stewartii.

FIG. 34. Peach bacterial spot on the leaves caused by Xanthomonas
pruni.

FIG. 35. Close-up of peach bacterial spot leaf infection.

FIG. 36. Peach bacterial spot (<u>Xanthomonas pruni</u>) fruit infection.

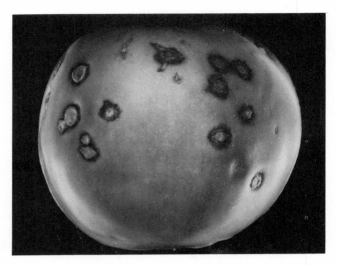

FIG. 37. Bacterial leaf and fruit spot of tomato caused by Xanthomonas vesicatoria.

of closely related hosts. A Phycomycete, Phytophthora fragariae Hickman,
causes the red stele root rot of strawberry. This fungus has eight different
races (reported in the United States) (46) based upon the resistance, susceptibility, or both of different strawberry varieties. For example, Blakemore,
a common strawberry variety, is susceptible to all races. Sparkle has
resistance to races A-1, and A-3; Surecrop has resistance to races A-1,
A-2, A-3, and A-4. Other varieties show multiple resistance. Another
species, P. infestans (Mont.) D By., which is infamous for the widespread
epiphytotics on potatoes in 1845-1847 in Europe, and contributed greatly to
the general famine in Ireland, has many races, greatly complicating the
breeding programs for developing resistant potato varieties.

    Stakman et al. (201) gives an excellent discussion of the varieties,
races, and biotypes of Puccinia graminis, the cause of black stem rust of
cereals and grasses. Within varieties of P. graminis some morphological
differences exist such as slight differences in size of the uredospore or in
shades or tints of color. In this case, the host may vary slightly. For
example, wheat, barley, and many wild grasses will be in one group, with
oats, and different wild grasses in a second group. When physiological
races of rust are grown on different varieties of one crop, race identification depends strictly upon the host reaction. For example, infection type 0
denotes that the host is "immune," with only small flecks of dead host tissue. Infection type 1 shows a very resistant host response with minute rust
pustules surrounded by dead areas. As we progress through types 2 and 3

FIG. 38. Bacterial soft rot (Erwinia carotovora) of Chinese cabbage.
Two diseased and one healthy head after 2 months in storage at about 40° F.

FIG. 39. Bacterial wilt of cucumber caused by Erwinia tracheiphyla.

FIG. 40. Fire blight disease (<u>Erwinia amylovora</u>) on an apple shoot showing exudation.

FIG. 41. Bacterial canker of tomato caused by <u>Corynebacterium</u>
<u>michiganense</u>.

to infection type 4, the host is very susceptible, with pustules large in size
and often united (60). A variety of the fungus <u>Puccinia graminis</u> known as
<u>P. graminis tritici</u> is known to have over 300 races distinguished only by
their infectivity on different wheat varieties. The development of physiological races definitely complicates the development of resistant varieties and
hence disease prevention.

III. SYMPTOMS

An important aspect of plant disease control is a basic understanding
of what is causing a disease. Thus, we have discussed fungal and bacterial
pathogens in some detail to establish their locus in the scheme of life.
Fungal spores and bacteria are largely microscopic and even though they
may have infected a leaf, stem, or fruit, we cannot see them. When infection has progressed sufficiently so that early visible symptoms appear we
know that a disease relationship has been established. The period of time
between the start of an infection until the first visible symptom is called the
<u>incubation period</u>. This period will vary depending upon such factors as

moisture, temperature, age, variety, condition of the plant, and so forth. For example, under optimum conditions the apple scab disease has an incubation period of about 10 days. The fire blight disease will appear in 3 to 4 days. One should remember that the pathogen infects the plant and that the incubation period refers to disease development.

It is not important here to describe the life cycle of a pathogen which involves the genetics of spore development, and so on. However, one should have an opportunity to understand a typical disease cycle that will indicate the series of developments of the disease through the winter, spring, summer, and fall months (Figs. 42 through 44).

In the apple scab disease (Fig. 42), the fungus winters over saprophytically in dead leaves. During this period perithecia are developed (Fig. 11). With the first spring rains, ascospores are discharged and are windblown to the young apple leaves just expanding from the buds. The dead

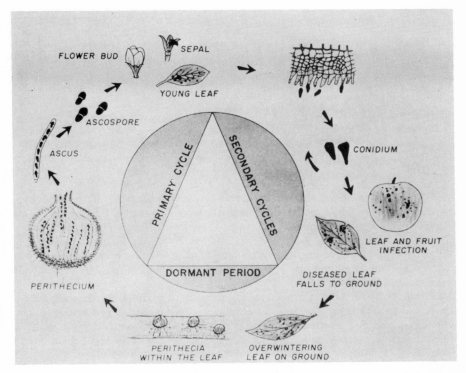

FIG. 42. Disease cycle of the apple scab fungus. (Courtesy of E. J. Klos, Michigan State University, East Lansing, Michigan.)

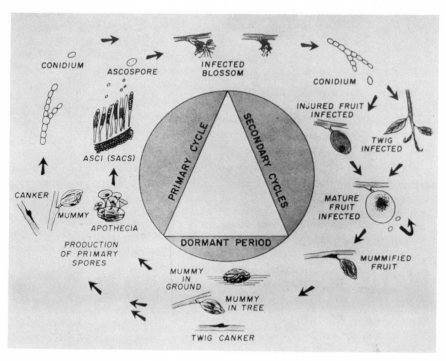

FIG. 43.   Disease cycle of the brown rot fungus of stone fruits.   (Courtesy of E. J. Klos, Michigan State University, East Lansing, Michigan.)

leaves are considered the source of the primary inoculum.   The ascospore forms a peg (appressorium), attaching the spore to the leaf.   Then a germ tube or hypha grows into the epidermal layer of the leaf and infection has occurred.   As infection develops, an olivaceous lesion becomes visible.   A microscopic observation shows that this is an acervulus containing hundreds of conidia borne on vertically arranged conidiophores (Fig. 14).   The conidia are the source of secondary inoculum that produce secondary infections. With the first rain the conidia are released and carried by wind and rain to new leaves where they infect.   Even though new acervuli are developed from these secondary infections, new conidia are referred to as secondary rather than tertiary, quaternary, and so on.   Some diseases like apple scab have repeating secondary inocula over most of the growing season.

Thus, we determine or diagnose a disease by the symptoms.   At least, they are the first key to diagnosis.   Heald (93) developed a rather detailed description and outline of plant disease symptoms.   These will be presented briefly with certain modifications:

1. <u>Discoloration or change of color</u> from the normal (possibly the result of virus infection or physiological abnormality).
   A. Pallor—pale green, yellow (etiolation, chlorosis), silvering, and albinism (Fig. 45).
   B. Color spots or areas—white, gray, red, purple, and so on.
   C. Autumnal or spring coloration.
2. <u>Shot-hole</u>—perforation of the leaf. Some fungi may do this but shot-hole is very likely the result of viral or bacterial infection or chemical injury (Fig. 34).
3. <u>Wilting</u>—"damping-off" of seedlings, or wilting due to drought or excessive soil moisture (Figs. 46 and 47). Recovery may occur under these conditions.
4. <u>Necrosis</u>—death of parts. Blight due to vascular nonfunction from bacterial or fungal infections (Fig. 48). Leaf spots cause localized tissue to be killed (Figs. 49 through 52). Bacterial leaf spots are usually angular (Fig. 35); fungal leaf spots are more likely to be circular.

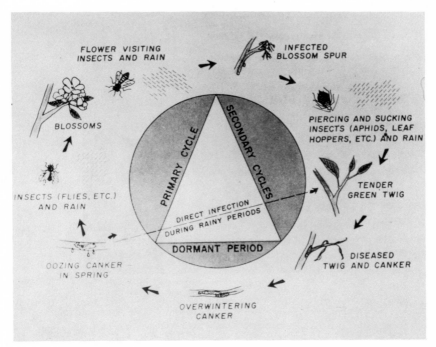

FIG. 44. Disease cycle of the fire blight bacterium (<u>Erwinia amylovora</u>) on apple and pear. (Courtesy of E. J. Klos, Michigan State University, East Lansing, Michigan.)

FIG. 45. Aster yellows of carrot caused by a mycoplasma. Healthy
carrot, right.

FIG. 46. Damping-off of cucumber seedlings. Healthy seedling in middle.

FIG. 47. Cucumber seed treatment showing seed decay and damping-off control. (Left) seed untreated; (right) seed treated with a fungicide.

5. <u>Dwarfing or atrophy</u> (hypoplasia). Sometimes whole plants are dwarfed or only infected parts show this symptom (Figs. 53 and 54). Fungi, bacteria, nematodes, viruses, or mycoplasmas may cause this condition.

6. <u>Hypertrophy</u> (hyperplasia). Increase in size of cells, or abnormal multiplication of cells (Fig. 55). Caused by fungi, bacteria, insects, or herbicide injury.

7. <u>Transformation of organs or replacement of organs</u> by new structures. Mostly the result of fungi, e.g., certain downy mildews, smuts and white-rusts. Some mycoplasmas, such as aster yellows (Fig. 45), may abort flowers into leaflike structures.

8. <u>Mummification.</u> Shrivelling, dehydration, and the production of mummies. Mostly caused by fungi; but some bacterial rots also mummify.

9. <u>Dropping of leaves, blossoms, fruits, or twigs</u>. Due to fungi, bacteria, adverse weather conditions, chemical injury, and so forth.

10. <u>Production of malformations, galls</u>, and so on. Mostly caused by fungi and bacteria although certain nematodes and many insects come under this category.

11. <u>Exudation</u>—oozing of gum and other exudates. Bacteria, fungi, insects, and certain viral infections cause this condition.

FIG. 48. Late blight of tomato caused by the fungus Phytophthora infestans.

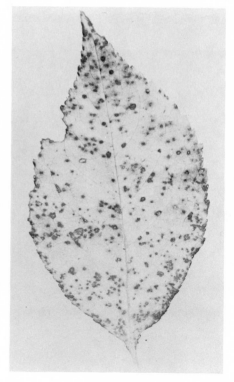

FIG. 49. Cherry leaf spot caused by Coccomyces hiemalis.

FIG. 50. Strawberry leaf spot caused by <u>Mycosphaerella fragariae</u>.

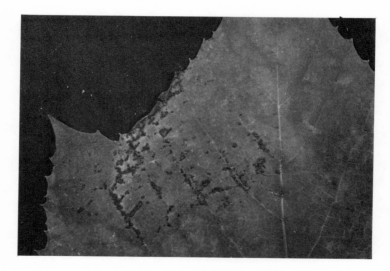

FIG. 51. Downy mildew (<u>Plasmopara viticola</u>) on upper surface of grape leaf.

FIG. 52. Apple scab (Venturia inaequalis) on fruit.

FIG. 53. Gibberella ear rot of corn caused by the fungus Gibberella zeae.

FIG. 54. Soybean mosaics caused by several viruses: (middle) common mosaic; (right) yellow mosaic; (left) healthy leaf.

FIG. 55. Plum pockets (<u>Taphrina communis</u>) showing hypertrophy of leaves and twigs.

12. Rotting. Various types of rot may occur (Figs. 56 and 57). Bacteria are noted for soft slimy rots (Fig. 38) but some fungi (e.g.,
Rhizopus, Phytophthora) are also capable of a soft watery rot.
Some fungi cause a firm to leathery or even a hard rot. Fungus
rots may show macroscopic masses of mycelial threads and various spore-forming structures. But, bacterial rots seldom show
visible evidence of the organism, except for droplets of exudate.

FIG. 56. Strawberry fruit rot. A mixture of Rhizopus stolonifer and
Botrytis cinerea showing mycelial mass on fruit surface.

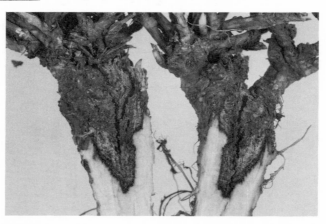

FIG. 57. Crown rot of alfalfa caused by Leptosphaeria pratensis.

IV. LOSSES CAUSED BY PLANT DISEASES

It is doubtful that many Americans are greatly concerned about present losses of food crops to disease or any other cause. We are blessed with a plentiful food supply, providing we can afford the price. Plant diseases, perhaps more than any other factor except climate and soil, qualify the locale in which certain crops and their varieties can be profitably grown. Yet it is extremely difficult to assess losses caused by them. For example, the Midwest could have a multimillion dollar pear industry if it were not for the fire blight disease (Figs. 40 and 44). The weather is optimum for the growth and development of the Bartlett pear, the "cadillac" of pear varieties, which is very susceptible to fire blight. It is also optimum for the development of the disease. Thus, the pear industry is primarily confined to the Pacific Coast states where fire blight, because of weather conditions, can be contained economically.

Many high quality peach varieties cannot be grown commercially because of their susceptibility to bacterial leaf spot (Figs. 34 through 36). There are many such examples throughout all our agricultural crops. This type of loss cannot be estimated but one cannot ignore the profound influence plant diseases have in crop production in addition to calculable losses.

Major diseases that attack the staple crops, such as wheat, corn, soybeans, potatoes, and others are of greatest concern. Epiphytotics in the chief centers of production could produce scarcity, with serious national and international effects.

It is important to realize that plant disease situations are always changing. No matter how up-to-date our information, there is always a new problem. A variety developed for resistance to one pathogen may be extremely susceptible to a previously unimportant organism, as was shown with the development of Helminthosporium (Victoria) blight on oat varieties with Victoria-Richland parentage (221).

The Agricultural Research Service (ARS), in cooperation with the Statistical Reporting Service, Economic Research Service, Soil Conservation Service, and Forest Service of the USDA, has summarized the losses in agriculture over a 10-year period, 1951-1960 (133). The following information has been obtained from this publication. Losses are essentially divided into production and marketing areas. Marketing in this case would include harvesting, grading, transit, storage, and so on. Thus, such losses have two kinds of economic effects: (a) they increase the cost of production; and (b) they reduce the quantity and, in some instances, the quality of products. (See tables 1, 2, and 3.)

Reducing disease losses would not only enable the farmer to lower his production costs but would also allow him to produce more food as the rapidly expanding world population requires maximum yields on decreasing acreages of top-quality cropland.

The estimates presented by the ARS indicate not only preventable reductions in production but also, in some cases, those not avoidable with present technical knowledge. The estimates of annual dollar loss do not mean that the farmer's cash income would have been increased to that extent if losses had not occurred. Increased production is usually followed by decreased unit values so that total farm income might not change appreciably unless the demand for the product increased simultaneously with production. Thus, losses must be interpreted as losses to the public rather than to farmers.

TABLE 1

Estimated U.S. Average Annual Losses in Value Caused by Plant Diseases and Air Pollution to Various Groups of Crops During Production and Cost of Controlling Diseases, 1951-1960 (133)

Crop group	Average annual loss[a] ($1,000)
Field crops (corn, cereals, soybeans, etc.)	1,890,836
Alfalfa and all other hay plants	614,766
Forage seed crops	23,584
Pasture and range plants	193,935
Fruit and nut crops	223,505
Vegetable crops	290,389
Ornamental plants and shade trees	14,099
All crops: loss from air pollutants	325,000
Total	3,576,114
Cost of controlling diseases	115,800
Grand total	3,691,914

[a]Data as estimated for the period 1951-1960 by the Statistical Reporting Service.

In the above table, field crops include 23 different crops of which the percentage losses vary per crop from a minimum of 3.0 (rye) to a maximum of 28.0 (peanuts).

Corn showed a 12.0% loss from potential production. This amounted to 0.527 billion dollars (1960), the maximum dollar loss of any field crop. More than 50 diseases occur to some extent on corn. The leaves, stalks, ears, and roots, may become infected. Generally losses are due to lowered grain quality, and to decreased value of fodder and yield.

TABLE 2

Estimated Average U.S. Annual Losses Caused by Specific Corn
Diseases for the Period 1951-1960

	Average annual loss (%)[a]
Stalk rots (fungi)	3.0
Helminthosporium leaf blights	2.3
Seedling blights	1.6
Root rots	1.3
Ear rots	1.2
Smut	0.7
Bacterial leaf blight	0.5
Physoderma brown spot	0.2
Others	1.2
Total	12.0

[a]Based on estimates of the Statistical Reporting Service (133).

The 3.0% loss from stalk rots is an example of the difficulty in making accurate estimates. In Illinois in 1965 many fields showed 50% down (lodged) stalks at harvest time. Premature stalk dying and breakage resulted in lightweight, chaffy ears and rotting of grain in contact with the soil. The overall State loss due to stalk rots was estimated at 20%, one of the highest on record. Yet Illinois corn farmers still harvested the first billion-dollar crop in any state's history. Ideal growing weather (frequent, well-spaced rains and favorable temperatures from planting to harvest) produced a bumper crop. Even a 20% loss due to stalk rots still left most farmers with the highest yield and biggest money-maker in history.

Following corn in economic importance are wheat (14%), cotton (12%), oats (21%), soybeans (14%), tobacco (11%), and others.

Field crop diseases are controlled primarily by the widespread plant-ing of resistant varieties. Losses will continue to be large since sources of plant resistance are unknown for many diseases or are constantly being broken down by the development of new, more pathogenic races. The use of protective chemicals, outside of seed treatments, is not generally recom-mended on these crops because the returns per acre are not sufficiently large to justify them.

Losses due to diseases in fruit and nut crops were confined only to those grown commercially and were estimated at 223.5 million dollars (1960) per year. Perhaps we should add to this a 52.9 million dollar annual loss during transit and unloading of fruits. (Data based on information from inspection certificates from Pittsburgh Produce Inspection Service from 1957 to 1961.) Some fruits, based upon the largest dollar loss per annum, would be listed as follows:

TABLE 3

Fruit Loss from Potential Production[a]

	Percent	Value in $1,000
Grape	27	48,415
Oranges	12	38,560
Strawberries[b]	16	26,869
Peaches	14	19,805
Apples	8	17,791
Cherries	24	12,355

[a]Based on estimates of the Statistical Reporting Service (133).
[b]Basic data represent the quantity marketed and its value.

The remaining fruits and nuts are listed in order of their decreasing dollar loss: lemons, pears, pecans, walnuts, raspberries (largest loss of 38%), fresh prunes, almonds, apricots, blueberries, plums, cranberries, grapefruit, filberts (hazelnuts), and tung. Most of these crops depend rather heavily upon fungicides to control certain major diseases. Of the 27% loss of grapes, 21.3% was the result of virus diseases which cannot be controlled by fungicides. Citrus trees and fruit are damaged by a number of fungi but they can be kept under control fairly well. On the other hand, a number of virus diseases menace the very existence of the citrus industry.

These are estimated to destroy up to 1 million trees annually in Florida, the Gulf Coast, and California.

Straying from the statistical report, strawberries offer a good example of the effectiveness of fungicides. It was shown that captan was very effective in controlling the blossom blight stage of Botrytis fruit rot (102, 174, 175). Experiments showed that during a serious epiphytotic, use of captan during the blossom period more than doubled the yield. Later, similar experiments showed that when an epiphytotic did not occur, captan actually decreased the yield over an unsprayed check. In some way it interfered with the normal set of fruit. Since the estimated disease loss to strawberries is 25% and fruit rots account for 15% (133), most growers apply a fungicide during the bloom period, just in case. The damage to blossoms from Botrytis is normally much more serious than from the fungicide.

Vegetables are extremely perishable and are subject to an annual field loss of $275 million and a transit-storage loss of $67 million, which exceed those of fruit. Vegetables also depend heavily on fungicides for disease control.

One should realize that Phytopathology, Horticulture, Food Science, Chemistry, Agricultural Economics, and Transportation are all concerned with plant disease losses. If the population continues to grow as predicted, America cannot long withstand these losses. Efforts are being made to increase agricultural production in more heavily populated and underdeveloped countries through international programs. America will also need to develop more technical and efficient crop production programs to reduce disease losses and maintain a comparatively high per capita yield of food and fiber.

V. PLANT DISEASE CONTROL

Plant diseases may be controlled by nonchemical and/or chemical methods. Nonchemical implies the use of resistant varieties, various cultural practices, such as sanitation, proper tillage, fertilization, and weed and insect control. In addition, such measures are employed as the propagation of disease-free plants, seed certification, tuber indexing (potatoes), use of hot and cold temperatures both under wet and dry environment, gamma radiation, and legal control by means of inspection and quarantines.

Biological control of plant diseases is being studied. For example, Trichoderma sp., when introduced into soil, was shown to reduce certain root diseases caused by Armillaria mellea (Vahl) Quel. and Rhizoctonia sp. under greenhouse conditions (18). A symposium on the ecology of soilborne plant pathogens has been presented as a prelude to biological control (110).

Nonchemical disease prevention is of utmost importance and often is the only means of control. Many diseases, however, can only be reduced by these procedures and chemicals are required to supplement them. Plant pathologists feel that fungicides are used only to complement nonchemical methods, if such are available. However, fungicides are the only practical means of preventing many diseases.

Economics plays an important role in the use of fungicides. Specialized, high value crops, such as vegetables, fruits, and nuts, can afford chemical control. The agronomic crops, however, seldom can support widespread fungicide usage because of tremendous acreages and the comparatively meager per acre profit.

TABLE 4

Data Obtained From Three Textbooks, Dickson (60), Walker (225), and Anderson (6), which Assess the Importance of Various Methods of Disease Control

| | | Diseases controlled by | | | |
| | | Disease resistance | | | |
Class	Total diseases	Only	Also	Fungicides only or best	Other means
Field crops	307	230	10	33	40
Vegetables	179	52	29	47	77
Fruits	112	20	30	33	64
Total	598	302	69 1	113	181
Percent of total		50.5	11.5	18.9	30.2

Courtesy of A. L. Hooker.

The percentages of these columns total 111 due to duplication of diseases in different categories. Some diseases may require three or four different types of treatment, yet one might be considered the most effective. It indicates that about 19% of the diseases depend on fungicides for control. Thus, fungicides are important adjuncts to good disease control programs.

Basically a chemical that will destroy bacteria is called a bactericide; a chemical used for fungal control is called a fungicide. For expediency, the term fungicide will be used for all chemicals applied for controlling both bacteria and fungi.

It is not feasible to construct specific programs for individual crops in this chapter. Such programs would not be applicable to all areas of the United States. State Extension plant pathologists have capably developed recommendations for plant disease control in their respective states. In many cases, diseases vary within a state (e.g., Oregon or Washington) so several programs need to be developed for a particular disease within a state. Variations such as elevation, latitude, longitude, and weather conditions that accompany these variables, such as amount and distribution of rainfall, temperatures and air humidity, greatly influence disease problems. In the Midwest, diseases vary in kind and intensity north and south between Lake Superior and the Ohio River. Similar variations occur from east to west from Ohio to Missouri or Colorado.

The suggestions given for controlling the various diseases will be of a general nature with explanations given as to why various fungicides are used in a particular area.

VI. FUNGICIDE PERFORMANCE

Horsfall (103) classifies the performance of fungicides on the basis of protection and therapy (cure) of the host. Protection destroys the pathogen before it can attack the host. Therapy destroys the pathogen after it has invaded the host.

Protection is accomplished by two means: (a) The pathogen is destroyed by a residual film placed on the host (by spraying, dusting, dipping, and so forth). Many fungicides perform this function. Some are mostly residual in performance such as sulfur, zineb, and glyodin (176). They must be on the foliage ahead of the pathogen to nullify the disease most effectively. Other fungicides perform a double role of residual protection and partial therapy of established infections (4, 95, 115). The fungus is not killed but is inhibited from developing further. This includes a large group of popular fungicides such as captan, benomyl, dodine, ferbam, dinocap, bordeaux mixture, the many insoluble or fixed coppers, lime-sulfur, and others. (b) Protection by contact denotes that the fungicide destroys the pathogen before it reaches the host or infection has occurred. Horsfall (103) cites two examples, the use of sodium cresylate in early spring on dead apple leaves under the tree to destroy the ascigerous stage of the apple scab fungus (Fig. 11). This prevents the discharge of ascospores in the spring, thus protecting apple trees from primary scab infection. The second example is the use of a fungicide during the dormant stage of the peach tree to destroy the conidia of the peach leaf curl fungus. These spores winter in cracks and crevices of the bark and buds. It is possible that a fungicide that destroys conidia and prevents secondary infections may also be considered a contact protectant. Such spores are not in association with the second infection court and thus concur with the above definition.

Plant chemotherapy has been defined by Dimond et al. (103, 1952b) "as the control of plant disease by compounds that through their effect upon the host or pathogen, reduce or nullify the effect of the pathogen after it has entered the plant." This could be modified or appended as follows: "The effectiveness of fungicides used in chemotherapy depends on their capacity to inhibit or kill the invading pathogen without seriously affecting the tissue structure of the host." With this modification, chemotherapy occurs when a pathogen is inhibited but not necessarily destroyed.

Therapy is divided into three main categories (104):

1. <u>Systemic chemotherapy of systemic infection.</u> Verticillium wilts, Fusarium wilts, and Dutch elm disease are examples of vascular diseases (Figs. 18 and 19) that result from systemic infections. Since the bacteria causing fire blight disease of apple and pear travel through the vascular tissue (128) use of streptomycin for its control is employing systemic chemotherapy.

2. <u>Systemic chemotherapy of local infections</u> involves the use of a systemic chemical. Streptomycin and Terramycin have shown capacity to control or prevent some bacterial leaf spot and crown gall diseases. Horsfall (103) cites examples of fungal leaf spot control by the use of systemic fungicides.

3. <u>Topical chemotherapy</u> equals what has been referred to as chemical eradication of a pathogen from an established infection (103). Killing out any infection within a leaf such as powdery mildew, or "burning out" apple scab infection has been cited as examples (103). Adhering strictly to definitions is always difficult. Inhibition appears to be part of chemotherapy. Many commonly used fungicides inhibit rather than kill (177) and may be included in Horsfall's category of "topical therapy." The fungicides mentioned above as having a partial therapeutic effect, plus the residual action group, also belong in this category. Fungicides in this group which destroy the fungus, or pathogen, would not likely have more than limited residual action. Also, such fungicides can be expected to be more phytotoxic than those exhibiting strictly residual qualities.

Another characteristic of fungicides needs to be noted as regards their performance. Some chemicals may actually kill the pathogen and thus are truly fungicides. Others merely inhibit their growth without actual killing. These chemicals are called fungistats or as having fungistatic properties. This distinction is probably not as important as it may seem. Many true fungicides perform fungistatically below the fungicidal dosage. Also, a material may be fungicidal to one pathogen and fungistatic to another.

It should not be difficult to construct a classification of the common fungicides based upon the above discussion. This has been done realizing that there are many exceptions to every rule, but that a general assessment of fungicides (ignoring the exceptions) would have considerable value (Table 5).

Unfortunately, each fungicide is a separate entity in regard to a specific pathogen. For example, microfine wettable sulfur was thought to be primarily a residual fungicide. Recently it was shown that with the proper adjuvants it functions as a topical chemotherapeutant in destroying powdery mildew infections (103). Dodine is an effective contact fungicide against early infections of the apple scab fungus and destroys the fungus effectively because it invades only the superficial epidermal layer of the apple leaf (118). In contrast, the fungus Diplocarpon earliana (E. & E.) Wolf, which causes the leaf scorch disease of strawberry, infects the under surface of the leaf, and develops endogenously to produce its acervulus on the upper surface. Even though this fungus is very susceptible to the fungitoxicity of dodine, the fungicide functions only as a residual or contact protectant to destroy the spores before they can become established.

Another fungal pathogen Monilinia fructicola (Wint.) Honey, which causes the brown rot disease of stone fruits, presents still another problem. It infects the acid tissues of fruits and some fungicides are not effective under these conditions. Dodine loses its fungitoxicity as the medium becomes acid (27). The brown rot fungus is inhibited at very low concentrations, near 5 ppm, by dodine in spore germination tests, but will not affect fungal growth on the peach fruit at 1600 ppm.

Horsfall (103) discusses mobilization of a residual protectant and how such factors as spore concentration, particle size of the fungicide, absorption of the fungitoxicant by the spore, diffusion, water solubility, and so forth, are important. The host-fungicide-pathogen relationships determine how effective a fungicide appears in preventing a disease. An interesting observation by Rich (103) was that a compound very similar to dinocap was not fungitoxic on glass slides but was on leaves. He found that the cuticle of the leaf dissolves and retains the compound in a fungitoxic state. More work of this type should be done.

Most of the commonly used protectant fungicides are formulated in a particulate form. When applied, they are evenly distributed in a film over the surface of the plant. Spores must have water to germinate. This can be in the form of a droplet or film. Rainwater has a very low surface tension and normally wets and spreads readily over the plant surface. Thus, under most situations optimum for disease development, the fungicide and spore are connected in a water film or by a water-bridge.

TABLE 5

A Classification of Some of the Common Fungicides Indicating Their Characteristics as Observed From Field Experiments[a]

Fungicide	Residual	Contact	Topical therapy	Systemic	Fungi-cidal	Fungi-static	Phyto-toxic	High temp.	High moist.	Compati-bility
Benomyl	3	2	2	1	3	1	1	0	0	1
Bordeaux mixture	2	1	1	0	2	2	1	0	3	2
BOTRAN (dicloran)	3	0	2	0	1	3	0	0	0	0
Captan	3	2	2	0	3	1	0	0	0	1
Chloranil	3	0	?	0	?	?	0	0	0	?
Cu oxinate	3	1	1	0	3	1	0	0	0	1
Cycloheximide	3	2	1	0	3	0	3	0	0	?
DACONIL (chlorothalonil)	3	3	0	0	2	2	2	0	0	?
DIFOLATAN (captofol)	3	2	2	0	3	1	0	0	0	1
Dichlone	1	0	2	0	1	3	1	3	0	1
Dikar	3	1	1	0	2	2	0	0	0	1
DITHANE M-45 (mancozeb)	3	1	1	0	2	2	0	0	0	1
Dodine	3	2	2	0	3	1	0	0	0	0
DYRENE (analizine)	3	2	2	0	3	2	3	0	0	1
Ferbam	3	1	1	0	2	2	0	0	0	1

Folpet	3	2	3	0	3	1	1	0	0	1
Glyodin	3	0	0	0	2	2	0	0	0	0
Mercury	3	3	0	0	3	0	2	0	0	?
Copper	2	1	1	0	2	2	2	0	3	1
KARATHANE (Dinocap)	2	2	1	0	2	2	1	0	0	1
Lime–sulfur	3	3	2	0	3	1	2	3	2	2
Maneb	3	1	2	0	2	2	0	0	0	1
MANZATE D	3	1	1	0	2	2	0	0	0	1
Nabam	3	0	0	0	2	0	0	0	0	1
NIACIDE	3	1	2	0	2	2	0	0	0	0
Streptomycin	0	0	0	3	0	3	0	0	3	2
Sulfur	3	1	0	0	2	2	1	3	0	1
Thiram	3	0	0	0	2	2	0	0	0	1
TOPSIN M (thiophanate)	3	2	2	1	3	1	1	0	0	1
Zineb	3	0	0	0	2	2	0	0	0	1
Ziram	3	0	0	0	2	2	0	0	0	0

[a]Key: 0, No effect, or problem; 1, Lightly effective or responsive; 2, Moderately effective or responsive; 3, Strongly effective or responsive.

Butt (33) showed that captan would diffuse through a water film and that it was toxic in the diffusible state to spores of the pear scab fungus. He also demonstrated that captan was not absorbed and translocated through the leaf tissue. Powell (177) used a water-bridge, which amounted to a piece of filter paper 2 by 31 mm, with one end in a drop of water placed over a dried fungicide residue and the other end in a drop of spore suspension. It was shown that sulfur, captan, dichlone, dodine, and zineb would all diffuse through water to inhibit spores of Monilinia fructicola (Fig. 8). Time of diffusion does not appear to be an important factor, since these spores germinate in a 4-hr period. Neither nabam nor THIONEB (ethylenebis thiuram monosulfide) (135), a degradation product of nabam, would diffuse in a toxic state (178). When spores are in direct contact with these chemicals, they are inhibited. Experience has shown that when relatively insoluble compounds show water-diffusible toxicity, they usually have promise of an effective field fungicide. The mode of action of fungicides in the field is very closely associated with this function of mobility. Studies have shown a relationship between redistribution of copper fungicides and different types of leaf tissue. Cocoa leaves solubilized bordeaux residue more than either banana leaves or glass slides (101).

The layman commonly expresses, "that he applied a fungicide according-ing to directions yet received no control." The inoculum potential of plant pathogenic organisms is so great that given optimum conditions (e.g., ideal moisture conditions and temperature) a fungicide, normally considered effective, will not control; or, a plant considered resistant under normal conditions will become infected. Ferbam was used for years for cedar-apple rust control on Jonathan apples and control was excellent. Cedar-apple rust (Figs. 24 through 29) does not normally infect the resistant Golden Delicious, Delicious, and Winesap varieties. Thus, they were never sprayed with ferbam. In 1957, cedar-apple rust infections occurred in Illinois on all resistant varieties and also the Jonathan variety which had been treated with ferbam (179). This was good proof that ferbam was only as effective in controlling cedar-apple rust as the weather conditions permitted.

In another case, commercial peach growers normally spray a fungicide on peach trees during the dormant period for control of peach leaf curl. Leaf curl is normally controlled 100% by this practice. In 1957, a long period of cool, rainy weather during the budbreak of peaches, optimum for the growth of the peach leaf curl fungus, resulted in a very serious epiphytotic of this disease (179). Every leaf and fruit in southern Illinois peach orchards showed peach leaf curl damage regardless of the fungicide used and efficiency of the spray application. Thus, weather again proved to be the controlling factor in the efficiency of a fungicide program.

VII. THE POME FRUITS

The pome fruits include the apples, pears, and quince. They comprise a subfamily of the Rosaceae, which have similar morphological structures. They also have similar cultural characteristics. The pome fruits are grown only in temperate climates.

The quince will not be discussed here since it has little economic value except in some parts of Asia. Quinces have essentially the same disease problems as do apples and pears, and the apple schedule listed below should suffice.

A. Apples

The apple (Malus sylvestris) is grown abundantly in both the north and south temperate zones. The United States has an average production of over 150 million bushels. Of all fruits grown, none require the special attention of apples as far as disease control is concerned. Plant breeders are developing new cultivars which have resistance to many of the important diseases. Some of these have been introduced on the commercial market.

The following important apple diseases, with the scientific name of their respective pathogens, are those that must be controlled by fungicides for adequate crop yield and quality.

Apple blotch, Phyllosticata solitaria
Apple scab, Venturia inaequalis (Figs. 8, 11, 14, 42, and 52)
Bitter rot, Glomerella cingulata
Bot rot, Botryosphaeria ribis
Brooks spot, Mycosphaerella pomi
Cedar-apple rust, Gymnosporangium juniperi-virginianae (Figs. 24
    through 29)
Collar rot, Phytophthora cactorum
Fire blight, Erwinia amylovora (Figs. 40 and 44)
Flyspeck, Lepthothyrium pomi
Frog-eye leaf spot, also black rot of the fruit, Physalospora obtusa
Hawthorn rust, Gymnosporangium globosum
Northern anthracnose, Neofabrea malicorticus
Powdery mildew, Podosphaeria leucotricha
Quince rust, Gymnosporangium clavipes
Sooty blotch, Gleodes pomigena

There are numerous minor diseases that are kept under control by cultivar resistance, sanitation, roguing infected plants and pherpa, more than we realize, by fungicides used for the more important diseases.

TABLE 6

A Typical Apple Fungicide Schedule

Application and purpose	Time to apply	Percent concentration of fungicide[a]
Dormant. Fire blight, frogeye leaf spot, bot rot, blotch	While buds still closed in spring	Copper sulfate, 0.50
Prepink. Apple scab, powdery mildew	When buds show 1/16 to 1/2 in. green	DIKAR, 0.25 or dodine 65W, 0.031 plus micro-fine sulfur, 0.375
Tight Cluster. Apple scab, powdery mildew	Blossom buds exposed but not separated	Same as prepink
Pink. Apple scab, powdery mildew, and cedar-apple rust	All blossom buds pink and stems fully extended	DIKAR, 0.25 or dodine 65W, 0.031 plus micro-fine sulfur, 0.375 plus zineb 75W, 0.125
Early Bloom. Fire blight on blight-susceptible varieties (Figs. 40 and 44)	Just before the king or center blossom opens	Streptomycin, 50-100 ppm[b] plus zineb 75W, 0.25 if needed
	Continue streptomycin at 4- to 5-day intervals until the fruit has formed. On nonfruiting trees use as long as needed.	
Petal-Fall. Apple scab, blotch, cedar-apple rust, frogeye leaf spot	When 3/4 of petals have fallen (this exact timing more important for insects)	DIKAR, 0.25 or captan 50W, 0.125 plus zineb 75W, 0.125[c]
Cover Sprays. For all diseases under petal-fall plus quince rust, sooty blotch and fly-speck	The first 5 covers at 10-day intervals following petal-fall; additional covers applied as needed	Same as petal-fall

[a]0.125% equals 1 lb/100 gal of water, also 1 oz, to 6 gal of water, see Table 8.

[b]Label has instructions for amounts to use.

[c]Benomyl (BENLATE) 0.062% may be substituted for DIKAR.

The use of fungicides on apples is regulated mostly by the stage of development of the tree as the growing season progresses. In actual practice, a fungicide, insecticide and acaracide are added together in one spray. Many multipurpose or all-purpose mixes are on the market. These mixes are well-accepted by the homeowner who does not wish to work with the separate ingredients. Normally they are not as effective as the home-mixed spray and are somewhat more expensive.

A typical apple spray schedule (Table 6) is presented, realizing that many comparable programs could be developed with other fungicides. Substitutes for zineb include ferbam, thiram, and NIACIDE. DIKAR is an all-purpose fungicide and can be used instead of dodine and sulfur. Glyodin is sometimes added. Being a protectant fungicide and a good surfactant, it is often used at one-half strength in apple fungicide programs.

Whether to use DIKAR, sulfur, or dinocap for powdery mildew is always controversial. Dinocap is specific for powdery mildew and an additional full-strength fungicide is needed for other diseases. This may increase the cost above the use of sulfur. DIKAR is probably the best choice.

Apple cultivars vary in their susceptibility to chemical injury, and an overall program is not always the best fungicidally. It is usually balanced between the attributes of fungitoxicity, nonphytoxicity, compatibility with other chemicals and economy. In many orchards the spray schedules will vary with the cultivar. A comparison of six of the most common commercial apple cultivars in relation to chemical injury and disease susceptibility is shown in Table 7. See Table 8 on how to compute fungicide dosage.

TABLE 7

Differences in Susceptibility of Some Common Apple Cultivars to Chemical Injury, Apple Scab, Fire Blight, and Powdery Mildew

Variety	Chemical injury	Apple scab	Fire blight	Powdery mildew
Golden Delicious	VS[a]	S	R	MS
Delicious	R	VS	R	R
Jonathan	S	MS	VS	S
McIntosh	R	VS	R	MS
Winesap	R	S	R	R
Rome Beauty	R	S	S	S

[a]Key: VS, very susceptible; MS, moderately susceptible; S, susceptible; R, resistant.

In some regions powdery mildew may be more important than any other disease. In this case use DIKAR in a full schedule.

Spraying is kept to a minimum during the bloom period because of honey bees. The bees leave with the beginning of the petal-fall period and spraying can be resumed without danger of poisoning them.

If sprays are thoroughly applied through the third cover, apple scab (Figs. 8, 11, 14, 42, and 52), blotch, the rusts (Figs. 24 through 29), frogeye leaf spot, and powdery mildew should be controlled for the season.

TABLE 8

Chart for Determining the Proper Concentration of Fungicide to Use

Percent concentration of chemical	lb/100 gal of water	oz/3 gal of water[a]
0.062	0.5	1/4
0.125	1.0	1/2
0.250	2.0	1.0
0.375	3.0	1.5
0.500	4.0	2.0
0.625	5.0	2.5
0.750	6.0	3.0
0.875	7.0	3.5
1.000	8.0	4.0

[a]Generally 1/2 oz is equal to 1 tablespoon when a material has a specific gravity of 1.0.

Diseases to be watched for the remainder of the season are bitter rot, bot rot, sooty blotch, flyspeck, and, where prevalent, northwestern anthracnose. Black rot may develop if fruit has been injured by insects, hail, and so on, since it will only infect injuries. Bot rot is most prevalent during hot and dry seasons in old, poorly pruned trees that have an abundance of dead wood. Dead twigs and limbs are the source of inoculum for both black rot and bot rot. Bitter rot, found normally only south of 40° latitude, is a very serious disease and can cause a reduction in quality of apples in a comparatively short time. Sooty blotch and flyspeck are generally only serious in cool, moist seasons or sometimes only in the lower elevations of an orchard. If a fungicide is needed, folpet 50W, 0.25% is recommended. This fungicide is particularly effective for late summer diseases.

Northwestern anthracnose has been fairly easy to control since the advent of captan.   Captan, or a captan-zineb application just prior to harvesting is recommended.   A judicious pruning program to remove as many twig and branch cankers as possible will complement the fungicide program.

B. Pears

The pear industry has a value of approximately 35% that of the apple. The apple spray schedule should answer the main fungicide problems, except a prebloom treatment may or may not be required, depending on location.   In northern states the prebloom period continues over a longer period.

The greater part of the pear industry is confined to the Pacific Coast states where fire blight can be controlled economically.   Also, pears thrive fairly successfully in some of the northern states where temperatures are below the optimum for fire blight development.

In California, Oregon and Washington a mycoplasma, disseminated by the pear psylla, has killed thousands of pear trees.   Pear decline, as the disease is called, is responsible for three-fourths of the pear losses for the years 1959-1960 (133).   Resistant cultivars and better pear psylla control curbed this disaster during the 1960s.

Fire blight (Fig. 44), second in importance to pear decline, is fairly well controlled by streptomycin as indicated in the apple schedule.   Pears are sufficiently more susceptible than apples, however, that even streptomycin is not as effective as is needed.   The fire blight bacterium is developing resistance to streptomycin in some areas.   Thus, in the west, a judicious pruning program helps the effectiveness of fungicides.   Several states still recommend copper sprays.   Bordeaux mixture and copper-lime dust formulations are noticeably effective (34).   Pears are less susceptible to russeting than apples, and copper can be used without serious fruit injury.

Pear leaf blight is caused by a fungus, Fabrea maculata. It infects the twigs (where it winters over), fruit, and foliage.   When serious, it causes defoliation early in the spring.   This disease is easily controlled with almost any fungicide.   Ferbam 76W, 0.187 to 0.25% has given excellent results.

Pear scab, caused by Venturia pirina, is closely related to apple scab. At least two different races (6) infect different cultivars.   This poses a problem in some regions.   Pear scab is controlled by any fungicide that controls apple scab.   Scab is sometimes serious on Bartlett pears in

California (34), and strong sulfur or lime-sulfur sprays are suggested in prebloom sprays. Organic fungicides are used in the petal-fall and first cover periods.

Pear leaf spot is not serious but may develop late in the season to cause defoliation of the tree. It is easily controlled with a minimal fungicide program.

One of the serious storage diseases of pears is caused by the fungus Botrytis cinerea (6). Dicloran (Botran) has shown effectiveness in reducing this disease on other fruits (164). More recently, Benomyl (BENLATE) 50W has been widely used.

The use restrictions as of December 1, 1974 imposed upon the fungicides recommended on apples and pears by the Environmental Protection Agency are as follows: benomyl, no time limit; copper sulfate, exempt from restrictions; captan, no restrictions; dicloran (BOTRAN), not permitted; dodine, do not use within 7 days of harvest; folpet, no limitations on apples, not approved for pears; DIKAR and maneb ion (MANCOZEB) approved on apples but do not use within 21 days of harvest; dinocap (KARA-THANE), do not use within 21 days of harvest; sulfur, no time limitations; zineb, 15 days on apples, but do not apply within 7 days of harvest on pears; streptomycin, 50 days for apples and 30 for pears. For any other material used as an alternate, read label carefully. These regulations came into effect on December 1, 1974 and are subject to change at any time; therefore do not use any chemical without first reading the label. The current label carries the latest official information.

VIII. THE DRUPES OR STONE FRUITS

These fruits also are in the family Rosaceae, genus Prunus and are known commonly as almonds, apricots, cherries, nectarine, peaches, and plums (prunes). The olive is also a drupe but belongs to another family, Oleaceae. It will not be considered here since it has little economic value in this country.

A drupe, except for the almond, has a hard endocarp or seed, a fleshy edible mesocarp, and a thin skin or exocarp. The almond fruit has a fleshy mesocarp when young, like a typical drupe, but it soon becomes hard and dry to form the hull. The shell is the endocarp which protects the seed or edible portion of the fruit.

The stone fruits share many of the same fungal and bacterial diseases. In addition, each fruit has certain specific diseases. Thus, it is not possible to develop one overall schedule that would apply equally well to all drupes. Table 9 indicates the relationship between the common diseases and their occurrence (109) and their importance on stone fruits.

A. Almonds

Almonds survive with a minimal fungicide program compared to the other stone fruits. The following information has been obtained from California (151) where most of the almonds are grown in the United States.

1. <u>Dormant.</u> Control brown rot and Coryneum blight using a 0.75% to 1.00% solution of sodium pentachlorophenate. It acts as a contact fungicide and destroys the mycelium and spores that winter over on the tree.

2. <u>Popcorn bud stage.</u> (Before bloom). For brown rot control use any one of a number of fungicides such as a 10:10:100 bordeaux mixture, fixed copper according to the label, or one of the following organic fungicides: dichlone 50W, 0.093%; captan 50W, 0.25%; or maneb 80W, 0.25%. If Coryneum blight control is necessary omit the copper and use dichlone, captan, or ziram 75W, 0.25%.

3. <u>Full Bloom.</u> For brown rot control and, except for the copper sprays, use the same materials as the popcorn stage.

4. <u>Petal-fall.</u> For control of Coryneum blight (shot-hole), leaf blight (Hendersona rubi), and scab use captan 50W 0.25%.

No other fungicide is needed for the remainder of the summer. In the fall (late September to October) soil fumigation is used to control Armillaria root rot. The soil is fumigated with carbon bisulfide or methyl bromide prior to planting. The soil is worked as deeply as possible to expose old roots 1 to 2 months before planting. The old roots are removed along with as many small pieces of roots as can be collected. The soil is then leveled and dampened to a depth of 3 in. Holes are drilled at 18-in. centers 5 to 6 in. deep. Carbon bisulfide is added to each hole, which is tamped shut immediately. Methyl bromide is applied according to manufacturer's instructions. If complete eradication is not obtained, retreatment may be necessary in localized areas.

There are restrictions on the use of fungicides on almonds with the exception of copper fungicides. Read the label carefully for instructions since they are subject to change at any time.

TABLE 9

Importance of Several Diseases Common to Stone Fruits

Disease	Pathogen	Almond	Apricot	Cherries	Peach and nectarine	Plum (prune)
Brown rot (Figs. 8, 12, 43)	Monilinia fructicola	-[a]	xx	xx	xx	xx
	Monilinia laxa	xx	xx	xx	x	x
Bacterial leaf spot (Figs. 34 through 36)	Xanthomonas pruni	x	xx	x	xx	xx
Bacterial canker	Pseudomonas syringae	x	xx	x	xx	xx
Black knot	Dibotryon morbosum	-	x	x	-	xx
Coryneum blight	Coryneum beijerinckii	xx	xx	x	xx	x
Leaf curl[b]	Taphrina spp.	-	x	x	xx	xx
Leaf spot (Fig. 49)	Coccomyces spp.	-	x	xx	-	xx
Perennial canker	Valsa spp.	-	xx	-	xx	xx
Root rot	Armillaria mellea	-	xx	-	xx	xx
Scab	Cladosporium carpophilum	xx	x	xx	xx	x
Soft rot	Rhizopus stolonifer	-	-	xx	xx	-

[a]-, no report; x, has been reported; xx, a serious disease.
[b]Many diseases are caused by Taphrina (Fig. 55) and these vary in severity on different hosts. These will be mentioned in the text.

B. Apricots, Nectarines, and Peaches

These fruits require essentially the same schedule. Any variation would be the result of geographical change. Apricots are grown commercially mostly in California although Utah and Washington have new plantings (133). Apricots are also grown in the Midwest but late winter to late spring freezes are most hazardous to normal fruiting. In California, brown rot and Coryneum blight are very important and bacterial canker may be serious. Otherwise, there are minor differences in fungicide programs throughout the United States.

1. <u>Dormant</u>. For peach leaf curl and plum pockets (Fig. 55) use practically any commercial fungicide at full strength. In California, sources of primary inoculum for brown rot and Coryneum blight should be destroyed with a 0.75% to 1.0% solution of pentachlorophenate applied in late November or early December. This treatment also kills peach leaf curl spores.

2. <u>Early</u> and <u>Full-bloom</u> Periods. For brown rot blossom blight, Coryneum blight, Sclerotinia and Botrytis green fruit rots (in California) use benomyl 50W, 0.062%, bordeaux mixture, fixed copper, maneb 80W, 0.187%, captan 50W, 0.25%, ziram 75W, 0.25%, or dichlone 50W, 0.031% plus microfine sulfur, 0.625%. (NOTE: Do NOT use ziram or dichlone on nectarines.)

3. <u>Petal-fall</u> and <u>Shuck-split</u> Periods. These stages are about 7 to 10 days apart. For brown rot use benomyl 50W, 0.062% or microfine sulfur, 0.375% or captan 50W, 0.25%.

4. <u>1st</u> and <u>2nd Covers</u>. Apply benomyl 50W, 0.062% or microfine sulfur, 0.375% at 10- to 14-day intervals following the shuck-split for brown rot and scab control.

5. <u>3rd</u>, <u>4th</u>, and <u>5th Covers</u>. Not needed in arid climates. For brown rot and scab control use benomyl 50W, 0.062%.

6. <u>6th</u>, <u>7th</u>, and <u>8th Covers</u>. Not needed in arid climates. Apply 3, 2, and 1 weeks before harvest, respectively. Use benomyl 50W, 0.062% or captan 50W, 0.25% for brown rot.

7. <u>9th Cover</u>. Not needed in arid climates. For Rhizopus rot apply dicloran (BOTRAN) 50W, 0.125% 18 days, 10 days, and 1 day before harvest (or apply as a postharvest dip or wash. Dicloran is usually combined with captan 50W, 0.25% or benomyl 50W, 0.062% for broad-spectrum disease control.

If powdery mildew is a problem dinocap (KARATHANE) 22.5W, 0.062%, benomyl 50W, 0.062%, microfine sulfur or sulfur dust is applied when mildew is first seen or expected. Several applications at 10- to 14-day intervals may be required for control. WARNING: Do not use sulfur on apricots.

Many variables affect the spray programs. Cultivar susceptibility is most important. Peach scab is a serious problem on some cultivars, and a fungicide should be continued at 7- to 10-day intervals until within 30 to 35 days of harvest. This disease has a long incubation period—40 to 45 days. If the fungicide is stopped too early and warm rainy weather prevails, scab will develop, sometimes appearing 2 to 3 days ahead of harvest.

Brown rot (Figs. 8, 12, and 43) is a constant threat in moist climates and needs constant attention throughout the growing period. Rhizopus rot is serious only in rainy climates. Dicloran gives excellent inhibition of this fungus.

Bacterial spot (Figs. 34 through 36) is always a threat to fruit quality and some cultivars are not grown because it is most difficult to control. Some states recommend an 8:9:100 zinc sulfate-lime-water mixture at 5- to 7-day intervals from petal-fall to a month before harvest (77). Zineb 75W, 0.25% has shown some promise when applied with the occurrence of rainfall. A combination of dodine and captan has also shown promise, particularly in preventing fruit infection. It has a label and will probably be more used in the future. Read label instructions carefully, however, before using this combination.

In California, bacterial canker and decline is partially controlled by bordeaux mixture applied at the beginning and end of leaf fall. In light sandy soil in the San Joaquin Valley, successful control is achieved with the preplant soil fumigants Telone and D-D. It is suggested that this treatment be followed at regular 2-year intervals with postplant fall applications of DBCP (Fumazone, Nemagon), either as injection or flood weed control.

Valsa canker (Perennial canker) has caused damage to these fruits east of the Rocky Mountains. Infection occurs in any of a number of injuries, such as leaf scars from autumn leaf-drop, borer infestation, winter damage, pruning wounds, and so on (100). Low-vigor trees are more susceptible than trees with good vigor. Valsa canker may develop in orchard or nursery. There is much speculation as to whether fungicides help control this disease. A fall application of dichlone 50W, 0.062% after leaf drop is suggested. Pruning from bud-break to petal-fall allows the pruning wounds to heal quickly, and the dichlone spray during this period would be helpful. Keeping trees in a good state of vigor is beneficial. There is little cultivar difference in susceptibility to this disease. Rio Oso Gem and Sunhigh are very susceptible.

As of December 1, 1974 restrictions on the use of the various fungicides were as follows: captan has no time limitations; dicloran can be used within 1 day of harvest on apricots, peaches, and nectarines; benomyl and sulfur are exempt from restrictions; zineb cannot be used after petalfall; dodine can be used up to 15 days before harvest on peaches only; dinocap cannot be used after 45 days before harvest on peaches only. All fungicides are permitted use in the dormant spray.

Hydrocooling (immersing packed fruit in ice cold water) is used to control rot development in transit. This program has been adopted by most commercial peach growers.

C. Cherries

Cherries have a brown rot schedule similar to that of the other stone fruits. There is little difference between sour and sweet cherries, except that the latter have a more serious preharvest brown rot problem (22, 112, 212, 236).

1. Full-bloom. For brown rot and Botrytis blossom blight use benomyl 50W, 0.031% to 0.062%, captafol (DIFOLATAN) 4F, 1 to 2 pt/ 100 gal, folpet (PHALTAN) 50W, 0.187% to 0.25%, dichlone 50W, 0.062%, or microfine wettable sulfur, 0.625% or combine the last two at one-half the above dosages. Some states (22) also recommend a prebloom spray for cherries; the same fungicides as used in the full-bloom spray.
2. Petal-fall, 1st, 2nd, 3rd, and 4th Covers. The cover sprays are applied at 10- to 14-day intervals following petal-fall. Cherry leaf spot (Fig. 49) control is necessary through this period. Use the same fungicides as for full-bloom, or use dodine 65W.
3. Postharvest. Cherry leaf spot continues to infect all summer. Extensive defoliation may occur and weaken the tree. Use dodine 65W, 0.031% to 0.062% or benomyl 50W, 0.062%. The number of applications depends on the weather. Excessive rainfall increases the need for better protection. Two or three such sprays should be adequate.

Powdery mildew may be a problem on sour cherries (see note under apricots, nectarines, and peaches).

Fungicide use restrictions on cherries, as of December 1, 1974, are as follows: benomyl, captan, dodine, and sulfur have unlimited use; dichlone should not be used within 3 days of harvest; dicloran is restricted to within 1 day of harvest and zineb to 7 days.

D. Plums and Prunes

While there is a plum tree in almost every backyard in the United States, California and Michigan produce most of the commercial plum crop. California, Idaho, Oregon, and Washington produce the bulk of the prunes. Prunes are any of a number of species of plums, which contain a high sugar content and can be easily dried without spoilage. They are all plums while still on the tree, and production problems are essentially the same.

The following is a composite of schedules from a number of states (22, 112, 113, 151, 212) to give the reader an idea of the number of plant disease problems involved in raising plums.

1. Delayed Dormant. For control of black knot, the dormant spray should be delayed as long as possible before bud-break. Use a 10% lime-sulfur solution (10 gal made up to 100 gal; 1 qt to 2 1/2 gal of water), zineb 75W, 0.25%, or dodine 65W, 0.062%. Another excellent treatment is a 12:12:100 bordeaux mixture plus 3% plant spray oil (preferably a miscible type).

2. Pink. (When flower buds are pink and just ahead of early bloom.) For black knot, use zineb 75W, 0.25% or benomyl 50W, 0.062%. Several states suggest mixtures of benomyl 50W, captan 50W, zineb 75W, and dichlone 50W or wettable sulfur plus ferbam 76W each at about half-normal strength.

3. Bloom. For brown rot and black knot use dichlone 50W, 0.062%, benomyl 50W, 0.062%, or microfine sulfur, 0.075%, or a combination of each at one-half the above strength (see also under Pink).

4. Petal-fall. For brown rot and in some areas leaf spot and black knot (113), use ferbam 76W, 0.125% plus microfine sulfur, 0.375%, 0.25% captan 50W alone, benomyl 50W, 0.062%, or one of the mixtures listed under Pink.

5. Shuck-split to Shuck-fall. Usually 10 to 14 days following petal-fall. For control of leaf spot and brown rot, use the same as for petal-fall. If black knot is still prevalent, use zineb 75W, 0.25%.

6. 1st and 2nd Covers. Apply at 10- to 14-day intervals following shuck-fall. For leaf spot, use benomyl 50W, 0.062% or ferbam 76W, 0.25%, or, if leaf spot and brown rot control are desired, use captan 50WP, 0.25% or benomyl 50W, 0.062%.

7. Additional Covers. These are needed only if the weather is rainy and brown rot infections are expected. Use benomyl 50W, 0.062%, captan 50W, 0.25%, or a reduced mixture of benomyl-captan as often as needed.

As of December 1, 1974 the restrictions on the use of fungicide on plums is as follows: benomyl, captan, lime-sulfur, and sulfur have no restrictions; dichlone cannot be used within 3 days of harvest; ferbam 7 days, and zineb until petal-fall.

Brown rot is the most serious disease on plums (133) and accounts for almost one-half of total losses. Plum pockets, caused by a species of the fungus Taphrina, is second in importance, even though it is easy to control with a dormant application of almost any fungicide. Bacterial leaf spot is third in causing losses. This disease degrades the fruit, attacks the limbs and branches, resulting in perennial cankers, and in addition causes defoliation from leaf infection. There is no control except the use of cultivars which show resistance. Black knot is fourth in importance (133). It is a difficult disease to control because its etiology is a 2-year cycle. Care needs to be taken for two consecutive years to control the disease completely. If not controlled, black knot forms large black cankerous overgrowths on the twigs and limbs, gradually girdling and killing them or the entire tree (114).

IX. THE SMALL FRUITS

A. Strawberries

These are aggregate fruits, also of the family Rosaceae. The fruit, familiar to all of us, is described botanically as having numerous achenes (seeds) on a fleshy edible receptacle.

Rapid advances are being made in strawberry production. In the late 1940s the production of virus-free strawberry plants was initiated by the USDA (221). Sources of virus-free plants and the proper indicator plants were soon developed. Over a 3-year period, enough virus-free plants were propagated for distribution to growers. Several states, California, Oregon, Michigan, Arkansas, Maryland, and others initiated their own programs. Now all reputable nurseries with the cooperation of the USDA and the State Agricultural Experiment Stations sell only virus-free plants. This is a boon to the strawberry industry because virus-free plants are very vigorous and productive. A planting can be retained for a fairly long period in a virus-free state if aphids and wild strawberries are controlled (217). Some commercial producing areas replant each year. Others may retain the same planting for several years.

The development of cultivars with multiple resistance to the several physiological races of the red stele fungus has brought many excellent cultivars. Many of these have been selected because of their moderate to excellent resistance to other diseases, such as leaf spot (Fig. 50), leaf scorch, and blight. No resistance to the fruit rots (Fig. 56) has been shown, thus over 50% of the losses from the potential strawberry yields are caused from these diseases (133).

Diseases which respond to fungicide treatment are as follows:

Diseases	Pathogens
Leaf spot (Fig. 50)	Mycosphaerella fragariae
Leaf scorch	Diplocarpon earliana
Leaf blight	Dendrophoma obscurans
Stem-end rot	Gnomonia fructicola
Gray-mold fruit rot (Fig. 56)	Botrytis cinerea
Soft rot or leak (Fig. 56)	Rhizopus stolonifer
Powdery mildew	Sphaerotheca macularis

A typical strawberry spray schedule is normally set up with pounds of various fungicides recommended on a per acre basis. Since readers may have but a few plants such a recommendation may be useless. Thus, we suggest that to receive good coverage one should apply 100 to 200 gal of spray per acre. If the percentage of chemical to be used is given on a weight basis, then this concentration can be mixed in a gallon of water and the plants sprayed thoroughly.

1. The 1st and 2nd applications in northern areas are applied when the blossom clusters are first visible and just before bloom, respectively. A choice of benomyl 50W, 0.125%, captan 50W, 0.375%, or dodine 65W, 0.09% is suggested.

In warmer climates plants develop more rapidly. The first spray is at the appearance of the first blossoms (early bloom) (112, 113, 193).

2. Additional Covers. These are applied as needed (usually at 7- to 10-day intervals until harvest), based on weather conditions and on cultivars grown. On cultivars susceptible to leaf diseases another spray or more of dodine 65W, 0.09%, benomyl 50W, 0.125%, captan 50W, 0.375%, or thiram 65 or 75W, 0.375% may be required after harvest. Should powdery mildew appear, use dinocap 22.5W, 0.062%, microfine sulfur, 1.0%, or benomyl 50W, 0.125%.

If the main problem is gray mold, causing blossom blight or fruit rot (Fig. 56), use benomyl 50W, 0.125%, captan 50W, or thiram 65W, 0.375%. These can also be mixed, one half each at half strength in one spray. Dicloran 50 or 75W, 0.125% is also excellent for the control of fruit rots, especially Rhizopus (Fig. 56). This material can also be mixed with captan 50W if desired.

Between early bloom and harvest, fungicides are usually applied about every 7 to 10 days. Nonrestricted fungicides may be applied between pickings. After harvest and renovation one or more thorough applications of dodine 65W, 0.09%, benomyl 50W, 0.125%, or captan 50W, 0.375% is made to inhibit the development of leaf diseases and keep plants vigorous.

Captan, dodine, and thiram dusts formulated at 7.5% active are available. Also, a 7% copper dust is available. Dusts are usually applied at least as often as sprays and at about 40 lb/acre per application. These are most effective when applied in early morning or evening with low wind and light foliage dew. Under these conditions dusts have often been more effective than sprays in controlling diseases. Dusting equipment is lighter, causing little soil compaction, and material is applied much more rapidly. Generally, dusts are preferred over sprays for control of strawberry diseases.

As of January 1, 1975, the restrictions on the use of the above fungicides on strawberries is as follows: benomyl, captan, sulfur, copper, and dicloran have unlimited use; anilazene (DYRENE) is restricted to within 5 days of harvest; zineb 7 days of harvest; dodine 14 days; and thiram should not be used within 3 days of harvest unless the fruit is washed with water before marketing.

California (233) is concerned about gray-mold rot but questions the value of preharvest sprays of captan and thiram. They have adopted a postharvest cooling program which effectively reduces rot development and allows safe transit of the fruit over the country without breakdown. It is similar to the hydrocooling process used on peaches except cold air is used instead of cold water. However, fungicides will reduce fruit rot development during the preharvest period (26, 102, 112, 151, 175, 193, 217). Thus, it is suggested that both procedures be employed for best results.

B. Grapes

The grapes are classified botanically in the family Vitaceae, genus Vitis. They are characterized as climbing shrubs, having fruits called berries (a berry is a fruit developed from a several-carpellate, several-seeded ovary, having the seeds imbedded in a fleshy mass surrounded by a very thin covering). These are a very perishable fruit with an estimated annual loss from potential production of 27.0% due to diseases (133).

Three types of grapes are grown in the United States. In the Western states are the high quality European table grapes. In the eastern states or east of the Rocky Mountains, are grown the American bunch and Muscadine grapes. The European table grapes are susceptible to many diseases that do not attack them in the semi-arid climate of the Pacific Coast. They are far from disease-free, however, since many destructive viruses and mycoplasmas are prevalent. Leaf roll, fanleaf, and Pierce's disease are some of the most important. California and New York have embarked on virus-free programs in grapes that have greatly improved their industry in recent years. Important fungal diseases in the west are powdery mildew, black measles—also referred to as black mildew of the fruit—and dead arm disease. The latter is also prevalent in the east (12, 147).

East of the Rocky Mountains, the counties bordering Lakes Erie and Ontario, produce most of the American bunch grape. These fruits are subject to damage mainly by black rot, downy mildew (Fig. 51), dead arm, and powdery mildew. Muscadine grapes are native to the southeastern part of the United States and may be cultivated over most of this region. Though not immune to fungus diseases, they are markedly resistant to them as well as many other destructive agents of the American bunch grape. Black rot blossom infection is quite destructive and may need control. Bitter rot (see apple) may cause fruit damage. In general, this crop is grown with a minimum of fungicide spraying (1, 112, 113, 147, 151).

While the history of grape growing extends back to 6000 B.C. (6) there would not be a significant industry today were it not for the efficient use of fungicides. The following diseases are kept under control by such programs:

Disease	Pathogen
Anthracnose	Elsinoë ampelina
Black measles	Secondary effects of wood-rotting fungi
Black rot	Guignardia bidwellii
Dead arm	Cryptosporella viticola
Downy mildew (Fig. 51)	Plasmopara viticola
Gray mold	Botrytis cinerea
Powdery mildew	Uncinula necator

Grape fungicide schedules vary throughout the country primarily because of weather conditions and type of grape grown. Cultivar variation to disease susceptibility may also affect fungicide usage. Thus, explanations need to be given as an overall program is developed.

1. Dormant. Dinoseb (dinitrobutylphenol) is suggested for insect control and thought to have an effect in reducing inoculum sources of dead arm (183). In California, for dead arm and black measles, the fungicide used is sodium arsenite, containing an equivalent of 4 lb of arsenic trioxide, 0.75% during the dormant period, but at least 4 weeks after pruning (151). Many precautions should be noted (183). Some states recommend a dormant spray of copper sulfate (bluestone) or liquid-lime sulfur to control anthracnose.
2. Bud-swell. Captan 50W, 0.187% or folpet 50W, 0.25% is used to control dead arm in prebloom (113, 151).

3. <u>After growth starts.</u> In some states sprays are applied only once before bloom (23). In others, applications are made at 1/2 to 1 in. of shoot growth and just before bloom (159, 183). Some prefer one or more additional sprays with growth 4 to 6 and 7 to 8 or 10 in. long (1, 112, 183). In all cases, a variety of fungicides is suggested, e.g., ferbam 76W, captan 50W, zineb 75W, or folpet 50W all at 0.25%. Captan and folpet are particularly effective in preventing early infections of dead arm. Ferbam and zineb are good for early black rot infections.

4. <u>Postbloom.</u> The same materials may be used as mentioned above. Ferbam is excellent for black rot but leaves a black residue on the fruit if used late in the season. Thus, one can switch to captan in the later applications. Folpet is recommended in some states instead of captan (212). It is phytotoxic to grapes south of the 40th parallel (183).

Black rot, downy mildew, and powdery mildew are important diseases from this time until harvest. Anthracnose is controlled by the fungicides used for other diseases. Applications are made as needed. In moist climates sprays are applied if it rains. If no precipitation occurs, sprays are not needed.

Powdery mildew is a semi-arid disease and must be treated with special care under such conditions. Sulfur dusts at 8 to 10 lb/acre, or spraying with sulfur, 0.25% or benomyl 50W, 0.062% with a good wetting agent, has proved effective. Sulfur has two disadvantages: it causes injury when temperatures are above 85° or 90° F and some grape cultivars are sulfur-sensitive (1). Thus, it should be used with caution. In the east, powdery mildew is primarily a problem north of the 42nd parallel where the annual rainfall is less than that farther south. Here, insoluble or fixed coppers or bordeaux mixture is used as needed. Copper sprays are used sparingly on grapes since the advent of the organic fungicides, because they may greatly reduce yield (1, 180).

Gray mold is also a problem in the table grape industry. A dust containing 15% captan and 25% sulfur is used at 20 lb/acre. This program is usually started between July 1 and 15 and four applications are made at about 3-week intervals. Dusting must be done ahead of rainfall, and it is suggested if rain threatens, just to be safe (1). Benomyl 50W, 0.062% is effective against powdery mildew. It is usually combined with captan 50W, folpet 50W, or ferbam 76W to give broad-spectrum disease control.

The restrictions of the above fungicides on grapes as of January 1, 1975, are as follows: benomyl, captan, folpet, copper-lime mixes, and sulfur have no limitations; ferbam and zineb should not be used within 7 days of harvest; sodium arsenite and dinoseb are to be used only when plants are dormant.

C. Brambles

The brambles include the raspberries and blackberries. Botanically they are in the family Rosaceae, genus Rubus. They develop an aggregate fruit composed of many druplets. These fruits are grown throughout the temperate and subtropical climates.

About one-third of the loss in growing brambles results from diseases. Viral diseases account for one-third of these losses (133). As with other crops, fungicides have no effect on their prevention. Roguing infected plants is the only satisfactory means of control. Virus-free brambles are now available commercially.

There are many fungal and bacterial diseases, such as crown and cane gall, Verticillium wilt, orange rust, and cane blight, that are controlled only by sanitation, good cultural practices, roguing infected plants, and so on (44, 47).

Anthracnose, caused by the fungus Elsinoë veneta, and fruits rots, caused mainly by Rhizopus stolonifer and Botrytis cinerea, follow viruses in importance (133). A fungicide program is essential to prevent destruction from anthracnose. This will prevent other minor diseases such as spur blight and several leaf spots from developing. The schedule generally is as follows (24, 35, 44, 112, 113, 183, 218):

1. Delayed Dormant. Apply when the new leaves are 1/4 to 3/4 in. long. Use one of the three sprays: (a) dry lime-sulfur, 2 1/2%; (b) liquid lime-sulfur, 10%; or (c) a mixture containing plant spray oil, 3%; plus an 8:8:100 bordeaux mixture (copper sulfate in the form of bluestone; or blue vitrol, 1%; and hydrated lime, 1%).
2. 1st Prebloom. When new cane growth is 6 to 8 in. high use ferbam 76W, 0.25%, captan 50W, 0.25%, or bordeaux 3:3:100.
3. 2nd Prebloom. Apply just prior to blossoming using the same materials as in the first prebloom. Some states suggest an additional spray when new canes are 12 to 15 in. high (24).
4. Postbloom. Apply ferbam, captan or folpet, 0.25% sprays just after blossoming (petal-fall).
5. Harvest. If fruit rots are prevalent, use captan 50W, 0.25% as required.
6. Postharvest. Apply as needed to control leaf spots and anthracnose using ferbam 76W, 0.25%. As needed means if precipitation is prevalent and these diseases are seen to be developing. Red and black raspberries, particularly, may need additional fungicide protection from anthracnose during this period.

The fruit rots are mostly prevalent on raspberries. A good practice is to pick the fruit each morning as they mature during the harvest period.

Powdery mildew may develop in arid climates. Wettable sulfur sprays
or dusts are recommended (35).

The minimum days wait between last application and harvest, as of
January 1, 1975, is: copper, lime-sulfur, and oil materials are exempt
from tolerance; captan and folpet 0 days; ferbam 40 days.

D.   Currants and Gooseberries

These deciduous fruits belong to the family Saxifragaceae (herbs,
shrubs, and small trees), genus Ribes. The fruit is a true berry derived
from a compound ovary with seeds embedded in the flesh (186).

Currants and gooseberries are produced primarily in the north temperate zone where the climate is relatively cool and moist. Commercially
they are grown north of 42° latitude from the Northwestern Pacific Coastal
states to the Hudson River Valley in New York (41). They respond well to
irrigation in arid climates. Gooseberries can be grown successfully farther
south than currants. There are approximately 900 acres each of gooseberries and currants grown commercially in the United States (2, 41).

Plantings of these fruits are limited since they serve as the alternate
host of white pine blister rust caused by the fungus Cronartium ribicola.
The value of the white pine industry greatly exceeds that of currants and
gooseberries, thus the latter have been eradicated. Because of this disease,
neither the white pine or Ribes hosts can survive if they are within 1/4 mile
of each other. Varieties of Ribes vary in resistance to the white pine blister
rust disease but thus far no immunity has been observed.

Cane blight, caused by the fungus Botryosphaeria ribis, is most destructive to Ribes species but fungicides have not been as successful in preventing this disease as a good sanitation and pruning program.

Anthracnose, caused by the fungus Pseudopeziza ribis, infects both
currants and gooseberries and can be prevented by ferbam or captan sprays.
Powdery mildew caused by Sphaerotheca mors-uvae is very severe on gooseberries. It can be controlled by either sulfur or dinocap.

The spray schedule is rather minimal as compared to most fruits
(113, 183).

1. Green tip. When leaf tips first appear use (only on gooseberries)
    either wettable sulfur, 1%; or liquid lime sulfur, 5%; dry lime sul-
    fur, 2%; or dinocap 25W, 0.062% to protect against powdery mildew.
    Thorough coverage is essential.
2. Postbloom. Sprays should be applied immediately after bloom and
    again 2 to 3 weeks later. Use ferbam 76W, 0.25% or captan 50W,

0.25%. It may be necessary to continue protection against powdery mildew on gooseberries using either wettable sulfur, or dinocap as above.

E. Blueberries

Blueberries are a most interesting group of deciduous or persistent shrubs or bushes. They are classified in the family Ericaceae, genus Vaccinium. Childers (41) describes the many different types of blueberries and their cultivars, and discusses their culture. The commercial crop is confined largely to special areas in Massachusetts, Michigan, New Jersey, North Carolina, New York, and Washington. This is not a big industry, but when grown properly, blueberries are most profitable.

Blueberries are indigenous to the United States, and have developed resistance to most diseases. Even so, it is estimated that the crop is reduced 14.0% annually by diseases. Blossom and twig blight, caused by the fungus Botrytis cinerea, causes almost one-third of the annual loss (4.4%). Stem canker caused by the fungus Botryosphaeria corticis; mummy berry caused by Monilinia vaccinii-corymbosi; stem blight caused by Botryosphaeria dothidea; stunt, a mycoplasma; and bacterial cane blight caused by Pseudomonas sp. are listed according to their importance (133). There are also several leaf spots that vary in severity from one area to another. Double spot, for example, causes a 0.3% annual loss and is a serious problem in North Carolina.

While Botrytis is a cosmopolitan fungus and is considered the most serious disease of blueberries, it is sporadic in nature and only occurs when long periods of rain and warm weather prevail. Mummy berry and stem canker appear to be the most consistent maladies. Stunt, caused by a mycoplasma, is gradually increasing in importance. It causes a slow decline in vigor and eventual death of the plant. Stunt is controlled by removing infected plants so that no sprouts will develop from the old roots.

Fungicides play a small part in the blueberry industry. Clean cultivation, pruning infected and dying shoots, roguing dying plants, and other forms of sanitation are the most important means of keeping many of the important diseases at a minimal level.

A few spray schedules have been developed. One such program utilizes either anilazene (DYRENE) 50W or ferbam 76W, 0.25% just after bloom, 7 to 10 days later, and following harvest, three sprays at 14-day intervals (158, 160). In another program such fungicides are not used (113).

As of January 1, 1975, the minimum days wait between the last application and harvest for blueberries are: captan 0; anilazene 14; and ferbam 40.

In addition, plant breeding programs continue to develop more resistant cultivars. It is hoped that chemical sprays can always be kept at the present minimal level.

F. Cranberries

The cranberry (Vaccinium macrocarpon) is a low-growing evergreen shrub with trailing vines. It has a close botanical relationship to the blueberry, even belonging to the same genus (Vaccinium) and subgenus Oxycoccus. It, like the blueberry, is indigenous to the northeastern area of North America, along the coastal parts, extending from Nova Scotia to North Carolina, and west to Wisconsin, Oregon, and Washington. In the southern latitudes, the disease problems increase and it becomes difficult to maintain a planting. Cranberries grow in bog or marsh land; thus their distribution is very restricted. The leading states for cranberry production listed in order of their importance are (41): Massachusetts, Wisconsin, New Jersey, Washington, and Oregon. Massachusetts produces almost one-half of the world's supply.

Disease losses (133) for the period 1951-1960 amounted to an average of 8.7% annually. The fruit rots cause the most damage before and after harvest.

There are eight species of fungi associated with the fruit rot complex; Guignardia vaccinii, Acanthorhynchus vacinii, Glomerella cingulata vaccinii, Godronia casandrae, Diaporthe vaccinii, Sporonema oxycocci, Pestalotia vaccinii, and Ceuthospora lunata. All do not attack at one time. They have widely different growth characteristics. Godronia and Sporonema may start to grow at 32° F with an optimum temperature between 59° and 68° F. Diaporthe, Glomerella, and Guignardia begin to grow between 35° and 40° F with an optimum between 59° and 86° F. Acanthorhynchus starts growing at 60° to 70° F with an optimum at about 85° F (221). Symptoms of the various fruit rots are not easy to distinguish from one another so that only a trained plant pathologist or experienced grower could identify them. Another interesting point is that the important rot pathogens usually become established early in the season, at blossom time, but do not develop until the fruit approaches maturity.

Fungicides are used to control these diseases in some areas (Massachusetts, New Jersey) but not in others. A fungicide application about mid-bloom and two additional sprays 10 to 14 days and 28 days after midbloom are usually adequate. Maneb, zineb, captafol, and ferbam have shown efficacy in preventing fruit rots. Fungicides have been shown to reduce pigment content of the fruit (70) but this is not considered a deterrent to their use for disease prevention. An interesting aspect of the use of fungicides on cranberries is their effect in alleviating preharvest frost injury (238).

Minimum days wait between last application and harvest, as of January 1, 1975, are: captafol, 50 days; ferbam, 28 days after midbloom; maneb, 30 days; and zineb, midbloom.

False blossom, a serious mycoplasmal disease of cranberry, is controlled by destroying its vector, the blunt-nosed leafhopper.

X. THE NUTS

Nuts include the chestnuts, butternuts, walnuts, hickory, pecans, filberts (hazelnuts), beechnuts and acorns. The peanut is considered a pod, corresponding to the fruit of bean or pea (186).

A. Filberts

The filberts are large shrubs or small trees cultivated for edible nuts or ornament. They are classified botanically in the family Betulaceae, as Corylus maxima and are grown commercially in Oregon and California.

Losses due to disease average about 4% annually (133). Bacterial or filbert blight, caused by the bacterium Xanthomonas corylina is present in all plantings. This pathogen attacks the buds, leaves, branches, and trunks of the trees. It rarely attacks the nuts directly but reduces the size of the crop. The most serious phase of filbert blight is trunk girdling and killing of trees up to 4 years of age. Two general methods for control of this disease are suggested (137, 138). A fungicide program is applied and pruning tools are sterilized with 70% rubbing alcohol. The fungicide program involves the use of a 6:3:100 (0.75% copper sulfate, 0.37% lime) bordeaux mixture, Tribasic copper sulfate, 0.75% or KOCIDE 101, 0.5 to 0.75% in late August or early September (before the first heavy fall rains) (137, 138). Should unusually heavy rains occur, a second application is made when about three-fourths of the leaves have fallen. A dust composed of monohydrated 25% copper sulfate, 50% hydrated lime, 21.5% talc, 2% bentonite and 1.5% light mineral oil may be substituted for or supplement the spray (137, 138).

There are many other diseases that attack filberts, but sanitation, pruning, and good cultural practices will effectively keep them under control.

B. Pecans

The family Juglandaceae includes the pecans, hickory, walnut and butternuts. They are characterized as timber trees of eastern North America and eastern Asia with hard tough wood and handsome foliage. Some are ornamental and some species are grown for their edible nuts. The pecan, Carya illinoensis, is native to the south central part of the United States. The commercial industry extends from South Carolina through Georgia, Florida, and westward to Texas, Oklahoma, and Arizona (88).

It is estimated that diseases cause a 21 percent crop reduction annually (133). The important fungus diseases and their pathogens are: scab (Cladosporium effusum) caused 9.8% of the loss, brown leaf spot (Cercospora fusca), downy spot (Mycosphaerella caryigena), and leaf blotch (Mycosphaerella dendroides).

Since scab is the most important disease, spraying is done to prevent its development and this is usually sufficient to prevent other diseases from occurring (10, 73, 145, 160). Dodine 65W (CYPREX), 0.062% to 0.125%, DU-TER 47.5W, 0.062%, or benomyl 50W, 0.062% are most commonly used in pecan spray schedules. Occasionally an alternate is suggested in a 4:1:100 (0.50% copper sulfate, 0.125% lime) bordeaux mixture in the first two sprays and a 6:2:100 (0.75% copper sulfate, and 0.25% lime) bordeaux in the remaining covers (73). Five to eight applications of a fungicide are sometimes required on scab-susceptible varieties as follows: first prepollination, when buds are bursting and the first leaves are showing; second prepollination, when first leaves are half grown; first cover, when tips of small nuts are brown; and four or five more covers at 3- to 4-week intervals. Some areas omit the second prepollination spray and the last cover or two.

Rosette is a physiological disease of pecans caused by a zinc deficiency. It is corrected by applying zinc sulfate, 0.25% to the foliage. On soils that do not fix zinc, use 1/2 to 1 lb of zinc sulfate for each year of tree age.

C. Walnuts

Two species of walnuts are cultivated for commercial production: the black walnut, Juglans nigra, has a forest tree range in the eastern one-half of the United States. It is not grown commercially to any extent, but mostly as an avocation. Enough nuts are harvested each year from native trees to reach the commercial market. The English walnut (also called Persian walnut) is indigenous to western Asia (not England). It is cultivated, especially in California, Oregon, and Washington (88).

Walnuts are estimated to have an 18% annual loss from diseases. Bacterial blight caused by Xanthomonas juglandis causes an average annual loss of 15%. This disease affects the nuts but may also produce leaf lesions and infect the blossom clusters in early spring. Black line, a disease of unknown etiology, causes an estimated average annual loss of 3% (133).

Bacterial blight is most effectively reduced by applications of a fungicide timed at the early prebloom, late prebloom, and early postbloom (139). Sprays should contain a 4:2:100 (copper sulfate, 0.50%, lime, 0.025%) bordeaux mixture plus summer oil, 0.125%; tribasic copper sulfate (approximately 50% copper), 0.375%; plus a suitable spreader-sticker, or Kocide 101, 0.25% plus superior type plant oil. In years when rainfall during the infection period is light, two spray applications in late prebloom and early postbloom are usually sufficient. Four to six dust treatments are an alternative to spraying. A suggested mix includes 15% monohydrated copper sulfate, 30% hydrated lime, 10% dusting sulfur, 41 1/2% talc, 2% bentonite, and 1 1/2% light mineral oil. Dusts are applied at approximately weekly intervals, beginning in the early prebloom stage. The first two applications require 1 1/2 lb of dust per tree; thereafter, 1 3/4 lb per average-sized tree (139). California (11) formerly suggested two prebloom applications of either fixed copper, used according to the label, or streptomycin 50 ppm. Streptomycin 1000 ppm in a dust was also suggested at 30 to 40 lb/acre. In wet locations, frequent bordeaux mixture or fixed copper sprays are suggested during bloom and early growth stages, especially just before rain (151).

Almost 100 diseases have been reported on walnuts (109). In addition to bacterial blight and black line, the following diseases and their respective fungal pathogens are important (221): branch wilt, Hendersonula toruloidea; melaxuma (black sap), Dothiorella gregaria; blotch, Gnomonia leptostyla; ring spot, Ascochyta juglandis; and Armillaria root rot, Armillaria mellea. Most of these diseases are controlled by methods other than fungicides.

XI. THE CITRUS FRUITS

The citrus fruits are composed of grapefruit, lemon, lime, orange, tangerine, and tangelos. They are in the family Rutaceae, genus Citrus, and are characterized as small to medium sized trees that grow in semitropical climates and produce edible fruits. The citrus fruit is a type of berry called a hesperidium because of its thick, leathery rind, numerous oil glands, and thick juicy portion composed of several wedge-shaped locules (186). In the United States, commercial production is in California, Texas, Arizona, and Florida. This has been a rapidly expanding industry for the past 40 years and now exceeds all other fruits in total production (2, 41).

The average annual loss from diseases of grapefruit, lemons, and oranges is 2%, 25%, and 12%, respectively (133). While there are a number of diseases of viral, mycoplasmal, fungal, bacterial, and physiogenic origin, not many respond to fungicide treatment. In arid regions, citrus spray programs are primarily for insect and mite control (36). However, there are many important diseases that attack citrus but are kept under control by such means as sanitation, disease-free nursery stock, pruning, scarifying cankers and disinfecting them, and so on (123).

In the rain belt, five fungal diseases require special attention (68). These diseases and their pathogens are as follows: anthracnose (Glomerella cingulata), melanose (Diaporthe citri), scab (Elsinoe fawcettii), brown rot (Phytophthora spp.), and greasy spot (Cercospora citri-grisea). Neutral or fixed copper formulations, ferbam, and zineb are the commonly used fungicides. Copper, 0.10% actual (see label), used in the dormant period before appreciable spring growth has started, will supplement later sprays for scab control. A second spray of copper, when two-thirds of the petals have fallen, will aid in scab and anthracnose prevention. A third spray, 1 to 3 weeks after petal-fall, will prevent melanose and greasy spot. Additional sprays may be necessary between the postbloom and summer periods to control anthracnose. Between June and August, greasy spot may develop if such sprays are not continued.

Brown rot control is obtained by spraying the lower 6 ft of trees with neutral copper, applying the spray around the middle of August in groves where the disease has been troublesome in the past. If only occasional, spraying may be deferred until just after the first appearance of affected fruit.

Ferbam 76W, 0.125% is rated more effective than copper for scab control in the 2/3 petal-fall application but should not be used on the fruit which limits it in citrus programs. Zineb 75W, 0.125% will give some prevention of greasy spot but it is used primarily as an acaricide on rust mite. Neither ferbam nor zineb is compatible with copper.

Soluble sulfates of copper, zinc, and manganese are used as needed to control mineral deficiencies. Hydrated lime is combined with zinc and copper sulfates to prevent phytotoxicity. Manganese sulfate can be used up to 0.75% without causing injury, although the addition of lime does not detract from its effectiveness. The routine use of zineb as a fungicide controls zinc deficiency.

Citrus fruits are subject to many postharvest rots which may be prevented by fungicides. They aid in preserving the fruits in storage, transit, and in the market. Biphenyl was introduced in 1944. It has a tolerance of 110 ppm and has approval for use when impregnated in wrappers, pads, and liners. Disease development is greatly reduced under refrigerated conditions (below 45° F). Immersion of fruit for 2 or 4 min in wash solutions of 115° to 120° F have been effective in eradicating infections (68).

XII. THE SMALL GRAINS OR CEREALS

The small grains or cereals include wheat, oats, barley, rye, and rice. Since the culture and disease problems of rice varies greatly from those of other small grains they will be treated separately. Dickson (60) has an excellent discussion of the diseases of these crops, their control, and where the crops are grown. The acres harvested, production, and yield of small grains in the United States in 1973 is given in Table 10.

A. Wheat

The cultivated wheats comprise several species, the most important commercially being common wheat (Triticum aestivum or T. vulgare), club wheat (T. compactum) and durum wheat (T. durum). Varieties of the common wheat represent the major acreages in most countries. Wheat is grown under a wide range of environmental conditions, principally for grain (60). The crop is grown in drier agricultural climates as well as humid areas of the world (2).

Wheat diseases cause large losses in yield and quality. Estimated average annual losses in the United States due to all wheat diseases for the 10-year period, 1951 to 1960, amounted to 14% of the crop, almost 174 million bushels, with an estimated value of almost $331.3 million (133).

Data collected annually in Illinois, over a period of 20 consecutive years, show that the average estimated annual loss from some 36 infectious wheat diseases is 18.5% or almost 8.5 million bushels (19).

B. Oats

The cultivated oats are of two species: the common oat (Avena sativa) and the red oat (A. byzantine). The common oat is grown in cool climates and comprises the largest acreage. The red oat varieties are grown chiefly in warmer climates (60).

Oat diseases cause large losses in the United States. The estimated average annual loss for the 10-year period, 1951 to 1960, amounted to 21% of the crop, over 299 million bushels, with an estimated value of almost $198.5 million (133).

C. Barley

The cultivated barleys are principally two species, the more common-ly grown six-rowed (Hordeum vulgare) barley and the two-rowed (H. distichon).

TABLE 10

Acreage, Production, and Yield of Small Grains in the United States
in 1973 (2)

Crop	Acres harvested (in thousands)	Production in bushels (in thousands)	Yield per harvested acre (bushels)
Wheat, all	53,875	1,711,400	31.8
Winter	38,407	1,269,653	33.1
Durum	2,974	84,860	28.5
Other spring	12,494	356,887	28.6
Oats	14,110	663,860	47.0
Barley	10,527	424,483	40.3
Rye	1,038	26,398	25.4

It is grown chiefly for grain, although it is an important forage crop in certain sections (61).

Barley diseases cause large losses in both yield and grain quality. Estimated average annual losses in the United States for the 10-year period, 1951 to 1960, amounted to 14% of the crop, almost 56.5 million bushels, with an estimated value of $55.5 million (133).

D. Rye

Cultivated rye (Secale cereale) is widely grown as a winter cereal in Europe, Asia, and, on lighter soils, in North and South America. Spring varieties of rye are grown, but their usefulness is limited. The extreme winter hardiness of rye permits growing of the higher-yielding winter varieties except in far northern climates (60).

Rye diseases are far less important than those of wheat, oats, and barley. The estimated average annual loss in the United States for the 10-year period, 1951-1960, amounted to 3% of the crop or 801 thousand bushels, with an estimated value of $928 thousand (133).

At present, due to relatively low cash returns per acre, fungicides are usually applied to cereals or small grains only as a protective (residual and contact) seed treatment. Some aerial spraying is done by helicopter or conventional aircraft, principally to control rusts (Fig. 23) and Septoria diseases, but this usage, relatively small at present, is rapidly gaining favor. Maneb 80W, mancozeb 80W, and zineb 75W are the fungicides in current use at the rate of 0.25%.

Table 11 lists the important diseases and causal organisms (fungi and bacteria) partially or completely controlled by seed treatment fungicides.

Treating small grain seed with a fungicide is cheap insurance for improving stands and grain quality, and for obtaining higher yields—especially if the soil is cold and wet following planting. Proper chemical cereal seed treatment returns to farmers about $100 for each $5 spent for fungicides.

The principal reason for treating seed with a fungicide is to control smut fungi (72) that infect seedling plants (Table 11). Loose smuts of wheat, barley, and rye (Ustilago tritici or U. nuda) are now controlled by a systemic fungicide seed treatment of carboxin (Vitavax).

A secondary benefit of seed treatment is the protection of germinating seeds and young seedlings from soilborne fungi that cause seed decay, seedling blights (damping-off) and root rots. These organisms are present in all soils and many are carried on the surface of seeds. In addition, fungicide treatments effectively control seedborne infections caused by scab (Fig. 10) and head blight fungi; anthracnose; Septoria diseases; certain leaf blotch diseases; seedborne root rots such as Rhizoctonia, Fusarium, and Pythium; and stripe of barley (Table 11). The fungicide may be applied as a dust, wettable powder, or liquid.

Seed can be treated on the farm using a cement mixer, oil drum, or barrel treater, spoonfed dust or liquid dripped into a grain auger loader as seed flows from a bin or truck, or application in the drill or planter box.

Commercial seed treatment is also widely available at country elevators, seed and feed houses, or seed processors after the seed is cleaned of weed seeds, light and shriveled kernels, smut particles and balls, chaff and other impurities. A number of machines have been devised for commercial seed treatment. These include the slurry, ready-mix, and mist-type treaters. Table 12 lists a number of seed treatment products for small grains together with their suggested rate, formulation, and methods of application. These products have been widely tested at agricultural experiment stations in the United States and Canada and represent the consensus of the principal states growing small grains.

All seed treatment materials are harmful if swallowed and may irritate the skin and mucous membranes. Treated seed should be carefully marked and not used for feed, food, or oil purposes even after it has been stored for months or years. Sacks, bins, grain auger loaders, wagons, or other containers used for treated seed should be thoroughly cleaned before using for other purposes.

All treated small grain seed must be prominently colored to avoid mixture with food or feed (Federal Food, Drug, and Cosmetic Act, Section

TABLE 11

Diseases and Causal Organisms of Small Grains—Wheat, Oats, Barley, and Rye—Partially or
Completely Controlled by Fungicide Seed Treatment

Small grain	Diseases	Cause[a]	Causal organism
Wheat, oats, barley, rye	Seed rot, seedling blights, seedling root rots, foot (crown) rots	F	Cochliobolus sativus, Gibberella and Fusarium spp.
Wheat, oats, barley, rye	Seedborne scab, head blight	F	Fusarium and Gibberella spp., Cochliobolus sativus
Wheat, oats, barley, rye	Anthracnose	F	Colletotrichum graminicolum
Wheat, oats, barley, rye	Septoria disease, speckled leaf blotch, black stem of oats	F	Septoria spp.
Wheat	Glume blotch	F	Septoria nodorum
Wheat, barley, rye	Bacterial black chaff, bacterial blight	B	Xanthomonas translucens
Oats	Halo blight, bacterial stripe blight	B	Pseudomonas coronafaciens Pseudomonas striafaciens
Wheat, barley	Basal glume rot	B	Pseudomonas atrofaciens
Wheat, oats, barley,	Covered smut (bunt of wheat and rye), covered smut (barley), semiloose smut (barley),	F	Tilletia spp. Ustilago hordei Ustilago nigra

TABLE 11 (Continued)

Small Grain	Diseases	Cause[a]	Causal organism
	flag smut (wheat and rye), covered smut (oats), loose smut (oats), stem or stalk smut (rye)		Urocystis tritici (U. agropyri) Ustilago kolleri (U. hordei) Ustilago avenae Urocystis occulta
Oats	Leaf blotch	F	Drechslera avenacea or Helminthosporium (Pyreno- phora) avenae
Wheat, barley, rye	Spot blotch, foot rot	F	Cochliobolus sativus or Bipolaris sorokiniana (Helminthosporium sativum)
Barley	Net blotch	F	Drechslera teres or Helmin- thosporium (Pyrenophora) teres
	Stripe	F	Helminthosporium gramineum
Barley, rye	Leaf scald	F	Rhynchosporium secalis, R. orthosporium

[a]Key: F, fungus or fungi; B, bacterium.

TABLE 12

Commonly Suggested Fungicide Seed Treatments for Wheat, Barley, Oats, and Rye

Fungicide[a]	Tradenames	Methods of application	Dosage and remarks
Captan and combinations	Captan 25 (25% captan)	Dust	3 oz/bu as planter box treatment on wheat; not for bunt control.
	Captan 80 (80% captan)	Slurry	2/3 oz/bu on wheat, barley, oats (not registered for rye) in slurry-type treater.
	Captan 88 (88% captan)	Slurry	1/2 oz/bu on wheat, barley, oats (not registered for rye) in slurry-type treater.
	Captan–Thiram (43% thiram)	Slurry	4.0 oz/cwt on wheat, 4.5 oz/cwt on barley (not registered for oats or rye) in slurry-type treater.
	EVERSHIELD SEED PROTECTANT CM (29.52% captan, 0.34% malathion)	Slurry	1 gal CM PROTECTANT concentrate/5 gal cold water, plus 7 oz EVER-SHIELD dye. Use 270 cc/cwt on wheat; 320 cc/cwt on barley; 420 cc/cwt on oats; 370 cc/cwt on rye in slurry-type treater. Not for bunt control, but gives some insect protection.

TABLE 12 (Continued)

Fungicide[a]	Tradenames	Methods of application	Dosage and remarks
Carboxin and combinations	VITAVAX-200 (37.5% carboxin, 37.5% thiram)	Wettable powder, slurry	3–4 oz/cwt on wheat and barley only. Controls smuts and seedling diseases. Apply by mist or slurry-type treater.
	VITAVAX 75% (34% carboxin)	Dusty, slurry, spray-liquid	2–4 oz/cwt on wheat and barley only. Controls all smuts. (Do not use treated crop for hay, straw or grain as food or feed. Do not graze.)
	VITAVAX-300 (37.5% carboxin, 37.5% captan)	Wettable powder, slurry	3–4 oz/cwt on wheat and barley only. Controls smuts and seedling diseases. Apply by mist or slurry-type treater.
Maneb	MANZATE (80% maneb)	Dust-rotary drum or dust attachment for mechanical loader or planter box dust	2 oz/bu on wheat, barley, rye (not registered for oats). Usually controls covered smut or bunt of wheat, barley and rye, but needs overcoat of VITAVAX for loose smut.
	Maneb 50%, or AGSCO DB GREEN or DB YELLOW, or COVER UP, or COVER UP PLUS, or GRANOL NM, or GRANOX NM (with HCB, 10%)		2 oz/bu on wheat, barley, oats, rye as planter box treatment. DB GREEN, GRANOL NM, and COVER UP PLUS have lindane 18.75% added for insect control. Keep maneb dry.

MANCOZEB or Maneb plus zinc ion coordinate	DITHANE M-45 SEED PROTECTANT (80% maneb plus zinc ion) or MANZATE 200 (80% maneb plus zinc ion)	Dust	1 1/3 oz/bu on wheat, barley, oats, rye. For dry treatment apply 2 oz/bu on wheat, barley, oats, rye. Seed may be pretreated in rotary drum, dust attachment, or in planter box. Keep dry. For slurry-type treatment use 230 cc/cwt on wheat, barley, rye and 288 cc/cwt on oats.
PCNB, TERRACLOR	TERRACOAT LT-2 (24% PCNB)	Spray-liquid	2 oz/bu on wheat, barley, oats (not registered for rye). Usually controls covered smut (bunt) and seedling blights but not loose smut. Treat at planting time or before.
TERRACLOR and TERRAZOLE	TERRACOAT L-205 (23.2% PCNB and 5.8% TERRAZOLE)	Spray-liquid	2 oz/bu on wheat, barley, oats (not registered for rye). Usually controls covered smut or bunt and seedling blights but not loose smut of wheat and barley.
TERRACLOR and TERRAZOLE	TERRACOAT SD-205 (20.0% PCNB and 5.0% TERRAZOLE)	Dust	3 oz/bu on wheat and oats, 4 oz on barley (not registered for rye). Usually controls covered smut or bunt and seedling blights, but not loose smut of wheat and barley. Apply in drill box at planting time.

TABLE 12 (Continued)

Fungicide[a]	Tradenames	Methods of application	Dosage and remarks
Thiram	ARASAN 50-RED (50% thiram)	Dust	2 oz/bu on wheat, barley and rye (not registered for oats). Will usually control covered smut or bunt of wheat, barley and rye but needs overcoat of carboxin (VITAVAX) for loose smut control.
	ARASAN 75 (75% thiram)	Dust	1 1/3 oz/bu on wheat, barley, rye (not registered for oats).
	ARASAN 70-S (70% thiram and 2% methoxychlor)	Slurry	1 1/3 oz/bu on wheat, barley, rye (not registered for oats).
POLYRAM	POLYRAM, 53.5%	Dust	2 oz/bu on barley and oats, 3 oz on wheat (not registered for rye). Treat at planting time or before.

[a]The fungicides listed have 1975 EPA label registrations with the limitations indicated. Other trade name materials may also have registration. The mention of specific products does not warrant or guarantee the standards of these products, nor does it imply approval of the materials listed to the exclusion of others that may be equally suitable.

408, November 4, 1964). Treated seed left over after planting may be held and planted the following year if it is not overtreated and if it is held in a dry, well-ventilated location and away from feed and foodstuffs. The moisture content of the seed should be held below 14%. It is good practice to have a germination test made on all holdover seed. Contamination of small grain seed with fungicide sometimes occurs through ignorance where a farmer failed to clean out a wagon or grain auger properly or accidentally mixed in some surplus treated seed with a wagon load of untreated grain sent to market for food or feed use.

The manufacturer's directions and precautions on the label should be carefully read and followed when handling seed treatment chemicals or using treated seed.

XIII. RICE

Cultivated rice varieties belong to a single species (Oryza sativa). This is a warm-climate cereal grown without irrigation (upland) and with irrigation (lowland). Both spring and intermediate to winter types are grown. Long-, medium-, and short-grain types are represented. The major commercial crop is grown under some form of irrigation or when the plants are partially under water for much of the growing season (9, 60). In 1973, rice was harvested from 2,170,200 acres and averaged almost 43 hundredweight (cwt) per acre, over 4 cwt less than in 1971 and 1972 (2).

Diseases are important in the economic production of rice. The estimated annual loss in the United States for the 10-year period 1951-1960 is 7%. This amounted to a loss from potential production of 3,874,000 cwt or 193,700 tons with an estimated value of $18.8 million (133). Losses from diseases in the more intensive rice-growing areas of Asia are frequently higher than those in the United States (162).

In extensive tests with rice, the same fungicides used to treat wheat, oats, barley, and rye seed (Table 12) have given excellent results as have products containing chloranil and dichlone (see under Corn). As with seed treatments on other crops, maximum benefits come from early plantings when the soil is cool and damp. Little benefit normally results from treating rice seed sown late in the season (94). The same precautions in handling these materials should be followed, as outlined above under Small Grains.

In certain humid areas where rice is grown intensively (e.g., Japan), blast (Piricularia oryzae)—by far the most economically important disease—is controlled by helicopter or conventional aircraft applications of antifungal antibiotics or other fungicides. Minor leaf and culm diseases such as brown leaf spot or Heliminthosporium blight (Table 13), Cercospora

TABLE 13

Rice Diseases in the United States and the Fungi that Cause Them,
Partially or Completely Controlled by the Same Seed Treatment
Fungicides Used for Small Grains

Disease	Causal organism
Seedling blights, kernel blights	Cochliobolus miyabeanus (Helminthosporium oryzae), Rhizoctonia solani, and species of Pythium and Fusarium (Gibberella)
Brown leaf spot or Helminthosporium blight	Cochliobolus miyabeanus (Helminthosporium oryzae)
Blast or rotten neck	Piricularia (Pyricularia) oryzae
Kernel smut	Tilletia horrida

or narrow brown leaf spot (Cercospora oryzae), sheath and culm blights
(Rhizoctonia spp.), and culm rot (Leptosphaeria salvinii = Helminthosporium
sigmoideum) are probably kept somewhat in check by the same fungicidal
applications.

XIV. COTTON

Cotton (Gossypium spp.) is a major fiber crop in warmer climatic
areas of the world and grown over a wide range of environments. In the
tropics, many species are perennial. The cultivated species and varieties
of cotton are commonly divided into four groups: (a) the wild Australian
species, (b) the wild and cultivated Asiatic species, (c) the wild American
species, and (d) the American semiwild and cultivated species.

The estimated average annual loss from potential production due to
diseases in 1951-1960 was 12% or 1.86 million bales valued at over $300.9
million (133). In 1973, cotton was harvested for lint on 11,995,200 acres,
yielding an average of 519 bales/acre (2).

Seedling diseases, which account for about 20% of the annual disease
loss (300,000 to 1/2 million bales), are caused by a complex of seed- or
soilborne fungi and bacteria (Table 14). Reduction in yield may be as high
as 15% (184). The seedling disease complx varies not only from year to
year—depending primarily upon the weather—but also from one field to the
next. During cold, wet seasons farmers may be forced to replant a second
or even a third time, resulting in additional losses of seed, fertilizer, time,

TABLE 14

Cotton Diseases and Their Causal Organisms that Are at Least Partially
Prevented by Fungicides

Disease	Cause[a]	Causal organism
Seedling disease complex		
(seed decay, damping-off,	F	Pythium spp., Fusarium spp.,
sore shin),	F	Rhizoctonia spp., Thielaviopsis basicola,
seedling blights,	F	Physalospora rhodina (Diplodia gossypina),
seed-borne anthracnose,	F	Glomerella (Colletotrichum) gossypii,
Ascochyta blight, and	F	Ascochyta gossypii,
bacterial blight	B	Xanthomonas malvacearum
Boll rots (primary)	F	Diplodia spp., Glomerella gossypii,
(many secondary fungi and bacteria involved)	F	Gibberella fujikuroi (Fusarium moniliforme), Aspergillus spp., Sclerotium sp.,
	B	Xanthomonas malvacearum
Rust	F	Puccinia stakmanii

[a]Key: F, fungus or fungi; B, bacteria.

labor and machinery operation. Replanted cotton is also commonly damaged by insects and weed competition (133), while poor seedling stands of
weakened plants may greatly reduce the efficiency of mechanization—particularly mechanical harvesting.

Seedling losses occur across the entire Cotton Belt, being greatest
in the Southeast where, on the average, seed quality is rather low (133).
As with other field and vegetable crops, research has shown that low-
quality seed when planted is affected by adverse conditions (cold—below
65° F, wet soil) and is also more susceptible to the seedling disease complex.

Table 14 lists the diseases of cotton that are partially to completely
controlled by fungicides.

Some fungicide spraying of the foliage using ground or air equipment,
is carried out in Texas, New Mexico, and Arizona to control southwestern
cotton rust (Puccinia stakmanii). Zineb 75W or maneb 80W (0.25% to
0.375%) is applied before the cotton is infected. Three applications are
usually suggested, at 12- to 14-day intervals, starting before the rains

begin in June or July. Zineb or maneb applications also help check such
minor leaf diseases as Alternaria leaf spot or blight (Alternaria spp.),
Cercospora or zonate leaf spot (Cercospora gossypina, C. althaeina),
Ascochyta leaf blight (Ascochyta gossypii), and areolate mildew or frosty
blight (Mycosphaerella areola).

Seed treatment is most effective after sulfuric acid is used to delint
the seed. Fungicides (see Table 12) are effective in increasing stands and
preventing seed decay. Captan, PCNB-captan, BUSAN 72 (60% liquid),
carboxin (VITAVAX) 75W, carboxin-thiram (1:1 dust), carboxin-captan (1:1
dust), PCNB (TERRACOAT LT-2, 24% liquid, TERRACOAT L-21, 34.2%
liquid), and thiram seed protectants are recommended by various states for
use on cotton seed. The same types of equipment are used to treat cotton
seed as are used for Small Grains. The dosages of seed treatment fungicides vary depending on whether the cotton seed is (a) fuzzy, i.e., not de-
linted, (b) acid-delinted, or (c) machine-delinted. As with other field and
vegetable crops, sowing top-quality, properly treated seed will (a) help
prevent seed decay and preemergent damping-off, (b) save cost of labor
and replanting, (c) improve emergence, resulting in better stands of more
uniform plants, (d) increase the average yield per acre, and (e) help prevent certain leaf and boll diseases (48).

Proper seed treatment, however, offers little protection against seedling blight organisms that attack just before or after seedling emergence.
Therefore, additional fungicides effective against species of Rhizoctonia,
Pythium, Fusarium, and other soilborne fungi must be placed in the seed
furrow and into the covering soil at the time of planting. Recommended
fungicides are applied by three methods: (a) in-furrow spray, (b) in-furrow
dust or granules, and (c) planter-box or hopper-box (Table 15). The in-
furrow spray method is the most effective while the planter-box treatment
is least effective, but the latter costs far less. In-furrow methods require
more fungicide per acre plus additional equipment (e.g., planter-mounted,
preemergence band sprayer). It is important that the seed and all soil
surrounding it (up to and including the surface soil) be properly treated.
The fungicide or mixture of chemicals should be thoroughly mixed with the
soil. Cotton soil fungicide research is an active, changing field.

It is impossible at present to suggest any one or a combination of fungicides that will guarantee increased stands and yields in all fields under
all environmental conditions throughout the Cotton Belt.

Experiment stations, USDA scientists, and chemical companies now
realize that the potential seedling disease complex in the soil should be
sampled prior to planting. Soil fungicide mixtures can then be suggested
depending on the complex of organisms present in the sampled field. Table
15 lists a number of fungicides recommended by Cotton Belt states (e.g.,
Alabama, Florida, Georgia, Louisiana, North and South Carolina) to reduce

TABLE 15

Fungicides and Combinations of Fungicides Suggested for Application to Soil to Control Seedling Diseases of Cotton

Fungicide (or proprietary name) and active ingredient	Method of application[a]	Rate per acre (a.i.)
PCNB (pentachloronitrobenzene) and ethazol (TERRAZOLE) - granular	I-F-D	1.3 to 2 lb
PCNB and ethazol - dust	I-F-D	1.3 to 2 lb
PCNB and ethazol - emulsion concentrate (EC)	I-F-S	1.3 to 2 lb
PCNB and ethazol (TERRACLOR SUPER X)	I-F-S	3/4 to 1 gal
PCNB 75 and captan 50W	I-F-S	1 1/2 to 2 lb and 2 lb
PCNB and captan (40:40)	I-F-S	2 1/2 lb
PCNB and captan or thiram (10:10)	I-F-D	10 to 15 lb
PCNB and captan or thiram (10:10)	P-b	3 to 10 lb[b]
Chloromeb (DEMOSAN 65W)	I-F-S	1.3 to 2 lb
Chloromeb (DEMOSAN 10D)	I-F-D	1 to 2 lb
Chloromeb (DEMOSAN 65W)	P-b	6.5 oz/100 lb[c]

[a]I-F-S = In-furrow spray; I-F-D = In-furrow dust; P-b = Planter-box or hopper-box. (NOTE: The planter-box method of application is not recommended when acid-delinted seed is used or when seed is planted more than 1 1/2 in. deep. Check label directions.)

[b]Pounds of fungicide per acre depends on the seeding rate which may vary from 15 to 100 lb/acre.

[c]Seed overcoat following treatment.

seedling disease losses. Additions and subtractions to this list can be expected yearly (68, 74, 160).

Where specific tradenames or manufacturers are mentioned, it is to be understood that they are not listed to the exclusion of similar and competitve products or firms. They are simply representative of the most generally available products in the United States.

XV.   PEANUT

The common peanut or groundnut (Arachis hypogaea) comprises several types and varieties based largely on seed size and color. The peanut is an annual legume, probably native to South America. It is cultivated for its edible seeds (and the oil derived therefrom) and for livestock feed and hay (60, 109). In 1973 almost 1.5 million acres were harvested for nuts, yielding an average of 2,323 lb of nuts per acre (2).

Estimated average annual losses caused by various peanut diseases, for the 10-year period 1951-1960, amounted to 28%, almost 500 million pounds, and worth an estimated $53.7 million (133). Principal losses from diseases in the production of peanuts are caused by Cercospora leaf spots (Cercospora (Mycosphaerella) arachidicola and C. personata = Mycosphaerella berkeleyi), stem rot or southern blight (Sclerotium rolfsii), and leaf rust (Puccinia arachidis) (111).

Fungicide seed treatments (Table 16) are relatively ineffective if the seed is of poor quality (on the small side, not fully matured, cracked or otherwise damaged, moldy or stained, not plump or well filled) and planted too deep in cool soil (below 70° F) that has not been well prepared. Most peanut seed is now treated by the processor. Treated seed should not be fed to man or animals under any circumstances.

Dusts and sprays to control leaf spots have increased peanut yields 300 to 1,000 lb/acre or more. Six applications of fungicides are usually suggested, spaced 10 to 14 days apart. Dusting or spraying should begin when the first spots appear on the older leaves at the base of the plant, and be continued until 2 weeks before expected harvest. Fifteen to 20 lb of dust per acre is suggested for the first application or two to insure good coverage. Larger amounts (20 to 30 lb) may be required as vine growth becomes more extensive. If a moderate to heavy rain occurs within 24 hr after application, a repeat dust or spray is suggested (210).

More care and precision is required when spraying than dusting. Liquid copper (CITCOP 4E, LCF 6, COPOLOID 6, FOR-COP 80, OXY-COP 8L, K-KOP 80), cupric hydroxide (KOCIDE 101, 0.250% or KOCIDE 404S, 1 to 2

TABLE 16

Peanut Diseases Partially or Completely Controlled by Fungicides, Their Causal Fungi,
Fungicides, and Methods Used in Their Control

Disease	Causal fungus	Fungicide	Method[a]
Seed decay	Rhizoctonia solani, Pythium spp., Aspergillus spp.,	Thiram, 50% or 75%	ST
Seedling blights	Penicillium spp., Fusarium	Captafol–captan, 30:30	ST
Root rots	spp., Macrophomina phase–olina (Sclerotium bataticola),	BOTRAN–captafol, 30%, 35% or 37 1/2% each or 60–20	ST
Peg rots	Sclerotium rolfsii, Rhizopus spp.	BOTRAN–captan, 30%, 35%, or 37 1/2% each	ST
Pod rots		Maneb–captan, 30% or 37 1/2% each	ST
Crown rot		Captan 18.75%, maneb 18.75%, PCNB 10%, and ethazol 2.2% (TSP 4–WAY)	ST
		Captan, 75% or 80%	ST
		PCNB–ethazol (TERRACOAT SD–205)	ST
Leaf spots (and leaf rust)	Cercospora arachidicola, C. personata, Alternaria sp., Leptosphaerulina arachidicola	Benomyl (0.031–0.062%)	S
		Sulfur (see label)	D, S
	Puccinia arachidis	Copper sulfate and sulfur (see label)	D, S

TABLE 16 (Continued)

Disease	Causal fungus	Fungicide	Method[a]
		Chlorothalonil (0.125-0.25%)	S
		Liquid copper (see label)	S
		KOCIDE 101 (0.250%) or KOCIDE 404S (1-2 qt/acre)	S
		Metiram (0.125-0.250%)	D, S
		DU-TER (0.046%)	S
		MANCOZEB (0.125-0.250%)	D, S
		COPPER COUNT, S and NC (see label)	S
Stem rot or southern blight	Sclerotium rolfsii	PCNB (see label)	D, S[b]
		PCNB-captan (see label)	D, S

[a]ST = seed treatment; D = dust or granules; S= spray.

[b]Stem rot is now largely controlled in most peanut-growing areas by rotation with corn, cotton, small grains or grain sorghum, deep covering of surface organic matter at the time of seed bed preparation, flat cultivation, keeping soil out of the row (nondirting), and good control of leaf spot diseases (111, 210). PCNB or PCNB-captan is usually applied in a 12-in. wide band centered over the row at early pegging as dust, granules or spray; about 10 to 15 lb of technical chemical(s) per acre.

qt/acre), ammonium carbonate (COPPER COUNT 1/2 to 3/4 gal/acre;
COPPER COUNT S and NC also contain sulfur, 1/3 to 3/4 gal), benomyl
50W (0.031 to 0.062%), chlorothalonil (BRAVO 6F), 1 to 1 1/2 pt/acre,
copper-sulfur dusts (MICRO-SPERSE also contains landplaster), copper-
sulfur liquid (FUNGI-SPERSE), 1 to 2 gal/acre, metiram (POLYRAM) 80W
(0.125% to 0.250%), DU-TER, 47.5% W (0.046%), and mancozeb (DITHANE
M-45, MANZATE 200) 80W, 0.125% to 0.250% are currently recommended
spray materials (Table 16). Ten to 50 gal of water per acre are required to
obtain good coverage with ground sprayers, 3 to 5 gal for materials being
applied by air. A spreader-sticker is usually added to reduce surface tension and increase adhesiveness.

XVI. SMALL-SEEDED LEGUMES

(Soybeans, alfalfa, sweet clovers, clovers, lespedeza,
various field beans and peas, lupines, vetches, Crota-
laria, velvetbeans, cowpea, birdsfoot-trefoil, kudzu)

These widely-grown, small-seeded legumes are grouped together because fungicides are not commonly a part of their production, at least in
the northern half of the United States. In the South, seed treatments are
commonly recommended on certain legumes where seed quality and germination are low. Increases in stand of 20% from proper seed treatment is not
uncommon. In many cases, however, increased stands have not resulted in
higher forage or seed yields. Since certain types of legume seed are easily
damaged, great care must be taken during harvest, cleaning, bagging,
other handling, and so forth. It is not uncommon for the field germination
of soybean seed to decrease following seed cleaning and fungicide treatment
due to seedcoat and embryo damage.

As with other crops, maximum benefits from seed treatment occur
when cool, wet weather follows planting and emergence is delayed.

Seed treatment chemicals are applied to control seedborne seed decay
and seedling blight fungi (Table 17).

Seed treatment fungicides are formulated for dry treatment of seed
(dust) to be applied by commercial dust treaters, by home made rotary
drums, or by mixing directly with the seed in a planter box. Some are
wettable powders that can be adapted to slurry treaters. Only a few are
liquids.

Another reason the seed of small-seeded legumes is not routinely
treated with a fungicide is that a bacterial inoculant, specific for each legume, is commonly coated on the seed before planting to ensure proper root
nodulation, necessary for "nitrogen-trapping." In the past, the seed

TABLE 17

Diseases of Small-seeded Legumes, Their Causal Fungi, and Seed
Treatment Fungicides Used in Their Control

Disease	Causal fungi[a]	Fungicide[b]
Seed decay seedling blights, seedborne anthracnose, leaf, stem and pod spots or blights, etc.	Pythium spp., Phytophthora spp., Aphanomyces euteiches, Glomerella or Colletotrichum spp., Ascochyta and Mycosphaerella spp., Cercospora spp., Diaporthe spp., etc.	Thiram, 50% or 75% Captan, 25%, 50%, 75%, or 80% Chloranil, 96% Captan-thiram, 43% captan and 43% thiram Captan-Moly, 25% captan, and 4.75% molybdenum EVERSHIELD CM SEED PROTECTANT, 29.52% captan, and 0.34% malathion Chloroneb, 65% PCNB-ethazol (TERRACOAT SD-205, 20% PCNB, 5.0% TERRAZOLE; TERRACOAT L-205, 23.2% PCNB, 5.8% TERRAZOLE)

[a]These fungi do not attack all small-seeded legumes. Many fungal species are limited to a single species or genus of legume.
[b]Lindane or malathion is added to certain formulations of thiram and captan to protect the seed against insect feeding.

treatment fungicides, when improperly applied, often reduced the effectiveness of the inoculant by killing the bacteria.

XVII. TOBACCO

The cultivated tobaccos comprise two species, Nicotiana tabacum, the common tobacco and major commercial crop, and N. rustica. The latter species is the principal tobacco cultivated for smoking in certain parts of Europe and Asia. These two annual (or rarely perennial) herbs hybridize readily and can be crossed with a number of additional species of Nicotiana indigenous to the Western Hemisphere. A number of other species are cultivated for ornament, and one species—tree tobacco, Nicotiana glauca—is the source of the insecticidal alkaloid anabasine (60, 109, 134). In 1973, tobacco was harvested from 885,000 acres and yielded an average of 1,962 lb/acre (2).

Tobacco is one of the five crops in the United States that annually returns to farmers more than $1 billion. In addition, federal, state, and local taxes total more than $3.5 billion each year.

During the period 1951-1960, estimated annual losses in tobacco due to disease were 11% in the United States, almost 243.4 million pounds and worth an estimated $132.2 million (133). Losses during this 10-year period were sharply reduced by the development and widespread use of tobacco varieties resistant to wildfire (Pseudomonas tabaci), the tobacco mosaic virus (Fig. 1), black shank (Phytophthora nicotianae var. parasitica), bacterial or Granville wilt (Pseudomonas or Xanthomonas solanacearum), Fusarium wilt (Fusarium oxysporum var. nicotianae), blue mold or downy mildew (Peronospora tabacina), black root rot (Thielaviopsis basicola), and Verticillium albo-atrum (2, 30, 109, 134). Other nonchemical control measures include crop rotation, stringent sanitation measures, use of cover crops, insect, nematode and weed control practices, improved soil drainage and balanced fertility (134, 156, 222).

At present, fungicides are used in the production of tobacco in the United States in four ways: (a) seed treatment, (b) disinfestation of plant bed soil, (c) plant bed dusts and sprays, and (d) field dusts or sprays (Table 18).

Few pathogens are carried in or on tobacco seed. For this reason, seed treatment is not generally practiced. It is much more convenient for the tobacco farmer to buy certified or disease-free seed from a reputable dealer than for him to grow, clean, treat, and test his own seed for germination (134).

TABLE 18

Operation, Diseases Controlled, Causal Organisms, and Prevention Treatments Used to Control Tobacco Diseases in the United States

Operation	Diseases controlled and causal organisms	Preventive treatment
Sterilization of plant bed soil before seeding	Damping-off (Pythium spp., Rhizoctonia solani, Fusarium spp., Phytophthora spp.), collar rot (Botrytis cinerea, Sclerotinia sclerotiorum), black root rot (Thielaviopsis basicola), black shank (Phytophthora nicotianae var. parasitica), angular leaf spot (Pseudomonas angulata), wildfire (Pseudomonas tabaci), blue mold (Peronospora tabacina), blackleg (Erwinia carotovora), Olpidium seedling blight (Olpidium brassicae), sore-shin (Sclerotium bataticola, Rhizoctonia solani), anthracnose (Colletotrichum destructivum), Verticillium wilt (Verticillium albo-atrum), bacterial or Granville wilt (Pseudomonas solanarum).	Steam, methylbromide[a], chloropicrin[b], SMDC[c] or DMTT[d] (see label instructions for use of chemicals)
Plants up; apply dusts or sprays once or twice weekly until transplanting to field	Damping-off or bed rot, blue mold, anthracnose, frogeye leaf spot (Cercospora nicotianae), Ascochyta leaf spot (Ascochyta phaseolorum)	Ferbam, zineb, POLY-RAM, or maneb (0.062% to 0.5%; see label) or fixed copper (0.25% to 0.5%)
	Angular leaf spot, wildfire (Pseudomonas angulata, P. tabaci)	Streptomycin spray, 200 ppm; drench, 100 ppm (see text)

| Plants set in field and growing (primarily shade-grown tobacco in wet seasons) | Blue mold, brown leaf spot (Alternaria tenuis), frogeye leaf spot (Cercospora nicotianae or capii), anthracnose | Zineb, maneb, or POLYRAM (see label) |

[a]Methyl bromide is sold as DOWFUME MC-2, BROZONE, PANOBROME, PANOBROME CL, PESTMASTER METHYL BROMIDE, PICRIDE, PROFUME, TRIZONE, GREAT LAKES BROMO-O-GAS, WEEDFUME, NEMASTER, MBC FUMIGANT, and others.

[b]Chloropicrin is sold as LARVACIDE 100, PICFUME, TRI-CLOR, and so forth. LARVACIDE SOIL FUMIGANT contains 93.5% chloropicrin and 6.5% EDB; NEMEX is a 50:50 mixture of chloropicrin and dichloropropenes. Chloropicrin is trichloronitromethane or tear gas.

[c]SMDC, containing 32.7% sodium N-methyldithiocarbamate, is sold as STAUFFER VAPAM SOIL FUMIGANT.

[d]DMTT, 3,5-dimethyltetrahydro-1,3,5,2H-thiadiazine-2-thione, is sold as MYLONE 25W, 50W, 85W, Dust 50; SOIL FUMIGANT M; MILLER MICRO-FUME 25D (MYLONE), and so forth.

Soil disinfestation using steam or a soil fumigant (Table 18) is a practical necessity to control diseases in commercial tobacco-growing areas. Methyl bromide is the preferred soil fumigant where steam is unavailable (134). Disinfestation, properly done, also controls such soilborne pests as nematodes, and certain plant-infecting viruses, insects and germinating weed seeds (30).

Depending on the frequency and amount of rainfall and high humidity, bed covering, ventilation system, tobacco-growing region, and so on, several sprays or dusts in the plant bed are required to control a number of fungal pathogens, especially the very important blue mold fungus (Table 18). Dithiocarbamate fungicides (ferbam, zineb, mancozeb, metiram) are preferred. On very young plants—the size of a dime—the usual recommended dosage is 0.062%. The percentage is gradually increased as the seedlings age to 0.250% or even higher (74, 134, 160). Ferbam and zineb dusts or sprays are generally preferred on very young plants, while maneb and metiram (POLYRAM) are more effective on older seedlings and after transplanting in the field. Fixed copper (0.25% to 0.5%) fungicides are sometimes substituted for the dithiocarbamates if wildfire and angular leaf spot (Pseudomonas angulata) appear in the seed bed (134). Streptomycin is widely used to control the latter two bacterial diseases. A 200 ppm solution is applied as a spray at the rate of 5 gal to 300 ft of bed, applied weekly for 5 weeks, and beginning when plants are in the two-leaf stage. For a drench, a 100 ppm streptomycin solution is used at the rate of 10 gal to 300 ft of bed. Streptomycin is sometimes mixed with a dithiocarbamate fungicide to give control of both fungal and bacterial pathogens. In some areas, the wildfire and angular leaf spot bacteria are now resistant to streptomycin (134).

A blue-mold warning service was established in 1945, operated by the USDA and cooperating states (134, 148). All tobacco-growing areas are thus kept informed of the progress and extent of the disease through releases by all types of mass media. A similar warning system was created in 1961 by Coresta (Centre de Cooperation Pour Les Recherches Scientifiques Relatives au Tabac) with correspondents in most European and Mediterranean countries. Any occurrence of blue mold is reported by cable to the service headquarters in Paris which then informs, by cable, the countries most threatened and warns all correspondents by air mail. A local blue-mold forecasting and spray warning service is also being used successfully in Australia (134).

Field spraying with fungicides is not normally practiced in the United States, except in wet seasons on shade-grown tobacco. Blue mold is generally the only important field disease requiring fungicide applications; maneb (mancozeb) is preferred (134). Field sprays probably also aid in checking brown leaf spot (Alternaria tenuis), frogeye leaf spot (Cercospora nicotianae or C. apii), and anthracnose.

XVIII. FIELD OR DENT CORN

All cultivated forms of corn, or maize, are represented by the single species, Zea mays. Dent corn (Z. mays var. indentata) comprises the major acreage in the United States and flint corn (Z. mays var. indurate) in South America, Europe, and Asia (60). In 1973, dent corn was harvested from 61,760,000 acres for grain (average yield 91.4 bushels/acre), 569,000 acres for forage, and 8,764,000 acres for silage (average yield 12.5 tons/acre) (2).

Corn production in the United States is reduced each year by more than 25 diseases (45, 60, 214). Estimated annual losses in the United States for the 10-year period, 1951-1960, were 12% of the potential production, over 413 million bushels valued at $527.3 million, the largest figure of any crop grown in the United States. Generally, losses are due to lowered grain quality, decreased value of fodder, and decreased yield (45, 60, 133, 214).

In contrast to some diseases in other crops, the diseases of corn seldom become severe over very wide areas. Up to now, the production of corn in any given locality of the United States has not been limited by disease, except during the "southern corn leaf blight year" of 1970, where soil and weather conditions have been favorable for the crop, nor has it been necessary to stop growing corn over a wide area because of disease (45, 214).

Two ways to control corn diseases are the use of adapted, disease-resistant hybrids and the application of fungicides to the seed. Other practices such as spraying plants with fungicides, weed and insect control, crop rotation, proper seed bed preparation, destruction of diseased crop residues, adjustment of soil fertility by intelligent use of fertilizers, maintenance of good soil drainage, and a favorable pH tend to reduce losses from certain diseases, but have limited application (45, 60, 124, 214).

The incorporation of genetic resistance into agronomically desirable hybrids by breeding is the most efficient and permanent means of controlling corn diseases. For most diseases, the resistance of a hybrid is generally proportional to the number of resistant inbred lines that were combined to make the hybrid. Thus a double-cross hybrid comprised of four resistant inbred lines is likely to be more resistant than one possessing only a single resistant inbred line in its pedigree (45, 214).

Treatment of seed corn with a fungicide, such as captan, thiram, chloranil, or dichlone may control seed rots and seedling diseases but not other diseases. The fungi causing this disease complex include Pythium spp., Diplodia zeae, Gibberella zeae (Fig. 53), Fusarium moniliforme, Penicillium oxalicum, Aspergillus spp., Nigrospora oryzae, Rhizoctonia zeae, and R. bataticola (Fig. 5) (45, 124, 214). Thiram and captan protect

the seed from invasion by soil-borne pathogens during the critical early
stages of germination. However, these materials have little if any effect
on seed-borne pathogens already established within the seed prior to planting. Essentially all seed corn is now treated by the processor before the
grower buys it (45, 214). Corn is especially susceptible to seedling blight
when the pericarp or seed coat is broken and not protected by a fungicide
(125).

Spraying corn plants with maneb, mancozeb, or zineb fungicides to
control fungal leaf blights (caused primarily by Helminthosporium turcicum,
H. carbonum, and H. maydis) is successful in southern Florida where this
practice is commercially feasible because it is applied to high value,
market-garden sweet corn (45, 214). More and more seed production
fields of dent corn are also being sprayed each year in the midwest. Usually two or three applications are made at 7- to 10-day intervals, using
1 1/2 to 2 lb/acre.

XIX. SORGHUMS, SUDANGRASS, JOHNSONGRASS, BROOMCORN

The cultivated sorghums are largely annuals of the species Sorghum
vulgare. The varieties of sorghum grown in the United States fall into four
more or less distinct groups: (a) Sorgos (sweet or saccharine sorghums),
grown for hay, fodder, silage, and syrup; (b) grain sorghums (including
hegari, kafir, milo, feterita, durra, kaoliang, shallu, and darso), grown
mostly for grain but also for forage and silage; (c) broomcorn (S. vulgare
var. technicum) cultivated for its brush fiber; and (d) grass sorghums con-
sisting of Sudangrass (S. vulgare var. sudanense) and the related perennial
Johnsongrass (S. halepense) grown for forage (60, 127, 205). In 1973, a
total of 15,940,000 acres of sorghum were harvested for grain (average
yield 58.8 bushels), 2,137, 000 acres for forage, and 11,995,200 acres for
silage (average yield 11.4 tons/acre) (2).

Diseases cause relatively large losses in these crops. Four general
types may be recognized: (a) those that reduce stands by rotting the seed
or by killing the seedlings; (b) those that attack the leaves and decrease the
value of the plants for forage; (c) those that attack only the heads and prevent normal grain formation; and (d) those that cause root or stalk rots and
prevent normal development and maturity of the entire plant (60, 82, 127,
133, 187, 205).

Reduction in yield of grain sorghum in the United States as a result
of disease was estimated at 9% for the 10-year period 1951-1960. This resulted in a loss from potential production of 33,866,000 bushels valued at
$34 million. Estimated losses for this same period for silage and forage
sorghums is also 9%, amounting to 1,493,000 tons valued at $8.55 million
(133).

Disease losses in sweet sorghum occur regularly, reducing the total yield of syrup. Losses from all diseases reduced the total annual yield an estimated 15% during the period 1951-1960. This amounted to 463,000 gal of syrup valued at $987,000 (133).

The planting of adapted, resistant hybrids and varieties, crop rotation with nongrass species, destruction or clean plow-down of crop residues, planting high quality seed, balanced soil fertility, insect, nematode and weed control, and proper management to ensure adequate soil moisture and avoid high soil temperatures, are important control measures (60, 82, 127, 187, 205).

Seed treatments are applied to control the same genera and species of fungi that attack germinating corn seeds. Sorghums are of tropical origin; therefore the seeds germinate best at a temperature (70° F or above) considerably higher than the soil temperatures that usually prevail at planting time in the United States. This subnormal soil temperature exposes the seeds to attack by various seed- and soilborne fungi (see under corn). These fungi, which may rot the seed or kill the young seedlings before or after emergence, can be controlled effectively by treating the seed with one of several fungicides, e.g., thiram[*], captan[**], dichlone[†], chloranil[††], captan-maneb[§], PCNB-ethazol[§§], or metiram[#] (60, 82, 127, 187, 205). These fungicides also generally control covered kernel smut (Sphacelotheca sorghi) and loose kernel smut (Sphacelotheca cruenta). Most seed processors in the United States now treat their commercial sorghum seed with one of these chemicals or a fungicide-insecticide combination (82, 127, 187, 205). Seed-treatment materials containing both a fungicide and an insecticide (e.g., diazinon, lindane, methoxychlor) are generally superior to those containing only a fungicide for improving stands by higher germination, and greater seedling vigor—especially on glume-free seed (127, 187, 205). These materials not only control seedling diseases and kernel smuts but also insects that attack the seed, such as wireworms, seed-corn maggots, and the thief or kafir ant (127, 187). The cost of this seed treatment is

---

[*]Thiram is sold as THIRAM SEED PROTECTANT, ARASAN 42-S, 70-S, SF-X, 50-RED and 75 SEED PROTECTANT. Thiram is also sold in combination with Captan as CAPTAN-THIRAM 43-43W.
[**]Captan is sold as CAPTAN 75 SEED PROTECTANT, CAPTAN 80W, ORTHOCIDE 75 SEED PROTECTANT, MILLER'S CAPTAN 75.
[†]Dichlone is sold as PHYGON SEED PROTECTANT, PHYGON-XL, CHIPMAN DICHLONE, and others.
[††]Chloranil is sold as SPERGON, SPERGON-SL, NIAGARA SPERGON SEED PROTECTANT, and others.
[§]Captan-Maneb is sold as AGROSOL S, CAPTAN-MANEB SEED PROTECTANT.
[§§]PCNB-ETHAZOL is sold as TERRACOAT L-205 and SD-205.
[#]Metiram is sold as POLYRAM.

about 1¢ to 15¢/acre, and average increases of 15 bushels of grain and 30% of forage per acre are not uncommon.

It is advisable to treat sorghum seed every year, because seed treatment helps control certain seedborne diseases other than smuts (e.g., bacterial stripe (Pseudomonas andropogoni), bacterial streak (Xanthomonas holcicola), bacterial spot (Pseudomonas syringae or P. holci), rough spot (Ascochyta sorghina), zonate leaf spot (Gleocercospora sorghi), target spot (Helminthosporium sorghicola), and anthracnose (Colletotrichum sp.), and also improves stands and yields (82, 127, 205).

XX. FLAX

Common flax (Linum usitatissimum) consists of two types: fiber and seed, which differ considerably in type of plant growth. Fiber-type varieties develop tall slender stems, with a high content of good-quality fiber, and bear seeds of low oil content. Seed-flax varieties develop shorter stems, which commonly branch more and usually bear larger seeds of higher oil content (60).

Flax is adapted to a wide range of environmental conditions, but does best in cool climates. It is chiefly grown in Texas and California as a winter crop and in northern states (primarily the Dakotas, Montana, Minnesota, northern Iowa, and Wisconsin) as a summer crop. Over half of the nation's flaxseed is produced in North Dakota. Seed-flax varieties are more generally grown in the drier areas whereas fiber flax is grown in humid sections (60).

The 1973 flax seed crop in the United States was estimated at 16,437,000 bushels, approximately one-half the 1960-1964 average. Harvested acres in 1973 were 1.725 million and the 1973 flax seed crop averaged 9.5 bushels (2).

Estimated average annual losses caused by flax diseases in the United States for the period 1951-1960 are 10%. This loss from potential production amounted to 3,757,000 bushels of flax seed valued at $11.6 million (133). Pasmo (Septoria linicola = Mycosphaerella linorum), rust (Melampsora lini), and Fusarium wilt (Fusarium oxysporum f. lini) are the most destructive flax diseases in the United States (50, 60, 67, 133, 221), causing an estimated annual loss of 6% (60, 133). These diseases are largely controlled by growing resistant varieties. A number of soilborne and seedborne fungi occasionally cause seed rot and seedling blights that result in poor stands of weakened plants. Some of the more important of these fungi include: Pythium spp., Thielaviopsis basicola, Alternaria spp., Polyspora lini, Fusarium spp., Helminthosporium sativum, Coniothyrium olivaceum, and Rhizoctonia spp. These organisms invade seedling tissues under favorable environmental conditions and cause severe damage to stands when mechanically-injured seed is used. Yellow-seeded varieties are more susceptible than brown-seeded types. Seedling blights are estimated to cause an annual loss of 1.5% in the United States (133).

Flax seed for planting was commonly treated with the same fungicides used for small grains (Table 12) (60). It is also generally recommended that farmers sow high-quality, crack-free flax seed. Treatment can be applied anytime before planting. Experiment stations often report yield increases of several bushels per acre for a few cents investment in a seed treatment chemical.

XXI. FORAGE GRASSES

Annual cash income from the use of grasslands in the United States is several billion dollars. Six of every 10 acres are better adapted to the production of hay and pasture crops than to most other purposes. Total area of grasslands (ranges, pastures, and haylands) in the United States aggregates more than a billion acres. This area is equal to six states the size of Texas or 18 states as big as Iowa. About 90% of this grasslands area is suitable only for the production of forage (61).

There are approximately 1,500 species of native and introduced grasses in the United States. About 100 of these grasses are very important to agriculture. Of this number, 50 are humid region grasses and 50 are adapted to subhumid conditions (3, 90, 106).

Scientists have divided the grass family into a number of tribes. Each tribe has its own characteristics of structure and seed arrangement (3, 61, 90, 106). The following discussion includes only the more important grasses used for forage on the range, in pastures, or as hay. For additional information on the characteristics and adaptability of these grasses, consult the texts of Donahue and Evans (61), Harlan (90), Ahlgren (3), and Hughes et al. (106). In 1973, a total of 62,190,000 acres of grass were harvested for hay and the average yield was 2.16 tons/acre (2).

On the basis of life history, grasses may be classified as (a) summer annuals, (b) winter annuals, and (c) perennials. The millets are summer annuals, common ryegrass is a winter annual. There are two types of perennials: (a) the bunchgrasses, and (b) the sod formers. Timothy (Phleum pratense), orchardgrass (Dactylis glomerata), and crested wheatgrass (Agropyron cristatum) are examples of bunchgrasses. Sod-forming grasses include smooth bromegrass (Bromus inermis), big bluestem

(Andropogon furcatus), Kentucky bluegrass (Pao pratensis), Bermudagrass
(Cynodon dactylon), and buffalograss (Buchloe dactyloides) (3).

Tables 19 and 20 give estimated average annual losses due to disease
for several important forage pasture and range plants (133).

The forage grasses are attacked by about the same range of disease-
producing organisms as the cereals (Table 11). The importance of disease
varies with the area of the country, kind of grass, production in pure stands,
nurseries or seed fields, and its use in pastures or on the range (60, 106,
133, 199). Control measures naturally differ greatly under these varied
conditions of growth. Many of the diseases have been studied intensively in
certain regions and not at all in others (60). Representatives of these
groups of pathogens attack plants at any stage of development, from seed
germination to maturity (60, 133).

Some of the more important diseases include bacterial blights (Pseudomonas and Xanthomonas spp.); seed decay, seedling blights, foot (crown)
and root rots (Pythium spp., Gibberella and Fusarium spp., Helminthosporium sativum (Cochliobolus sativus), Rhizoctonia solani, Colletotrichum
graminicolum); powdery mildew (Erysiphe graminis); ergot (Claviceps spp.),
downy mildews (Sclerospora spp.); take-all (Ophiobolus graminis); leaf spot,
blights, and blotches (Helminthosporium spp., Scolecotrichum graminis,
Selenophoma spp., Piricularia grisea, Ascochyta spp., Septoria spp.,
Stagonospora spp., Cercosporella spp.); leaf scald (Rhynchosporium spp.);
anthracnose (Colletotrichum graminicolum); leaf, culm, head, and kernel
smuts (Ustilago spp., Urocystis spp., Entyloma spp., Sorosporium spp.,
Tilletia spp., Sphacelotheca spp.); stem (Figs. 21 and 22), leaf, and stripe
rusts (Puccinia spp.); and snow molds (Fusarium nivale, Typhula spp.)
(62, 79, 109, 199).

Some organisms, such as the seedling blight and root rot fungi and
stripe smut, inhibit seed germination, kill seedlings, and reduce stands or
attack and retard the growth and development of older plants. Seed yields
are reduced and weakened grass plants often fail to withstand adverse conditions that normally would not seriously affect their growth and survival.
Leaf-spotting fungi cause leaves to wither and die prematurely. Foliage
diseases weaken plants and reduce their productivity, palatability, and
nutritive value. Fungi causing head smuts and ergot reduce natural reseed-
ing on ranges. In addition, cattle grazing on ergot-infected grass heads are
sometimes poisoned (60, 106, 133, 199). Losses from disease organisms
in irrigated regions sometimes approach those in humid regions (133).

Certain seedborne organisms are controlled by sowing certified,
disease-free seed or treating the grass seed with thiram, maneb, HCB,
PCNB, or captan fungicides; the same ones used on small grains, corn,
and sorghum seed (see Tables 11 and 12 and under corn).

TABLE 19

Estimated Average Annual Losses Due to Disease for Several Important
Forage Grasses in the United States, 1951–1960 (133)

Grass	Percent loss	Quantity of seed (lb)	Value
Bluegrass, Kentucky	8	1,749,000	$525,000
Bromegrass, smooth	8	1,054,000	126,000
Fescue, tall	4	1,365,000	236,000
Orchardgrass	8	1,056,000	169,000
Ryegrass	6	7,726,000	505,000
Timothy	4	1,620,000	171,000
Grasses			
Miscellaneous	5	4,196,000	623,000
Turf	12	1,979,000	607,000
Total			$2,962,000

TABLE 20

Estimated Average Annual Losses Due to Disease for Pasture and
Rangeland Plants in the United States, 1951–1960 (133)

Type pasture or range	Percent loss	Quantity (acres)	Value
Cropland pastures	9	6,387,000	$ 76,645,000[a]
Forest land pastures	3	7,839,000	15,677,000[b]
Grassland pastures and range	5	33,871,000	101,613,000[c]
Total			$193,935,000

[a]Computed at $12/acre.
[b]Computed at $2/acre.
[c]Computed at $3/acre.

Other important control measures include: growing adapted, disease-resistant varieties; seeding a mixture of forages; rotation with nongrass crops (where feasible); avoiding close grazing, clipping, and leaving a heavy mat of hay on the grass in damp weather; carefully controlled burning of dead grass; destruction of weed grasses by cultural or chemical means; and clean plowdown or early removal from the field of any hay crop from severely infected fields (60, 106, 137, 199).

Protective applications of maneb or mancozeb, 3 lb/acre, are sometimes used to control rusts in grass fields grown for seed production (137). Grass treated with these fungicides should not be fed to livestock.

XXII. HOPS

The common hop (Humulus lupulus) is a twining perennial vine of Europe, its fruit is used in brewing; it is also grown for ornament and cultivation (109). Commercial production of hops is chiefly in California, Idaho, Oregon and Washington (2, 109). Production was 55,769,000 lb in 1973, about 20% above average. The yield per acre was 1,744 lb harvested from 31,400 acres (2).

The average annual loss in the United States due to hop diseases, for 1951-1960, is estimated at 13%. This loss in potential production is calculated at slightly more than 7 million pounds valued at $3,318,000 (133). Virus diseases cause the greatest losses in hop yields (10%). Downy mildew (Pseudoperonospora humuli), the most damaging and widespread fungus disease of hops, caused an estimated 3% loss. Verticillium wilt (V. albo-atrum) is not as yet a serious hazard to hop production in the Pacific North-west, but it is expected to become more destructive in the future (2). Hop diseases are controlled largely by growing resistant varieties, planting of certified disease-free roots, and roguing infected plants or plant parts. Sprays or dusts of zineb and copper-lime dust are suggested at weekly intervals on downy mildew-susceptible varieties when mildew is present and wet or humid weather prevails (137, 140). Treating the planting hole with SMDC (STAUFFER VAPAM SOIL FUMIGANT) is a recommended control where Verticillium wilt-infected plants have been located (137).

XXIII. CASTORBEANS

The castorbean (Ricinis communis) is a member of the Euphoriaceae or spurge family. In temperate zones it is a gigantic annual herb, grown

THE ROLE OF FUNGICIDES IN CROP PRODUCTION

for its seeds—which yield approximately 50% castor oil—and for ornament (25, 109). Castorbeans became a cash crop in the High Plains area of Texas during the late 1950s, after high-yielding dwarf varieties and efficient harvest-hullers were developed. Acreage expanded from a few hundred in 1957 to over 10,000 in 1959 (25).

The estimated average annual loss in the United States due to diseases for the 10-year period, 1951-1960, is 11%. This amounted to 3,369,000 lb of castorbean seed valued at $184. The most destructive diseases are alternaria leaf spot (Alternaria ricini) and capsule molds (Alternaria ricini, Botrytis sp. = Sclerotinia ricini = Botryotinia ricini). Capsule drop has been particularly severe in humid areas east of Texas. Some varieties in Mississippi have prematurely dropped up to 70% of their capsules (133).

Seed rot and seedling blights, caused by various soil- and seedborne fungi, including Alternaria ricini, Fusarium spp., Rhizoctonia solani, and Pythium spp., cause an estimated average annual loss of 1% (109, 133). This seedling disease complex can be greatly reduced by sowing high quality seed (high test weight and germination, plump, free of cracks and disease organisms) treated with a thiram or captan seed protectant fungicide suitable for corn or sorghum. Planting at the recommended depth, 2.5 to 3 in. deep, in warm moist soil is also recommended (25).

XXIV. SUGARCANE

Sugarcane (Saccharum officinarum) belongs to the grass family, Gramineae, which includes about 5,000 species of plants (120, 142). The name sugarcane is applied both to the species from which many cultivated sugarcanes are derived, as well as to the cultigen in which this species and others (e.g., S. sinense, S. barberi, S. robustum, and S. spontaneum) have been combined by interspecific crossbreeding and selection (60, 109, 120). All are robust perennial grasses, native to southern and southeastern Asia and neighboring islands (Melanesia). One authority places the origin of sugarcane as a useful food material at from 8,000 B.C. to 15,000 B.C. By 6,000 B.C. it had become established through most of Melanesia, Indonesia, and also in India and China (120).

Sugarcane is grown chiefly in the Gulf States (from Florida to Texas; chiefly in Louisiana) and in Hawaii for sugar, syrup, feedstuffs, and various byproducts (109). In 1973, the crop was grown for sugar and seed on 748,300 acres and yielded an average of 36.8 tons/acre (2).

Sugarcane diseases decrease growth of the crop and reduce the sucrose content and yield of sugar (120, 133). Diseases are a primary factor in the rise and fall of cane varieties (120).

The average annual loss of sugarcane in the United States to disease, 1951-1960, is estimated at 23%, slightly over 1.9 million tons, and valued at $13,521,000. Virus diseases—Ratoon stunting, mosaics, and chlorotic streak—cause the greatest losses (14.5%), followed by red rot (Physalospora tucumanensis or Colletotrichum falcatum) at 4.8% and 1.7% due to soil pathogens (complex of Pythium spp., Phytophthora erythroseptica and P. megasperma, Rhizoctonia spp., Marasmium stenophyllus and M. sacchari, Physalospora tucumanensis, Fusarium spp., Gibberella fugikwia, Melanconium sacchari, Physalospora rhodina, Plectospira gemmifera, Sclerotium rolfsii, Clathrus columnatus, Ithyphallus rubicundis, and Thielaviopsis paradoxa or Endoconidiophora paradoxa. Important foliar diseases include brown stripe (Cochliobolus stenospilus or Helminthosporium stenospilum, eyespot (Helminthosporium sacchari), and pokkah boeng (Fusarium moniliforme) (60, 108, 109, 120, 133, 142).

Disease-resistant cultivars or hybrid selections control most commercially important sugarcane diseases. Propagation by clones, chiefly culm bud cuttings (often called seedpieces or setts), facilitates the use of good-quality, disease-resistant selections free of disease. Quarantines are effective in preventing further distribution of viruses, mycoplasmas, and other organisms carried in the cuttings (60, 108, 120, 142). Other major disease control methods include roguing out diseased plants; early harvest and plow-out orders; disinfection of cane-cutting knives or blades of planting and harvesting machines by heat, alcohol or "Lysol"; balanced fertility, based on soil and tissue tests, with high nitrogen; grassy weed control; improved soil drainage; and hot-water (50° to 52° C for 20 min to 3 hr) or hot-air (54° to 58° C for 8 hr) treatment of seed clones (60, 108, 120, 142). Temperature control is extremely critical as cultivars differ in their response to heat treatment.

XXV. SUGARBEET

While beets (Beta vulgaris) have been cultivated for table use and stock feed for many centuries, their value for sugar production was not widely recognized until about the middle of the 18th century. The first beets analyzed contained 2% to 6% sugar (sucrose), but by selecting plants for high sugar content, many crops now reach 15% to 20% or more of the root weight (14, 15, 107, 121, 129).

Beet sugar did not appear on the market until 1803 and then in small quantities (14, 15). Commercial production of sugarbeets in the United States on a major scale dates from 1870 when the first sugar factory was built in Alvarado, California (121). Today the sugarbeet supplies about

40% of the world's sugar (14). Sugarbeets were harvested from 1,222,200 acres in the United States in 1973, yielding 20.1 tons/acre (2).

The average annual loss in the United States to sugarbeet diseases for the 10-year period, 1951-1960, was estimated at 16%. This amounted to 2,420,000 tons of beets valued at $27,254,000 (133).

Diseases are a serious hazard to sugarbeet production and cause losses in all areas. Virus diseases (primarily curly top, beet yellows, western yellows and mosaic) were responsible for an average annual loss of 9%—practically all being west of the Rocky Mountains (133).

In sugarbeet-growing areas east of the Rocky Mountains, Cercospora leaf spot (C. beticola) caused an estimated annual loss of 3%, black root (primarily Aphanomyces cochlioides, but also Pythium spp., Rhizoctonia solani, Phoma betae), a loss of 1%. Both these fungus diseases reduce yield and quality (133).

Rhizoctonia (R. solani) and other crown and root rots (Fusarium spp., Aphanomyces cochlioides, Helicobasidium purpureum (Rhizoctonia crocorum), Phoma betae, Phymatotrichum omnivorum, Phytophthora drechsleri, Pythium spp., Rhizopus spp., Sclerotium rolfsii, Sclerotinia sclerotiorum) are responsible for the severe loss of mature roots before harvest, the estimated average annual loss being 2%. These losses fluctuate with weather conditions and cropping practices (15, 133).

Fungicides are used on sugarbeets in warm, humid areas to control Cercospora leaf spot where leaf spot–resistant cultivars are not grown due to low yield and reduced sugar content, or where the level of resistance does not hold up in prolonged, warm, moist weather. Where leaf spot(s) (also including Ramularia and Alternaria leaf spots) are serious enough to warrant the expense of spraying or dusting, a number of fungicides can be used. Applications should begin when the disease is first observed in the field. Higher rates are used when disease is severe. The fungicides recommended generally by several states are: benomyl (BENLATE) 50W, 0.046 to 0.062%; triphenyltin hydroxide (DU-TER) 47.5W, 0.046%; thiabendazole (MERTECT 360, 60W and MERTECT 340, 42 FLOWABLE), 0.038 to 0.076%; fixed and soluble copper fungicides (e.g., OXYCOP 8L, Tribasic copper sulfate, Basic copper sulfate, C-O-C-S, TC-90, SOL-KOP-10)—see label; maneb (DITHANE M-22, MANZATE D), 0.187 to 0.250%; mancozeb (DITHANE M-45, MANZATE 200), 0.187 to 0.250%; and metiram (POLY-RAM) 80W, 0.187 to 0.375%. Benomyl and thiabendazole should be applied at 14- to 21-day intervals, DU-TER every 10 to 14 days and the remainder at 7- to 10-day intervals. Label precautions should be read and followed. All of the fungicides listed have given good control when applied at proper dosages and at the right intervals, using hydraulic or air-blast ground spray equipment or airplane. High pressure, ground equipment requires 40 to 100 gal of spray per acre, air-blast machines about 20 to 40 gal, while

airplane application is successful with 5 to 7 gal of water per acre (17, 132).
Increases in yield, ranging from 2 to 7 tons/acre, may result from proper
use of fungicides.

Sprays are superior to dusts and thorough coverage is essential.
Applications should start when leaf spot is first evident about midseason.
The number of applications needed to control the disease depend on weather
conditions and the relative prevalence of the disease in the local area (14).

Seed treatment with a protective fungicide containing dichlone,
DEXON-PCNB, thiram, or captan helps prevent seed decay and damping-
off (called seedling blight or black rot) (42, 132, 151, 225).

XXVI. POTATO

The Irish or white potato (Solanum tuberosum) is a member of the
nightshade family Solanaceae, and is indigenous to South America. It is
closely related to the tomato, eggplant, pepper, tobacco, and such well-
known weeds as horsenettle, buffalo-bur and Jimson-weed. Normally a
perennial, it is grown as an annual (89, 107, 121).

The potato is one of the world's great food crops, combining high
yield, low cost, and a nutritious palatable food (89, 107, 121). It is also
used as a livestock feed, silage, production of starch, alcohol, dextrins
and glucose or "corn syrup" (32, 89, 166, 188). The crop was grown on
1,303,100 acres in 1973 and yielded an average of 228 hundred weight per
acre (21).

Potatoes are susceptible to a relatively large number of diseases.
Estimated average annual losses in the United States, caused by various
diseases, for the period 1951-1960, amounted to 19%. This loss from potential production equals 45,052,000 cwt valued at $89,527,000. Late
blight (Phytophthora infestans), with an average annual loss of 4% from
foliage blight and tuber rot, is destructive in many major producing areas
in the Central and Atlantic states, and at times in some sections of California (Fig. 7). Verticillium wilt (V. albo-atrum), causing an estimated loss
of 3%, is most serious in the western states. Leaf roll, a virus disease,
reduces both yield and quality (annual loss 3%). Less severe but important
losses result from scab (Streptomyces scabies), 1.5%; early blight (Alternaria solani), 1%; Rhizoctonia black scurf and sprout canker (R. solani)
(Fig. 16); Fusarium wilt and stem-end rot (Fusarium spp.), 0.5%; blackleg
(Erwinia phytophthora or E. atroseptica), 0.5%; and bacterial ring rot
(Corynebacterium insidiosum), 0.5%. Most of these diseases occur
wherever potatoes are grown, but are especially prevalent in the North

Atlantic, North Central, and Pacific Coast states. Since viruses (latent mosaic, 1% percent; mild mosaic, 1%; rugose mosaic, 1%; spindle tuber, 0.2%) are carried in seed tubers, these diseases occur wherever potatoes are grown. Losses result from decreased yield and quality as well as from increased cost of production due to (a) planting certified seed and (b) keeping plants comparatively free from infection (133).

Fungicides are used in potato production as treatments of cut "seed," as foliar sprays or dusts, and as disinfectants in the storage house. A variety of fungicides are recommended by the various states as potato seed treatments. There are only a few locations, however, where more than 50% of the acreage is treated. Where seedpiece diseases are of minor importance, growers have learned that seed treatment is not an economical practice (168). In general, fungicide seed treatment may reduce seedpiece decay losses (primarily due to Pythium spp., Fusarium spp., and Erwinia spp.) when cut seed is held before planting or is planted in cold, wet soil. It may also prevent the introduction into noninfested soils of the surface-borne organisms, inciting such diseases as scab, Rhizoctonia, Verticillium wilt, and blackleg (169).

Some of the fungicides recommended as seedpiece dusts or dips include captan (7.5% or 15% dust), dichlone (PHYGON) 50W, mancozeb (DITHANE M-45 or MANZATE) 8% dust, maneb (7 1/2% or 8% dust), metiram (POLY-RAM) 7% or 7 1/2 dust, zineb (DITHANE Z-78 DUST, AGSCO 8% ZINEB), and so on. It is suggested that poor quality seed and seed that is to be held after cutting be treated (132). The cut potatoes are dusted and planted as soon as possible. Dusts are applied with a drum-type treater using 1/2 to 1 1/2 lb of dust to each 100 lb of seed. Dip or wet spray treatments are not as popular as dusts. Dip solutions also lose chemical efficiency rapidly when contaminated with soil from dirty seed tubers (54, 68, 78, 86, 91, 132, 168, 169, 223). Where blackleg and the closely related bacterial soft rot (Erwinia carotovora and E. aroideae) are problems, the use of streptomycin at the rate of 100 ppm in combination with captan or other seedpiece treatment fungicide is recommended (132). Streptomycin may cause poor stands under certain conditions (29). Where cut potato seed is dipped, it should be planted immediately after treatment.

For both early and late blight control in humid areas, applications of fungicides are recommended to start when potato plants are 2 to 8 in. high (or before infection occurs) and continue throughout the growing season. Thus all new plant foliage growth will be protected soon after it develops. Thorough coverage is essential for success. This is accomplished by (a) using the correct volume, pressure, and recommended amount of fungicide per acre, and (b) making sure the sprayer is functioning properly (e.g., check cleanliness of nozzles, wear and position of nozzles, height of spray boom, accuracy of pressure gauge, agitation in spray tank, and so forth). With commonly used high pressure-high volume hydraulic sprayers, it is recommended that 75 to 100 gal of spray be applied at 350 to 400 psi. Under severe disease conditions, or after vines fill the row, the volume should be increased to 100 to 125 gal/acre. Low pressure-low volume hydraulic sprayers, used effectively by some growers, generally utilize a pressure of 80 to 125 psi and a volume of 35 to 60 gal/acre. Regardless of the sprayer or gallonage of spray used, it is important to apply the recommended amounts of fungicide per acre (29, 68, 78, 132, 169).

Modern potato fungicide spray programs are based on whether late blight is present in the field or immediate area and if conditions for blight are favorable (cool and moist) or not favorable (warm and dry):

1. If late blight is present or conditions for blight are favorable, apply per acre, every 4 to 6 days:
   Maneb or maneb and zinc ion coordination product (mancozeb) 80W, 2 lb or
   Chlorothalonil (BRAVO) 6F, 1.5 pt or
   Captafol (DIFOLATAN) 4F, 3 pt or
   DU-TER 47.5%, 7 oz or
   Metiram (POLYRAM) 80W, 2 lb
2. If conditions are not favorable for late blight, apply per acre, every 7 days:
   Maneb or mancozeb 80W, 1 1/2 lb or
   Chlorothalonil 6F, 1.5 pt or
   Captafol 4F, 2.5 pt or
   DU-TER 47.5W, 7 oz or
   Metiram 80, 1 1/2 lb.

Sprays are superior to dusts, especially when late blight occurs frequently (29, 54, 68, 78, 86, 91, 167, 169, 223, 224). When a field contains new late blight infections and harvest is near, the vines should be killed immediately to help prevent tuber infection. Tops should be entirely dead before digging.

If only dusting is possible, maneb, mancozeb, or metiram (7% to 10% active) is recommended as follows:

1. If late blight is present in the field, or conditions are favorable for blight (cool and moist), apply 35 to 45 lb of dust per acre every 4 to 5 days.
2. When conditions are not favorable for late blight (warm and dry), apply 20 to 35 lb of dust per acre every 6 to 7 days.

If possible, dust when vines are damp and the air is still. Although sometimes giving inferior disease control, dusts are used because of the low cost of labor and machinery, ease of operation, and speed of application (168).

For more economic control of late blight, many potato growers spray according to a late blight forecasting program, a cooperative state-USDA program. For infection to occur, late blight needs at least 10 straight hours of temperatures below 70° F and relative humidity above 91% (224). In Florida, growers are urged to spray following 8 consecutive days when the weekly daily temperatures range from 50° to 77° F and the 10-day total rainfall is more than 1 in. When this forecast method is used, intervals between late blight sprays may vary from 5 to 14 days or more, depending on the severity of the disease and the duration of periods when temperatures and rainfall favor its development (68). After the potato crop is removed from storage, the area is thoroughly cleaned and the walls, floors, and ceiling washed or sprayed with a disinfectant. See under Carrot (page 158) for suggested disinfectants and additional information.

XXVII. SWEETPOTATO

The sweetpotato (Ipomoea batatas), a member of the Convolvulaceae or morning-glory family, is a tuberous-rooted perennial. It is regarded as a native of tropical and subtropical South America. The sweetpotato was carried to the islands of the Pacific very early, and records indicate that it was used as food in prehistoric times in both tropical America and some islands of the Pacific. Columbus found the crop grown by the Indians in Cuba. Later Spanish explorers found it growing in Mexico and South America. Sweetpotatoes were probably cultivated in Virginia as early as 1610 (121, (155, 207). The crop was grown on 113,200 acres in 1973 and yielded an average of 109 hundredweight per acre (2).

Field losses in sweetpotatoes from diseases are due primarily to stem rot or Fusarium wilt (F. oxysporum f. batatas) and black rot (Ceratocystis Endoconidiophora fimbriata). Stem rot causes losses in yield by killing 10% to 50% of the crop wherever susceptible varieties are grown in the United States. Black rot reduces yield in all areas where sweet potatoes are grown. Losses occur in the seedbed, field, in transit and marketing, and in storage. These two diseases are estimated to cause an average annual loss in the United States of 12%. This is two-thirds the loss of 18% due to sweetpotato diseases for the 10-year period 1951-1960. This loss from potential production is estimated at 2,173,000 cwt valued at $8,930,000 (133).

Less important diseases include yellow dwarf (virus), estimated to reduce yields by 2%, and internal cork (virus), which causes greater losses (1.5%) in storage than as a field disease. Leaf spot (Septoria bataticola), scurf (Monilochaetes infuscans), and soil rot or pox (Streptomyces ipomoea) are estimated to incite a combined loss of 2.5%. These three diseases all reduce yields, lower eating quality, and are found throughout sweetpotato-growing areas of the United States (99, 133). Scurf and soil rot also greatly reduce market value (99).

Fungicides are used in sweetpotato production as (a) root or sprout dips prior to planting, (b) for pasteurization of seedbed soil, (c) postharvest sprays or dips, and (d) for fumigation of the storage house. Diseases partially controlled by a seed-rot or sprout-dip include black rot, scurf, stem-rot or Fusarium wilt, Java black-rot (Diplodia theobromae or D. tubericola), and foot-rot (Plenodomus destruens). Disease-free bedding roots are dipped for 30 sec or so in one of the following fungicides: thiram 75W (1 lb/5 gal water); thiram (ARASAN 42S), 1 pt/5 gal; VANCIDE 51 (3/4 pt/5 gal), or thiabendazole (MERTECT 340 F), 1/2 pt/7.5 gal. Treated roots should not be used for food or feed (54, 160, 223).

The soil or sand in the propagative bed is "sterilized" and made free of scurf, stem rot, black rot, foot rot, and nematodes, using steam, or by fumigation with chloropicrin, VORLEX, or methyl bromide (75, 76, 160).

Many sweetpotatoes shipped to market are now treated with a postharvest fungicidal dip or spray before packaging to control decay, particularly Rhisopus soft rot (R. stolonifer), the principal cause of spoilage during transit and marketing. BOTRAN 50W or 75W, 0.125% to 0.187%, is recommended as a treating solution (54, 126, 160, 223). BOTRAM is also useful as a root dip prior to bedding to control soft rot (54).

Sweetpotato roots should be stored only in a thoroughly cleaned and disinfected storage house. A number of disinfectants or fumigants are used. Copper sulfate (blue vitriol), 0.5%, is thoroughly sprayed on the walls, floor, ceiling, bins, storage crates, and baskets (136). The house should be dried out before storing potatoes (75).

Another method is to fumigate the storage house with (a) a mixture of potassium permanganate and formaldehyde, using 3 pt of commercial 40% formalin and 23 oz of potassium permanganate for every 1,000 ft^3 of storage space. The house should be kept tightly closed and locked for at least 24 hr then opened and ventilated thoroughly for at least 2 weeks (99, 136). (b) One-half pound or 6 fl oz of chloropicrin are used. (c) Burn 0.35 to 0.7 lb of 100% sulfur per 1,000 ft^3 of storage space (160). The house must be airtight and moistened thoroughly with water for 2 days in advance of fumigation.

Other important disease control measures include (a) selection of disease-free seed roots and vine cuttings—certified if possible, (b) a rotation of 3 to 5 years between sweetpotato crops, (c) careful handling to keep root injuries at a minimum, (d) prompt curing for 1 to 2 weeks at a temperature of about 85° F and 85% to 90% relative humidity, (e) storage at 55° to 60° F and a humidity of 85% to 90% following curing, (f) planting of disease-resistant varieties, (g) weed and insect control, (h) harvest during warm, dry weather before frost kills vines, and (i) soil rot or pox does not normally develop in soils below pH 5.2 (52, 76, 99, 150, 153, 207).

Sweetpotato leaf diseases, primarily white-rust (Albugo ipomoeae-panduratae), rust (Coleosporium ipomoeae), Septoria leaf spot (Septoria bataticola), and leaf blight (Phyllosticta batatas) are not serious enough to warrant fungicide sprays or dusts (207).

XXVIII. SOLANACEOUS FRUITS

A. Tomato

The tomato (Lycopersicon esculentum and L. pimpinellifolium) is a tender, warm-season annual that requires a relatively long season to produce profitable yields. Like the potato, eggplant, pepper, and husk tomato (Physalis), it is a member of the Solanaceae. It is one of the most popular and important vegetables. The tomato is grown in nearly all home gardens and by a large percentage of market gardeners and truck growers. It is a major forcing crop in greenhouses in the northern part of the United States, Europe and many other countries. The tomato is the most important processing crop in the United States, being made into soups, conserves, pickles, catsups, sauces and other products. Few human foods lend themselves to so great a variety of uses as does the tomato. It is one of the most popular salad vegetables in the raw state, but is also commonly served baked, stewed, fried and as a sauce with various other foods (207, 215, 216). In 1973 tomatoes were harvested on 431,400 acres and the average yield was 16 tons/acre (2).

The diseases that cause the most severe losses in fresh market tomatoes are leaf spots and blights, virus diseases, and fruit rots (Table 21). The most important wilt is caused by Verticillium (V. albo-atrum) (Figs. 18 and 19), which affects the fresh market crop to some extent in Florida, California, Utah, certain areas of the Midwest, Northeast and the Middle Atlantic states (42, 64). Losses are now dropping as more growers turn to the newer Verticillium wilt-resistant varieties. Defoliation associated with foliage diseases reduces yields and exposes fruit to sunscald (42, 64, 133). Bacterial wilt causes minor losses in Florida and other southern states (68).

Some leaf spots and blights occur only on the foliage and stems, but other pathogens that infect leaves also cause rotting of the fruit. Septoria leaf spot (S. lycopersici) and gray leaf spot (Stemphylium solani) are the chief diseases that affect only the leaves. Leaf mold (Cladosporium fulvum) causes minor losses. Early blight (Alternaria solani) (Fig. 20) and late blight (Phytophthora infestans) (Fig. 7) cause both leaf blighting, stem rot

TABLE 21

Estimated Average Annual Losses in the United States for the
Period 1951-1960 Caused by Specific Tomato Diseases (133)

Disease (causal agent)	Fresh market (%)	Processing (%)	Green-house (%)	Trans-plant (%)
Tobacco mosaic (virus)	5.0	1.0	7.0	
Tobacco streak (virus)			1.0	
Curly-top (virus)	1.0	3.0		
Bacterial spot (Xanthomonas vesicatoria)(Fig. 37)	1.0	1.0		2.5
Bacterial wilt (Pseudomonas solanacearum)	0.5			
Gray leaf spot (Stemphylium solani)	3.0	2.0		
Early blight (Alternaria solani) (Fig. 20)	1.0			3.0
Alternaria collar rot (A. solani)				1.0
Late blight (Phytophthora infestans) (Fig. 48)	0.5			0.5
Septoria leaf spot (S. lycopersici)	0.5			
Leaf mold (Cladosporium fulvum)	0.5		8.0	
Anthracnose (Colletotrichum phomoides)		4.0		
Rhizoctonia stem canker (R. solani) (Fig. 16)				1.0
Verticillium wilt (V. albo-atrum) (Figs. 18 and 19)	3.0	4.0		
Fusarium wilt (F. oxysporum f. lycopersici) (Fig. 19)	1.0	2.0		
Blossom-end rot (physiogenic) (Fig. 4)	1.0			
Others	3.0	5.0	4.0	
Totals	21.0	22.0	20.0	8.0

and fruit rot; but late blight, always a major threat to tomatoes, is most injurious as a fruit rot. Most of these diseases are widespread in humid areas east of the Continental Divide. Leaf mold, however, occurs chiefly in the humid South Central and South Atlantic states, while gray leaf spot may cause injury from the humid Southeast into the North Central and Middle Atlantic states (42, 64, 133).

Some fruit rots, such as anthracnose (Colletotrichum phomoides), soil rot (Rhizoctonia solani), and buckeye rot (Phytophthora parasitica, P. capsici, P. drechsleri), are caused by fungi that do not seriously affect the leaves, and they commonly cause only minor losses to the fresh market crop except in wet seasons (42, 64, 133).

Viruses infect tomatoes wherever the crop is grown. The tobacco mosaic virus (TMV) (Fig. 1) causes some reduction in yield in most fields, especially those that are pruned and trained, but other viruses which are not as readily transferred mechanically cause only minor losses east of the Rocky Mountains except in southwest Texas, where some losses occur most years (42, 64, 133). A few, such as curly-top and spotted wilt, damage tomatoes only in regions where their specific insect carriers (vectors) are abundant (64).

Blossom-end rot (Fig. 4), a noninfectious disorder of tomato fruit due to a calcium deficiency and fluctuating soil moisture supply, causes some loss throughout the United States in both the field and the greenhouse (42, 64, 133, 163).

Diseases causing the most severe losses in tomatoes for processing are anthracnose in the East and Verticillium wilt in the West (Table 21). Some losses due to Verticillium wilt occur in the Middle Atlantic, Northeast, and North Central states. Virus diseases and nonparasitic disorders are just as prevalent on tomatoes for processing as on the fresh market crop (42, 64, 133).

Leaf mold and TMV continue to be the primary causes of fruit-production losses to greenhouse tomato growers (Table 21). Blotchy ripening and gray wall or internal browning of the fruit may cause severe losses (42, 64, 133, 163).

Diseases were estimated to cause annual losses of 8% to the southern plant-bed tomato industry during the period 1951-1960 (Table 21). Alternaria leaf spot and collar rot (stem canker) and bacterial spot (Xanthomonas vesicatoria) (Fig. 37) caused the greatest losses, mostly in Georgia, Florida, and Alabama. Rhizoctonia stem canker (R. solani) (Fig. 16) and late blight (Figs. 7 and 48) are other important diseases (42, 64, 133).

Fungicides are used in tomato production as (a) seed treatments; (b) seedbed and greenhouse fumigants, dusts, or soil drenches; and (c) field or greenhouse sprays or dusts. The diseases controlled by these treatments,

TABLE 22

Fungicides Commonly Used in Tomato Production in the United States

Type of treatment	Principal disease at least partially controlled (causal agent)	Chemicals used, application, and remarks (references)
1. Seed treatment (controls organisms carried in and on seed; protects seedlings until emergence	Bacterial spot, bacterial speck (Pseudomonas tomato), bacterial canker (Corynebacterium michiganense) (Fig. 41), seed decay and damping-off (Pythium spp., Rhizoctonia solani), Septoria leaf spot, anthracnose, early blight (Fig. 20) and collar rot, gray leaf spot, Phoma fruit rot (P. destructiva), nailhead spot (Alternaria tomato), Rhizoctonia stem canker (Fig. 16) and soil rot, Verticillium wilt (Figs. 18 and 19).	Hot water soak (122° F for 25 min) followed by a dust of thiram, 1 tablespoon/lb or 5.5 oz/cwt. Use of seed from healthy plants or purchase of certified seed is highly recommended. The hot water soak kills bacteria and fungi under the seedcoat; the dust protects against seedborne fungi causing preemergence damping-off (42, 64, 131, 150, 163, 213). To free seed of bacterial canker, ferment crushed fruit pulp at 70° F for 72 to 96 hr; soak freshly extracted seed in 0.8% solution of acetic acid; or soak dried seed in 0.6% acetic acid (42, 64, 163).
2. Seedbed, Greenhouse, and Hotbed Preplant Treatment (heat—as steam or electricity—is preferred to using a soil fumigant since all types of soilborne pests are killed when soil is held at 180° to 200° F for at least 30 min	Fusarium wilt (Fig. 19), Verticillium wilt, bacterial wilt, bacterial canker, bacterial spot, bacterial speck, Septoria leaf spot, early blight and collar rot, nailhead spot, Botrytis stem rot, Phoma rot, gray leaf spot, anthracnose, buckeye rot, southern blight (Sclerotium rolfsii), Sclerotinia stem rot	1. Methyl bromide (trade names: DOWFUME MC-2, BROZONE, PANO-BROME CL, PESTMASTER METHYL BROMIDE, KOLKER METHYL BROMIDE, PROFUME, MBC FUMIGANT, and so om). 2. Chloropicrin (trade names: LARVACIDE 100 and SOIL FUMIGANT, PICFUME, TRI-CLOR).

3. MIT (trade name: VORLEX—contains 20% methyl isothiocyanate and 80% chlorinated C_3 hydrocarbons including dichloropropenes.

4. SMDC (trade name: VAPAM SOIL FUMIGANT) contains 32.7% sodium N-methyldithiocarbamate.

Streptomycin (100 to 200 ppm) alone or mixed with fixed copper (see label) controls bacterial diseases. Streptomycin is ineffective in some areas due to resistant strains of spot bacterium. Dusts, sprays, or drenches of thiram, captan, ferbam, or zineb, 0.156%-0.25% are applied to control fungal diseases (42, 54, 64, 65, 131, 150, 160, 163). Applications are made weekly from seeding to almost transplanting time. The most popular treatment is a captan drench (3/4 to 2 tablespoons/gal of water or 3/4 to 2 lb of 50W in 100 gal, applied at the rate of 1 gal over every 100 ft² followed by weekly sprays, 0.025%.

Formulations of maneb, mancozeb, captafol, chlorothalonil, zineb, POLYRAM, anilazene, 0.25% to 0.375%, 25 to 40 lb dust per acre (see label for sprays). Gray mold is commonly more severe where maneb, mancozeb or zineb are applied regularly. Anilazene (DYRENE) is often added to maneb to

(S. sclerotiorum), damping-off, and Rhizoctonia stem rot

Bacterial spot, bacterial speck, gray leaf spot, Septoria leaf spot, late blight (Fig. 48) early blight and collar rot, anthracnose, damping-off, Phoma rot, Rizoctonia stem canker

Late blight, early blight, nailhead spot, Septoria leaf spot, gray leaf spot, leaf mold, anthracnose, soil rot (Rhizoctonia solani), buckeye rot, gray mold and ghost spot (Botrytis cinerea)

Seedbed Treatment (Fungicides applied as dusts, coarse sprays, or soil drenches to protect against seed rot and damping-off)

Field or Greenhouse Sprays and Dusts (Number of applications, 5 to 15, depending on weather conditions. A 5- to 10-day schedule is usually recommended, starting no later than 2 weeks

TABLE 22 (Continued)

Type of treatment	Principal disease at least partially controlled (causal agent)	Chemicals used, application, and remarks (references
after the first flowers open and continuing to harvest. Maneb, manco-zeb and chlorothalonil are the preferred fungicides. Sprays are much more effective than dusts.)		better control gray leaf spot and gray mold; but anilazine is weak against late blight. If bacterial spot is a problem, apply a mixture of fixed copper (2 lb actual copper per acre) plus maneb, mancozeb or zineb (2 lb per acre). Sprays or dusts are not usually needed in greenhouses where the relative humidity is kept below 90%, ventilation is ample, and heat is turned on when the outside temperature is below 60° F (52, 54, 64, 131, 150, 160, 163, 223). Captafol is registered for use on machine-harvested tomatoes only. When the relative humidity in greenhouses cannot be maintained below 90% growers use chlorothalonil (EXOTHERM, 1 bomb/100 ft^2 of floor area, or TERMIL, 1 tablet/400 ft^2 of floor area) to control gray mold and leaf mold. If either disease becomes established, the application should be repeated every 7 to 10 days (54). Spraying the soil surface under plants with maneb or mancozeb, 4 lb per acre, following the last cultivation, reduces losses from anthracnose and soil rot (131).

chemicals in common use, and information on amounts used per acre, frequency of application, and so on are given in Table 22.

In greenhouses, Botrytis stem rot or canker is controlled by (a) painting cankers with a paste containing equal parts of copper sulfate and hydrated lime, (b) spraying the lower stem with dicloran (BOTRAN 75W, 0.125%), or (c) applying a 50:50 mixture of pentachloronitrobenzene (PCNB) and captan as a dust or spray. The PCNB-captan dust (25 lb each of captan 50W and TERRACLOR 75W (75% PCNB) plus 50 lb of talc) mixture is applied in the row before planting and mixed in at the rate of 100 lb/acre, or 10 lb each of captan 50W and TERRACLOR 75W with 80 lb of talc are dusted at the rate of 30 to 35 lb/acre to the base of the plants only. A soil drench containing 2 lb each of captan 50W and TERRACLOR 75W per 100 gal of water may be used at the rate of 1/2 pt per plant. Captan 50W, 0.25%, is also used. The drench should be applied a week to 10 days after setting followed by one or two more applications depending on the weather (163).

B. Pepper

Cultivated peppers (Capsicum spp.) are of American origin, known from prehistoric remains in Peru. They were widely cultivated in Central and South America in early times, and were unknown in Europe before Columbus carried seed to Spain in 1493. The first peppers grown in the United States probably came directly from Europe rather than from Central or South America (207).

Considerable confusion persists regarding the taxonomy of cultivated peppers. Heiser and Smith (94) recognized four species, only two of which are grown in the United States, Capsicum annuum and C. frutescens. The other two species, C. pubescens and C. pendulum, are cultivated in Latin America. According to Thompson and Kelly (207), all principal cultivars of peppers grown in the United States, with the exception of tabasco, belong to C. annuum. In 1973, green peppers were harvested on 48,030 acres, with an average yield of 98 hundredweight (2).

Peppers are subject to a number of diseases that reduce both yield and market value of the fruit. Losses are principally due to leaf spots, fruit spots and rots, mosaics, damping-off (Rhizoctonia solani, Pythium spp.), downy mildew or blue mold (Peronospora tabacina), southern blight (Sclerotium rolfsii), ripe rot (Colletotrichum capsici and C. nigrum), blossom-end rot (physiogenic), and sunscald (physiogenic) (21, 42, 133, 207).

Leaf spots and fruit rots occur wherever peppers are grown in humid areas. Viruses (primarily producing mosaic-type diseases) are present in

all pepper-growing sections. Fusarium wilt (F. oxysporum var. vasinfectum) is most prevalent in the Southwest (133).

Estimated average annual losses caused by the various diseases of
green sweet peppers in a recent 10-year period are given in Table 23. The
average estimated annual loss of 14% or 450 cwt is valued at $3,752,000
(133).

Fungicides are used in pepper production as (a) seed treatments;
(b) seedbed preplant treatments, dusts, or soil drenches; and (c) field
sprays or dusts.

Seed treatment is by a hot water soak (30 min at 125° F). This eradi-
cative type of seed treatment is then followed by a protective dust treatment
of thiram, captan, or chloranil. Such treatment controls seed decay and
damping-off plus seedborne anthracnose, ripe rot, Phytophthora blight,
Cercospora leaf spot, and bacterial spot (21, 42, 150, 192, 203, 213).

Commonly recommended seedbed preplant treatments are the same
as those used for tomatoes, i.e., heat, methyl bromide, chloropicrin,

TABLE 23

Estimated Average Annual Losses in the United States for the Period
1951-1960 Caused by Specific Diseases of Green Sweet Peppers (133)

Disease	Causal agent	Average annual loss (%)
Bacterial spot	Xanthomonas vesicatoria	6.0
Ripe rot	Colletotrichum capsici, C. nigrum (C. piperatum)	2.0
Fusarium wilt	F. oxysporum var. vasinfectum	1.5
Cercospora ("frogeye") leaf spot	Cercospora capsici	1.0
Phytophthora blight	Phytophthora capsici	1.0
Tobacco mosaic	Virus	1.0
Cucumber mosaic	Virus	0.5
Potato mosaic	Virus Y	0.5
Tobacco etch mosaic	Virus	0.5
Total		14.0

MIT, and SMDC (42, 54, 68, 160, 195). Dusts and coarse sprays are also applied in the seedbed to protect seedlings against seed decay and damping-off. These fungicide treatments are the same as used for tomatoes.

Field sprays or dusts of zineb, mancozeb, or maneb are sometimes recommended for use after plants have attained almost full size (or when disease first appears) to control anthracnose, Cercospora leaf spot, downy mildew, Phytophthora blight, and various fungal fruit rots (anthracnose, ripe rot, Alternaria, Diaporthe, Phoma, and Phytophthora (21, 42, 54, 213, 223). If bacterial spot is still a problem after fruit are set, fixed copper is often added to the zineb or maneb as outlined under tomato. As for other vegetables and fruits, good results are obtained only if applications are begun before a disease is prevalent in a field or garden and if the spray or dust thoroughly coats the plants. Rates of fungicides per acre for peppers and the equipment used are generally the same as for tomatoes.

C. Eggplant

The eggplant (Solanum melongena) is a minor vegetable in the United States. In India, China, and the Philippines, however, it is grown more extensively than is the tomato (207). Also, methods for growing eggplant are the same as for tomato.

Eggplant diseases reduce both the yield and the quality of the fruit. The major diseases are fruit rots and leaf diseases, which occur wherever the crop is grown, but are most damaging in humid areas of the South. Verticillium wilt causes some losses in the north (42, 133).

Estimated average annual losses due to eggplant diseases for a recent 10-year period are given in Table 24. The average estimated annual loss is 12% or approximately 64,000 cwt valued at $317,000 (133).

Fungicides are used in eggplant production as for pepper and tomato, i.e., seed treatments; seedbed preplant treatments, sprays, dusts, or soil drenches; and field sprays or dusts. Seed treatment with hot water (122° F for 25 to 30 min) followed by a dust treatment with thiram or captan, the same as for tomato, controls seed decay and damping-off (Rhizoctonia solani, Pythium spp., Phomopsis vexans), and seedborne Colletotrichum fruit rot, Cercospora leaf spot, and Verticillium wilt (42, 54, 149, 150, 160, 192, 194, 203, 223). Seedbed preplant treatments using heat, MIT, SMDC, or chloropicrin control Verticillium wilt and other soilborne pathogens (149).

Seedbed sprays and dusts of ziram, captan, or ferbam (0.125% to 0.25%) or soil drenches of these fungicides (1 gal/20 ft^2 of bed area), applied over both seedlings and soil at 5- to 7-day intervals, gives good control of seed decay and damping-off, Phomopsis (foot rot or tip-over phase), Colletotrichum fruit rot, and Cercospora leaf spot (42, 54, 160, 203, 223).

TABLE 24

Estimated Average Annual Losses in the United States for the Period
1951-1960 Caused by Eggplant Diseases (133)

Disease	Causal agent	Average annual loss (%)
Phomopsis blight	Phomopsis vexans	6.0
Cercospora leaf spot	Cercospora melongenae	2.0
Sclerotinia stem rot	Sclerotinia sclerotiorum	1.5
Colletotrichum fruit rot	Colletotrichum spp.	1.0
Verticillium wilt	Verticillium albo-atrum	1.0
Yellows	Virus	0.5
Total		12.0

In the field, captan, zineb, or maneb (0.187% to 0.375%) give good
control of Phomopsis blight (leaf spot and fruit rot), anthracnose, Colletotrichum fruit rot, Cercospora leaf spot, early blight (Alternaria solani),
and possibly rust (Puccinia substriata), Asochyta leaf spot (A. lycopersici
or Diplodina lycopersici), and gray-mold fruit rot (Botrytis cinerea) (42,
54, 160, 223). Field sprays or dusts to control leaf spots and fruit rots
are not usually required except in humid areas of the South or in very rainy
seasons in northern states.

XXIX. CUCURBITS OR VINE CROPS

The cucurbits, or vine crops, belong to the same family, Cucurbita-
ceae. They include cucumber, muskmelon, watermelon, pumpkin, squash,
West Indian Gherkin, and citron. They are all tender annuals that thrive
only in hot weather. All have similar cultural requirements and share many
of the same diseases and insect pests (207).

All cucurbits suffer large losses to diseases not only through reduced
yields but also through decreased quality because of blemished or deformed
fruit. In many regions disease is the principal limiting factor in profitable
production of the crops. Estimated average annual losses due to diseases,
1951-1960, are given in Table 25 (42, 133, 207).

TABLE 25

Estimated Average Annual Losses in the United States for the Period
1951-1960 Caused by Specific Diseases of Cucurbits (133)

Disease (causal agent)	Cucumber FM[a]	G	P	Muskmelon	Watermelon
	Average annual loss (%)				
Anthracnose (Colletotrichum lagenarium)	3.5		0.5	1.0	5.0
Alternaria blight (A. cucumerina)	1.0		0.5	1.0	
Scab (Cladosporium cucumerinum)	0.5		2.5		
Cercospora leaf spot (C. citrullina)	1.0		0.2		
Downy mildew (Pseudopero-nospora cubensis)	3.5		0.5	2.5	0.5
Powdery mildew (Erysiphe cichoracearum (Fig. 15)	1.5	6.5	0.5	3.0	
Fusarium wilt or foot rot (Fusarium spp.)	0.3	0.2		0.3	2.0
Gummy stem blight (Myco-sphaerella melonis)				0.5	1.0
Bacterial wilt (Erwinia tracheiphila) (Fig. 39)	0.2	0.3	0.2	0.2	
Pythium fruit rots (Pythium spp.)	0.3		0.1		
Diplodia fruit rot (D. gossypina)	0.2				
Angular leaf spot (Pseudo-monas lachrymans)	2.0		1.5		
Cucumber mosaic (virus)	2.0	1.0	2.5		
Watermelon mosaic (virus)	1.0				1.0
Tobacco ringspot (virus)	1.0		2.0		0.5
Virus complex (primarily mosaics)				3.5	
Crown blight (cause ?)				4.0	
Total	18.0	8.0	11.0	16.0	10.0

[a]Key: FM, fresh market; G, greenhouse; P, pickling cucumbers.

A. Cucumber (Cucumis sativus)

The cucumber is an important vegetable crop in the United States, being grown in the home garden, in market gardens and truck farms, as a forcing crop in greenhouses, and for pickles (207). In 1973, cucumbers were harvested from 173,120 acres, yielding an average of 4.76 tons/acre (2).

The principal diseases in cucumbers grown for fresh market are caused chiefly by downy mildew (Pseudoperonospora cubensis), anthracnose (Colletotrichum lagenarium), angular leaf spot (Pseudomonas lachrymans), and the cucumber mosaic virus. Less severe losses are caused by scab (Cladosporium cucumerinum), powdery mildew (primarily Erysiphe cichoracearum) (Fig. 15), Alternaria blight (A. cucumerina), Cercospora leaf spot (C. citrullina), watermelon mosaic (virus), Fusarium wilt and foot rot (Fusarium spp.), bacterial wilt (Erwinia tracheiphila) (Fig. 39), Pythium fruit rots (Pythium spp.), and Diplodia fruit rot (Diplodia gossypina). Loss from downy mildew is confined almost entirely to the Atlantic Coast states where moisture is plentiful and the temperature moderately high. Anthracnose and angular leaf spot are widespread in humid regions east of the Continental Divide, while one or more virus diseases occur throughout most cucumber-growing areas. Scab is troublesome primarily in northern states where the moisture is high for at least short periods and temperatures, especially at night, are cool (42, 133).

The principal disease losses in cucumbers grown in the greenhouse are caused by powdery mildew and cucumber mosaic, with minor losses from bacterial wilt and Fusarium foot rot. All these diseases may occur and reduce yields wherever the crop is grown (133).

The major disease losses in cucumbers grown for pickling are caused by cucumber mosaic, scab, tobacco ringspot (virus), and angular leaf spot. Mosaic and bacterial wilt are especially prevalent in the North Central states. Scab is most damaging in the northern tier of states from Maine to Minnesota. Angular leaf spot is damaging in certain years in all humid regions east of the Rocky Mountains. Tobacco ringspot causes losses from Texas into Wisconsin (42, 133).

B. Muskmelon (Cucumis melo)

The muskmelon or cantaloupe is a popular home garden and market garden crop and is widely grown as a special crop in warmer sections of the United States, especially in hot, dry climates of the West and Southwest where irrigation is used (61, 207). Over 60% of the commercial crop is produced in California and Arizona (207). In 1973, cantaloupes were harvested from 93,100 acres, with an average yield of 121 hundredweight. In this same year, honeydew melons were harvested on 14,100 acres, with an average yield of 174 hundredweight (2).

Losses in muskmelons or cantaloupes are caused chiefly by crown blight, leaf diseases, and viral diseases. Downy and powdery mildews cause the most leaf damage with anthracnose (Fig. 17) and Alternaria blight or leaf spot causing some loss. Severe damage in the West is caused by a virus complex: watermelon mosaic, squash mosaic, cucumber mosaic, and curly top. The tobacco ringspot virus is serious in Texas (63, 133).

Muskmelon disease losses are regional. Crown blight and powdery mildew are major diseases in the West, especially the Southwest. Downy mildew is normally important only in humid areas in the Atlantic and Gulf Coast States (63, 133). Anthracnose, Alternaria leaf blight, and bacterial wilt are destructive to muskmelons only in humid areas east of the Continental Divide. Mosaics are much more damaging in Arizona, California, Texas, and Washington than in the North Central and eastern states. Curly top is largely confined to areas west of the Continental Divide (63, 133).

In another type of muskmelon, the honeydew and honeyball, diseases of major importance are crown blight (annual loss estimated at 6%) and powdery mildew (4%). Viral diseases, primarily watermelon mosaic, cucumber mosaic, squash mosaic, and cucurbit latent virus, cause a 4% loss due to stunting, reduced yields, and lowered quality from fruit blemishes and deformity (133).

C. Watermelon (Citrullus vulgaris) and Citron (C. vulgaris or C. vulgaris var. citroides)

The watermelon requires a longer and warmer growing season than most cucurbits and hence production is primarily in the southern states. In the 1950s, 70% of the watermelons produced in the United States came from Alabama, California, Florida, Georgia, South Carolina and Texas. Another 20% were produced in Arkansas, Arizona, North Carolina and Oklahoma. Indiana and Missouri were other important producing states (62). In 1973, watermelons were harvested from 241,300 acres and the average yield was 109 hundredweight/acre (2).

Anthracnose is the most widespread and damaging disease of watermelons in warm humid areas east of the Continental Divide. It may cause severe defoliation and fruit rot where resistant cultivars are not grown. Downy mildew is severe in humid areas in the Atlantic Seaboard and Gulf States; watermelon mosaic, primarily in the south and southwest; and tobacco ringspot in Oklahoma and Texas. Fusarium wilt is common and serious wherever susceptible varieties are grown in infested soil. Gummy stem blight (Mycosphaerella melonis) may be serious in the Southeast and Midwest in wet seasons (42, 62, 133, 207).

D. Pumpkin and Squash (Cucurbita pepo, C. moschata,
C. maxima, C. mixta, and C. ficifolia)

Pumpkin and squash are discussed together because of similar cultural requirements and the great confusion in terminology. Both crops are grown in nearly all areas of the United States as fresh vegetables by millions of home gardeners and thousands of commercial producers in the United States. Pumpkins and squashes are also processed in large quantities in the United States, for use mainly as pies. Both are important food crops in many other countries (207).

These crops are attacked by most of the same diseases that attack other cucurbits. It is seldom, however, that they are seriously damaged by disease, and control measures, except for seed treatment, rotation, and sanitary practices, are often omitted (42, 207).

Fungicides are used in vine crop production as seed treatments, seedbed preplant treatments and sprays, and field sprays or dusts. Seed treatment with thiram, captan, or chloranil is standard. Some of the better treatments with captan and thiram have the insecticide lindane added to control seed-eating insects, such as the seed-corn maggot. Seed is usually dust-treated by seedsmen (54, 130, 150, 213, 223). The combination eradicative-protective treatment controls fungi and bacteria, causing seed decay and preemergence damping-off (primarily Pythium spp. and Rhizoctonia solani (Figs. 46 and 47) and seedborne gummy stem blight or Mycosphaerella black rot, angular leaf spot, anthracnose, Alternaria leaf blight, scab, and Fusarium wilt or foot rot (42, 53, 62, 63, 86, 122, 130, 143, 146, 206, 207, 223). In growing transplants, captan 50W, 0.025%, is sprinkled or sprayed over the plants, using 1 gal/100 ft^2 every 5 days, starting with the first watering (150).

The use of a preplant broad-spectrum soil fumigant, e.g., VORLEX, in conjunction with a polyethylene mulch, is gaining favor with muskmelon and watermelon growers. This practice provides excellent fungal disease, nematode, soil insect, and weed control within the row; produces a much larger early yield—important in the northern states—and much greater total yield. Early plant survival and growth is better under a clear plastic film than with black film alone or a bare soil (144).

Damping-off of seedlings in the seedbed (Figs. 46 and 47) is controlled by spraying the soil at 5- to 7-day intervals after planting, using captan 50W, 0.25% applied at the rate of 1 gal to each 125 ft^2 of bed (42, 63, 130, 150).

Formulations of maneb, mancozeb, folpet (PHALTAN), metiram (POLYRAM), or zineb (2 to 3 lb/acre), chlorothalonil (BRAVO 6F), 2 to 3 pt/acre or captafol (DIFOLATAN 4F), 3 to 5 pt/acre as sprays, used alone or in alternating treatments, are applied at 5- to 10-day intervals in humid regions to greatly reduce losses from foliage and fruit diseases, such as downy mildew, anthracnose, scab, Alternaria leaf spot or blight, Cercospora leaf spot, Diplodia fruit rot or stem-end rot, Pythium fruit rots, and gummy stem blight or Mycosphaerella fruit rot (42, 53, 54, 62, 63, 122, 130, 143, 146, 150, 160, 206, 213, 223, 228). Check labels for restrictions and exclusions.

Sometimes a fixed copper (2 to 4 lb of a product containing 50% metallic copper per acre) is added to the maneb, mancozeb, chlorothalonil, captafol, or zineb to aid in the control of angular leaf spot, downy mildew, powdery mildew, Mycosphaerella black rot, and Choanephora fruit rot (C. cucurbitarum). Copper materials are sometimes injurious to cucurbits, especially young plants, and muskmelons in the blossoming state (42, 53, 54, 63, 122, 146, 223). Copper fungicides, however, are not effective against anthracnose, scab, or Alternaria leaf blight (42, 62, 63, 122, 146, 149).

Ziram and captan (2 to 4 lb/acre) are safer to use on young cucurbits and are commonly recommended early in the growing season to control anthracnose, scab, Alternaria leaf blight, and other fungal diseases of the foliage (42, 53, 63, 149, 194). Ziram and captan, however, are not as effective against downy mildew as are maneb, mancozeb, chlorothalonil, captafol, zineb, or copper compounds.

Powdery mildew is commonly controlled by using resistant varieties, i.e., muskmelons in the Southwest. Benomyl (BENLATE) 50W, 1/4 to 1/2 lb/acre, dinocap (KARATHANE) at the rate of 3/4 lb/acre of 22.5W or 30 lb/acre of 1% dust is often recommended (42, 54, 62, 122, 146, 149, 150, 206, 223). Dinocap may severely injure cucurbits, especially muskmelons, at high temperatures (63). It can be mixed with the other fungicides discussed above in the spray tank (54, 149, 223). One to three applications at 7- to 10-day intervals may be required to control powdery mildew (42, 54, 150, 160, 206), starting when disease appears. Liquid coppers (SOL KOP 10, TC-90) have provided excellent control of angular leaf spot when applied alone or with other fungicides (170). Check labels for the latest registrations.

Field application of fungicides is usually started before or shortly after vining commences and continues through the growing season until a week or so before harvest (42, 54, 63, 150, 160, 223). If a fixed-boom hydraulic sprayer is used, 150 gal/acre of spray is required late in the season. With low-gallonage equipment, commonly done with an airblast (mist) sprayer, 30 to 40 gal of spray are usually applied per acre (63). When fungicides are applied as dusts, 20 to 60 lb of a 5% to 10% active material are applied with a power duster (63, 143, 144, 146). Sprays are generally more effective than dusts (63, 194, 228).

Watermelons can be protected against Diplodia stem-end rot—primarily an in-transit disease—by treating the freshly cut stems with a disinfectant paste in the field or at the time of loading. A copper sulfate-starch paste is usually applied with a small brush. Commercially prepared pastes are available (42, 62).

Captan 50W, 0.50%, is recommended as a postharvest dip or spray to control several fruit rots of cucumber and muskmelon. Special formulations of captan (e.g., ORTHOCIDE FRUIT and VEGETABLE WASH, Captan 80W SPRAY-DIP) are also available for this purpose (194, 213).

XXX. THE CRUCIFERS

These vegetables are closely related hardy crops belonging to the mustard family or Cruciferae. All thrive in a relatively cool climate. Many are grown in the South as a winter crop and in other parts of the country as a spring or late summer to late-fall crop. The cultural requirements for all members of the group are similar and they share many of the same diseases (207).

The crucifers include the cabbage (Brassica oleracea var. capitata), cauliflower (B. oleracea var. botrytis), broccoli (B. oleracea var. botrytis), brussels sprouts (B. oleracea var. gemmifera), kohlrabi (B. oleracea var. gongylodes), kale (B. oleracea var. viridis), collards (B. oleracea var. acephala), Chinese cabbage (B. pekinensis and B. chinensis), turnip (B. rapa), rutabaga (B. napobrassica or B. campestris var. napobrassica), leaf mustard (B. juncea), giant curled and ostrich plume mustards (B. juncea var. crispifolia or B. japonica), black mustard (B. nigra), garden radish (Raphanus sativus), horseradish (Armoracia rusticana), cress (Lepidium sativum), watercress (Nasturtium officinale or Rorippa nasturtium-aquaticum), and seakale (Crame maritima) (109, 207). The major crucifers, the total acres harvested in 1973, with the average per acre yield in parentheses, is as follows: broccoli, 53,590 acres (63 hundredweight); Brussels sprouts, 5,590 acres (115 hundredweight); cabbage, 110,350 acres (218 hundredweight); and cauliflower, 30,480 acres (87 hundredweight) (2).

The crucifers as a group do not suffer as great losses to disease as do the cucurbits, tomato, potato, and other vegetables. Estimated average annual losses due to diseases during 1951-1960, for several important crucifers are given in Table 26. The losses are due to reduced yield, quality, and decay of edible parts in the field, storage, and transit (133).

A. Broccoli

Downy mildew (Peronospora parasitica) causes the greatest annual loss, estimated at 1%. Loss from potential production is 46,000 cwt valued at $369,000 (133).

B. Cabbage

Major diseases include Fusarium yellows (Fusarium oxysporum f. conglutinans), blackleg (Phoma lingam), clubroot (Plasmodiophora brassicae), watery soft rot (Sclerotinia sclerotiorum), black rot (Xanthomonas campestris), mosaic (virus) and downy mildew. Fusarium yellows may occur wherever susceptible cabbage varieties are grown (42, 133). Blackleg is present throughout the Temperate Zone where susceptible cabbage and other crucifers are grown extensively. Clubroot is widespread in northern states, while watery soft rot may be destructive wherever cabbage is grown. Black rot occurs chiefly in humid areas, and downy mildew of cabbage and other crucifers is most prevalent in coastal regions (42, 133). Estimated annual loss in the United States to cabbage diseases is 8% or 2,229,000 cwt valued at $3,973,000 (133).

C. Cauliflower

Important diseases in the United States include clubroot, black rot, blackleg, downy mildew and mosaic. Clubroot is widespread in humid

TABLE 26

Estimated Average Annual Losses in the United States for the Period
1951-1960 Caused by Specific Diseases of Crucifers (133)

Disease (causal agent)	Average annual loss (%)			
	Broccoli	Cabbage	Cauliflower	Kale
Downy mildew (Peronospora parasitica)	1.0		1.0	1.0
Fusarium yellows (F. oxysporum f. conglutinans)		2.5		1.0
Clubroot (Plasmodiophora brassicae)		1.5	1.0	1.0
Blackleg (Phoma lingam)		1.0	1.0	1.0
Watery soft rot (Sclerotinia sclerotiorum)		1.0		
Black rot (Xanthomonas campestris)		0.5	1.0	1.0
Mosaic (virus)			1.0	
Others	1.0	1.5	3.0	3.0
Total	2.0	8.0	8.0	8.0

regions. Mosaic of cauliflower and other crucifers occurs throughout the
United States. Losses are due not only to reduced yield, but also to head
discoloration, and curd damage (42, 133). Annual cauliflower losses in the
United States are estimated at 8% or 223,000 cwt, valued at $1,476,000
(133).

D. Kale

This crop is grown commercially principally in Virginia. Blackleg,
black rot, clubroot, downy mildew and Fusarium yellows are all major diseases. Disease losses reduce potential production in the United States by
an estimated 8% or 16,000 cwt worth $70,000 (133).

Fungicides are used in crucifer production as seed treatments; seedbed preplant treatments, dusts, sprays and soil drenches; in the transplant
water, and as field sprays and dusts.

Crucifer seed is routinely treated with hot water (122° F for 18 to 30
min, depending on the crop) followed by a dust treatment of thiram, captan, or
chloranil, to control seed decay and preemergence damping-off (primarily
Pythium spp. , and Rhizoctonia solani) plus seedborne blackleg, black rot,
downy mildew, Alternaria or black leaf spot (mostly A. brassicae), Fusarium yellows, anthracnose (Colletotrichum higginsianum), Cercosporella leaf
spot or white spot (C. brassicae or C. albomaculans), Cercospora leaf
spots (C. spp.), Mycosphaerella leaf spot or ring spot (M. brassicicola),
bacterial leaf spot of cauliflower (Pseudomonas maculicola), and powdery
mildew (Erysiphe polygoni) (40, 42, 53, 54, 58, 122, 150, 192, 207, 213,
223, 227). A relatively new treatment is the soaking of seed for 25 hr in a
thiram solution (0.2% active) at 86° F and then dry. The thiram treatment
does not kill black rot bacteria (150).

Cabbage, cauliflower, Brussels sprouts, broccoli and certain other
cruciferous crops generally are started in a plant bed (227). Treatment
of the soil before planting (using heat or a broad-spectrum soil fumigant) is
the same as outlined for tomato. Proper soil treatment, a 3- or 4- year
rotation, strict sanitation practices, and the sowing of disease-free seed
effectively aid in controlling bacterial leaf spot, clubroot, blackleg, black
rot, downy mildew, Rhizoctonia disease (R. solani), watery soft rot (Sclerotinia sclerotiorum), Mycosphaerella leaf spot and black root of radish
(Aphanomyces raphani) as well as nematodes, weeds, and soil insects (42,
53, 149, 213, 223, 227).

If damping-off and wirestem (Rhizoctonia solani, Pythium spp. , Phoma
spp. , Sclerotinia sclerotiorum) have been problems in cold-frame seedbeds
or in the field, a dust or spray (drench) of captan and PCNB (TERRACLOR)
before or after planting is recommended: (a) Before planting, a commercial

10:10 dust mix (5 lb/1,000 ft^2) or a homemade dust mixture of PCNB 20% and captan 7 1/2% to 10% (10 cups of each per 1,000 ft^2 or about 187 lb of TERRACLOR 20, and 181 lb of captan 7 1/2%/acre) is rototilled or otherwise worked evenly into the top 4 in. of soil; (b) After planting, a soil drench of PCNB 75W and captan 50W, 0.03% to 0.06% each (1/4 to 1/2 lb each in 100 gal of water) is applied with a sprinkling can or hose proportioner over 2,000 ft^2 of seed bed (5 gal/100 ft^2). Postplant treatment is not as satis- factory as the preplant one. Commercial PCNB-captan mixes are sold as TERRACAP, PCNB-captan, and ORTHOCIDE SOIL TREATER "X." PCNB- captan suspension is sprayed on both plants and soil (53, 58, 150, 194, 223). This treatment is not presently registered for collards, kale, or turnips (223). An alternative treatment is to apply PCNB 75W (7.5 lb/13,000 linear ft) in an 8-in. band prior to seeding. The PCNB is mixed into the top 2 in. of soil just before seeding (54).

An effective control of clubroot is the addition of PCNB to the trans- plant or setting water, using 1 to 3 lb of PCNB (TERRACLOR 75) in 50 gal of water (0.125% to 0.375%) and 1/3 to 3/4 pt per plant. WARNING: Do not use PCNB emulsion; it will "burn" the roots; and avoid using where carrots or parsnips will be grown within 3 years (42, 54, 150, 213). Another con- trol measure in infested soils is to make the soil alkaline (at least to pH 7.2) using hydrated lime. If the soil is already at pH 7.2 or higher, 1,500 lb of hydrated lime is harrowed or rototilled into an acre of soil at least 6 weeks before planting. Excellent control of clubroot has been obtained where liming has been coupled with PCNB in the transplant water (42, 53, 149, 150, 160, 191).

Sprays of maneb, mancozeb, fixed copper, or zineb (1 1/2 to 3 lb/acre) or chlorothalonil (BRAVO 6F) (1 to 2 pt/acre) are recommended at 5- to 10-day intervals, starting in the seedling stage in the seedbed or in the field at about midseason and continued until the heads or other edible parts start to form. These applications control downy mildew, Alternaria leaf spot, blackleg, and probably other minor fungal diseases of the foliage (42, 54, 122, 149, 150, 160, 213, 223, 227). Fixed coppers (especially KOCIDE, 2 lb 86SP per acre) is the suggested fungicide where black rot is a problem (150). Chlorothalonil is not registered for collards, kale, and turnips (223).

XXXI. THE BULB CROPS

The bulb crops are all hardy pungent herbs, belonging to the same genus, Allium, of the family Liliaceae. The onion is the most important member of this group of plants, being grown in all parts of the United

States. All bulb crops have similar cultural requirements and are attacked by most of the same diseases (207).

Members of this family include the onion (A. cepa), leek (A. porrum), welch onion or cibol (A. fistulosum), garlic (A. sativum), shallot (A. ascalonicum), and chives (A. schoenoprasum) (207). Onions were harvested on 106,560 acres in 1975, with an average yield of 228 hundredweight (219).

Estimated average annual losses caused by the various diseases of onions and shallots for the period 1951-1960 are given in Table 27.

Downy mildew (Peronospora destructor), Fusarium root and bulb rots (Fusarium spp.), pink root (Pyrenochaeta terrestris), purple blotch (Alternaria porri), and smudge (Colletotrichum circinans) reduce yields and cause greater losses in onions than do other diseases. Bulbs of infected plants are undersized and rot in the field, storage, or in transit (42, 133, 219, 226).

Smudge occurs only on white onions. Downy mildew is most destructive in cool moist climates. Major losses to the bulb and seed crops occur in California, Louisiana, Michigan, New York, and Oregon. In unusually cool, humid seasons it may become severe in the Midwest, Colorado, Idaho and Texas. Neck rot or gray-mold rot and leaf blights (Botrytis spp.) are

TABLE 27

Estimated Average Annual Losses in the United States
Caused by the Various Diseases of Onions and
Shallots for the Period 1951-1960 (133)

Disease (causal agent)	Average annual loss (%)	
	Onion	Shallot
Fusarium root rots (Fusarium spp.)	4.0	2.0
Neck rot (Botrytis spp.)	4.0	2.0
Pink root (Pyrenochaeta terrestris)	3.0	3.0
Purple blotch (Alternaria porri)	3.0	2.0
Downy mildew (Peronospora destructor)	1.0	2.0
Smudge (Colletotrichum circinans)	1.0	2.0
Yellow dwarf (virus)		4.0
Others	4.0	4.0
Total	20.0	21.0

most severe in the upper North Central and Northeastern United States. Pink root is most prevalent in the Rio Grande Valley and central California. Fusarium root and bulb rots and purple blotch occur wherever onions are grown extensively (42, 109, 133, 226).

Greatest losses in shallots are caused by yellow dwarf, an aphid-borne virus disease. Other losses are caused by the same diseases that attack onions (Table 27) (133).

The annual 20% loss of onions from potential production in the United States from diseases is estimated at 5,367,000 hundred pound sacks valued at $13,901,000. The estimated average annual loss of shallots (21%) amounts to 35,000 cwt worth an estimated $245,000 (133).

Fungicides are used in onion production primarily as seed and/or in-furrow treatments, field sprays or dusts.

Proper seed treatment with captan, thiram, or chlorothalonil (Bravo) controls seed rot and preemergence damping-off, smut, and seedborne pur-ple blotch (42, 54, 55, 86, 122, 130, 157, 213, 226). To protect seed for bulb production, better distribution of the dust is obtained by applying a methyl cellulose (METHOCEL) sticker to the seed first (42, 157, 226). Most seed treating is done for growers by seedsmen and custom treaters (157).

An in-the-furrow drip treatment of nabam (6 qt/acre of 22%) dribbled into the open furrow with the seed, immediately behind the planter shoe, using 75 to 125 gal of water per acre (low rate for moist soil; high rate for dry soil), or sprayed in the furrow using 25 to 40 gal/acre, has largely re-placed the older and more toxic formaldehyde (1 to 1 1/2 gal of 37% to 40% commercial formalin per acre) treatment to control smut and damping-off (42, 150, 157, 213). Mancozeb (DITHANE M-45, MANZATE 200), 3 lb of 80W per 100 gal and applied like nabam has proven effective. Some growers prefer to dribble granules containing thiram (1 to 1 1/2 lb actual per acre) into the seed furrow. The rates are calculated for a 15-in. row spacing and a V-shaped planting shoe. The smut fungicide is commonly combined with an insecticide (e.g., diazinon or DASANIT) to control maggots (55, 149, 150, 157, 213). Thiram should not be used on green or bunching onions.

Fungicides are applied in the field largely to control Botrytis blast or leaf blights, downy mildew, purple blotch, and, to a minor extent, post-emergence damping-off, smudge, and black mold (Aspergillus niger). The number of applications depends on the type of onion grown, whether the crop is started from sets or seed, temperature and moisture conditions, and the region. Spraying is begun in the seedling stage or when the plants are 5 to 8 in. tall and is continued at 3- to 10-day intervals until four to nine applica-tions have been made (42, 54, 55, 150, 157, 213, 226). Commonly recom-mended fungicides on a per acre basis include maneb, maneb and zinc ion (mancozeb) or zineb all at 2 to 3 lb, anilazine (DYRENE) 3 lb, and captafol

(DIFOLATAN 4F) 2.5 pt. If only Botrytis is present, captan 50W (3 to 4 lb) may be used with fairly good results (42, 157). A good spreader-sticker or wetting agent, 0.062% is needed to ensure proper coverage of the waxy foliage (55, 122, 130, 150, 157, 160, 213, 226). These fungicides are normally combined with one or more insecticides to control thrips and other insects (42, 150, 157).

Dusting is inferior to spraying and is suggested only as a last resort (42). Successful control with a 6.5% zineb dust has been reported by timing the applications to coincide with the formation of a fine film of moisture on the leaf surface at certain periods of the day (86).

White rot (Sclerotium cepivorum), pink root, and Fusarium bulb and root rots can be controlled where severe, and economically possible, with a broad-spectrum soil fumigant as outlined under tomato. Dicloran (BO-TRAN 75W) as an in-furrow treatment is effective against white rot. A broad-spectrum soil fumigant (e.g., VORLEX) also controls such other pests as the stem or bloat and root-knot nematodes, and mites. Fumigation is usually done in the late summer or fall while soil temperatures are still fairly warm (54, 157, 160).

XXXII. LETTUCE, ENDIVE, ESCAROLE, CHICORY

Lettuce (Lactuca sativa), endive and escarole (Cichorium endiva), and chicory (C. intybus) are annual crops belonging to the sunflower family, Compositae. While endive, chicory and escarole are minor crops, lettuce exceeds in farm value all fresh vegetable crops except potato and tomato. During the 1950s, the acreage for lettuce harvest fluctuated between 204,000 and 233,000 acres. The farm value fluctuated between $96 million in 1950 and about $142 million in 1957. These figures do not include the huge amount grown in home gardens, for roadside stands, and for local market acreage (61). In 1973, lettuce was harvested from 219,070 acres and the average per acre yield was 227 hundredweight (2).

Lettuce is also an important, as well as one of the oldest, greenhouse-forcing crops. It is well adapted to forcing during the colder months when light and temperature are less favorable for other vegetable-forcing crops (209).

Lettuce is subject to a large number of diseases, but only a relatively few cause serious losses under most conditions (42, 84, 207, 229). Field-grown lettuce is damaged severely by lettuce mosaic, a virus disease. Downy mildew (Bremia lactucae), which withers the leaves, is another cause of heavy losses. Big vein (virus) and tipburn, the latter a noninfectious

disease, often damage lettuce. Bottom rot (<u>Rhizoctonia solani</u>) and Sclerotinia rot, drop, or collar rot (<u>S. sclerotiorum</u>) decay both stems and leaves
(42, 84, 133, 207, 229).

Mosaic and downy mildew are widespread, particularly in western
growing areas, but aster yellows is found chiefly in the northwestern
states. Tipburn and bottom and Sclerotinia rots occur wherever lettuce is
grown (207).

Other diseases of occasional importance in specific areas are damping-
off (<u>Rhizoctonia solani</u>, <u>Pythium</u> spp. , <u>Botrytis</u> spp.), spotted wilt (virus),
anthracnose (<u>Marssonina panattoniana</u>), powdery mildew (<u>Erysiphe cicho-
racearum</u>), and gray mold rot (<u>Botrytis cinerea</u>) (42, 84, 133, 229).

The greatest losses from disease are due to decreased quality caused
by discoloration, deformities, and off-flavor (42, 84, 133, 207, 229).

The annual estimated loss (Table 28) from potential production of
field-grown lettuce, 1951-1960, amounted to 12% or 4,278,000 cwt valued
at $17,274,000 (133).

Disease losses in escarole during the same period were caused by
stem and leaf spots (primarily anthracnose caused by <u>Marssonina panattoni-
ana</u>, bacterial rots (<u>Pseudomonas</u> spp.), gray mold rot (<u>Botrytis cinerea</u>),
Cercospora leaf spot (<u>C. cichorii</u>), bacterial soft rot (<u>Erwinia carotovora</u>),
aster yellows (mycoplasma), and a number of minor leaf, stem, and root
diseases. Most of these occur in Florida, where the bulk of the crop is
grown. Losses are due not only to reduced yield but also to decreased
quality because of discoloration (133).

The annual estimated loss from potential production of escarole or
endive is 6% (Table 28) or 46,000 cwt valued at $214,000 (133). This crop
was grown on 8,400 acres in 1973, with an average per acre yield of 140
hundredweight (2).

Fungicides are used in lettuce production as seed treatments, soil
treatments, and foliage sprays or dusts.

Suggested seed treatment dusts to control seed rot and damping-off,
where stands have been poor in previous years, include captan, chloranil,
and thiram (42, 54, 56, 149, 213).

Soil sterilization is widely practiced for lettuce grown in cold frames,
hotbeds, and greenhouses to kill soilborne inoculum of fungi and bacteria
that incite leaf, stem, head, and root rots. Steam, where available and
application is feasible, is generally the preferred method of treatment
(54, 56, 209).

Damping-off and <u>Botrytis</u> in the seed bed can be controlled by spraying leaf lettuce 7 days after transplanting, using dicloran (BOTRAN) 75W,

TABLE 28

Estimated Average Annual Losses in the United States
Caused by Various Diseases of Field-grown Lettuce
and Escarole (Endive) for the Period
1951-1960 (133)

Disease (causal agent)	Average annual loss (%)	
	Lettuce	Escarole
Mosaic (virus)	4.0	
Downy mildew (Bremia lactucae)	3.0	
Tipburn (noninfectious)	1.5	
Big vein (virus)	1.0	
Aster yellows (mycoplasma)	1.0	1.0
Damping-off (Botrytis spp., Rhizoctonia solani, Pythium spp., etc.)	0.5	
Others, including bottom (R. solani) and Sclerotinia (S. sclerotiorum) rots	1.0	
Stem and leaf spots (several fungi and bacteria)		2.0
Bacterial soft rot (Erwinia carotovora)		2.0
Miscellaneous		1.0
Total	12.0	6.0

2.7 lb/acre. The treatment is repeated when plants are 50% mature. Do
not spray later than 14 days before harvest (150). An alternative treatment
is ferbam 76W (1 lb/gal; 1 gal/100 ft^2) (54, 160).

Botrytis gray mold is troublesome under glass or in the earliest field
lettuce. Dicloran (1 oz/yd of 4% dust) is worked into the top 3 in. of soil
2 to 3 days before planting, or dicloran is sprayed (2.7 lb of 75W per acre)
several days before thinning and repeated 7 and 14 days later (150).

Dicloran also gives some control of bottom rot (Rhizoctonia solani)
and drop (Sclerotinia sclerotiorum) using 2 lb/100 gal of 75W. Spray at
the first appearance of disease and repeat at 7- to 10-day intervals (160).
The spray is applied as a 12-in. band over plants so as to wet the foliage

when plants are 2 to 3 in. tall and again 20 days later. This treatment is doubtful on high organic (muck) soils (54).

The fungicides for downy mildew control are listed in Table 29. Start spraying before the disease appears and make an application every 4 to 7 days in cool, damp weather. Do not spray within 7 days of harvest. A zineb dust (6.5% to 8% active) is sometimes used instead of a spray (42, 56, 122, 149). Where lettuce rows are spaced 18 in. apart, 150 to 200 gal of spray per acre are suggested.

In Florida, zineb 75W, 0.25% or maneb 80W, 0.187% is recommended to control Alternaria leaf spot of lettuce. Applications should be repeated at 4- to 5-day intervals (68).

Powdery mildew can be controlled by applications of a sulfur dust, starting when the disease first appears, provided temperatures are high enough to volatilize the sulfur. Dusting with sulfur will not control downy mildew (84).

TABLE 29

Fungicides and Suggested Dosages for Preventing Diseases
of Various Vegetable Crops

Fungicides[a]	Amounts to use per acre (lb)
Maneb 80W	2-3
Mancozeb 80W	2-3
Zineb 75W	2-3
Captan 50W	2-4
Chlorothalonil (BRAVO) 75W	2-3
Metiram (POLYRAM) 80W	2-3
Fixed copper, 48%-53% metallic copper	2-4
Anilazene (DYRENE) 50W[b]	1 1/2-3
Thiram 75W[b]	1
Ferbam 76W[b]	1-2
Ziram 76W	2

[a]The fungicides are to be used singly except when indicated and are not necessarily listed according to their effectiveness.
[b]Suggested mostly for celery; check label restrictions.

XXXIII. UMBELLIFEROUS CROPS

A. Carrot

The carrot (Daucus carota var. sativa) is a very popular root vegetable of the Umbelliferae or parsley family. It is widely grown in home
gardens and commercially for fresh market, processing, or storage (20,
207). In 1973, carrots were harvested on 82,500 acres and the average
yield was 260 hundredweight per acre (2).

Carrot is subject to a rather large number of diseases, but only a
relatively few cause serious losses under most conditions (42, 207). Most
of the disease losses in carrots during the period 1951-1960 were caused
by bacterial blight and root scab (Xanthomonas carotae), estimated at 2%
annually, Alternaria (A. dauci) and Cercospora (C. carotae) leaf blights
(2%), and yellows caused by the aster yellows mycoplasma (2%). Losses
caused by minor diseases, primarily due to other viruses, and field or
storage root rots were also estimated at 2%, making a total of 8%. This
annual loss from potential production was estimated at 1,284,000 cwt valued
at $3,845,000 (133).

Bacterial blight occurs chiefly in California and other western states.
It occurs on seeds, foliage, flowers, and roots. Alternaria and Cercospora
blights, which attack the foliage and floral parts, occur wherever carrots
are grown. Aster yellows (Fig. 45) is most destructive in the western half
of the United States. It not only reduces yield but also decreases the value
of roots because of their atypical shapes and sizes (42, 109, 133).

Fungicides are used in carrot production as seed treatments, foliage
sprays and dusts, postharvest dips, and as disinfectants in the storage
house.

Treating of carrot seed with hot water (15 to 20 min at 122°F or 10
min at 126°F), followed by a protective dust of captan or thiram, helps control bacterial blight, Alternaria blight, seed rot, and damping-off. The hot
water treatment is made in areas where bacterial blight has been destructive or is expected. A protective fungicide dust is often applied just before
planting only if there have been past problems in procuring satisfactory
stands or if the hot water treatment has been applied (42, 122, 150, 213).

It is also optional whether or not to spray the carrot seedlings as
soon as they emerge and at 5-day intervals as long as they require protection against postemergence damping-off (primarily caused by Pythium spp.,
Rhizoctonia solani, Alternaria dauci) and early-appearing leaf blights.
Use either ziram 76W or captan 50W, 0.25% when early protection is required (42).

In humid regions, fungicide sprays and dusts are commonly and frequently applied (e.g., maneb, mancozeb, chlorothalonil, zineb) in the field

to control <u>Alternaria</u>, <u>Cercospora</u>, and <u>Stemphylium</u> (<u>S. radicinum</u>) leaf blights as well as to reduce storage losses from root decays. Applications are usually suggested to begin when plants are 3 to 8 in. tall, or when leaf spots are first evident, and continued at 7- to 10-day intervals as long as the plants require protection. In very wet weather the intervals are shortened. Thorough coverage with a fungicide is essential (Table 29) (42, 54, 122, 139, 150, 160, 213, 223). If bacterial blight is a problem, fixed copper (e.g., tribasic copper (KOCIDE) is preferred (213).

After each carrot or other root crop is removed from storage, the area is thoroughly cleaned and the walls, floor, and ceiling washed or sprayed with 1.0% to 2.0% copper sulfate or formaldehyde solution (37% to 40% commercial formalin, 1.25%). This removes all contaminated carrot refuse and inoculum. If old crates, baskets, or other storage containers must be used, they should be dipped in a formaldehyde solution and aired thoroughly before storage time (42, 185).

Postharvest losses from bacterial and fungal root rots (primarily <u>Erwinia carotovora</u>, <u>Sclerotinia</u> spp., <u>Botrytis cinerea</u>, <u>Rhizoctonia solani</u> and <u>R. carotae</u>, <u>Centrospora acerina</u>, <u>Rhizopus</u> spp., <u>Fusarium</u> spp., <u>Phytophthora</u> spp., <u>Pythium</u> spp., <u>Mucor</u> spp., <u>Penicillium</u> spp., <u>Aspergillus</u> spp., <u>Typhula</u> spp., <u>Gliocladium aureum</u>, <u>Geotrichum candidum</u>, <u>Thielaviopsis basicola</u>, <u>Zygorhynchus</u> spp.) are reduced by (a) precooling immediately after digging and topping, (b) carefully removing all diseased, mechanically injured or wet roots before packaging or placing in storage, and (c) maintaining a temperature as close to 32° F as possible without freezing, and with a high humidity (95%) to prevent shriveling. Free moisture in the package, in transit, storage and during marketing must be avoided (42, 150, 185, 197).

Off-flavors result when carrots are stored with apples, pears and other fruits; a bitter taste results from exposure of the carrots to the ethylene gas given off by the fruit in the respiration process. Bitterness may also occur in carrots affects with the aster yellows mycoplasma (150, 197).

If postharvest rots have been a problem, growers may dip freshly harvested carrots in BOTRAN 75W, 0.125% (150).

B. Parsnip

The parsnip (<u>Pastinaca sativa</u>), like the carrot, is a member of the Umbelliferae, grown for its edible, fleshy taproot. It is one of the less important vegetable crops. Although a biennial like carrot, it is grown as an annual. The parsnip is adapted to much of the United States, but commercial production is practically limited to the northern states from New England to the Pacific Coast. About 60% of the commercial acreage is grown in Massachusetts, New York, Illinois, Pennsylvania, Washington, and California (13, 207).

Diseases can be serious in parsnip culture. Roots affected with disease and nonparasitic defects are discarded in packing the roots for market and channeled elsewhere for livestock feed. Only fancy roots are marketed. Thus, the loss from disease is essentially at the farm level and familiar only to growers (85).

If the parsnip crop is overwintered for spring harvest, there is more loss from disease. There is less disease among roots harvested in late summer than in the autumn season. As for other crops, excessive and persisting moist weather and poorly drained soil are important factors contributing to parasitic diseases (85).

Important diseases of parsnips include yellows, caused by the aster yellows mycoplasma; fungal leaf spots and blights (Cercospora pastinacae, Cercosporella or Ramularia pastinacae, Itersonilia perplexans, Phomopsis diachenii); and crown and root rots which develop in both field and storage (primarily Itersonilia perplexans, Phoma spp., Sclerotinia sclerotiorum, Rhizopus spp., Penicillium sp., Rhizoctonia solani, Botrytis cinerea, Fusarium spp., Centrospora acerina, Erwinia carotovora, Pseudomonas pastinacae) (42, 85, 109, 194, 197). Loss of foliage from leaf spot diseases seriously impairs the quality and size of the roots. These diseases are serious in wet seasons and on land with poor air and soil drainage (85).

Fungicides are not as widely used in parsnip production as they are for many major vegetable crops. Seed treatment with thiram or captan has often been recommended to prevent seed decay and damping-off (42, 122, 150). Fixed or other coppers, maneb, mancozeb, or zineb are suggested to control leaf diseases and parsnip canker (Itersonilia perplexans) (Table 29) (42, 150, 160, 213). Applications are begun in midsummer or when leaf lesions are first evident and are continued throughout the season (October) to near harvest.

Control of postharvest root rots before and during storage is the same as for carrots (85, 149, 150, 194). Sulfur dioxide gas is also used as a disinfectant. The fumigant is obtained by burning 5 lb of brimstone sulfur to 10,000 ft^3 of storage. Burning is done on a fireproof base or the ground floor of the cellar to prevent fire. This fumigant corrodes metal (85).

A better method of fumigation is with formaldehyde gas generated by combining 1 lb of potassium permanganate crystals and 1 qt of commercial formalin to 1,000 ft^3 of storage. Containers of 25 gal capacity to each 10,000 ft^3 are used. The required amounts of material are prepared for each container. The permanganate crystals are added to the formaldehyde in the pails, beginning at the farthest pail and then to the others toward the exit. The gas is confined for 24 to 48 hr. A moist atmosphere and a room temperature above 60° F contribute to the toxicity of the gas (85).

C. Celery

Celery (Apium graveolens var. dulce), a member of the family Umbelliferae, and grown for its edible leafstalks, is second to lettuce as a salad crop in both value and popularity. A large part of the celery grown in the United States is consumed raw, but considerable quantities are used in vegetable juices, soups, stews and as a cooked vegetable. In England some celery is canned (207). In 1973, celery was harvested from 34,080 acres, yielding an average of 501 hundredweight per acre (2).

Celeriac (Apium graveolens var. rapaceum), also known as turnip-rooted celery, is a form in which the leaves are borne on a thickened, turniplike crown that is the edible part of the plant (197).

Celery is subject to a large number of diseases caused by fungi, bacteria, and viruses. Most of the major diseases are widespread and occur wherever the crop is grown. Others are more limited in distribution or destructiveness (42, 133, 197, 207).

The more serious diseases include the leaf blights, Fusarium yellows or wilt (F. oxysporum f. apii), virus diseases, and nonparasitic disorders. The major leaf diseases are early (Cercospora apii) and late (Septoria apii) blights, but bacterial leaf spot or blight (Pseudomonas apii) causes some losses. Cucumber mosaic (virus) and common or western aster yellows (mycoplasma) are other important diseases. Major nonparasitic diseases are blackheart (caused by a lack of calcium) and cracked stem (due to boron deficiency) (Table 30). Western aster yellows, which is limited to the western United States is damaging chiefly in California. In the northeastern United States, cucumber mosaic is the most serious virus disease. Losses are due to reduced yield as well as blemishes and bitterness that reduce quality (42, 133, 197, 207).

The estimated average annual loss in the United States due to celery diseases is 17% (Table 30) or 2,955,000 cwt, valued at $10,866,000 (133).

Fungicides are used in celery production as seed treatments, soil fumigants or disinfectants, foliage dusts or sprays, and disinfectants in the storage area.

Celery and celeriac seed less than 2 or 3 years old are commonly soaked in hot water (118° F for 30 min) to destroy seedborne early, late, and bacterial blights, followed by a protective dust treatment of captan, thiram, or chloranil. The dust treatment protects against seed decay and damping-off organisms (primarily Pythium spp., Rhizoctonia solani, Sclerotinia spp., Fusarium oxysporum f. apii) in the soil. Seed 2 years or older does not require the hot water soak, since seedborne blight organisms are killed in less than 24 months. The seedbed soil should be sterilized, as outlined for tomatoes, before planting to control damping-off and other

TABLE 30

Estimated Average Annual Losses in the United States Caused by
Various Celery Diseases for the Period 1951-1960 (133)

Disease (causal agent)	Loss (%) from potential production
Early blight (Cercospora apii)	4.0
Late blight (Septoria apii)	4.0
Cucumber mosaic (virus)	2.5
Western aster yellows (mycoplasma)	2.5
Bacterial leaf spot (Pseudomonas apii)	1.0
Blackheart (nonparasitic)	1.0
Cracked stem (nonparasitic)	1.0
Fusarium yellows (F. oxysporum f. apii)	1.0
Total	17.0

soilborne disease (49, 54, 150, 160, 169, 202). A seed treatment for controlling seedborne, early and late blights is to soak new seed in thiram
solution (0.2% active) for 24 hr at 86° F (150).

In the seedbed, damping-off, early and late blights are controlled by
spraying every 7 days after emergence, using various fungicides (Table 29).
At least 3 gal of spray should be applied per 1,200 ft^2. Maneb and mancozeb may injure young celery seedlings in the seedbed (49, 149, 150).

In the field, both early and late blights as well as Rhizoctonia stalk
rot (R. solani) are controlled by regularly spraying every 7 to 10 days
throughout the season (Table 29). In rainy weather the spray interval may
have to be reduced to 4 or 5 days (42, 49, 150, 160, 202). A commercial
spreader-sticker should be added to all sprays (202). Thiram 75W controls
Rhizoctonia stalk rot or crater rot.

If bacterial blight is also a problem, fixed copper or soluble ammoni-
ated copper (SOL KOP-10, 2 lb) may be combined with half strength maneb
or anilazine. Copper zinc chromate, 47.5W plus ziram 38% (MILLER
658-Z), 3 to 4 lb of the formulation per acre, gives good control of the three
major leaf blights (223). Dusts containing maneb, zineb, anilazine, or fixed

copper may also be used when there is little or no wind and the foliage is damp. It is suggested that copper compounds be applied only every 14 days when the weather is warm and dry and/or bacterial blight is checked (42, 68, 150, 160). Some states also suggest applications of streptomycin (17% W), 200 ppm, be applied to the plant bed only at 4- to 5-day intervals (160).

Careful handling and storage as outlined for carrot will control post-harvest decays of the leafstalks. Bacterial soft rot (Erwinia carotovora) and gray mold rot (Botrytis cinerea) are primarily transit and storage diseases (197).

XXXIV. ASPARAGUS

Garden asparagus (Asparagus officinalis var. altilis) is a perennial, dioecious herb, 4 to 10 ft tall, in the Liliaceae or lily family. It is widely grown in home gardens and for local fresh market sale throughout the United States. Commercial production—fresh, canned, and frozen—is principally in California, New Jersey, Washington, Illinois, South Carolina, Michigan, Pennsylvania, Maryland, Massachusetts, Delaware, and Iowa (109, 197, 207, 208). In 1973, asparagus spears were harvested from 115, 380 acres, with an average yield of 22 hundredweight (2).

Green asparagus is preferred on most markets in the United States, although a considerable part of the processed product is blanched or white. The demand, however, has been increasing for greeen asparagus for canning and freezing (207).

Asparagus is subject to a number of diseases, but rust, caused by the fungus Puccinia asparagi, is the only one of great economic importance (42, 133, 208). Average annual losses in the United States were estimated at 7% for the period (1951-1960) (133). The rust fungus develops on the stems (fern growth) that form after the cutting season is over, causing the tiny needlelike branches to fall. Seriously affected plants are weakened or killed. Injury to the top growth slows the storage of food reserves in the fleshy roots, reducing the size and number of shoots produced the following year. The marketable portions of the plants, however, are uninfected. Rust occurs throughout the United States, but is most severe in humid, high-rainfall regions (42, 116, 133, 197, 204, 208, 237). Fusarium stalk wilt and root decay (F. oxysporum f. asparagi) also cause some losses (about 1% each). This annual loss of 9% from potential production is esti-mated at 347,000 cwt, valued at $4,073,000 (133).

Fungicides are used in asparagus production primarily as sprays or dusts. These are applied only to seedlings and to fern growth in bearing plantations after the spears are harvested. Suggested fungicidal sprays include maneb, mancozeb, and metiram 80W or zineb 75W applied at 2 to 3 lb/acre on a 7- to 10-day protective schedule, starting right after the end of harvest and continued to about midsummer (54, 122, 130, 150, 160, 213, 223, 237). Airplane applications have been successful when fungicides were used at 3 lb/acre (54).

XXXV. OTHER VEGETABLE CROPS

The remaining vegetable crops are routinely dusted or sprayed with fungicides. Where feasible, some states suggest using sprays or dusts of sulfur, to control powdery mildews and rusts, and fixed coppers to prevent bacterial blights and other diseases. Organic fungicides, e.g., maneb, mancozeb, chlorothalonil, metiram, are suggested for a few diseases, such as downy mildews and anthracnose. Applications should normally start when disease is first evident or expected. A summary of these fungicide recommendations are given below.

1. Okra (Gumbo). Weekly sprays of wettable sulfur 2.0% or dusts of 325 mesh sulfur for powdery mildew (Erysiphe cichoracearum) (68).
2. Pea, Garden. Powdery mildew (Erysiphe polygoni) is controlled by sulfur (see above). Applications should start when disease is first evident and be repeated at 10- to 14-day intervals, or often enough to keep the disease under control. It may be necessary to adhere to a strict program. (WARNING: Sulfur may cause injury if applications are made to plants when wet or when the temperature is above 90°F) (68, 86, 160).
3. Garden Beans (Dry, Snap, Pole, Lima). Powdery mildew (Erysiphe polygoni) and rust (Uromyces phaseoli var. typica) are important diseases. Maneb, mancozeb, chlorothalonil, sulfur, and zineb sprays or dusts are effective against rust (54, 57, 122, 150, 160, 213, 223). Sulfur or dinocap (KARATHANE) may be used against powdery mildew (42, 57, 150, 160, 223). The first application should be made as soon as any evidence of powdery mildew or rust is seen. The second application should follow within a week or 10 days (57, 76, 150, 160).

Spread of bacterial blights (Xanthomonas phaseoli, Pseudomonas phaseolicola, P. syringae) and anthracnose (Colletotrichum lindemuthianum) can be checked by timely sprays of fixed copper (especially KOCIDE). Weekly dusts or sprays of zineb, maneb, mancozeb, captan, thiram, ferbam, or ziram are effective against anthracnose (Table 29). However, fungicides are not ordinarily needed if disease-free seed, rotation, and strict sanitation is practiced (42, 54, 57, 130, 150, 160, 213, 223).

Downy mildew of lima bean, caused by the fungus Phytophthora phaseoli, is largely controlled by the use of resistant cultivars and by sowing disease-free seed. Maneb, mancozeb, zineb, and fixed copper are effective (Table 29). The same materials may be used if stem anthracnose (Colletotrichum truncatum) and pod blight (Diaporthe phaseolorum) are damaging. Copper sprays have been reported to cause foliage and pod injury and should not be used on fresh market limas (42, 57, 76, 150, 220). Dusts usually give less control than sprays.

Watery soft rot (Sclerotinia sclerotiorum) or white mold, root rot, and damping-off are controlled by applying PCNB (TERRACLOR) 75W, 2.5 lb in 15 to 20 gal of water applied in the furrow as a band spray (8,400 ft of row) at planting. The amount for bush beans covers 14,500 ft. Other chemicals useful against watery soft rot include benomyl 50W (1/2 to 2 lb/ 100 gal) and BOTRAN 75W, 3 to 4 lb/100 gal sprayed or dusted at the first appearance and repeated at 7- to 14-day intervals (54, 150, 160, 213, 223).

Benomyl 50W, 0.125 to 0.250%, is also suggested for Botrytis when it first appears. The spray is repeated at peak bloom (54, 160, 223). Botran does not have label approval for lima or dry beans.

Chloroneb (DEMOSAN) 65W, 1 to 1.5 lb in 10 to 20 gal applied in the seed furrow (14,500 linear ft), gives good protection against damping-off and soilborne root rot fungi (54, 68, 160).

4. Spinach. Downy mildew (Peronospora effusa) can be controlled by spraying with fixed copper, maneb or zineb or dusting with maneb or zineb (Table 29). One to three applications are suggested, 7 to 10 days apart, during cool wet periods (42, 54, 150, 160, 213, 223). These fungicides may also aid in controlling white rust (Albugo occidentalis) and anthracnosese (Colletotrichum spinaciae and C. spinacicola) (42, 160, 223). Powdery mildew is controlled by sulfur dust, 7.5 to 30 lb/acre when disease first appears and repeated at 7- to 10-day intervals (160). Damping-off is checked by applying 25 to 30 gal of captan 50W, 0.75% or captan 10% dust (25 to 30 lb/acre), in the seed row at planting (40,800 ft of row) (54, 150, 213).

5. Rhubarb. Leaf spot (primarily Ascochyta rhei), where serious, can be controlled by spraying with captan (213).

6. Garden (Red) or Table Beets. Cercospora leaf spot (C. beticola) is controlled as outlined under Sugarbeet.

XXXVI. SEED TREATMENT FOR OTHER VEGETABLES

Seed treatments are routinely applied by seedsmen to certain types of seed (beans, beets, peas) while others (anise, parsley, dill) are not treated. Seed of minor vegetable crops is normally treated only where poor stands have resulted in past years. Label directions should be carefully followed. Proper application protects seed and seedling against seed decay, preemergent damping-off and other seed-borne diseases. Table 31 lists commonly recommended seed treatment fungicides. Treatment normally gives maximum benefits when the soil is cold and wet or when emergence is otherwise delayed (42, 192, 194).

It is necessary to use fungicides within restrictions approved by the Food and Drug Administration and the Federal Environmental Protection Agency (Table 32). Tolerances are given in parts per million and the number of days between the last application at normal rate and harvest or they give the date of last application that will keep residues within tolerances set by the FDA (Table 33). A fungicidal usage may require such restrictions as "do not apply after first blooms appear" or "do not apply after edible parts form."

Growers must follow a spray program that will assure the production of vegetables with no excessive residues. Vegetables marketed with residues exceeding FDA tolerances may be injurious to consumers, may be confiscated, and may cause the grower to be brought to court. Growers have nothing to fear from the law so long as they use fungicides according to the label only on the crops specified, in the amounts specified and at the time specified. The label is the latest and most up-to-date information. Before using a fungicide, read the label carefully.

TABLE 31

Seed Treatment Materials, Diseases Controlled, Time, and Remarks Concerning Other Vegetables
(42, 68, 122, 130, 149, 150, 160, 194, 213, 220, 223)

Crop	Diseases controlled	Treatment[a]	Time and remarks
Anise, chervil, caraway, dill, fennel, parsley, salsify	Seed rot, damping-off	3 or 4	Any time
Beans, garden	Seed rot, damping-off, stem blights root rots	3, 4, or 5	Any time
Beet, Swiss chard, mangel, mangold	Seed rot, leaf spot, damping-off	3, 4, or 5	Any time
Coriander	Bacterial blight	Hot water then 3, 4, or 5	Hot water soak (30 min at 127° F) just before sowing
Okra (gumbo)	Seed rot, damping-off	3, 4, 5, or 6	Any time
Pea, lentil	Seed rot, foot and root rots, damping-off, Ascochyta and Mycosphaerella blights, Fusarium wilts, bacterial blights	3, 4, 5, or 6	Any time
Spinach	Seed rot, damping-off, anthracnose	3, 4, 5, or 6	Any time. If downy mildew is a problem, soak seed in hot water (25 min at 122° F) then apply fungicide.

[a]3, captan; 4, thiram; 5, chloranil; 6, dichlone.

TABLE 32

Fungicide Uses for Vegetables, Approved by the EPA, October 1, 1975[a],[b]

Crop	Benomyl (BENLATE), 0.2-15 ppm	Captan (D) (See ppm below)	Chlorothalonil, 0.1-15 ppm	DIFOLATAN 0.1-15 ppm	Anilazene (DYRENE), 10 ppm	Maneb, 4-10 ppm / Maneb with zinc salt	Mancozeb or zinc ion maneb[c] (See ppm below)	Zineb, 4-25 ppm
Asparagus	..	root dip	..	..	..	A[d]	(0.1 ppm), A	A
Beans (dry, lima, snap)	14[e], B (snap only) 28	(25 ppm), pp, 0[e]	7[e], B (snap only)	..	..	4 on limas or snap	..	7
Beet, garden	..	(2 ppm-root, 100 ppm-greens), 0, pp	..	..	..	..	..	7 (tops)
Broccoli	..	(2 ppm), pp	0[e]	..	..	3 or trim and wash	..	7
Brussels sprouts	..	(2 ppm), pp	0	..	..	0	..	7
Cabbage	..	(2 ppm), pp	0	..	..	7	..	7
Cantaloupe (muskmelon)	0	(25 ppm), 0, ph[d], pp	0	0[e]	0	5	(0 ppm in edible parts), 5[e]	5
Carrot	..	(2 ppm), 0	0	..	..	0	(2 ppm) 7, B (tops)	7 (tops)
Cauliflower	..	(2 ppm), pp	0	..	..	0	..	7
Celery	7	(50 ppm), 0, pb	7	..	0	strip and wash, 14	(5 ppm), 14	strip and wash, 14

Chinese cabbage	..	7 (0.5 ppm cob and kernel)	..	..	..	..	7
Corn, sweet and pop	..	(15 ppm), Fodder and forage	0, B	..	14, B[f]	(2 ppm), 10, B, pp	0, B, D
Cucumber	0	(4 ppm), 5	5	0	0	(25 ppm), 0, ph, pp	5
Eggplant	..	..	0	..	..	(25 ppm), 0, ph, pb	0
Endive, escarole	..	..	10 and wash	..	..	..	10
Kale, collards	..	..	10 and wash	..	..	(2 ppm), pp	10
Kohlrabi	..	..	0	..	..	..	(halfgrown)
Lettuce	..	..	10 (strip and wash)	..	..	(100 ppm), 0	10
Mustard greens	..	..	10 and wash	..	..	(2 ppm), pp	7
Onion	0	(0.5 ppm dry), 7	0	0	..	(50 ppm green, 25 dry), 0, ph	10
Peas	..	..	..	..	..	(2 ppm), pp	10, D
Pepper	..	..	0	..	..	(25 ppm), 0, pb, pp	0
Potato, Irish[d]	(1.0 ppm), 0	(0.1 ppm), 0, D	0	0	(25 ppm), 0, ph	0 and seed, D, pp
Pumpkin	..	..	0	0	0	(25 ppm), 0, pp	0

TABLE 32 (Continued)

Crop	Benomyl (BENLATE), 0.2–15 ppm	Captan (D) (See ppm below)	Chloro-thalonil, 0.1–15 ppm	DIFO-LATAN, 0.1–15 ppm	Anila-zene (DYRENE), 10 ppm	Maneb, 4–10 ppm Maneb with zinc salt	Mancozeb or zinc ion maneb[c] (See ppm below)	Zineb, 4–25 ppm
Radish	..	..	..	..	0	..	..	0
Rhubarb (greenhouse)	..	(25 ppm), 0	..	..	..	..	..	..
Spinach	..	(100 ppm), 0, pp	..	..	..	10 and wash	..	10
Squash	0	(25 ppm), 0, pp	0	..	0	5	(4 ppm), 5	5
Sugarbeet[d]	21	0	..	..	..	10, B, D, 14, (no feeding restrictions)	(65 ppm tops), 14 (2 ppm roots), 10, B, 14, no feeding restrictions	
Swiss chard	..	0	..	..	..	..	..	7
Tomato	..	(25 ppm), 0, pp	0	0g	0	(4 ppm), 5, C	(4 ppm), 5	5

Turnip, rutabaga	··	(2 ppm), pp	··	··	··	7 and wash	··	(7 ppm), 7–tops
Watermelon	0	(25 ppm), 0, pp	0	0	5	5	(0 ppm edible parts), 5e	5

[a] No tolerances have been set for these fungicides on dill, horseradish, okra, parsley, and parsnip.

[b] The following abbreviations are used: A, Postharvest application to ferns only or to young plantings that will not be harvested; B, Do not feed treated tops or forage to livestock; C, To avoid damage, do not use on tender young plants; D, Do not use treated seed or seed pieces for feed or food; M, Maneb; Z, Zineb; pb, Plant bed treatment; ph, Postharvest spray or dip; pp, Preplant soil treatment.

[c] Maneb and zinc ion are sold as DITHANE M–45 and MANZATE 200.

[d] Tolerances are not needed for pesticides applied only to the foliage and not translocated to the tubers or roots.

[e] Number indicates number of days between last application and harvest; 0, up to harvest.

[f] Do not apply if crop is to be used for processing.

[g] Machine harvest only.

TABLE 33

Label Information on Fungicides of Less General Use

Fungicide (tolerance)	Crops and use restrictions
Dicloran (BOTRAN)	Beans (snap)—white mold. 2 days to harvest. Do not feed forage to livestock. Greenhouse tomato—to harvest. Do not drench seedlings or newly set transplants. Carrot—postharvest dip or spray, see label. Garlic, Onion—soil application before seeding or spray to soil around sets or bulbs. Do not plant spinach as followup crop in treated soil. Leaf lettuce (greenhouse)—14 days[a] (do not apply to wilted plants or seedlings). Head lettuce—14 days. Celery—7 days. Cucumber (greenhouse)—see label. Rhubarb (greenhouse)—3 days). Potato—14 days (do not feed to livestock). Sweetpotato—root dip and plant bed treatment. NOTE: Do not plant tomatoes as a followup crop in treated soil. Do not use spent roots for food or feed. Postharvest spray or dip as directed.
Copper, fixed, neutral, and basic (including Bordeaux mixture)	Exempt if used with good agricultural practices. Not exempt if used at time of or after harvest. See label.
Fenaminosulf (DEXON)	Cleared only for seed-treatment use on Beans, Beets, Corn, Cucumbers, Peas, Spinach, Sugarbeets. Do not use treated seed for food, feed, or oil. Slurry seed treatment for planting in light soils or soils high in clay or organic matter.
Dinocap (KARATHANE)	Cantaloupe (Muskmelon), Cucumber, Honeydew melon, Pumpkin, Squash, Watermelon—7 days. For control of powdery mildew only.
Ethazol (TERRAZOLE, TRUBAN, KOBAN)	Seed treatment: Beans, Peas, Sugarbeets.
Metiram (POLYRAM) (0 ppm)	Cantaloupe, Cucumber, Potato, Sugarbeets—no time limitations. Celery—14 days. Tomato—5 days. Potato—seedpiece treatment. Do not feed Sugarbeet tops to meat or dairy animals. Celery—strip, trim, and wash. Postharvest application to Asparagus ferns.

TABLE 33 (Continued)

Fungicide (tolerance)	Crops and use restrictions
PCNB (TERRACLOR, BRASSICOL, FUNGICLOR) (0.1 ppm)	Beans—base of plants before blossoming, soil and seed treatment at planting, or foliar spray. Do not feed treated Bean vines to livestock. Do not apply after first bloom. Broccoli, Brussels sprouts, Cabbage, Cauliflower—transplant solution (3/4 pt per plant) or row treatment before transplanting. Pepper, Potato, Tomato—soil treatment at or before planting. Tomato (greenhouse)—transplant solution (1/2 pt of 0.2% per plant). Garlic—soil and seed treatment at planting.
Streptomycin (0.25 ppm)	Celery, Pepper, Tomato—plant beds only (200 ppm spray); Potato—seedpiece treatment only (100 ppm dip or dust). Soak cut seed pieces less than 30 min. Beans—seed treatment for halo blight control. Do not use treated seed for food or feed.
Sulfur, lime, and lime-sulfur	Exempt when used with good agricultural practices. See label.
Thiabendazole (MERTECT)	Sweetpotato—"seed" root treatment. Do not use treated pieces for food or feed. Potato "seed" tubers only (1,500 ppm—20 sec dip).
Thiram, TMTD (0.5-7 ppm)	Onion—Furrow treatment. Celery—7 days (strip, trim, and wash). Sweetpotato—preplant root dip (1,200 ppm). Tomato—10 days for leaf spots and fruit rots. Seed treatment: Beans, Beets, Broccoli, Brussels sprouts, Cabbage, Cantaloupe, Carrot, Cauliflower, Corn, Kale, Collards, Cucumber, Eggplant, Endive, Kohlrabi, Lettuce, Okra, Onion (bulb, seed, and set), Peas, Pepper, Pumpkin, Radish, Spinach, Squash, Swiss chard, Tomato, Turnip, Watermelon. WARNING: Do not use treated seed for food, feed, or oil.

[a]Number of days between last application and harvest.

A FINAL WORD

There are many different approaches to the subject of this chapter. We have chosen to define Plant Pathology, describe its history, and briefly discuss the causal agents of plant diseases. This sparse entry into this intricate science paves the way to discuss the control measures which, in this instance, involve the fungicides with emphasis on their performance under field conditions.

Rather than present a general discussion of the role of fungicides in crop production, we chose to be specific and discuss the importance of each crop, losses resulting from disease damage, and fungicide programs to prevent such losses. We are aware that new fungicides are constantly being studied and are entering the various programs of plant health. The recommendations suggested here may be outdated in 1 or 10 years hence, depending on future progress. In general, most of our suggestions have been in effect for 5 to 20 years, thus we feel that new developments will not appear too rapidly. In any case we have presented a template on broad fungicide usage, which is not likely to be modified for many years.

REFERENCES

1. L. E. Adams, G. L. Gubb, Jr., C. M. Haesler, T. H. Obourn, and D. H. Peterson, 1974 Grape Disease, Insect, and Weed Control, Pa. Agric. Exp. Stn. Rep. Serv., 1974.

2. Agricultural Statistics—1974, Washington, D.C., U.S. Government Printing Office, 1974, 619 pp.

3. G. H. Ahlgren, Forage Crops, Ed. 2, New York, McGraw-Hill, 1965, 536 pp.

4. J. A. Albert and J. W. Heuberger, Plant Dis. Rep., 45:759-763 (1961).

5. C. J. Alexopoulos, Introductory Mycology, Ed. 2, New York, Wiley, 1962, 613 pp.

6. H. W. Anderson, Diseases of Fruit Crops, New York, McGraw-Hill, 1956, 501 pp.

7. P. A. Ark, Plant Dis. Rep., 37:404-406 (1953).

8. J. G. Arthur, Bot. Gaz., 10:343-345 (1885).

9. J. G. Atkins, USDA Farmer's Bull. 2120, 1958.

10. G. L. Barnes, Okla. State Univ. Bot. Plant Pathol. Processed Ser. P-482, 1964.

11. M. M. Barnes, Ca. Exp. Stn. Serv. Leaf. 80, 1964.

12. A. H. Bauer, Pa. State Agric. Stn. Ext. Leaf. Circ. 513, 1963.

13. J. H. Beattie and W. R. Beattie, USDA Leaf. 154, 1938.

14. Beet Sugar Handbook, San Francisco, California, Western Beet Sugar Producers, Inc., 221 pp.

15. Beet Sugar Technology, R. A. McGinnis (ed.), New York, Reinhold, 1954, 574 pp.

16. E. A. Bessey, Morphology and Taxonomy of Fungi, New York, Blakiston's, 1950, 791 pp.

17. H. L. Bissonette, N.D. Ext. Circ. A-455, 1964.

18. D. E. Bliss, Phytopathology, 41:665-683 (1951).

19. G. H. Boewe, Ill. Nat. Hist. Surv. Circ. 48, 1960.

20. V. R. Boswell, USDA Leaf. 353, 1963.

21. V. R. Boswell, S. P. Doolittle, L. M. Pultz, A. L. Taylor, L. L. Danielson, and R. E. Campbell, WSDA Agric. Inf. Bull. 276, 1964.

22. J. L. Brann, Jr., P. A. Arneson, and G. O. Oberly, Tree-Fruit Production Recommendations for Commercial Growers, Cornell Ext. Ser., 1975.

23. J. L. Brann, Jr., J. P. Tomkins, and P. A. Arneson, Grape Pest Control Guide, New York Coop. Ext. Serv., 1975.

24. J. L. Brann, Jr., P. A. Arneson, and J. P. Tomkins, New York Bramble Pest Control Culture Guide, New York Coop. Ext. Serv., 1974.

25. R. D. Brigham and B. R. Spears, Tex. Exp. Stn. Bull. B-954, 1960.

26. A. N. Brooks and E. G. Kelsheimer, Fla. Agric. Exp. Stn. Bull. 629, 1961.

27. I. F. Brown and H. D. Sisler, Phytopathology, 50:830-839 (1960).

28. H. P. Burchfield and A. Goenaga, Contrib. Boyce Thompson Inst., 19:141-156 (1957).

29. O. D. Burke, Pa. Ext. Circ. 349, 1960.

30. Burley Tobacco, The American Tobacco Co., 1958.

31. T. J. Burrill, Am. Nat., 15:527-531 (1881).

32. W. G. Burton, The Potato: A Survey of Its History and Factors Influencing Its Yield, Nutritive Value, and Storage, London, Chapman and Hall, 1948, 319 pp.

33. J. D. Butt, Rep. Agric. Hort. Sec. Res. Stn., 127-135 (1955).

34. California Experimental Station Service Leaflet 71, 1964.

35. California Agricultural Experiment Station Extension Service Leaflet 75, 1964.

36. G. E. Carman, Spray Program for California Citrus Fruits, Univ. Ca. Citrus Exp. Stn. Agric. Ext. Serv., 1964.

37. D. Cation, Ann. Rep. State Hort. Soc. Mich., 83:21 (1953).

38. D. Cation, Plant. Dis. Rep., 41:1029 (1959).

39. D. Cation, Phytopathology, 42:57 (1953).

40. J. N. Chand, E. K. Wade, and J. C. Walker, Plant Dis. Rep., 47:94-95 (1963).

41. N. F. Childers, Modern Fruit Science, Ed. 4, Somerville, New Jersey, Somerset Press, 1969, 912 pp.

42. C. Chupp and A. F. Sherf, Vegetable Diseases and Their Control, New York, The Ronald Press, 1960, 693 pp.

43. N. G. Clark, A. F. Hains, D. J. Higgins, and H. A. Stevenson, Chem. Ind., 572-573 (1960).

44. H. Cole, Pa. State Univ. Coll. Agric. Exp. Serv. Circ. 523, 1965.

45. A Compendium of Corn Diseases, St. Paul, Minnesota, The American Phytopathological Soc., 1973, 64 pp.

46. R. H. Converse, Phytopathology, 57:173-177 (1967).

47. R. H. Converse, USDA Agric. Hand. 310, 1964.

48. Cotton Diseases - How to Recognize and Control Them, Council and National Cotton Council, 1953.

49. R. S. Cox, Fla. Agric. Exp. Stn. Tech. Bull. 598, 1958.

50. J. O. Culbertson, Adv. Agron., 6:144-182 (1954).

51. H. S. Cunningham and E. G. Sharvelle, Phytopathology, 30:4 (1940).

52. S. H. Davis, Jr. and S. R. Race, Jr., N.J. Ext. Leaf. 335-A, 1966.

53. S. H. Davis, Jr. and S. R. Race, Jr., Rutgers Ext. Leaf. 331-B, 1966.

54. S. H. Davis, Jr. and J. K. Springer, 1975 Pesticides for New Jersey, N. J. Ext. Serv., 1975.

55. S. H. Davis, Jr. and F. C. Swift, Rutgers Ext. Leaf. 340, 1964.

56. S. H. Davis, Jr. and F. C. Swift, Rutgers Ext. Leaf. 341, 1964.

57. S. H. Davis, Jr. and F. C. Swift, Rutgers Ext. Leaf. 330, 1963.

58. S. H. Davis, Jr. and F. C. Swift, Rutgers Ext. Leaf. 332, 1963.

59. A. E. Dimond, J. W. Heuberger, and J. G. Horsfall, Phytopathology, 33:1095-1097, 1943.

60. J. G. Dickson, Diseases of Field Crops, Ed. 2, New York, McGraw-Hill, 1956, 517 pp.

61. R. L. Donahue and E. F. Evans, Englewood Cliffs, Prentice-Hall, 1973, 441 pp.

62. S. P. Doolittle, A. L. Taylor, L. L. Danielson, and L. B. Reed, USDA Agric. Inf. Bull. 259, 1962.

63. S. P. Doolittle, A. L. Taylor, L. L. Danielson, and L. B. Reed, USDA Agric. Hand. 216, 1961.

64. S. P. Doolittle, A. L. Taylor, and L. L. Danielson, USDA Agric. Hand. 203, 1961.

65. J. C. Dunegan, R. A. Wilson, and W. T. Morris, Plant Dis. Rep., 37:604-605 (1953).

66. I. M. Febler and C. L. Hammer, Bot. Gaz., 110:324 (1948).

67. H. H. Flor, USDA Yearbook of Agriculture, Washington, D.C., U.S. Government Printing Office, 1953.

68. Florida Plant Disease Control Guide, Fla. Ext. Serv., 1974.

69. J. H. Ford, W. Klomperens, and C. L. Hammer, Plant Dis. Rep., 42:680 (1958).

70. F. J. Francis and B. M. Zuckerman, Proc. Am. Soc. Hort. Sci., 81:288-294 (1962).

71. D. E. H. Frear, Chemistry of the Pesticides, Ed. 3, New York, Van Nostrand, 1955, 300 pp.

72. S. G. Fushtey, Ontario Can. Dep. Agric. Bull. 524, 1957.

73. W. N. Garrett and V. R. Coleman, Univ. Ga. Ext. Circ. 519, 1963.

74. H. R. Garriss and J. C. Wells, N.C. Ext. Leaf. 125, 1967.

75. H. R. Garriss and J. C. Wells, N.C. Agric. Ext. Leaf. 113, 1966.

76. H. R. Garriss, N.C. Plant Pathol. Inf. Note 103, 1963.

77. Georgia Agricultural Station Extension Service Bulletin 630, 1966.

78. C. J. Gilgut and H. E. Wave, Massachusetts Pest Control Chart for Potatoes, Mass. Ext. Serv., 1966.

79. M. C. Goldsworthy, E. L. Green, and M. A. Smith, J. Agric. Res., 66:277-291 (1943).

80. R. N. Goodman, Am. Fruit Grower, 73:7 (1953).

81. R. N. Goodman, Mo. Agric. Exp. Stn. Res. Bull. 540, 1954.

82. Grain Sorghum Handbook, Kan. Ext. Circ. 494, 1974.

83. R. A. Gray, Plant Dis. Rep., 39:567-569 (1955).

84. R. G. Grogan, W. C. Snyder, and R. Bardin, Univ. Ca. Circ. 448 (1955).

85. E. F. Guba, Univ. Mass. Agric. Stn. Bull. 522, 1961.

86. Guide to Chemical Control of Plant Diseases in Colorado, Colorado State Ext. Serv., 1974.

87. J. M. Hamilton and M. Szkolnik, Plant Dis. Rep., 41:293-300 (1957).

88. Handbook of North American Nut Trees, R. A. Jaynes (ed.), The Northern Nut Growers Association, 1969, 421 pp.

89. E. V. Hardenburg, Potato Production, Ithaca, New York, Comstock Publishing Co., 1949, 270 pp.

90. J. R. Harlan, Theory and Dynamics of Grassland Agriculture, Princeton, New Jersey, Van Nostrand, 1956, 281 pp.

91. M. R. Harris and D. H. Brannon, Wash. Ext. Bull. 553, 1962.

92. J. B. Harry, R. H. Wellman, F. R. Whaley, H. W. Thurston, Jr., and W. A. Chandler, Contrib. Boyce Thompson Inst., 15:195-210 (1948).

93. F. D. Heald, Manual of Plant Diseases, Ed. 2, New York, McGraw-Hill, 1933, 953 pp.

94. C. B. Heiser, Jr. and P. G. Smith, Econ. Bot., 7:214-227 (1953).

95. J. W. Heuberger and R. K. Jones, Plant Dis. Rep., 46:159-162 (1962).

96. J. W. Heuberger and P. C. Poulos, Plant Dis. Rep., 37:81-83 (1953).

97. J. W. Heuberger, S. H. Davis, Jr., L. P. Nichols, and L. D. Buchler, Phytopathology, 37:9 (1947).

98. J. W. Heuberger and J. G. Horsfall, Phytopathology, 29:303-321 (1939).

99. E. M. Hildebrand and H. T. Cook, USDA Farmer's Bull. 1059 (1959).

100. E. M. Hildebrand, Cornell Univ. Agric. Exp. Stn. Mem. 276 (1947).

101. E. C. Hislop, Ann. Appl. Biol., 57:475-489 (1966).

102. N. L. Horn, Plant Dis. Rep., 36:309-310 (1952).

103. J. G. Horsfall, Principles of Fungicidal Action, Ed. 2, San Fran-
      cisco, California, Chronica Botanica Co., 1956, 279 pp.

104. W. S. Hough and A. Freeman Mason, Spraying, Dusting, and Fumi-
      gating of Plants, Revised Ed., New York, MacMillan, 1955, 726 pp.

105. F. L. Howard and M. B. Sorrell, Phytopathology, 33:1114 (1943).

106. H. D. Hughes, M. E. Heath, and D. S. Metcalfe, Forages: The
      Science of Grassland Agriculture, Ed. 2, Revised, Ames, Iowa,
      Iowa State University Press, 1953, 723 pp.

107. H. D. Hughes and D. S. Metcalfe, Crop Production, Ed. 3, New
      York, MacMillan, 1972, 627 pp.

108. C. G. Hughes, E. V. Abbott, and C. A. Wismer, Sugar Cane Diseases
      of the World, Vol. 2, New York, Elsevier, 1964, 354 pp.

109. Index of Plant Diseases in the United States, USDA Agric. Hand. 165,
      1960.

110. International Symposium on Factors Determining the Behavior of
      Plant Pathogens in Soil, K. F. Baker and W. O. Snyder (eds.), San
      Francisco, California, University of California Press, 1965, 571 pp.

111. C. R. Jackson and D. K. Bell, Ga. Exp. Stn. Res. Bull. 56, 1969.

112. B. F. Janson and R. L. Miller, Ohio State Ext. Serv. Bull. 506,
      1975.

113. A. L. Jones, D. Ramsdell, W. W. Thompson, A. J. Howitt, and
      J. Hull, Mich. State Univ. Ext. Bull. 154, 1974.

114. A. L. Jones, Mich. State Univ. Ext. Bull. E-714, 1971.

115. R. K. Jones, J. W. Heuberger, and J. D. Bates, Plant Dis. Rep.,
      47:420-424 (1963).

116. R. P. Kahn, H. W. Anderson, P. R. Hepler, and M. B. Linn, Ill.
      Agric. Exp. Stn. Bull. 559 (1952).

117. H. L. Keil, Agric. Chem., 20:23-24 (1965).

118. G. W. Keitt and L. K. Jones, Wisc. Res. Bull., 73:1-106 (1927).

119. W. A. Kellerman and W. T. Swingle, Kan. Agric. Exp. Stn. Bull.
      12, (1890).

120. N. J. King, R. W. Mungomery, and C. G. Hughes, Manual of Cane
      Growing, Revised Ed., New York, Elsevier, 1961, 542 pp.

121. M. S. Kipps, Production of Field Crops, Ed. 6, New York, McGraw-Hill, 1970, 790 pp.

122. J. E. Klinker, D. L. Matthew, R. W. Samson, J. P. Coleman, G. F. Warren, G. E. Wilcox, and L. Hafen, Purdue Univ. Mimeo. 1D-56, 1964.

123. L. J. Klotz, Ca. Agric. Exp. Stn. Ext. Serv. Circ. 396 (revised), 1960.

124. B. Koehler and J. R. Halbert, Univ. Ill. Agric. Exp. Stn. Bull. 354, 1930.

125. B. Koehler, Univ. Ill. Agric. Exp. Stn. Bull. 617, 1957.

126. L. J. Kushman, W. R. Wright, J. Kaufman, and R. E. Hardenburg, USDA Marketing Res. Rep. 698, 1965.

127. R. W. Leukel, J. H. Martin, and C. L. Lefebvre, USDA Farmer's Bull. 1959, 1960.

128. S. Lewis and R. N. Goodman, Phytopathology, 55:719-723 (1965).

129. J. G. Lill, USDA Farmer's Bull. 2060, 1964.

130. M. B. Linn, Ill. Ext. Circ. 802, 1964.

131. M. B. Linn and W. H. Luckman, Ill. Ext. Circ. 912, 1965.

132. E. H. Lloyd, Jr. and R. L. Kiesling, North Dakota Plant Disease Control Guide, N.D. Ext. Serv., 1975.

133. Losses in Agriculture, USDA ARS Agric. Hand. 291, 1965.

134. G. B. Lucas, Diseases of Tobacco, Ed. 2, New York, Scarecrow Press, 1965, 778 pp.

135. R. A. Ludwig and G. D. Thorn, Plant Dis. Rep., 37:127-129 (1953).

136. J. M. Lutz and J. W. Simons, USDA Farmer's Bull. 1442, 1955.

137. I. C. MacSwan and P. A. Koepsell, Oregon Plant Disease Control Handbook, Corvallis, Oregon, 1976.

138. I. C. MacSwan and P. A. Koepsell, 1975 Oregon Plant Disease Control Handbook, Ore. State Univ. 1975, 214 pp.

139. I. C. MacSwan, Ore. State Univ. Ext. Circ. 646, 1974.

140. O. C. Maloy, Disease Control Guides for Washington, 1966.

141. H. Martin, Ann. Appl. Biol., 19:98-120 (1932).

142. J. P. Martin, E. V. Abbott, and C. G. Hughes, Sugar Cane Diseases of the World, Vol. 1, New York, Elsevier, 1961, 542 pp.

143. N. E. McGlohon, Ga. Ext. Leaf. 67, 1967.

144. N. E. McGlohon, Ga. Ext. Leaf. PP. No. 2C-3, 1966.

145. N. E. McGlohon, R. J. Ledbetter, U. L. Diener, and G. H. Blake, Ala. Ext. Serv. Auburn Univ. Circ. 598, 1962.

146. N. E. McGlohon, Ga. Ext. PP. No. 2C-2, 1967.

147. J. R. McGrew and G. W. Still, USDA Farmer's Bull. 1893 (Revised), 1972.

148. P. R. Miller, Plant Dis. Rep., 32:160-166 (1948).

149. P. A. Minges, A. A. Muka, A. F. Sherf, and R. F. Sandsted, Cornell Misc. Bull. 76, 1967.

150. P. A. Minges, A. A. Muka, R. F. Sandsted, A. F. Sherf, and R. D. Sweet, Vegetable Production Recommendations, Cornell Ext. Serv., 1975.

151. W. D. Moller, D. H. Hall, A. H. McCain, and A. O. Paulus, California Study Guide for Agricultural Pest Control Advisers on Plant Diseases, Ca. Ext. Serv., 1972.

152. J. Monteith, Jr. and A. S. Dahl, U.S. Golf Assoc. Greens Sec. 12, 1890.

153. R. E. Motsinger, Ga. Ext. Leaf. 47, Revised, 1972.

154. A. E. Murneek, Phytopathology, 42:57 (1952).

155. National Sweet Potato Collaborators, 1960, 64 pp.

156. W. C. Nettles, F. H. Smith, C. A. Thomas, and D. A. Benton, Clemson Ext. Bull. 109.

157. A. G. Newhall and W. A. Rawlins, Cornell Ext. Bull. 1018, 1958.

158. New Hampshire Agricultural Extension Service Bulletin 151, 1964, 9 pp.

159. L. W. Nielson, N.C. Plant Pathol. Inf. Note 87, 1962.

160. North Carolina Agricultural Chemicals Manual, N.C. Ext. Serv., 1975.

161. J. M. Ogawa, S. D. Lyda, and D. J. Webber, Plant Dis. Rep., 45:636-638 (1961).

162. G. W. Padwick, Commonwealth Mycological Institute, 1950, pp. 1-198.

163. R. E. Partyka and L. J. Alexander, Ohio State Univ. Ext. SB-16, 1965.

164. A. O. Paulus, Botran Symposium, Kalamazoo, Michigan, The Upjohn Co., 1965.

165. U. S. Pickering, J. Chem. Soc., 91:1988-2001 (1907).

166. Potato Handbook, Potato Assoc. Am., New Brunswick, N.J., 1960, 80 pp.

167. Potato Fungicide Spray Schedule, Pa. State Univ. Plant Pathol. Ext., 1967.

168. Potato Handbook, Potato Assoc. Am., New Brunswick, N.J., 1957, 79 pp.

169. 1975 Potato Recommendations for New York State, Cornell Ext. Serv., 1975.

170. H. S. Potter, R. Harrison, and B. Hodgin, Mich. Res. Rep. 62, 1967.

171. D. Powell, Phytopathology, 36:572-573 (1946).

172. D. Powell, Trans. Ill. State Hort. Soc., 96:60-61 (1962).

173. D. Powell, Am. Fruit Grower, 86:36 (1966).

174. D. Powell, Plant Dis. Rep., 36:97-98 (1952).

175. D. Powell, Plant Dis. Rep., 38:209-211 (1954).

176. D. Powell, Plant Dis. Rep., 44:176-178 (1960).

177. D. Powell, Ill. Res., 4:8-9 (1962).

178. D. Powell, Unpublished data.

179. D. Powell, Trans. Ill. State Hort. Soc., 91:25-28 (1957).

180. D. Powell, Trans. Ill. State Hort. Soc., 90:139-141 (1956).

181. D. Powell and H. W. Anderson, Trans. Ill. State Hort. Soc., 77:488-493 (1943).

182. D. Powell, A. Khettry, P. J. Sesaki, G. E. Brussel, Plant Dis. Rep., 42:493-498 (1958).

183. D. Powell and R. H. Meyer, Ill. Agric. Ext. Serv. Fruit Leaf. 4, 1965.

184. J. T. Presley, USDA Farmer's Bull. 1745, 1954.

185. W. E. Rader, Cornell Univ. Agric. Exp. Stn. Bull. 889, 1952.

186. W. W. Robbins, T. E. Weir, and C. R. Stocking, A Textbook of General Botany, Ed. 2, New York, Wiley, 1957, 578 pp.

187. W. M. Ross and O. J. Webster, USDA Agric. Inf. Bull. 218, 1964.

188. R. N. Salamon, The History and Social Influence of the Potato, London, Cambridge University Press, 1949, 685 pp.

189. A. J. Salle, Fundamental Principles of Bacteriology, Ed. 5, New York, McGraw-Hill, 1961, 812 pp.

190. V. Sauchelli, Ind. Eng. Chem., 25:363-367 (1933).

191. A. F. Sherf, Cornell Ext. Bull. 1130, 1964.

192. A. F. Sherf, Cornell Ext. Bull. 1128, 1964.

193. M. C. Shurtleff, S. M. Ries, R. Randell, and C. C. Zych, Ill. Ext. Serv. Fruit Leaf. 1 (Revised), 1975.

194. M. C. Shurtleff, How to Control Plant Diseases in Home and Garden, Ames, Iowa, Iowa State University Press, 1966, 649 pp.

195. M. C. Shurtleff, D. P. Taylor, J. W. Courter, and R. Randell, Ill. Ext. Circ. 893 (Revised), 1969.

196. E. F. Smith, Zentbl. Bakt. Abt. II, 1:364-373 (1895).

197. M. A. Smith, L. P. McCulloch, and B. A. Friedman, USDA Agric. Hand. 303, 1966.

198. W. H. Snell and E. A. Dick, A Glossary of Mycology, Cambridge, Harvard University Press, 1957, 171 pp.

199. R. Sprague, Diseases of Cereals and Grasses, New York, The Ronald Press, 1950, 538 pp.

200. R. Sprague, Proc. Wash. State Hort. Assoc., 45:47 (1949).

201. E. C. Stakman and J. G. Harrar, Principles of Plant Pathology, New York, The Ronald Press, 581 pp.

202. F. C. Swift and S. H. Davis, Jr., Rutgers Ext. Leaf. 334, 1963.

203. F. C. Swift and S. H. Davis, Jr., Rutgers Ext. Leaf. 342, 1961.

204. F. C. Swift and S. H. Davis, Jr., Rutgers Ext. Leaf. 333, 1963.

205. S. A. J. Tarr, Sorghum Diseases, London, The Commonwealth Mycological Institute, 1962, 380 pp.

206. C. A. Thomas and F. H. Smith, Clemson Ext. Circ. 534, 1950.

207. H. C. Thompson and W. C. Kelly, Vegetable Crops, Ed. 5, New York, McGraw-Hill, 1957, 611 pp.

208. R. C. Thompson, USDA FB 1646.

209. R. C. Thompson, S. P. Doolittle, and T. J. Henneberry, USDA Agric. Hand. 149, 1958.

210. S. S. Thompson, Univ. Ga. Ext. Leaf. 25, 1972.

211. W. H. Tisdale and I. Williams, Disinfectant, U.S. Patent 1,972,96 , 1934.

212. Tree Fruit Production Recommendations for Pennsylvania, Pa. Ext. Serv., 1975.

213. O. C. Turnquist, J. A. Lofgren, and H. L. Bissonnette, 1974 Weed, Insect, and Disease Control Guide for Commercial Vegetable Growers, Minn. Ext. Serv., 1974.

214. A. J. Ullstrup, Corn Diseases in the United States and Their Control, USDA Agric. Hand. 199 (Revised), 1974.

215. USDA Consumer and Marketing Services AMG 48, 1966.

216. USDA Consumer and Marketing Service AMG 51, 1966.

217. USDA Farmer's Bull. 2140 (Revised), 1974.

218. USDA Farmer's Bull. 2208 (Revised), 1975.

219. USDA Statistical Rep. Serv.,

220. USDA Summary of Registered Agricultural Pesticide Chemical Uses, Ed. 2, Sppl. 1, 2, and 3, 1964-67.

221. USDA Yearbook of Agriculture, USDA House Document 122, 1953.

222. W. D. Valleau, E. M. Johnson, and S. Diachun, Univ. Ky. Circ. 522-A, 1963.

223. Virginia Plant Disease Control Guide, Va. Ext. Serv., 1975.

224. E. K. Wade, Wisc. Ext. Circ. 424, 1963.

225. J. C. Walker, Plant Pathology, Ed. 3, New York, McGraw-Hill, 1969, 819 pp.

226. J. C. Walker and R. H. Larson, USDA Agric. Hand. 208, 1961.

227. J. C. Walker, USDA Agric. Hand. 144, 1958.

228. T. W. Whitaker and G. W. Bohn, Econ. Bot., 4:52-81 (1950).

229. T. W. Whitaker, E. J. Ryder, and O. A. Hills, USDA Agric. Hand. 211, 1962.

230. F. Wilcoxon and S. E. A. McCallan, Contrib. Boyce Thompson Inst., 3:509-538 (1931).

231. F. Wilcoxon and S. E. A. McCallan, Phytopathology, 20:391-417 (1930).

232. F. Wilcoxon and S. E. A. McCallan, Contrib. Boyce Thompson Inst., 9:149-159 (1938).

233. S. Wilhelm, Ca. Agric. Exp. Stn. Serv. Circ. 494, 1961.

234. H. F. Winter, Plant Dis. Rep., 46:560-564 (1962).

235. H. F. Winter and H. C. Young, Plant Dis. Rep., 37:463-464 (1953).

236. Wisconsin Agricultural Experiment Service Special Circular 103, 1966.

237. W. J. Zaumeyer and H. Rex Thomas, USDA Agric. Hand. 225, 1962.

238. B. M. Zuckerman, Plant Dis. Rep., 45:253-254 (1961).

Chapter 2

NEMATODE DISEASES OF PLANTS AND THEIR CONTROL INCLUDING THE IMPORTANT ROLE OF NEMATICIDES

J. N. Sasser
Department of Plant Pathology
North Carolina State University
Raleigh, North Carolina

I. INTRODUCTION

Few advancements in agriculture have had so profound an effect on crop production as the development and widescale use of nematicides. For centuries, man has been plagued by the ravages of nematodes, microscopic organisms that feed on plant roots, buds, stems, crowns, leaves, and even the developing seed. The various symptoms of nematode attack, and the effects of specific nematodes on host tissues, depend upon the host and the nematode, but the result to the grower is less profit. The visible deleterious effects of nematode attack are reduced yields and poorer quality of such crops as potato, rice, vegetables, fruits, tobacco, cotton, and peanuts. Highly prized ornamentals and turf show a slow decline and unthriftiness, or death.

In addition to the visible and measurable effects nematodes have on plant responses, they also have other effects which are less spectacular. Plant roots damaged by the feeding activities of nematodes are not efficient in the utilization of available moisture and nutrients in the soil. This is evidenced by the fact that a common symptom of nematode attack is nutrient deficiency, showing up in the foliage even when there is an adequate supply of these nutrients in the rhizosphere.

Root necrosis, root pruning, root galling, and cessation of root growth are common symptoms of nematode attack on plant roots. Roots, weakened and damaged in this manner, are easy prey to many types of saprophytic bacteria and fungi that invade the roots and accelerate root rot. Frequently plants, developed for resistance to certain pathogenic bacteria and fungi, succumb to attack by these organisms in fields where nematodes are not controlled. Also, many important soilborne virus diseases of crop plants are transmitted by plant-parasitic nematodes. The discovery of these interrelationships between nematodes and other soil-inhabiting pathogens has been an important development of recent years.

Prior to discussing the specific role of nematicides in crop production, it may be helpful to present some general background information concerning the Science of Nematology. With a brief discussion of each of the following topics, perhaps the reader will better appreciate the nature and magnitude of nematode problems and the necessity for their control.

A. Early History of Nematode-Induced Diseases

We have known about nematodes for many centuries, but the first plant parasite to be discovered was the wheat gall nematode (Anguina tritici) in 1743 when Needham (15) described this serious nematode disease of wheat and other grains (Fig. 1). This nematode is a damaging parasite of wheat, and if not controlled it can cause serious losses wherever it occurs. A nematode that is more universally recognized, the root-knot nematode (Meloidogyne spp.) (Fig. 2), was discovered and described in some detail in 1855 by the Rev. M. J. Berkeley (2). The various species now comprising this genus are considered by many nematologists to be the most important group known, perhaps because of their worldwide distribution and extensive host range.

Nematodes were not regarded as serious plant pathogens until the latter part of the 19th century, when the sugarbeet nematode (Heterodera schachtii) and what is now known as the potato root nematode (H. rostochiensis, commonly referred to as the golden nematode in the United States),

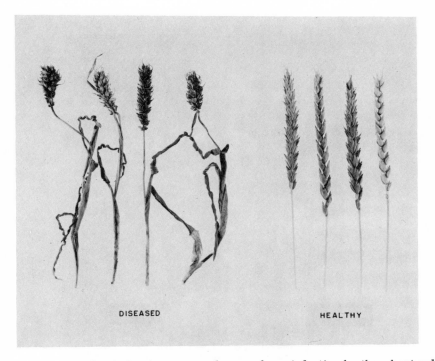

FIG. 1. Wheat showing severe damage from infection by the wheat gall nematode, Anguina tritici. (Photo by courtesy of Marvin Williams.)

began to cause serious losses to the sugarbeet and potato crops, respectively, in Europe. Scientists began to study these organisms and to evaluate the effects of crop rotation as well as chemicals (carbon bisulfide) applied to the soil, in reducing crop damage. Some of the early work on nematodes in this country was conducted at Alabama Polytechnic Institute (now Auburn University) by Atkinson (1) and in Florida by Neal (14). These investigators worked primarily on the root-knot nematode, and the accuracy of their observations and descriptions is astounding. Bessey (3) also worked extensively on this group. These scientists working in the 1880s and early 1900s, recognized the damage these pests caused but perhaps were not aware of the many additional types present in the soil, which are also of great economic importance. Cobb (24) pioneered the nematological work in the U.S. Department of Agriculture, followed by Christie, Steiner, Thorne, Chitwood, Taylor, Courtney, and others. These scientists, along with Linford, Godfrey, Newhall, and others working at universities, described many new species and characterized the type of damage caused by certain forms and conducted many experiments on control. Support for nematological research prior to 1945, however, was not very great, and

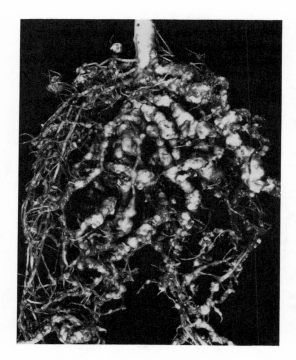

FIG. 2. Root-knot of bean caused by <u>Meloidogyne incognita.</u>

scientists trained in nematology were few in number. Even so, a vast amount of basic knowledge was accumulated pertaining to the biology of nematodes and their destructive nature.

B. Recent Developments Indicating the Importance of Nematodes in Crop Production

A new era in nematology began in the early 1940s, an era character- ized by several significant developments all of which have indicated the importance of nematodes in crop production. It would seem appropriate to enumerate certain of these developments, which have brought about sub- stantial support for what was little more than a museum science prior to these events.

The first of these developments, and perhaps the most important, was the discovery of D-D soil fumigant (1,3-D) in 1943, a relatively cheap nematicide which could be used effectively and economically on a field basis. The significance of this discovery lies in the fact that growers, as well as research workers, could compare in the field crops grown in nematode- infested soil with those grown in soils in which most of the nematodes had been killed. In many cases crop response to fumigation was so spectacular that growers accepted the practice long before it was recommended by experiment station workers. The annual use of nematicides has grown to a multimillion dollar industry, and it appears that their use will be greatly expanded in the future.

A second development was the discovery of the potato root eelworm (golden nematode of potatoes) in the important potato-producing region on Long Island, New York, in 1941. The seriousness of this pest in Europe was already known, and the threat or possibility of this nematode becoming established in the major potato growing states of this country was a matter of grave concern. The Federal Government, as well as the State of New York, appropriated sizeable sums of money to be used in the containment and study of this pest.

A third development had to do with the demonstration of the importance of several ectoparasitic species, belonging to the genera Belonolaimus, Dolichodorus, Xiphinema and Trichodorus, in the early 1950s (6, 7, 20, 22). Prior to this work, it was generally assumed that a nematode had to enter the roots or other parts of plants to cause injury, and little attention was paid to forms not found within the plant tissues. These ectoparasites are now recognized as major pests of many crops and some of the most devastating effects on root systems are now known to be caused by these and other genera that were considered of little economic importance prior to 1950.

A fourth development had to do with the demonstration of the role of certain nematode species in increasing the incidence of certain major diseases of tobacco, such as black shank, caused by Phytophthora parasitica var. nicotianae, and Granville wilt caused by Pseudomonas solanacearum. These diseases were a serious threat to the tobacco industry in North Carolina and other southern states, and although varieties resistant to these diseases had been developed, it was becoming increasingly evident that in the presence of certain pathogenic nematodes, the plants were no longer resistant. Our knowledge in this phase of nematology has expanded rapidly in recent years, and this subject will be treated in some detail in another section of this chapter.

A fifth development of great significance was the discovery of the causal organism of the spreading decline disease of citrus in Florida (23). The citrus industry in Florida, comparable to the tobacco industry in North Carolina in terms of the economy of the state, was being threatened by a rapidly spreading disease for which no control was known. Perhaps no other nematode has received as much publicity by the newspapers as has Radopholus similis, the causal agent of spreading decline. Again, a greatly expanded program was initiated by the State of Florida and the U.S. Department of Agriculture. Emphasis has been on preventing spread and on research designed to control or, if possible, eliminate the nematode from the soil.

A sixth development in the rapidly growing science of nematology was the recognition, on the part of many scientists in the Southeast, of the need for cooperative research on nematode diseases and the subsequent initiation of a regional research project on nematodes, designated S-19. Because of the shortage of trained nematologists, a series of workshops was held and various publications were written as aids to those engaged in research dealing with nematodes. Several institutions initiated graduate courses in nematology. The idea of cooperative regional projects extended to the Northeastern states, where the project is known as NE-34, and likewise to the North Central and Western states and all of the states have organized cooperative regional programs. These research projects center around problems common to the cooperating states with each state, contributing to a particular phase of the overall project. These projects are funded from Federal sources but are administered by State experiment stations.

A seventh development had to do with the formation of the Society of European Nematologists and the subsequent independent publication of Nematologica, an international journal of nematological research. Also the organization in 1962 of the Society of Nematologists was initiated, an outgrowth primarily from the American Phytopathological Society and the publication of the Journal of Nematology commenced in Jan. 1969. Each of these two societies now have several hundred members and hold meetings each year. These organizations have given some autonomy to those engaged in the science of nematology, which is slowly but surely assuming its rightful place in the scientific family.

C. Estimates of Losses Caused by Nematodes

It would be difficult indeed, if not impossible, to determine the economic losses caused by nematodes to agricultural and horticultural crops on a worldwide or even national basis. For many countries, few or no studies have been made to determine the prevalence and extent of damage caused by pathogenic nematodes and where surveys have been made, presence of nematodes has not always been correlated with crop performance. Studies made in the United States, Great Britain, The Netherlands, and other nations, however, leave little doubt concerning the destructive nature of plant-pathogenic nematodes and their role in crop production.

Control efforts have been directed primarily toward those crops the grower depends upon for his income—tobacco, cotton, peanuts, citrus, potatoes, sugarbeets, ornamentals, and many others. For these crops, growers are spending millions of dollars annually for nematicides to control nematodes and in most cases are receiving in return several dollars for every one dollar expended. This practice must be profitable, since growers historically have not adopted and continued practices not profitable for them.

For some of these crops, there are reliable estimates of damage caused each year, but for many crops where the damage is not spectacular and control measures are not taken, there is little information. Even for these low value per acre crops, such as corn, soybean, pasture, and others, the damage will probably amount to from 1% to 5% of crop value. For higher acre value crops, such as ornamentals, citrus, tobacco, sugar cane, pineapple, turf, certain vegetables, and fruit crops, the damage may range from 5% to 20%. When all crops are included, an average annual loss of perhaps as much as 10% of the total crop production occurs, especially when the many indirect affects of nematode damage are taken into consideration. Although these are only estimates, the cumulative data substantiating losses and their overall importance are overwhelming. Improved methods of estimating disease losses will increasingly reveal that nematodes are even more important than presently considered by nematologists.

II. CHARACTERISTICS OF NEMATODES

A. Gross Morphology and Anatomy*

1. General Structure of a Nematode

Nematodes are triploblastic, bilaterally symmetric, unsegmented, pseudocoelomate animals. Their shape is more or less cylindrical, sometimes fusiform, pear-shaped, or otherwise modified, particularly in the adult female. The mouth opening is generally anterior and is usually surrounded by lips bearing sensory organs. The mouth is followed by a mouth cavity or stoma, an esophagus, intestine, and a rectum terminating in a ventral, terminal, or subterminal anus in females, or a cloacal opening in males. The body is covered with cuticle. There are usually no external appendages, but appendages do occur in rare forms. The body wall is composed of a hypodermis or epithelium, situated beneath the cuticle, and a single layer of muscles. Sexes are usually separate. The male reproductive system opens directly into the rectum, forming a cloaca, while the reproductive system of the female has a separate opening, the ventrally situated vulva. Excretory and nervous systems are present, but there are no specialized organs of circulation or respiration.

2. External Characters

Nematodes, or roundworms, are generally vermiform animals of long cylindrical shape, circular in cross-section. There are two general types of body form, the fusiform and the filiform, the latter being less common. Other variations are the short, plump, pyriform or oval shape assumed by females of Heterodera and Meloidogyne. Most attain a length of 1 to 3 mm. Males are nearly always smaller than the females.

Nematodes, in general, lack coloration, being transparent or of a whitish or yellowish tint conferred on them by the cuticle. The body is not divisible into definite regions and lacks a distinct head, although this term is sometimes applied to the anterior end. The ventral surface of nematodes is identifiable by the presence in the midventral line of the excretory pore, the gonopore in the female, and the anus. In most forms the excretory pore is located proximal to the base of the esophagus. The vulva is usually situated in the posterior body half. The anus or cloacal opening of the male lies near the posterior end.

---

*Refer to Plates I through IV at the end of this chapter. Drawings provided through the courtesy of Dr. Hedwig (Hirschmann) Triantaphyllou.

The oral opening is surrounded by six lips in many genera, but others show great modifications from this pattern. In the region of each lateral lip, but behind the anterior face, there is a pair of sense organs very characteristic of nematodes, the amphids. The amphidial openings are cuticular depressions and consist of a gland and nerve endings and are presumably chemoreceptors. In most plant-parasitic forms, the pore-like opening of the amphid cannot be seen except by a study of en face preparations or with the aid of a scanning electron microscope.

The general body surface of nematodes may be smooth but very often is marked by a regular series of transverse striations. These striations are often interrupted by the lateral fields, which are quite prominent in some forms but rather inconspicuous in others. In addition to the transverse striae, there may also be longitudinal striations. In the tail regions of some males, extensions of the cuticle often form lateral alae (caudal bursa) employed in copulation and which generally bear genital papillae. Near the posterior end of many nematodes there is a pair of cuticular pouches, resembling the amphids, called phasmids, which are probably sensory in function. They are located in the lateral fields, generally in the tail region or just above it. Each phasmid consists of a short duct opening on the surface of the cuticle and leading inward to a small unicellular gland. In some cases the gland and duct may degenerate, leaving only a surface papillae as evidence of their former existence.

3. Body Wall

The body wall of nematodes consists of cuticle, hypodermis (epidermis, subcuticle), and muscle layer. The cuticle is a noncellular layer extended inward at the mouth, excretory pore, vulva, and anus. It is intimately connected with and, undoubtedly, is a product of the hypodermis. Histologically, it consists of several layers which are reducible to three kinds of material: the cortex, the matrix, and the fiber layers. The cortical layer consists of a dense material of the nature of a keratin and is resistant to solvents and to digestion. The matrix layer consists of, or contains, a fibroid named matricin, rich in sulfur. The innermost part of the cuticle consists of two or three fiber layers of very dense connective tissue running in different directions in adjacent layers. These fiber layers consist chiefly of collagen.

The hypodermis is a syncytial layer that bulges into the pseudocoel at four places to form four longitudinal ridges termed the longitudinal chords, mid-dorsal, midventral, and lateral in position. The nuclei of the hypodermis are confined to the chords. The nerves and excretory canals (when present) are in these chords. Some forms may have more than four chords.

The musculature of the body may be divided into two general types: the somatic musculature and the specialized muscles. The somatic musculature is the general muscular layer of the body wall and is composed of a single layer of obliquely arranged, more or less spindle-shaped cells attached to the hypodermis throughout their lengths. Specialized muscles, apparently of the same origin as the somatic musculature, are limited to some particular part of the body, such as labial muscles, somatoesophageal muscles, somatointestinal muscles, rectal muscles, and copulatory muscles.

4. The Digestive System

The mouth leads into the buccal capsule, variable in size, shape, and degree of differentiation in different nematodes. The buccal capsule and its stiffenings often exhibit a triradiate arrangement, corresponding to that of the esophagus. In some nematodes, especially rhabditoids, the buccal capsule is divisible into three sections: an anterior chamber enclosed by the lips, the vestibule or cheilostom; a middle and longest and most sclerotized portion, the protostom; and a small terminal chamber, the telostom. The walls of the various parts of the stoma are called rhabdions and are termed cheilorhabdions, protorhabdions, and telorhabdions, respectively.

In certain groups of nematodes, the buccal capsule is armed with a conspicuous protrusible spear or stylet, used to puncture plants and animal prey. This may be formed, as in tylenchoids, by the coming together of the sclerotizations of the buccal capsule, so that it constitutes a buccal stylet. This is necessarily hollow and forms the path of food intake. This structure is called a stomatostylet. In other forms, as in dorylaimoids, the spear represents an enlarged tooth that originates in the esophagus wall. This type is called odontostylet. All plant-parasitic forms have stylets.

The structure of the esophagus varies in different nematode groups and is, therefore, an important taxonomic character. It is a tube lined by a thin cuticle and covered by a membrane. The esophagus lumen is triradiate, being extended into three symmetrically arranged longitudinal grooves that partially divide the esophagus wall into three sectors, one dorsal and two ventrolateral. Three salivary glands are typically imbedded in the esophageal wall, one dorsal and two ventrolateral. In most plant-parasitic nematodes, the esophageal glands are single cells, often with conspicuous nuclei. The main cuticularized duct of each gland opens, often by way of an ampulla, into the lumen of the esophagus. It is usual for the dorsal gland to open much farther forward than the ventrolateral glands. In some tylenchoids, the glands protrude from the esophagus wall into the pseudocoel. Esophagi may be quite diversified due to feeding habit. Certain things, however, are constant: (a) triradiate luman, (b) cuticular lining, and (c) glands.

The esophagus may be connected with the intestine through a short structure termed the esophageal-intestinal valve, which often projects some distance into the lumen of the intestine. This functions as a valve which impedes the flowing back into the esophagus of food contained in the intestine. Like the esophagus, it is lined with cuticle.

The intestine is composed of a single layer of epithelial cells. It is usually a straight tube, in contrast to the reproductive organs which may be reflexed or coiled.

Posteriorly, the intestine leads into the rectum, which is lined by an invagination of the body cuticle and opens at the anus. The rectal glands, generally three in number, open into the rectum, one dorsal and two subventral. In the male, the sperm duct enters the ventral wall of the rectum. The rectum is thus in whole, or in part, a cloaca in male nematodes. In both sexes the rectum empties posteriad on the ventral surface of the body.

5. The Reproductive System

Nematodes, as a rule, are dioecious, existing as separate males and females. Males are readily distinguished externally from females by the presence of copulatory spicules. Other distinguishing features, which may or may not be present, are smaller size, curvature of the posterior end, and presence of bursae, genital papillae, and other accessory copulatory structures.

Nematode gonads are of tubular shape, varying greatly in length, and may be straight, sinuous, reflexed, or coiled back and forth. The male gonad consists of a single testis or paired testes. A single testis is usually present, and this extends anteriorly. However, two testes occur in many nematodes, and these are oppositely oriented, except in Meloidogyne where they are parallel. The terms diorchic and monorchic are convenient for referring to the two-testes or one-testis condition, respectively. When two testes are present, one usually extends forward and the other backward in the body cavity, but they join medially into the vas deferens, which runs posteriorly ventral to the intestine and finally narrows down to an ejaculatory duct that opens into the rectum to form a cloaca.

Male nematodes, with few exceptions, are provided with copulatory spicules lodged in and secreted by spicule pouches. Often the spicules are accompanied by an accessory piece, the gubernaculum, which is a sclerotization of the dorsal wall of the spicule pouch.

The female reproductive system may be paired or single, the organ lying in the body cavity alongside the intestine, either outstretched or reflexed. The terms monodelphic and didelphic are convenient for indicating the single or double condition of the female tract, respectively. Each gonad consists essentially of an ovary, containing developing eggs, and a tubular portion which, in many forms, is divisible into an oviduct and a uterus. The uterus joins on to the vagina, which opens on the ventral body surface at the vulva.

6. Nervous System

The most easily recognizable part of the nervous system is the nerve ring, which encircles the isthmus region of the esophagus. Associated with it are a number of ganglia, and, according to Chitwood and Chitwood (5), six nerves are directed anteriad from the nerve ring. The four submedial have three chief distal branches, the two lateral have only two branches. These branches innervate sensory organs of the anterior extremity (sensory papillae or setae). The amphids are innervated from the amphidial nerves which originate in the lateral ganglia of the nerve ring. There are a dorsal, a ventral, four submedian, and one, two, or three pairs of lateral nerves situated in the chords of the hypodermis, which proceed posteriorly from the nerve ring. The paired postanal lateral sensory organs (phasmids), the ventral supplementary organs of males, and the genital papillae are all innervated by branches from one or the other of the main nerves. The nerves themselves and their branches cannot be seen without special methods for demonstrating them or by means of sections. In routine microscopic examinations, the only part of the nervous system observed is the nerve ring, which surrounds the esophagus, and this is often difficult to see.

7. The Excretory System

The excretory system presents a varied picture in the phylum as a whole. It is simplest in the Class Adenophorea (Aphasmidia) where there is a single ventral excretory cell, or renette, which opens through an excretory pore on the midventral line in the region of the esophagus by way of a short to long duct. In the Class Secernentea (Phasmidia), there are two lateral excretory canals imbedded in the lateral chords of the hypodermis throughout most of the body length. They are connected anteriorly and ventrally by a transverse canal, thus forming an H or U shape. A duct, variable in length, connects the transverse duct with the excretory pore. The terminal excretory duct is cuticularly lined in the Secernentea and can be observed in routine microscopic examinations. In the Adenophorea, the terminal excretory duct is not lined with cuticle (except in some plectinae), thus making it difficult to see. There may or may not be two special cells (the renette) associated with the transverse duct.

In a few genera, including Dorylaimus, no excretory system has been found. Considerable excretion through the digestive tract may occur in all nematodes.

8. Circulatory and Respiratory Systems

Organs or structures associated with circulatory and respiratory systems are not known. Movement of fluids of the body cavity apparently is involved in both circulatory and respiratory processes.

B. Kinds of Nematodes and General Feeding Habits

There are several hundred species of plant-parasitic nematodes known to attack crop plants, and virtually every crop has its complement of nematode parasites. All parts of plants are attacked by nematodes, including roots, stems, trunks, buds, leaves, flowers, and even seeds. The majority of plant-parasitic forms are root feeders that spend their entire life either in the soil or in roots or other plant parts below ground, such as tubers, rhizomes, bulbs, or stems. Some of these enter the root completely where they mature and lay eggs within the root or while they are still attached to it. These are called endoparasites and include such important genera as the root-knot nematodes (Meloidogyne spp.), the lesion nematode (Pratylenchus spp.), and the cyst nematodes (Heterodera spp.).

Many soil forms feed without entering the roots and are called ectoparasites. In general, ectoparasitic species are larger than the endoparasites and have a longer and more powerful stylet for penetrating root tissues than do the endoparasites. Well-known and destructive genera of ectoparasites include the dagger nematode (Xiphinema spp.), the sting nematode (Belonolaimus spp.), the stubby-root nematode (Trichodorus spp.), and the awl nematodes (Dolichodorus spp.). Some forms are semiecto or endoparasites, in that they partially penetrate the root with the anterior part of their bodies but rarely enter the host roots completely. These include the spiral nematodes (there are several genera in this group), the ring nematodes (Criconemoides spp.), the sheath nematode (Hemicycliophora spp.), and the pin nematodes (Paratylenchus spp.). The overall size of these genera and the length of their stylets is intermediate between the endo- and ecoparasites. A diagrammatic and hypothetical drawing illustrating several important types of plant-parasitic nematodes feeding on a single rootlet is shown in Fig. 3. Note the different sizes and shapes of the nematodes, feeding sites, and types of root damage caused by various genera.

Some genera are classified as aerial pathogens since they infect and damage primarily the above-ground tissues. These above-ground pathogens

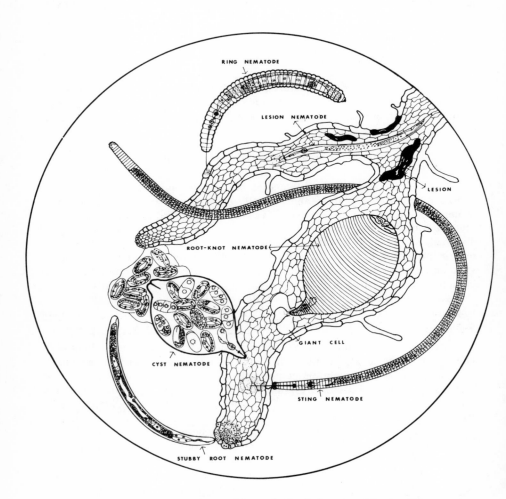

FIG. 3. A diagrammatic and hypothetical drawing illustrating several important types of plant-parasitic nematodes feeding on a single rootlet. Note the different sizes and shapes of the nematodes, feeding sites, and types of root damage caused by the various genera. (Courtesy of R. P. Esser.)

may carry out their life cycle in the soil or in the shallow surface layers, often in host plant residues. When suitable host plants develop and favorable conditions prevail, these aerial pathogens then ascend the plant or attack the growing seedling and mature above ground. The alfalfa stem nematode (Ditylenchus dipsaci), the leaf nematodes (Aphelenchoides spp.), and the seed gall nematode (Anguina tritici) are well known examples.

C. Typical Life Cycle of Plant-Parasitic Species

There are four larval stages and four molts in the life cycle of plant-parasitic and soil nematodes. The egg develops into the first stage larva within the egg shell. The first molt occurs within the egg shell and the second stage larva emerges free in the soil. Upon feeding, the second, third, and fourth molts with their corresponding third, fourth, and fifth or adult stages occur. Between each molt there is further growth and development of the nematode, with the development of the reproductive systems in each sex being the most obvious. Upon maturity, and in the presence of a favorable host, the female lays eggs and the life cycle is repeated.

Many species are bisexual and reproduction occurs only after copulation and fertilization of egg by male sperm. In other forms the male may be absent or rare and eggs develop without fertilization; reproduction occurs parthenogenetically or the female gonads may produce both egg cells and sperm cells, thus functioning as an hermaphrodite. For those forms in which both males and females occur in about equal number, the male is generally necessary and functional. Males and females are of the same general shape and appearance, differences being primarily in the posterior portions where the reproductive organs of the male, namely the spicules, bursa, and other copulatory organs, give them a different appearance from the females. In a few species, pronounced sexual dimorphism occurs and the adult female is either lemon-shaped, pear-shaped, kidney-shaped, or otherwise modified; whereas the male retains the eel-like shape common for most species of nematodes.

D. Distribution

Plant-pathogenic nematodes have been found in every region or country where surveys have been made. Information on the known distribution of various species is most complete in those countries where nematodes have been studied extensively and where they are important factors in crop production. For example, while records are far from complete, information concerning the kinds and distribution of nematodes in the United States, Great Britain, The Netherlands, Canada, and Germany, is much more advanced than in India, Africa, Latin America, China, and certain other countries. Although many plant-pathogenic genera of nematodes are world-wide in distribution, records indicate that certain genera and species have limited distribution.

Various factors of the environment, such as temperature and soil type, undoubtedly play a major role in nematode distribution and population levels attained. Some species are highly host-specific and consequently population increase for those species occurs only when suitable hosts are grown. In general, nematodes appear to be more prevalent and attain higher population levels in the warmer regions of the world than in the more northerly regions. The lower populations in the cooler regions are no doubt due to the effect of cold temperature on nematode activity and repro-duction rates as well as a lack of available food for the nematode to feed upon. For example, in Florida where crops grow the year round, soil nematode problems are much more serious than in Canada or Alaska. Wherever there are plants, however, parasitic nematodes of some type are likely to be found in association with their roots. The kinds, population levels, and degree of damage are important differences when one region is compared with another.

III. PLANT-NEMATODE RELATIONSHIPS

A. Gross Symptoms of Nematode Infection of Plants

Plant-pathogenic nematodes cannot reproduce without actually feeding on live tissues of host plants and, therefore, are obligate parasites. In the absence of suitable hosts, plant-pathogenic nematodes will gradually exhaust the stored energy reserves in their bodies and a great majority of the members of a population will die in a matter of months. Some forms, because of various protective or survival mechanisms, may remain viable for several years. Such forms, however, are in the minority and most will not survive long in the absence of a food source in the form of higher plants.

The plant parts attacked vary depending upon the nematode species, the host plant, and the stage of development of the nematode. The greatest number of plant-pathogenic forms feed on the fibrous root systems of plants or other underground fleshy parts, such as tubers, rhizomes, bulbs, and corms. In addition to those that feed on underground parts, there are those that feed on above-ground portions of the plant. Such forms, although they survive in the soil, infect developing young buds of such plants as alfalfa, clover, wheat, strawberry, and other plants and may feed externally between the bud leaves or they may enter the shoots, stems, or leaves directly and feed endoparasitically in these above-ground parts. These bud, stem, and leaf pathogens include species of the genus Ditylenchus, Aphelenchoides and Anguina. These three genera, in addition to being highly specialized aerial pathogens, also have the remarkable capacity to go into a state of cryptobiosis (latent life) when the host plant becomes unfavorable for continued feeding. This phenomenon is an important survival mechanism for the nematode and makes it more difficult to control.

1. Above-Ground Symptoms of Nematode Attack

Above-ground symptoms of nematode damage, excluding certain aerial pathogens, are not generally distinguishable from those caused by other soil pathogens or adverse conditions of the environment, such as low fertility and moisture. Generally, root damage caused by pathogenic nematodes is reflected in the above-ground portions of the plant. This may be reduced top growth, resulting in lower yields as well as poorer quality for such crops as tobacco, cotton, corn, citrus, soybeans, fruit, and certain vegetables where the harvested crop is above ground.

Other general symptoms include chlorosis of foliage, general unthriftiness, and wilting during the hot part of the day. Some typical above-ground effects of nematodes on plants are shown in Figs. 4 and 5. Above-ground symptoms caused by certain aerial pathogens are rather specific, such as the crown and stem swellings of alfalfa caused by Ditylenchus dipsaci (Fig. 6), the leaf, stem, and seed galls of Plantago aristata (Fig. 7), caused by Anguina plantaginis, and the leaf crinkling and distorted seed heads of wheat caused by Anguina tritici (shown earlier in Fig. 1).

2. Below-Ground Symptoms of Nematode Attack

Below-ground symptoms of nematode damage are also difficult to distinguish from those caused by other pathogenic organisms or by conditions of the environment. Because of this, if nematodes are suspected, it is wise to determine the kinds and population levels present in the soil or root tissue before making a diagnosis. Some nematodes, however, because of their size, can be seen attached to the roots of the diseased host plants, in which case there is little question concerning the causal organisms. The soybean cyst nematode (Heterodera glycines), shown in Fig. 8, and the golden nematode of potatoes (H. rostochiensis) (Fig. 9) are good examples. Known symptoms resulting from the feeding of various pathogenic nematodes, include an overall reduction in the root system, root pruning, lesions, and stubby roots as shown in Fig. 10, as well as root swelling, proliferation, and curling. These gross effects, not only greatly influence the development of the above-ground portions of the plant, but also reduce yield and quality of the underground portion of the crop which is harvested, such as potatoes, beets, peanuts, bulbs, and various vegetables. Examples of damage caused to certain of these crops are shown in Figs. 11 and 12.

B. Pathological Effects of Nematode Infections

The pathological effects of nematode-feeding on crop plants range from that of simple mechanical injury, caused by migration of the nematode between or through plant cells, to complex host-pathogen interactions involving physiological changes in host tissues resulting from toxic substances injected into the plant by the nematode and perhaps substances produced by the plant because of the presence of the nematode.

Root-galling caused by Meloidogyne spp. is a well-known host response and involves the production of so-called giant cells in the vascular tissues of susceptible plants. The feeding of the second stage larvae of this nematode

FIG. 4. Lesion nematode (Pratylenchus brachyurus) damage to tobacco. (Courtesy of C. J. Nusbaum.)

apparently results in the dissolution of cell walls, the contents of which coalesce, reorganize, and become three to six large cells, each with several nuclei. The stimulus from the digestive fluids of the nematode, perhaps enzymatic and hormonal in nature, also causes an increase in the number and size of other cells in the vicinity of the feeding site, resulting in the root swelling. Figure 13 illustrates the typical histological changes resulting from infection by the root-knot nematode, showing the adult female, egg mass, giant cells in vascular tissue, and the increase in size and number of cells in adjacent areas of the root. The responses of plants to infection by this nematode are related to plant resistance and susceptibility, and although the host-parasite relationships for this genus have been studied extensively, a complete understanding of the mechanisms of injury involved are lacking.

Lesion nematodes (Pratylenchus spp.) have an entirely different effect on host cells, in that the cells are killed by the feeding of the nematode and consequently small lesions or necrotic areas develop. Such lesions range from those barely visible to the unaided eye to almost complete discoloration and destruction of the root system. Unlike root-knot nematodes, which feed in vascular tissue, lesion nematodes are cortical feeders. As they destroy plant cells, secondary root-rotting organisms present in the soil begin to invade these necrotic areas and accelerate and magnify the root-rot complex.

The digestive secretions of some nematodes apparently contain a pectinase which dissolves the middle lamella of cells in the area where it is feeding and causes them to separate. The stem nematode (Ditylenchus dipsaci) is an example. Hypertrophy and separated cells cause the soft, puffy, water-soaked appearance of affected tissues. Apparently dissolution of the middle lamella is necessary for survival of the nematode since the nematode does not enter or is unable to survive in plants where feeding does not result in dissolution of the middle lamella.

Certain ectoparasites feed at root tips and apparently suppress mitosis. The stubby-root nematode (Trichodorus christiei), when it feeds on root tips, causes cells in the apical meristem to stop dividing and the root stops growing. Slight hypertrophy may also result but there is no discoloration or necrotic lesions and punctures made by the nematode are scarcely detectable. The physiological and biochemical basis for this arrest of root-tip growth is not known, but the effects on plant growth are very severe.

Other pathological effects caused by nematodes include root-pruning and root proliferation. Although each of these effects will vary with the nematode, host plant, and other organisms present, all have the net result of stunted plant growth, lower yields and quality of most crops, and less return to the grower.

FIG. 5. Red ring disease of coconut palm, caused by Rhadinaphelen-chus cocophilus. (A) Healthy; (B) diseased; (C) cross-section of diseased tree trunk showing ring of infected tissues.

FIG. 6. The effect of the stem nematode <u>Ditylenchus dipsaci</u> on alfalfa. (Left) healthy, (right) infected.

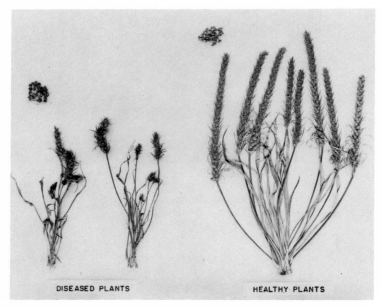

FIG. 7. Diseased and healthy plants of <u>Plantago aristata</u>. Leaves, stems, and seed heads of diseased plants are infected with <u>Anguina plantaginis</u>. (Photo by courtesy of Marvin Williams.)

FIG. 8. Soybean roots severely infected with the soybean cyst nematode, Heterodera glycines. Note white cysts attached to roots. (After Endo Sasser, Phytopathology, 48:571-574.)

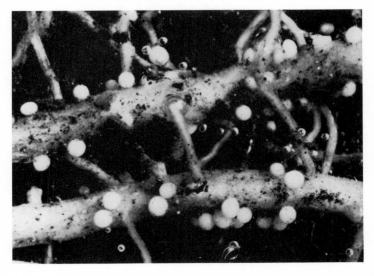

FIG. 9. Potato roots infected with the golden nematode, Heterodera rostochiensis. Note white cysts attached to roots. (Courtesy of W. F. Mai.)

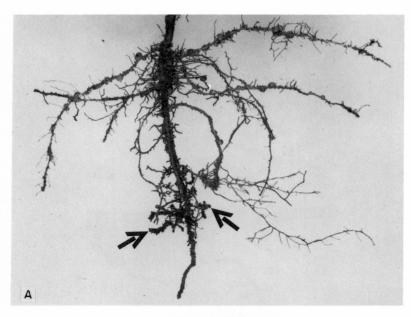

FIG. 10. Severe root damage indicated by arrows caused by the sting nematode, Belonolaimus longicaudatus. (A) Peanuts, (B) corn.

FIG. 11. Healthy potato (A) compared with diseased potato, (B) infected with the root-knot nematode <u>Meloidogyne incognita.</u>

FIG. 12. Sugar beet showing severe root-knot caused by the root-knot nematode <u>Meloidogyne incognita</u>. (Courtesy of H. R. Garriss.)

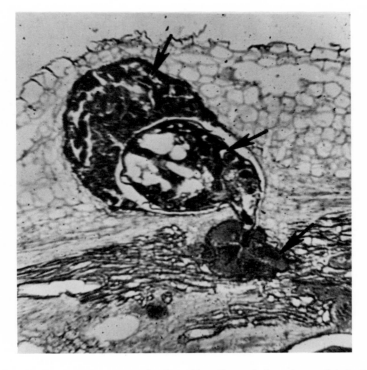

FIG. 13. Photomicrograph of section through tomato root infected with the root-knot nematode <u>Meloidogyne incognita</u>. Arrows indicate from left to right, egg mass, body of female nematode and giant cells. (Courtesy of R. G. Owen and C. J. Nusbaum.)

IV. INTERRELATIONSHIPS OF NEMATODES WITH FUNGI, BACTERIA, VIRUSES, AND OTHER NEMATODES IN CAUSING PLANT DISEASES

One of the most significant developments in nematology and plant pathology during the last decade has been the demonstration of many interactions between nematodes and other soil-inhabiting pathogens—fungi, bacteria, and viruses—in causing plant diseases. Performance of several crop varieties selected for resistance to specific fungal or bacterial pathogens, in nematode-infested fields, figured heavily in this discovery. Differences in growth of these disease-resistant varieties in fumigated and nonfumigated soils were surprisingly large. Controlled greenhouse tests proved conclusively that certain plant-pathogenic nematodes greatly enhance development of diseases caused by fungi and bacteria in plants resistant to these pathogens. Specific examples are the tobacco varieties developed for resistance to the black-shank fungus (Phytophthora parasitica var. nicotianae) and the Granville wilt bacterium (Pseudomonas solanacearum), and cotton and tomato varieties resistant to the Fusarium wilts interacting with root-knot nematodes.

Although various theories of the role of the nematode have been advanced, mechanisms that render the nematode-infected plants susceptible to fungi and bacteria have not been elucidated. In some bacterial diseases, it is thought that endoparasitic nematodes, as they invade roots, merely provide avenues for entrance by bacteria. In some nematode-fungus-plant interactions, the role of the nematode is known to be more complex than simple wounding of the plant, since wounds are not required for infection of plants by the fungi involved. Recent investigations indicate that host physiology is altered by nematode infection of plant tissue. For example, the black-shank fungus develops more vigorously and extensively in tobacco root tissues galled by root-knot nematodes than in nongalled roots of resistant varieties.

Synergistic relationships between phytopathogenic fungi and nematodes in increasing the severity of plant diseases have also been demonstrated. For example, the incidence and severity of wilt of egg plant, caused by the soilborne fungus Verticillium dahliae were increased in the presence of a lesion nematode Pratylenchus penetrans, and likewise the number of nematodes within egg plant roots was increased in the presence of the fungus. In addition to increasing the rate of nematode reproduction, fungus-infected roots are more attractive to and are more readily invaded by certain nematodes than are noninfected roots. These relationships are little understood and research is needed to determine the influence of individual environmental factors affecting the development of pathogens and disease expression in plants exposed to various combinations of organisms simultaneously.

Information on the relationship between two or more nematodes in causing a plant disease is fragmentary. One relationship that must exist is that of competition. It is not unusual to find that certain species are primary in a given disease situation; however, whether or not a given nematode is the predominant species in causing disease in a plant will depend on factors such as the host, the initial population level of that nematode species as well as other nematodes, relative reproductive rates of the species involved, soil type, and other environmental factors. Furthermore any advantage one nematode may have over another in a given situation is likely to be temporary and may change with the planting of a different crop and as the environment is otherwise modified.

Soil treatments that kill nematodes have been shown to prevent certain diseases caused by "soilborne" viruses. When certain nematodes were allowed to feed on virus infected plants and were then transferred to healthy plants, the healthy plants became diseased. Most of the soilborne viruses were really "nematode-borne" (transmitted) viruses. These fall into two distinct groups—the spherical or polyhedral shaped (NEPO-Viruses) and the rod or tubular shaped (NETU-Viruses). The NEPO-Viruses are transmitted only by species of Xiphinema and Longidorus, whereas the NETU-Viruses are transmitted by species of Trichodorus. Examples of the NEPO-Viruses include tomato ringspot, tomato black ring, tobacco ringspot, and the grapevine fanleaf. The NETU-Viruses include pea early browning and tobacco rattle. Nematodes of some species retain the ability to transmit viruses for several weeks after feeding on an infected plant, indicating that the viruses are probably carried internally in the nematodes. The fanleaf disease of grapes was the first virus disease shown to be transmitted by nematodes. Nematode species in three genera, Xiphinema (dagger nematodes), Trichodorus (stubby-root nematodes), and Longidorus (needle nematodes) are now known to transmit some 20 viruses, many of economic importance, which together attack a wide range of host plants.

An understanding of interrelationships among various kinds of soil-inhabiting plant pathogens is extremely important. Most agricultural soils are infested with plant-parasitic nematodes of one to six or eight genera, as well as various phytopathogenic fungi, bacteria, and viruses. Root systems of plants are often attacked simultaneously by several different soil-inhabiting organisms, and diagnosis of disease resulting from multiple infection is impossible. Such diseases are frequently classified as root-rot complexes, indicating that several organisms are involved in causing disease and that it is difficult to assess the relative roles of the different organisms present. Since multiple infections of plant roots by several pathogens is common, it is important to gain a thorough understanding of the contributions of each component pathogen and the factors favoring their individual activity before complex interrelationships can begin to be understood.

The breeder who is developing plant varieties resistant to specific diseases needs to be aware of plant disease complexes caused by interacting pathogens and, if possible, incorporate resistance to each of the pathogens involved.

Recognition that nematodes interact with other pathogens in causing disease has, to some extent, promoted nematode control by use of nematicides. Growers may elect to accept plant yield losses caused by some nematode species, especially where the direct damage is slight, but they cannot afford the losses resulting from nematodes interacting with other organisms that cause such destructive diseases as black shank and Granville wilt of tobacco and Fusarium wilt of tomato and cotton. In many instances control of nematodes by soil fumigation results also in effective control of other diseases that would become economically damaging in the presence of the nematode.

This recognition of the interaction between nematodes and other soil-borne pathogens suggests that nematode infections alter host physiology to a greater extent than previously assumed. The nature of changes rendering host tissues more susceptible is not now known but it is at least recognized, and this is the first step toward progress in this area.

At present soil nematicides, including the multipurpose chemical soil treatments, are cheaper and more effective than soil fungicides or soil bacteriocides, and for this reason nematode control is currently the cheapest means of control for many disease complexes. If less expensive means for control of soil fungi or bacteria, or multiple resistance factors in plants are developed, the situation could change.

V. PRINCIPLES AND METHODS OF NEMATODE CONTROL

A. Nematode Control Through Land Management and Cultural Practices

Plant-parasitic nematodes can be controlled to varying degrees by land management and cultural practices. These include (a) fallowing (keeping the land free of all plant growth), (b) flooding, (c) growing winter cover crops, (d) crop rotation, (e) time of planting, (f) organic manuring, (g) removal or destruction of infected plants, (h) trap and antagonistic crops, (i) nutrition and general care of host, and (j) sanitation and use of nematode-free planting stock.

The specific principles involved in control of nematodes by land management and cultural practices differ; however, all are based on the inability of nematodes to survive, multiply, and cause disease under the conditions imposed upon them by the use of these practices. Most of the practices discussed below reduce nematode populations gradually over a period of weeks, months, or even years, as opposed to rapid kill such as that obtained with heat or toxic chemicals. Furthermore, control is a relative term, and satisfactory economic control may not be achieved by any single practice listed below. On the other hand, a combination of several practices may bring about economic control.

1. Fallow

This is the practice of keeping land free of all vegetation for varying periods by frequent tilling of the soil either by discing, plowing, or harrowing, or application of herbicides to prevent plant growth. At least two principles of nematode control are represented by this practice. The first, and perhaps the most important, is starvation of the parasite. Plant-parasitic nematodes are dependent upon living hosts for the food necessary to develop to maturity and reproduce, and thus are obligate parasites. In the absence of a host plant the nematode will die after the stored food in the body has been depleted. Some of the cyst nematodes (Heterodera spp.) can survive as unhatched eggs or dormant larvae in cysts in the soil in the absence of a host for at least 14 years, but these are exceptions. Nematodes of most plant-parasitic species probably would not survive in upper soil layers for more than 12 to 18 months and their populations are greatly reduced in the first 6 months. Length of survival will vary with depth in the soil of the nematodes and is also influenced by whether or not the nematodes are in pieces of roots from previous crops or are free in the soil.

The second principle involved in fallowing is death through desiccation. Although there are some exceptions, nematodes of most species, depending on stage of development, will die if exposed to the drying action of the sun and wind. When fallow land is tilled frequently to destroy vegetation, the surface strata of soil is exposed to the drying and heating effects of the wind and sun. Fallow is especially effective in areas of low rainfall or in areas where rainfall is seasonal, thus resulting in long periods, perhaps 6 months or more, of dry conditions.

There are several objections to the practice of fallowing: (a) the operations necessary to maintain lands completely free of vegetation are difficult and expensive; (b) fallowing is a poor conservation practice, often resulting in erosion by wind and water, and is likely to impair the physical structure of the soil; and (c) fallow land does not contribute to farm income.

2. Flooding

Flooding of fields to control nematodes has not been widely accepted. Results from early investigations indicated that flooding for 12 to 22 months is required to rid soil of root-knot nematodes (Meloidogyne spp.). Where water is plentiful and level land can be taken out of production for long periods, flooding may be a useful control practice. Certain crops like rice, for example, can grow under flooded conditions and experiments have shown that rice, seeded in water and kept flooded for 4 to 6 weeks, had only a trace of the white tip disease caused by Aphelenchoides oryzae, whereas rice drilled and flooded after the seedlings were 3 to 4 in. tall had 60% disease. In this case the nematode is seedborne rather than soilborne. This is a special situation, however, since flooded fields are not generally planted with crops.

The principles of control involved in flooding are not understood. Presumably, if suitable host plants for certain species are eliminated by flooding, the nematodes may die from starvation. For other species, low levels of oxygen may be the limiting factor. Chemicals lethal to nematodes, such as butyric and propionic acids, hydrogen sulfide, and perhaps others, often develop in flooded soils of low pH containing large amounts of rapidly decomposing organic matter. Flooding of rice fields in Louisiana has given good control of certain species parasitic on rice. It should be remembered, however, that nematodes are essentially aquatic and some species persist in saturated soils. A disadvantage of flooding is the possibility of introducing new pests with the flood water.

3. Use of Cover Crops

Cover crops are commonly grown in the winter as a soil conservation measure and to provide forage for livestock, or in the summer between rows of widely spaced crops, such as fruit trees. Populations of some nematode species may decline but others undoubtedly increase on cover crops. The reduction in population is probably due to resistance of that crop to the particular nematode and conversely, any increase would be due to susceptibility of the cover crop to the species that increased. Low winter temperatures might also be a limiting factor when populations of parasitic species do not increase, even though the cover crop grown is susceptible. With some sedentary endoparasites the trap-cover principle may be operative, in which case the larvae enter the roots, develop to an immobile stage, but fail to develop into adults; thus they are trapped within the root tissues.

The addition of organic matter, which usually results from the plowing-under of green manure crops, increases the population and predacious activity of nematode trapping fungi, predacious nematodes, and of the internal parasites of nematodes. Also, nematicidal substances, such as butyric acid, are formed during the decomposition of cover crops, such as rye and timothy. The importance of these substances in biological control of plant-parasitic nematodes, however, is little understood and although some

investigators are optimistic about biological control, no practical control
of nematodes by predacious fungi, toxic substances resulting from decomposition of crop residues, or other biological agents has been reported.

4. Crop Rotation

The use of crop rotation to reduce nematode populations is without
question the most effective and most widely used land management practice.
This practice was used by growers long before its significance as a means
of nematode control was understood. To be effective, crops that are unfavorable hosts for the nematode to be controlled must be included in the rotation sequence. Some of the more serious nematode pathogens, such as the
golden nematode of potatoes (Heterodera rostochiensis), the soybean cyst
nematode (H. glycines), the stem nematode of alfalfa (Ditylenchus dipsaci),
and some species of the root-knot nematode (Meloidogyne spp.), are relatively host-specific, which makes selection of unfavorable hosts not too
difficult. Furthermore, where one of these nematodes occur, it is usually
the predominant species and growth of a resistant crop for 2 to 4 years will
greatly decrease the population of that species by starvation. Growth of a
resistant crop for only 1 year is generally inadequate. Two resistant crops
between susceptible crops may give fair control; however, 3 or 4 years,
and with some nematodes 7 or 8 years, are necessary for effective control.
Figures 14 and 15 illustrate the increased growth of certain important crops
resulting from crop rotation.

Although crop rotation is widely used and is effective in nematode
control, it has some important limitations. First, the degree of control is
based on the levels of resistance in the crops used in the rotation and the
number of years between susceptible crops. Also, populations of other
species of nematodes may occur on the alternate crop. Furthermore, the
nonhosts or resistant crops grown in the rotation may be of low value and
consequently contribute little to farm income.

5. Time of Planting

Certain pathogenic nematodes are inactive during the winter months
because low temperatures inhibit their activities. For example, in California sugarbeet production in fields infested with sugarbeet nematode (Heterodera schachtii) is much higher if the beets are planted in January or February, than if planted in March or April. The root-knot nematode seldom
damages the spring potato crop in North Carolina since the crop grows at
temperatures too low for much activity by the nematodes. Potatoes planted
in the late spring, on the other hand, make most of their growth during the
hot summer months, and are harvested in the fall. Under these conditions,

FIG. 14.   The effect of preceding crop on peanuts growing in field heavily infested with the root-knot nematode, Meloidogyne hapla. (Left) Preceding crop, cotton; (right) preceding crop, soybeans. (Courtesy of J. L. Perry.)

root-knot can be a serious problem unless otherwise controlled. In the British Isles the golden nematode may be controlled effectively by early planting of early potato varieties. Apparently, potato roots develop and grow at a lower temperature than that favorable for the development of nematodes.

6. Adding Organic Manures

Several investigators have found a reduction in the population levels of plant-pathogenic nematodes following the addition of organic manures to soil. In most cases there was increased activity of microorganisms in the soil following these treatments and reductions in nematode populations were assumed due to the build-up of nematode-destroying organisms in the soil. However, in most cases the actual factors responsible for killing the nematodes were not determined.

7. Host Refuge Removal or Destruction

Some annual crops such as tobacco will continue to live for several weeks after harvest is completed because of sucker growth. This growth

FIG. 15. The effect of crop rotation on the control of root-knot nematodes (<u>Meloidogyne incognita</u>) on tobacco. (A) Tobacco following peanuts;
(B) continuous tobacco. (After Sasser and Nusbaum, Phytopathology,
<u>45</u>:540–545.)

is sufficient to keep the root systems active by growing and consequently parasitic nematodes present in the roots continue to reproduce. One or two additional generations may develop between the end of harvest and the time the plant is killed by frost. Experiments have shown, however, that if the stalks are cut soon after harvest and the root system of the plant is then turned out and exposed (Fig. 16), the population of nematodes, especially root-knot, is reduced through the drying action of the sun and wind. Two control principles are utilized in this practice: (a) the host plant is destroyed by cutting the stalk and uprooting the plant, thus preventing further reproduction of the nematode, and (b) large numbers of nematodes concentrated in the soil around the root system and in the roots are killed by desiccation.

8. Use of Trap Crops and Antagonistic Crops

    Early investigators employed the trap crop principle in attempts to control certain species of the cyst and root-knot nematodes, by planting highly susceptible crops in infested fields and allowing the crop to grow only long enough for the second-stage larvae to enter the roots and begin their development toward adults. Since only the second-stage larvae of nematode species in both of these genera are infective, any development beyond the second stage would render the nematode immobile, and death would occur if the crop was destroyed before the nematodes matured. This

FIG. 16. Field scene showing tobacco stubble exposed to the drying effects of the sun and wind. (Courtesy of F. A. Todd.)

practice undoubtedly is effective in reducing soil populations of larvae of certain species, but it also could result in an even higher population if for some reason destruction of the crop was not accomplished before the nematodes completed their development and reproduced. In fact the population could increase several fold above the original infestation if reproduction occurred. Another disadvantage is the expense of planting and destroying a crop that brings in no revenue. The trap crop method of nematode control, while theoretically feasible, is rarely used commercially.

A more effective approach to the use of trap crops is to plant crops that are highly susceptible to invasion by the nematode but resistant to the development of larvae to adults. In this case the crop does not have to be destroyed but can be harvested or used as a cover crop and turned under as green manure. Crotolaria is a crop that has been successfully used for this purpose in reducing the population of certain species of root-knot nematodes.

In recent years considerable interest has centered around plants whose roots apparently give off chemicals toxic to nematodes, which reduces the population level of some nematode species. African marigolds (Tagetes sp.) and Asparagus officinalis L. are examples of such plants. While the use of antagonistic crops has reduced the populations of certain nematode species under some conditions, little is understood concerning principles involved and such practices have not as yet been developed to the point of giving effective nematode control.

9. Influence of Nutrition and General Care of Host

The deleterious effects of nematode damage to certain crops can, to some degree, be offset by proper nutrition, moisture, and protection from adverse conditions, such as drought or cold, which place the plant under stress. For this reason greenhouse plants can usually tolerate much higher populations of nematodes than field grown plants. Practices that tend to lessen the damage caused by nematode attack are irrigation, mulching to conserve moisture, fertilization, protection of plants on cold nights, and the control of other root and foliage diseases caused by pathogens other than nematodes. It should be pointed out, however, that these are only delaying tactics and that in time the nematode population will reach such proportions as to cause serious damage. The rapidity of disease development and the magnitude of the damage will depend on the host and nematode species involved, the resistance or tolerance of the host, and on various factors of the environment which favor or deter the development of the disease.

Some recent investigations have shown that soil population levels of several nematode species may be differentially altered by host nutrition and similarly, that disease development and severity are more pronounced in infected plants deficient in one or more of the essential nutrient elements.

It has also been shown that nematode infection may cause either an increase or decrease in concentration of one or more elements in leaf or root tissue. The application of such information to fertilization programs designed to minimize damage caused by nematodes to crop plants is just beginning.

10. Sanitation and Use of Nematode-Free Planting Stock

The various land management and cultural practices discussed above reduce nematode populations in fields to varying degrees. Most of these measures, however, have limitations and the degree of control is erratic, and sometimes factors actually responsible for the reduction in nematode populations are not fully understood. On the other hand, sanitation and the use of nematode-free planting stock are sure and effective means of nematode control. Cost of these practices is negligible, yet many growers continue to use nematode-infected transplants or seedpieces of such crops as tomato, pepper, strawberry, peach, sweetpotato, tobacco, potato, as well as bulbs, corms, rhizomes, and tubers of many other plants.

Examples of nematode-infested seeds and plants are alfalfa seed infested with stem nematode (Ditylenchus dipsaci); wheat seed, infested with wheat gall nematode (Anguina tritici); rice seed and strawberry and chrysanthemum plants, infested with species of bud and leaf nematode (Aphelenchoides spp.). Furthermore, nursery planting stock harbors numerous nematode parasites and is shipped all over the world. Although pathogenic nematodes are already widespread, indiscriminate use of nematode-infected plants and plant parts introduces, into many fields, species that do not already occur there, and therefore complicates control measures. Furthermore, nematodes introduced in this manner are in a favorable position for rapid development since they are already in host-plant tissues. The greatest damage, however, is probably not to the plants on which the nematodes were introduced but on plants grown in subsequent years in the newly infested field.

B. Nematode Control by Resistant Hosts

The somewhat selective character of nematode infestation, their slow spread, their persistance in soil and the relatively high cost of physical and chemical control, make the breeding of resistant and tolerant varieties of commercially important crops an economically attractive approach. Most of the available information indicates that resistance to nematodes is dependent on physiological factors of a rather complex nature.

Practical results have been obtained in breeding for resistance to several major nematodes. For example, resistance to certain species of the root-knot nematode (Meloidogyne) has been found in tomatoes, peppers, lima beans, common beans, soybeans, sweetpotatoes, tobacco, cotton, lespedeza, peach, fig, rose, gardenia, corn, and grape.

Development of a soybean variety highly resistant to the soybean cyst nematode (Heterodera glycines) is another example of a highly successful program in plant breeding. Progress has also been made in developing potato varieties resistant to the golden nematode, Heterodera rostochiensis, clovers and alfalfa varieties resistant to the stem nematode (Ditylenchus dipsaci). Commercial varieties of cotton, lima beans, and soybeans resistant to the root-knot nematode and varieties of alfalfa, oats, and barley resistant to the stem nematode are grown extensively.

Research in the development of varieties of other crops resistant to specific nematodes is in progress and undoubtedly will result in the availability of many additional crop varieties possessing high resistance to certain pathogenic nematodes. In many instances, resistance to even a single nematode species may be very beneficial, especially if the species for which the crop has resistance is the primary one present. In most fields, however, there are many species of parasitic nematodes, representing several genera. Without multiple resistance, of which there are few examples, the crop grown may be damaged considerably by those species present. Nevertheless, there are many instances in which a single species is primary and varieties of crop plants resistant to that particular species may prove very useful.

C. Nematode Control by Biological Means

Certain fungi are known to capture and kill nematodes in the soil. Known collectively as nematode-trapping fungi, these microorganisms are able to consume worms of microscopic dimension. Some are Phycomycetes and at least one is a Basidiomycete, but many are Fungi Imperfecti of the order Moniliales. Arthrobotrys, Dactylaria, Dactylella, and Trichothecium are the genera most commonly represented. Some fungi capture nematodes by adhesion, but many employ, for this purpose, specialized organelles that include networks of adhesive branches, stalked adhesive knobs, nonconstricting rings, and constricting rings. Although these organelles vary greatly in forms, they trap nematodes by either adhesion or occlusion.

Whatever the species of nematode-trapping fungus or the trapping mechanism involved, the fate of the nematode is similar. It struggles for a time and then appears dead. The surface of the nematode is penetrated and the fungus hyphae ramify throughout the carcass, digesting and absorbing its contents. Under favorable conditions, large numbers of nematodes may be captured, especially by those fungi that form networks of hyphal loops. The actual cause of death of a trapped nematode is uncertain. It may be due to mechanical damage and exhaustion during its struggle or it may be killed by a toxin produced by the fungus.

Biological control of nematodes in field soils has interested nematologists, mycologists, and other biologists for several decades. There is little doubt concerning the ability of predacious fungi to capture and destroy nematodes in large numbers under laboratory or small pot tests. In field tests, however, results have not been as encouraging and the successful application of this principle of control appears to be far in the future. If fungi are to be exploited for the control of nematode diseases, it is not enough to have them present in the rhizosphere—their growth and predacious activities must be sustained. While the adding of organic matter may provide such encouragment, effects of organic matter on soil are numerous and activation of the food chain in favor of predacious fungi is yet to be established experimentally.

D. Nematode Control by Physical Factors

Excellent control of plant-pathogenic nematodes can be achieved by using various physical factors. Heat, in various forms, is very effective and is the most widely used. When applied to soil, usually in the form of steam, it is effective also against other pathogens and weed seeds. Temperatures necessary for effective control will depend on (a) other biological agents to be controlled; (b) soil condition, that is, whether the soil is dry or moist, in a stationary mass or being tumbled; and (c) exposure time.

Hot-water treatment of plants infected with nematodes is also a widely used and accepted method of nematode control. The application of this method on nematode-infected plants is based on the fact that essential enzymes in nematodes are inactivated at temperatures near 50° C; however, if properly applied a short exposure to heat will not destroy the plant enzymes. This method has become of great importance in treating various bulb crops—iris, daffodils, and gladiolus—against the stem nematode. Strawberry plants, peach and citrus seedlings, and various ornamentals are also commonly hot-water treated to kill nematodes within the roots. Each plant-nematode combination has its own temperature-time requirements and must be done fairly precisely, or control will not be satisfactory. Specific details of treatment have been worked out for several crops and can be found in the selected references.

E. Nematode Control by Chemicals

The use of nematicides on a field basis was not possible until the early 1940s, when an effective and economical fumigant was discovered. Prior to this development, only greenhouse and nursery soils were freed of nematodes, either by steam or by the use of chloropicrin, methyl bromide, and other highly effective but expensive and difficult-to-apply chemicals. Such tests or use of chemicals to control nematodes did not provide the growers and research workers with spectacular differences in growth resulting from the use of nematicides under field conditions. Growers and research workers at first were reluctant to accept the fact that nematodes could cause such serious damage to their crops. Fields previously abandoned or classified as "dead" or "worn out" were put back into production. Yields increased and the quality of many crops was greatly improved. Repeated field tests and farmer demonstration plots convinced all concerned of the ravaging effects of nematodes when not controlled, and farmers accepted the practice in many areas even before experiment station scientists were free to recommend it. In essence a new industry was born, and while only a few hundred acres were treated in 1943, the acreage treated grew rapidly. Now several hundred thousand acres are treated annually and the industry has grown to a multimillion dollar business. Several materials have been developed and improvements in methods of application have come about as a result of much research. In many parts of the country, soil treatment has become an established farm practice, much like that of applying insecticides or herbicides.

1. Purpose of Treating Soil with Nematicides

The primary reason for applying nematicides to the soil is to reduce the population of plant-pathogenic nematodes to a level that will result in increased yield and quality of crops grown. Crops are most vulnerable to the effects of nematode attack while they are young seedlings trying to get established. Nematicides applied prior to planting provide a rhizosphere relatively free of nematodes and the crop is able to develop a good root system. Depending on the reduction of the nematode population, the crop may then attain maturity before the residual nematodes multiply to an injurious extent. Although the primary reason for the treatment of soils with nematicides is to reduce damage caused directly by nematodes, it is not the only benefit resulting from treatment. As discussed earlier, nematode control may result in control of several other diseases caused by soil-inhabiting organisms present in the soil, namely pathogenic fungi, bacteria, and viruses. Also, by controlling the parasitic nematodes, which usually cause various types of lesions, galls, or other malformations on roots, various saprophytic fungi and bacteria, which require wounds or necrotic tissue for food, are unable to become established.

Nematode control also results in more efficient use of moisture and mineral elements in the soil. Roots of plants damaged by the feeding activities of nematodes cannot utilize the moisture, present in the soil, efficiently and plants infected with nematodes often will show symptoms of wilting in the same field where other plants growing in fumigated soils do not wilt. Also, mineral elements, instead of being taken up by the plant will tend to accumulate in the soil or leach out and therefore are wasted.

Another benefit derived from nematode control is that of increased stands and a more uniform growth of the crop. This is helpful because of fertilization and cultivation practices, harvesting operations, and marketing of the crop. For example, in the production of woody ornamentals, if the crop is uniform in size, the nurserymen can sell the entire planting at one time and then put his land back into production of another crop. Without this uniformity in growth, the inferior plants either have to be moved or left until they reach a salable size.

Still another important benefit resulting from chemical treatment, depending on the material used, is weed control. Several of the nematicides in use are multiple purpose, and not only control nematodes but control weeds as well. This results in tremendous savings in labor costs in the production of tobacco plants, pine seedlings, and ornamentals, to mention a few.

A benefit derived from nematode control, that is perhaps underestimated, is protection of the overall investment in the crop. Growers customarily fertilize, cultivate, spray for insects and disease control, and irrigate; yet if nematodes are not controlled, money spent on these practices may, to a large extent, be wasted since the crop yields and quality may be greatly affected. On the other hand, if nematodes are controlled, a healthy and fibrous root system develops and the crop has an excellent chance to grow to maturity and make a profit for the grower.

2.   Factors to Consider Prior to Use of Nematicides for Nematode Control

a.   Value and Susceptibility of Crops to be Grown.   The crop to be grown is important in that it must be of sufficient economic value to make treatment profitable. Where nematodes are a limiting factor in crop production, many crops, such as tobacco, peanut, cotton, vegetables, potatoes, citrus, strawberry, and ornamentals, are of high enough value per acre to justify the cost of treatment. Acre values for some crops, however, like corn, soybeans, and small grains, may not be high enough to justify the average expenditures of $15 to $300/acre for nematicides, depending on crop, chemical used, dosage, and purpose of treatment. For a high percentage of the acreage treated, the purpose is to control nematodes only, and the average cost is in the range of $15 to $20/acre.

Secondly, the susceptibility of the crop to be grown to the particular nematodes present should be known. If the crop to be planted is not susceptible, little or no benefit will be derived from the treatment. In fact some stunting of growth could occur due to the chemical. Because of this the species should be determined and if past knowledge indicated that the crop to be planted was not susceptible, treatments should not be applied. The crop to be grown would also influence the choice of nematicide to use since some nematicides are very phytotoxic to certain groups.

b. _Nematode(s) to be Controlled._ Some species are more difficult to control than others and may require higher dosages of certain nematicides, or a particular nematicide, more effective against the nematode present, may be available.

c. _Soil Type, Preparation, Moisture, and Temperature._ Diffusion of volatile compounds is definitely influenced by soil type. Clay particles or excessive organic matter can absorb the compound and reduce the dispersion of the chemical in the soil. In clay soils pore spaces are smaller than in sandy soils and are likely to be blocked by excessive moisture, resulting in incomplete fumigation, especially if the nematicide is highly volatile or short-lived in the soil. On the other hand, sandy soils contain large pores and the problem of sealing the surface to prevent loss of vapor is more difficult.

Soil preparation is very important for best results with most chemicals. In order for the chemical to be effective, it must penetrate the entire soil mass for the depth required. Treatment of soils that are in poor tilth and those that contain substantial crop residues will produce discouraging results in that the chemical will not penetrate all the soil mass and/or will be absorbed by the undecayed organic matter. Furthermore, nematodes remaining within this undecayed plant detritus may escape the action of the toxicant.

Soil moisture and temperature are important factors in that diffusion of gases are dependent on both. Conditions of moisture and temperature, which are optimum, will vary with soil type, chemical used, and other factors. In general soils should not be treated when extreme conditions prevail, that is, wet or dry, cold or hot.

3. Methods of Application and Dosage

The method of application and rate will depend primarily on the job to be done. There is also the consideration of the nature of the crop to be grown, that is, annual versus perennial. For example, if the rows are spaced more than 2 ft apart, row application is just as effective as overall application and the cost is approximately one-half. For crops planted in rows closer than 2 ft apart and for crops, where an unusually high degree of control is required, the overall method of application may be desired. For crops, where it is desired to obtain control of other organisms in addition to nematodes, the dosage may be increased, another chemical or combination of chemicals may be selected, and the method of application may be that which would confine the chemical to the soil mass for longer periods of time. This can be done with packers, water seals, or covers.

For perennial crops, dosages recommended are usually at least double the amount for annual crops. The reason for this is the need for a higher degree of control as well as deeper penetration, as the perennials are deeper rooted and grow over a period of many years.

Other factors that influence method of application include volatility of chemical and formulations, that is, granular, solutions, or emulsions, seed treatment, or root soak.

In summary, the selection of the nematicide to be used in any given situation is dependent on (a) the crop to be planted, (b) nematodes to be controlled, (c) overall job to be done, (d) equipment available, (e) availability of nematicides, and (f) costs and other factors. Some nematicides are highly phytotoxic to certain crops even at very low rates per acre. Some are very ineffective against certain nematode species. Some nematicides are also effective against weeds, fungi, bacteria, and insects, and control of these may be very important for the crop to be grown. All things being equal, the grower should choose the least expensive nematicide which can be applied with the equipment available.

F. Commercial Nematicides

A list of chemicals used for nematode control is presented in Table 1. The most common and widely used nematicides at the present are volatile and have a fumigating action in the soil. These are referred to as soil fumigants and include chloropicrin (Larvacide, Picfume); methyl bromide (Dowfume MC-2); 1,3-dichloropropene and 1,2-dichloropropane mixture (D-D Mixture, Dowfume N, Vidden D); dichloropropenes mixture (Telone); ethylene dibromide [1,2-dibromoethane] (EDB, Dowfume W-85, Bromofume, Nephis); and dibromochloropropane [1,2-dibromo-3-chloropropane] (Fumazone, Nemagon). These basic chemicals may also be used in various combinations or with other chemicals to increase the spectrum of activity, beyond that of nematode control, to include soil insects, fungi, and weeds. For example, Trizone, a wide-range soil fumigant is a mixture of methyl bromide, chloropicrin, and propargyl bromide. Brozone is a mixture of methyl bromide, chloropicrin, and petroleum hydrocarbons. Another mixture with wide-range activity, Vorlex, is composed of 1,3-dichloropropene and 1,2-dichloropropane and methyl isothiocyanate. Dorlone, a mixture of ethylene dibromide and 1,2-dichloropropene, has proven highly effective for controlling a wide range of nematode types.

Chloropicrin and methyl bromide, or chemical mixtures containing these materials, are used primarily to treat soil for greenhouse use, seed or plant beds, nursery beds, or other soils used to produce high-acre value crops. Because of their high vapor pressures, soil covers are required to retain these chemicals in the soil long enough to effect kill of pathogenic organisms and weed seeds. The higher cost of these chemicals and the necessity of covers make treatment of large acreages impracticable.

TABLE 1

A List of Chemicals in Use for Nematode Control, Indicating Names and Designations, Structural and Empirical Formulas, Manufacture and Use[a]

Names and designations	Structural and empirical formulas	Manufacturer and use[b]
1. Chloropicrin Trichloronitromethane Larvacide Picfume	Cl_3CNO_2 CCl_3NO_2	Dow C. Fum. Morton S. Fum.
2. Bromomethane Methyl bromide	CH_3Br CH_3Br	Amer. Pot. C. Fum. Dow S. Fum. Frontier Great Lakes Mich. Chem.
3. 1,3-dichloropropene and 1,2-dichloropropane mixture D-D Mixture Dowfume Vidden D	$CHCl=CHCH_2CL$ + $ClCH_2CHCH_3$ with Cl $C_3H_4Cl_2$ + $C_3H_6Cl_2$	Dow Nem. Shell S. Fum.

TABLE 1 (Continued)

Names and designations	Structural and empirical formulas		Manufacturer and Use[b]	
4. Dichloropropenes mixture	$ClCH=CH-CH_2Cl$	$C_3H_4Cl_2$	Dow	Nem.
Telone				S. Fum.
5. Ethylene dibromide	$BrCH_2CH_2Br$		Amer. Pot.	C. Fum.
1,2-dibromoethane ethylene dibromide		$C_2H_4Br_2$	Dow	Nem.
			FMC	S. Fum.
Bromofume			Great Lakes	
Nephis			Mich. Chem.	
			Montecatini	
6. Dibromochloropropane	$\overset{\displaystyle Br}{\underset{\displaystyle |}{BrCH_2CHCH_2Cl}}$		Dow	S. Fum.
1,2-dibromo-3-chloro-pane		$C_3H_5Br_2Cl$	Shell	Nem.
Fumazone				
Nemagon				
7. Methyl isothiocyanate	CH_3NCS	C_2H_3NS	Morton	S. Fum.
Vorlex				

8. Natham

Sodium methyldithio-carbamate

Vapam

$$CH_3-NH-\overset{\overset{\displaystyle S}{\|}}{C}-S-Na$$

$C_2H_4NNaS_2$

Stauffer — Nem. / S. Fum.

9. TCTP

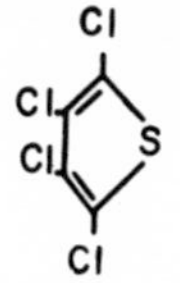

Tetrachlorothiophene

Penphene

C_4Cl_4S

Pennsalt — S. Fum.

10. Prophos

0-ethyl S,S-dipropyl phosphorodithioate

Mocap

$$C_2H_5O-p-\overset{\overset{\displaystyle O}{\|}}{}(SCH_2CH_2CH_3)_2$$

$C_8H_{19}O_2PS_2$

Mobil Chem. — Ins. / Nem.

11. Dichlorofenthion

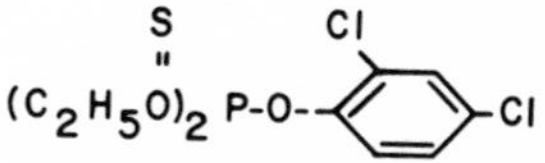

0-2,4-dichlorophenyl 0,0-diethyl phosphorothioate

V-C 13

V-C 13 Nemacide

$C_{10}H_{13}Cl_2O_3PS$

Mobil Chem. — Acar. / Ins. / Nem.

TABLE 1 (Continued)

Names and designations	Structural and empirical formulas	Manufacturer and use[b]	
12. Thionazin		Amer. Cyan.	Ins.
0,0-diethyl 0-2-pyra-zinyl phosphorothioate	$(C_2H_5O)_2P-O-$ (with S double bond to P) pyrazine ring		Nem.
Nemafos			
Nemaphos	$C_8H_{13}N_2O_3PS$		
Zinophos			
13. Aldicarb		Union Carb.	Acar.
2-methyl-2-(methylthio) propionaldehyde 0-(methylcarbamoyl) oxine	$CH_3-S-CCH=N-O-C-NH-CH_3$ with CH_3 groups and O double bond		Ins.
			Nem.
Temik	$C_7H_{14}N_2O_2S$		
14. Fensulfothion		Chemagro	Ins.
0,0-diethyl 0-p-(methylsul-finyl) phenyl phosphorothioate	$(C_2H_5O)_2P-O-$ phenyl ring $-S-CH_3$ (with S double bond to P, and O double bond to S)	Farbenfab-rikan, Bayer, A.G.	Nem.
Terracur-P	$C_{11}H_{17}O_4PS_2$		
Dasanit			

No.	Name / Chemical name / Trade names	Structure	Formula	Manufacturer	Uses
15.	Phorate 0,0-diethyl S-(ethyl-thio)-methyl phosphorodithioate Thimet	$\overset{S}{\overset{\|}{(C_2H_5O)_2P}}-S-CH_2-S-C_2H_5$	$C_7H_{17}O_2PS_3$	Amer. Cyan.	Acar. Ins. Nem.
16.	Ethyl parathion 0,0-diethyl 0-p-nitro-phenyl phosphorothioate Alknon Niran Thiophos	$\overset{S}{\overset{\|}{(C_2H_5O)_2P}}-O-\bigcirc-NO_2$	$C_{10}H_{14}NO_5PS$	Amer. Cyan. Amer. Pot. Monsanto Shell Stauffer Sumitomo Velsicol	Acar.
17.	Demeton Mixture of 0,0-diethyl S- (and 0)-2-(ethylthio) ethyl phos=phorothioates Systox	$\overset{O}{\overset{\|}{(C_2H_5O)_2P}}-S-CH_2CH_2-S-CH_2CH_3$ $\overset{S}{\overset{\|}{(C_2H_5O)_2P}}-O-CH_2CH_2-S-CH_2CH_3$	$C_8H_{19}O_3PS_2$	Chemago Farbenfab-riken, Bayer, A.G.	Acar. Ins. Sys. Nem.

TABLE 1 (Continued)

Names and designations	Structural and empirical formulas	Manufacturer and use[b]	
18. Diazinon		Geigy	Nem.
0,0-diethyl 0-(2-iso-propyl-4-methyl-6-pyrimidinyl) phosphorothioate			Ins.
Sarolex			
19. Carbofuran		FMC Corp.	Sys. Ins.
2,3-dihydro-2,2-dimethyl-7-benzo-furanyl methylcar-bamate			Nem.
Furadan	$C_{12}H_{15}NO_3$		

[a]Adapted from E. E. Kenaga, Commercial and Experimental Insecticides, Bull. Entomol. Soc. Am. 12(2): 161–217 (1966).

[b]Key: Acar., Acaricide; C. Fum., Commodity or space fumigant; Ins., Insecticide; S. Fum., Soil fumigant; Nem., Nematicide; Sys. Nem., Systemic nematicide; Sys. Ins., Systemic insecticide.

The increased spectrum of activity, however, makes the use of these materials profitable in plant beds and in nurseries where conditions for disease development are favorable and where labor costs for weed control are high.

The first really useful nematicides for field scale applications were D-D mixture and ethylene dibromide introduced in 1943 and 1945, respectively. These fumigants in various formulations are primarily nematicides and are used for large scale application in the field for nematode control. Because of their phytotoxicity, treatments are applied from 2 to 4 weeks before planting.

The search for less phytotoxic and less volatile compounds resulted in the development of dibromochloropropane. This compound is 8 to 16 times as active as D-D mixture and of such a low phytotoxicity that it can be applied to certain growing crops, or at time of planting, without damage to the plants.

Vorlex (chlorinated C_3 hydrocarbons including dichloropropenes, dichloropropane, and related chlorinated hydrocarbons, 80%; methyl isothiocyanate, 20%) controls nematodes and at rates considerably higher than that required for nematodes, will control certain weeds, fungi, and soil insects.

In addition to the halogenated hydrocarbons mentioned above certain of the dithiocarbamates and related compounds have proven effective as nematicides. Metham or Vapam (sodium N-methyl dithiocarbamate) is the most widely used and in addition to nematodes, controls fungi, bacteria, insects, and weeds.

Tetrachlorothiophene (Penphene), a low-volume nematicide has been recommended for use on some crops.

The organic phosphates constitute a group of nonfumigant compounds which have shown considerable promise as nematicides, although they are primarily insecticides. This group of nematicides because of low volatility do not diffuse in the soil to any appreciable extent and therefore dispersal requires good soil incorporation or they must move through the soil in rainfall or irrigation. At first, well-known phosphate insecticides were tested for nematicidal activity and some were found to be effective. Dichlorofenthion (O,O-diethyl-O-2,4-dichlorophenyl-phosphorothioate) was one of the first marketed organophosphate nematicides. While still used to a limited extent for nematode control, its greatest use at present is for chinch bud control on lawns. Parathion and demeton, well-known and widely used insecticides, are somewhat "specialty products" with reference to nematode control, in that their primary usage are as foliar sprays against the foliar nematode of chrysanthemum (Aphelenchoides ritzemabosi).

Thionazin (O,O-diethyl-O-2-pyrazinyl-phosphorothioate) has been shown to be effective against several species of parasitic nematodes when applied to the soil and is also used as a root dip. A combination of thionazin

plus Phorate (O, O-diethyl S-ethylthiomethyl phosphorothioate) is being
used in some areas. Thionazin, developed primarily because of its nematicidal properties, also controls certain insects. Phorate, developed as an
insecticide, also controls certain nematodes. The combination appears to
be more effective than equivalent amounts of either chemical alone.

Prophos (O-ethyl S, S-dipropyl phosphorodithioate) is used for the
control of soil nematodes, and insects. Mocap has little or no fumigant or
systemic activity.

Diazinon (O, O-diethyl O-(2-isopropyl-4-methyl-6-pyrimidinyl) phosphorothioate) is used primarily for the control of nematodes and insect pests
of turf and lawns.

The organic phosphate nematicides have the advantage of high nematicidal action with low phytotoxicity and most of them can be applied at time
of planting. They also have a moderate residual action but this is not always
desirable because of high mammalian toxicity. Furthermore, it may limit
their application to ornamental and other nonfood crops, although certain
phosphate nematicides have been cleared for use on some food crops.

G. Nematicide Usage and Effects on Crop Production

Although it would be difficult to appraise fully the value of nematicides
in crop production and therefore their economic return, it is apparent that
the use of these materials has had significant effects on crop production.
Since the development of practical field nematicides in the early 1940s,
their usage has grown from a few hundred pounds in 1943 to in excess of 40
million pounds by 1967. Grower acceptance of fumigation has been an outstanding testimony of the value of this practice in crop production. Largest
acreages of crops treated have been those of relatively high-acre value and
those centered around certain industries, such as pineapple, sugarcane,
citrus, tobacco, cotton, peanuts, ornamentals, and commercial vegetables.

Increases in per acre values resulting from the use of nematicides
may range from a few dollars to several hundred, depending on the crop
and the severity of the problem. An example, for which fairly reliable data
are available, is the increased returns to tobacco growers in North Carolina resulting from the use of nematicides. For example, since 1954 the
annual increased returns to tobacco growers in North Carolina resulting
from fumigation, after subtracting cost of materials, has ranged from $25
to $30 million. Based on these figures, the increased returns from tobacco
alone in North Carolina during the past 13 years, has amounted to from
$325 million dollars to $390 million dollars. Figure 17 illustrates the
increased growth of tobacco grown in treated soil compared with that grown
in infested soil. The untreated plot shown in Fig. 17 yielded 1598 lb/acre

and the acre value was $1121.00. Yield from the treated plot was 2259 lb/acre and the value was $1548, or an increase in value per acre of $427.00. The cost of treatment was approximately $15.00. Another illustration of improved crop growth resulting from fumigation is shown in Fig. 18. Similar results could be given for such crops as citrus, cotton, pineapple, vegetable crops and ornamentals. Not only are yields affected, but sometimes equally important is the improved quality of the crops grown. This is especially true where the harvested crop is in direct contact with the nematode or is actually infected by it. Such crops include sugarbeets, sweetpotatoes, Irish potatoes, peanuts, carrots and certain bulb crops. These crops may be so distorted and malformed that the grower either fails to harvest the crop, or has to accept a very low price. In such cases, he may break even or actually lose money on the crop.

Although use of nematicides is a well established and profitable farm practice, growers should use them with discretion. They are obviously of no value in fields where nematodes are not a problem—in fact, there would be a good chance of the chemical having a retarding effect on plant growth. A knowledge of the nematode species present and the level of infestation is an important safeguard against needless treatment with nematicides. Even for those fields known to be heavily infested, dosages should be kept at that level which is required to do the job because of possible residue build-up in the soil over a period of years resulting from repeated applications. Treatment of the same field for several years in a row should be avoided when possible since this could lead to the development of populations resistant to the toxic effect of the chemical. Rotations and other cultural practices combined with use of nematicides should make the development of resistant strains of the pathogen less likely.

FIG. 17. Photograph showing (left) tobacco growing in nonfumigated soil; (right) fumigated soil. (Courtesy of F. A. Todd.)

FIG. 18. Photograph of peanuts growing in fumigated and nonfumigated soil. (Left) fumigated with DBCP and (right) control.

Research in development of nematicides has continued to increase
over the past two decades and it is very likely that many new and different
kinds of nematicides for various purposes will be developed.

VI. FUTURE RESEARCH NEEDS

Few sciences have developed more rapidly than that of Nematology in
the last two decades. This expansion in our knowledge has come about
primarily because of increased emphasis on the training of professional
nematologists, resulting in a rapid increase in the number of scientists
engaged in nematological investigations. It is futile to discuss future research needs without pointing out the necessity of making provision for the
adequate training of scientists capable of doing the research needed in the
field of nematology. In spite of the fact that a great deal of change has
taken place with reference to the training of nematologists during the past
decade, adequate programs are few in number and recognition and concern
for this important field of biology are developing much too slowly in light of
the ravaging effects nematodes are having on food and fiber production
throughout the world. The only hope for progressing at a rate commensu-
rate with need is to establish centers of excellence, or institutes, staffed

and supported so as to provide broad and "in-depth" training to a large
number of U.S. and foreign students who will be doing research on nematode diseases and their control in the decades immediately ahead. So much
of the information obtained to date, although extremely important, has been
relatively easy to secure, but major advances in the future will be few and
retarded unless intensive measures are taken with reference to the recruitment and training of nematologists.

Assuming that something can be done to provide for an accelerated
and comprehensive program in the training of nematologists, research into
the broad aspects of the nematode and its environment, an understanding of
which is essential before major advances in nematode control are made,
can be initiated.

Plant-parasitic nematodes have evolved and adapted themselves to
their present state of parasitism over eons of time and the changes that have
occurred, whether sudden or gradual, have been associated with a variety
of biological and environmental factors. Certain species have evolved or
have adapted more rapidly and to a greater degree than have others, while
some have developed an almost unlimited capacity for parasitism and survival. The evolutionary process responsible for the development of a new
species or of resistance-breaking biotypes are not well known. Present
methods and principles of control, other than chemical, are based upon a
knowledge of the etiology of the species of nematode to be controlled, its
feeding habits, parasitic stages, host range, life cycle, reproductive
capacity, pathogenic stages, pathogenic variation, and its response to various factors in the environment, including the host, upon its parasitism,
reproduction, and survival. Such knowledge must include the identification
of its host, that may potentially exert a selection pressure upon the particular nematode population in specific environments, the susceptibility of
various crop plants to attack by various species of nematodes, and the
effects of desiccation, temperature, flooding, fallow (deprivation of food
source), and many other factors in the environment, upon its development.
Many of the widely used methods and principles of control really are not
thoroughly understood. Frequently the results from a given control practice are erratic and unpredictable, even from such a familiar control as
crop rotation.

An area of limited knowledge is the inherent physiological factors
which enable nematodes to withstand and adapt to the many extremes of the
environment induced by the intimate and constant association of the parasite with both the soil medium and the living host. The soil itself is not
static and changes occur constantly by the presence of other microorganisms.
Such changes include the kinds and concentrations of salt that result from
amendments added to the soil or by the uptake of these materials by crop
plants or by the various processes in the soil which change and consume
these various chemical components and modify the form and concentration

of soil compounds. The influence of plant roots, whether it be by exudates given off by active root systems or by the organic matter released by death of roots and plant detritus is not as yet clearly understood. Changes in temperature, moisture, gases, and the major impact of various tillage methods, induce tremendous changes within the soil over relatively short periods of time. In virgin lands where the soil is well protected by dense growth, is undisturbed by tillage or by additions of inorganic salts, or removal of crop growth, or by cultivation, the biotic activity of the soil reaches a steady state equilibrium, and pathogens, where present, are not subject to such extremes to which they have to endure or adapt. When genetic and physiological changes in the pathogen occur under these pristine conditions, they are usually minor and infrequent.

In the future man will necessarily continue to depend upon the soil for his food and fiber production. The kinds of nematodes present, their population levels, and their pathogenicity to agricultural crops will indubitably be influenced by the particular crops grown and these various factors of the environment. To minimize the damage caused by nematodes, it is essential to assess and understand those factors which affect the survival and development of the parasite. These factors are multiple and the interactions between the host, the parasite, and the environment are complex. Such studies demand a thorough knowledge of the organism—its morphology and anatomy, its physiology and metabolism, its cytology and its mode of reproduction, and, particularly, how these are influenced or modified by the conditions of the environment imposed upon the population.

Answers to these questions will make it possible to accelerate the development of resistant varieties, and to modify the physical and biological characteristics of the soil and the plant phase of the nematodes' environment, and thereby to decrease or to eliminate the adverse effect on economic crops caused by parasitic nematodes. Similarly knowledge of the metabolism of parasites will lead to the development of contact and systemic chemical nematicides, which can be applied to destroy or inactivate parasitic nematodes that live either free in the soil or within plant tissues.

PLATES I through IV

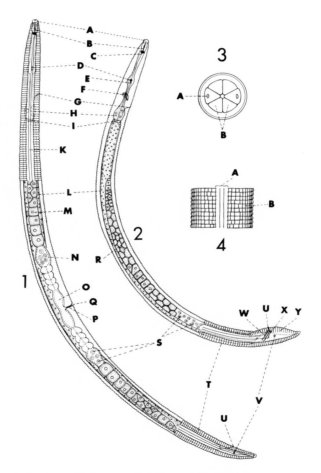

PLATE I. Tylenchorhynchus claytoni Steiner, 1937.

PLATE I.

1. Female
   A, lip region
   B, stylet
   D, median bulb of esophagus
   G, excretory pore
   H, basal bulb of esophagus
   I, cardia
   K, lateral field
   L, intestine
   M, ovary
   N, seminal receptacle
   O, uterus
   P, vagina
   Q, vulva
   S, sperm
   T, annulation of the cuticle
   U, anus
   V, phasmids

2. Male
   A, lip region
   B, stylet
   C, dorsal esophageal gland orifice
   D, median bulb of esophagus
   E, subventral esophageal gland orifice
   F, nerve ring
   G, excretory pore
   H, basal bulb of esophagus
   I, cardia
   L, intestine
   R, testis
   S, sperm
   T, annulation of the cuticle
   U, cloacal opening
   V, phasmids
   W, spicule
   X, gubernaculum
   Y, caudal bursa

3. Face view
   A, amphid
   B, lips

4. Cuticle detail
   A, lateral field with four incisures forming three ridges
   B, annules subdivided by longitudinal striations

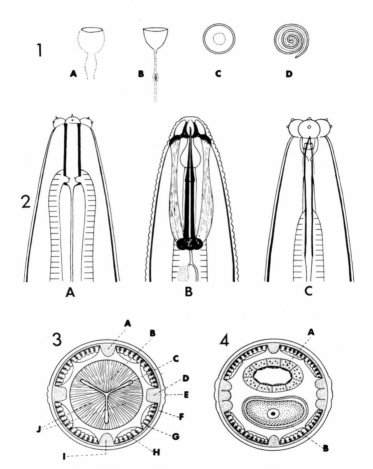

PLATE II. Amphidial openings; variations of the stoma; cross-section through esophageal region, and cross-section through regions of reproductive organs in female.

PLATE II.

1. Amphidial Openings
   A, cyathiform type (Trilobus)
   B, cyathiform type (Dorylaimus)
   C, circular type (Monhystera)
   D, spiral type (Choanolaimus)

2. Variations of the Stoma
   A, cylindrical stoma (Rhabditis)
   B, stomatostyl (Rotylenchus)
   C, odontostyl (Dorylaimus)

3. Cross-Section Through Esophageal Region
   A, dorsal chord
   B, dorsal sector of esophagus
   C, esophagus lumen (subdorsal radius)
   D, lateral chord
   E, lateral field
   F, cuticle
   G, hypodermis
   H, somatic musculature
   I, ventral chord
   J, the two subventral sectors of esophagus

4. Cross-Section Through Region of Reproductive Organs in Female
   A, intestine
   B, ovary with unfertilized egg (oocyte)

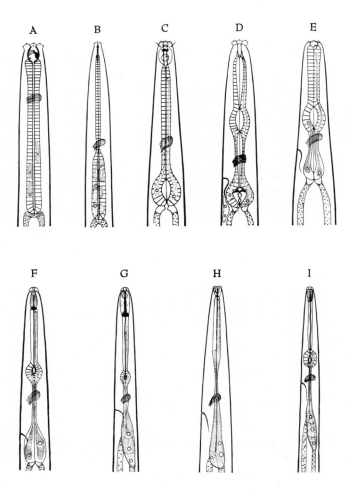

PLATE III. Different types of oesophagi.

PLATE III.

A, cylindrical (Monochus)

B, dorylamoid (Dorylaimus)

C, bulboid (Ethmolaimus)

D, rhabditoid (Rhabditis)

E, diplogasteroid (Diplogaster)

F, G, H, tylencoid (Tylenchorhynchus, Rotylenchus, Neotylenchus)

I, aphelenchoid (Aphelenchus)

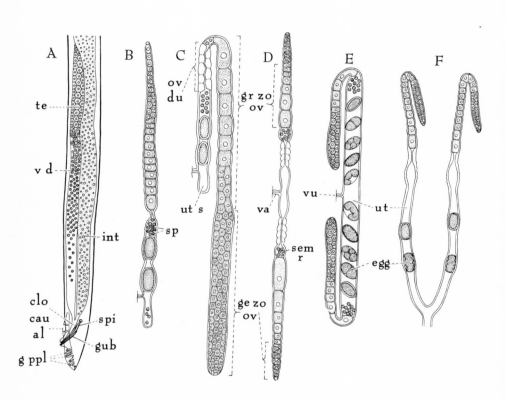

PLATE IV.   Reproductive systems of nematodes.

PLATE IV.

A, male reproductive system (<u>Rhabditis</u>, monorchic)

B, C, D, E, F, different types of female reproductive systems

B, single gonad, directed anteriad, ovary outstretched
(<u>Ditylenchus</u>, monodelphic, prodelphic)

C, single gonad, directed posteriad, ovary reflexed
(<u>Panagrolaimus</u>, monodelphic, prodelphic)

D, paired gonads, branches opposed, ovaries outstretched
(<u>Tylenchorhynchus</u>, didelphic, amphidelphic)

E, paired gonads, branches opposed, ovaries reflexed
(<u>Rhabditis</u>, didelphic, amphidelphic)

F, paired gonads, directed anteriad, ovaries reflexed
(<u>Meloidogyne</u>, didelphic, prodelphic)

> cau al = claudal alae
>
> clo = cloaca
>
> g ppl = genital papillae
>
> ge zo ov = germinal zone of ovary
>
> gr zo ov = growth zone of ovary
>
> gub = guvernaculum
>
> int = intestine
>
> ov du = oviduct
>
> sem r = seminal receptacle
>
> sp = sperm
>
> spi = spicule
>
> te = testis
>
> ut s = postvulvar uterine sac
>
> ut = uterus
>
> v d = vas deferens
>
> va = vagina
>
> vu = vulva

REFERENCES

1. G. F. Atkinson, Ala. Agric. Exp. Stn. Bull., 1:1-54 (1889).

2. M. J. Berkeley, Gardener's Chronicle, 14:220 (1855).

3. E. A. Bessey, Root-knot and Its Control, U.S. Dep. Agric. Bull. 217, 1911, 89 pp.

4. F. E. Caveness, A Glossary of Nematological Terms, Yaba, Nigeria, The Pacific Printers, 1964, 68 pp.

5. B. G. Chitwood and M. B. Chitwood, An Introduction to Nematology, Revised Ed., Baltimore, Md., Monumental Printing Co., 1964, 213 pp.

6. J. R. Christie and V. G. Perry, Science, 113:491-493 (1951).

7. J. R. Christie, A. N. Brooks, and V.'G. Perry, Phytopathology, 42:173-176 (1952).

8. J. R. Christie, Plant Nematodes—Their Bionomics and Control, Gainesville, Fla., University of Florida Press, 1959, 256 pp.

9. J. M. Good and J. Feldmesser, The future role of chemicals for controlling plant nematodes, in Pest Control by Chemical, Biological, Genetic and Physical Means—A Symposium, American Association for the Advancement of Science (Section 0 on Agriculture), Dec. 26-31, 1964, Montreal, Canada, ARS 33-110, July 1966, pp. 52-58.

10. J. B. Goodey, Soil and Fresh Water Nematodes, New York, Wiley, 1963, 544 pp.

11. L. H. Hyman, The Invertebrates, III, Acanthocephala, Aschelminthes and Entoprocta, New York, McGraw-Hill, 1951, 572 pp.

12. A. E. Kehr, Current status and opportunities for the control of nematodes by plant breeding, in Pest Control by Chemical, Biological, Genetic and Physical Means—A Symposium, American Association for the Advancement of Science (Section 0 on Agriculture) Dec. 26-31, 1964, Montreal, Canada, ARS 33-110, July 1966, pp. 126-138.

13. E. E. Kenaga, Bull. Entomol. Soc. Am., 12:161-217 (1966).

14. J. C. Neal, The Root-knot Disease of the Peach, Orange, and Other Plants in Florida Due to the Work of Anguillula, U.S. Department of Agriculture, Division of Entomology, Bulletin 20, 1889, 31 pp.

15. T. Needham, Philos. Trans. R. Soc., 42:634-641 (1944).

16. Nematology Fundamentals and Recent Advances with Emphasis on Plant Parasitic and Soil Forms, J. N. Sasser and W. R. Jenkins (eds.), Chapel Hill, North Carolina, University of North Carolina Press, 1960, 494 pp.

17. A. C. Papavizas, Biological methods for the control of plant diseases and nematodes, in Pest Control by Chemical, Biological, Genetic and Physical Means—A Symposium, American Association for the Advancement of Science (Section 0 on Agriculture) Dec. 26-31, 1964, Montreal, Canada, ARS 33-110, July 1966, pp. 82-94.

18. J. E. Peachey and M. R. Chapman, Chemical Control of Plant Nematodes, Technical Communication No. 36 of the Commonwealth Bureau of Helminthology, St. Albans, Commonwealth Agricultural Bureaux, Farnham Royal, Bucks, England, 1966, 119 pp.

19. F. C. Peacock and J. E. Peachey, Adv. Chemother., 2:1-22 (1965).

20. V. G. Perry, Proc. Helminth. Soc. Wash., 20:21-27 (1953).

21. D. Pramer, Predaceous fungi and the biological control of nematode pests, in Proceedings, Thirteenth Annual Meeting of the Agricultural Research Institute, Oct. 12-13, 1964, Washington, D.C., NAC-NRA, 1964, pp. 47-54.

22. A. F. Schindler, Nematologia, 2:25-31 (1957).

23. R. F. Suit and E. P. DuCharme, Plant. Dis. Rep., 37:379-383 (1953).

24. G. Thorne, Principles of Nematology, New York, McGraw-Hill, 1961, 553 pp.

25. H. B. A. Welle, Physostigmine and Structural Analogues as Direct and Systemic Nematicides, Utrecht (Nederland), Schotanus and Jens Utrecht NV, 1964, 187 pp.

Chapter 3

PEST CONTROL IN ORNAMENTAL
AND LANDSCAPE PLANTINGS

Richard H. Gruenhagen
Department of Plant Pathology and Physiology
Virginia Polytechnic Institute and State University
Blacksburg, Virginia

255

I. INTRODUCTION

In our world where over half the population goes to bed hungry every night and starvation and malnutrition are, in many areas, the order of the day, it is quite easy to understand why great emphasis is placed on control of the pests that jeopardize our food supplies. In the more enlightened areas of our civilization, however, we have realized that there is another facet to the problem. While food feeds the body, the beauty of nature in trees, shrubs, ornamentals, and lawns feeds the spirit. As with food and fiber crops, these aesthetic contributors to our living space are continually subject to the ravages of diseases, nematodes, insects, and weeds. Man has always had to wage relentless battle against those pests that threaten his food supply; but, until recently, little effort was made to control pests that destroy the beauty of areas in which we live, work, and play.

With the development of modern agricultural technology over the past century, the number of farms has decreased and those remaining have increased in size; thus, fewer people are needed to produce food. A vast migration to suburban and urban areas has occurred to supply goods and services. These people employed in industry generally work only a 40-hr week, and even this is being reduced. This has given people more leisure time and, consequently, the use of parks, forests, and public recreation areas has increased. In addition, such spare hours have encouraged home, neighborhood, and community landscaping. All garden clubs and many civic organizations endorse such efforts to preserve and enhance natural beauties, for it is well-recognized that ugly environments demoralize people.

Such programs to beautify homes are now becoming national in scope. In 1965 President Lyndon B. Johnson stated before the United States Congress,

> For centuries Americans have drawn strength and inspiration
> from the beauty of our Country. It would be a neglectful genera-
> tion indeed, indifferent alike to the judgment of history and the
> command of principle, which failed to preserve and extend such
> a heritage to its descendants.

Machinery is now operative at various levels of government to implement programs for the beautification of the Nation. City planners and construction engineers are now concerned with, in addition to blueprints of brick and mortar, the planting of lawns, flowers, shrubs, and trees in areas where people live and work in order to encourage the soul of man.

Today homeowners devote many hours beautifying their home grounds in order to make their outdoor living the more enjoyable (Figs. 1 through 3). Regardless of how diligently people work to landscape their homes, they must inevitably wage an untiring battle against pests that destroy the fruits of such labors. Vast forest fires may make national news, but few people realize that diseases and insects destroy more timber in the United States annually than does fire. Parks, national forests, or roadside picnic areas lose value rapidly if the trees and plants are destroyed by pests, or if camping areas become overgrown with poisonous plants and consequently become uninhabitable.

Although the natural beauty of our landscape may be damaged by many courses, such as fire, vandalism, and human irresponsibility, this chapter will explore the problems of control of the more subtle pests that threaten our flowers, ornamentals, trees, lawns and personal comfort. These

FIG. 1. Good civic planning and planting. (Photo: American Association of Nurserymen.)

FIG. 2. Carefully maintained home grounds are a pleasure to the owner and a credit to the neighborhood. (Photo: American Association of Nurserymen.)

FIG. 3. A place for rest and relaxation. (Photo: American Association of Nurserymen.)

include diseases, nematodes, insects and weeds which constantly threaten the aesthetic values we hope to preserve in our outdoor areas. Before discussing these pests, however, we will review briefly the kinds and characteristics of ornamental plants and the problems which beset their use.

II. ORNAMENTALS IN HISTORY

Many new flower and ornamental shrub varieties are introduced each year, and some hybrids may differ radically in form and color from their ancestors; yet, of the ornamental plants grown today, many have been known and prized since recorded history began. Interesting and detailed accounts of plant history may be found in books by Moldenke (22) and Quinn (27), and only a few examples need be mentioned here.

Boxwood, a stately plant traditional in the southeastern United States, is recorded in Greek mythology. [*]

Boxwood was noted by Pliny, 1st century A.D., as a prized ornamental shrub. From its hard wood the Greeks made small chests and boxes and referred to it as buxus; when boxwood moved to England, the name adapted from the Roman became box. Actually, boxwood probably originated in China and found its way into western Asia along ancient trade routes. The Bible contains references to the box plant. [**]

The Japanese dwarf boxwood probably was introduced into the United States about 1860 and now is used and prized widely as an edging and small hedge broadleaf evergreen.

In the 18th century the renowned botanist Linnaeus, while classifying plants, found reference to a brightly flowering plant from Arctic Lapland

---

[*]The legend tells how Apollo, God of the Sun, chased a wood nymph through field and woodland; and she, in desperation, called upon Diana for help. The Goddess of the Moon and the Chase, taking pity on the nymph, changed her into a bush with lustrous flowers. Apollo, thus thwarted, in rage tore all the flowers from the bush. Diana, whose power did not exceed Apollo's, could only replace the large beautiful blooms with tiny inconspicuous florets; but she bestowed upon the bush lustrous green leaves which would never die. From this everlasting habit of boxwood foliage comes the species name, sempervirens.

[**]"...The glory of Lebanon shall come unto thee, the fir tree, the pine tree, and the box together..." (Isaiah 60:13)

said to flourish only in dry places; from the Greek azaleos (dry), he took
the name azalea. Many legends and customs are associated with azaleas. *

Flowering dogwood (Cornus), well known in England in the 16th century, was called hounds tree, dogberry tree, or dogs tree. Possibly these
names were associated with the tree because the berries and bark were
extracted in hot water and used to treat mangy dogs. The American Indian
also used a similar preparation in his medicine. Subsequently, it was
found that dogwood bark contains the valuable antimalarial pharmaceutical
quinine. The short shrubby habit of the dogwood was also subject of a
legend. **

Pyracantha, or Firethorn, well known in early English gardens and
hedgerows, was prized not only for its beauty but also for its long sharp
thorns, which made it impenetrable. Lodeman (20), in 1629, described it in
detail.

Forsythia, also found in early England, was first known as goldenrain
tree but later was given its present name, after William Forsythe, who imported it from China for the Royal Gardens at Kensington.

Holly, now classified as genus Ilex, has been known under many names
for over 2,000 years. European holly, originally known as aquifolium, was
put into the genus Ilex, with the species name aquifolium, by Linneaus.

Many plants mentioned in the Bible are today planted as ornamentals
in our parks and gardens, including cedar of Lebanon, Andorra Juniper,
Boston ivy, hawthorn, pin oak, narcissus, crocus, Christmas rose, star-
of-Bethlehem, passion vine, lily, goldenrain tree, anemone, bayberry,
redbud and burning bush. The goldenrain tree was designated one of five
official memorial trees to be planted on the tombs of scholars; the lily became the emblem of purity and grace; the mulberry tree considered the
wisest of all trees; and in Exodus, the Lord spoke to Moses from a burning
bush. Although the identity of the bush remains speculative, a logical
explanation suggests the illusion of fire was created by the profuse blooms
of a crimson-flowered mistletoe, Loranthus acaciae, which flourishes on
an Acacia species common in the Holy Land.

---

*One interesting tale related to Nieuw Amsterdam, old Dutch New York.
Whitsuntide was celebrated with a 3-day festival called Pinxter Frolics,
from the Dutch Pinckster, meaning the 50 days ending on Whitsunday Pen-
tecost. During the festival, young people strolled through the woods and
plucked boughs of azalea blooms—the Pinxter Flower.
**Legend relates that the then tall and stately dogwood tree provided the
wood to build the cross upon which Christ was crucified. So distressing
was this, the dogwood resolved never to grow large enough to be used
again for a cross; and to commemorate the tragedy, its flowers grow in
a cross and bear the imprints of the nails to this day.

III. GARDEN HISTORY

An extensive discussion on the history of ornamental gardening is beyond the scope of this chapter and is also only indirectly relevant. A few pertinent highlights, however, will serve to relate the art of ornamental landscaping and decoration to the practical businss of pest control.

The garden of primitive man was strictly a necessity where meager crops grew to feed his family, where struggle for survival was paramount and no aesthetic values were considered. Only when nomadic peoples began to settle down in small villages did organized gardening evolve. All such early gardens were solely concerned with producing both food crops and medicinal herbs.

Early Egyptian gardens started the aesthetic trend in the cultivation of trees and herbs. These gardens were frequently near temples and became meeting places for the priest-physicians. It has been noted, also, that the grapevine was an early favorite plant in these gardens.

Although the Greeks did little to advance gardening, they did lay out formal gardens containing many fragrant flowers thought to keep the air pure and ward off disease. Much Greek mythology and superstition could be found in these old gardens with their grottoes and statues.

The Roman gardens were more elaborate and displayed the window box feature, still widely used today. The medieval monastery garden tended toward the classical Roman style and, small in size, strongly stressed decorative flowers and medicinal herbs.

The Renaissance gardens revealed an awakening interest in natural beauty. English medieval gardens followed this trend by cultivating many natural wild plants. As gardening developed, however, the formal garden emerged. One can imagine the pest problems that occurred as intensive types of growing practice ensued.

For centuries oriental gardens have fascinated landscape artists because of the mystery and symbolism latent in their design. Many such modern decorative plants, such as the peony, the peach blossom, and the chrysanthemum, are derived from ancient oriental gardens. These gardens of the East were an integral part in the lives of the Indian, the Chinese, the Japanese and the Persian; and many dedicated hours were devoted to their maintenance.

Throughout the study of early gardens, one can perceive the laborious efforts of those who tended and husbanded such gardens; yet it is also clear that superstition and ignorance prevailed with respect to methods of pest control. Today we not only have this heritage of beauty but also have modern chemistry to aid us in its preservation.

IV. PESTS ARE NOT NEW

Despite those who contend that plant pests are relatively new, arising from man's modern cultural practice, particularly his use of fertilizers and pesticides, the fact is that plant pests are recorded in the earliest history of mankind. Fossil remains indicate that disease and insect pests were ravaging plants on earth millions of years before man himself emerged.

Both Aristotle and Theophrastus wrote of the "mildew, blighting, and blasting" of crop plants, terms still in use today to describe certain disease symptoms. Many references to plant disease occur in the Bible.*

By the 18th century attention was directed at the impact on social economy. Even before the precise cause of plant disease was known, Zallinger (36), in 1773, described diseased plants in some detail. In 1833 Unger (33) wrote a definitive textbook, now a classic, on the broad subject of plant diseases. In the mid-19th century DeBary (12) and Berkeley (1) established and clarified the microbial nature of many plant diseases.

Insects, too, were long recognized as pests of plants, livestock, and man. As described in Exodus, the gardens of ancient Egypt were beset with pests.**

Numerous other historical references tell how the cankerworm and the caterpillar laid waste to the vines and barked the fig trees. In 1915 Forbes (13) commented,

> The struggle between man and insects began long before the dawn of civilization and has continued without cessation to the present time, and will continue, no doubt, as long as the human race endures. ...We commonly think of ourselves as the lords and conquerors of nature, but insects had thoroughly mastered the world and taken full possession of it long before man began the attempt.

Paleozoologists have established that insects have lived on earth for over 50 million years, while humans emerged from 500,000 to 1,000,000 years ago. In view of this long tenure of insects on earth, it is indeed ridiculous to contend that present-day insect problems stem primarily from the use of insecticides.

---

*"I smote you with blasting and with mildew and with hail in all the labors of your hands, yet ye turned not to me, sayeth the Lord." (Haggai 2:17)
**"...and the locust went up all over the land of Egypt and rested in all the coasts of Egypt...and they did eat every herb of the land and all the fruit of the trees which the hail had left..." (Exodus 10:14:15)

Weeds also have long been recognized as pests to man, and many passages in the Bible allude to the damage they cause.* Nettles, briars, thorns, wormwood, mustard and other noisome and harmful plants are mentioned in the Scriptures, providing presumptive evidence that ancient people were fully aware of these undesirable forms of vegetation.

Poisonous plants were known before man began to cultivate crops, for ancient hunters tipped their weapons with extracts of poisonous plants to kill game for food. Even today, in isolated areas of the world, primitive natives dip darts and arrows into plant poison to aid in killing game or foe. The first accurate compilation of the poisonous and medicinal properties of plants was made by Dioscorides in his manuscript, Of Medicinal Matters (14). Centuries, however, elapsed before writings dealing specifically with poisonous plants appeared, possibly because during this period poisoning had become an art with its secrets closely guarded. Interest in the scientific recording of objective information on the therapeutic and toxic properties of plants reappeared in the 17th century, for during this and the next century a number of treatises were published which dealt exclusively with poisonous and medicinal plants.

Woodville, in his text, Medical Botany (35) published between 1790 and 1794, reveals the paucity of reliable scientific information on plant poisons and he states,

It is a lamentable truth that our experimental knowledge on many of the herbaceous samples is extremely defective for, as writers on the Materia Medica have usually done little more than copy the accounts given by their predecessors, the virtues now ascribed to several plants are wholly referable to the authority of Dioscorides.

In the United States experimental work on poisonous plants began in the latter half of the last century. The establishment of the land grant and veterinary colleges and the founding of the Department of Agriculture in the 1860s provided the impetus needed for such research.

Ancient man struggled with an unfriendly nature to wrest his food and shelter; and though today we are still plagued by pests that are essentially similar to those that threatened our forebears, we now have effective means for controlling virtually all predators. We can thank modern technology, particularly chemistry, for the development and use of pesticides, which make our lives easier, safer, and more pleasant.

---

*"Let thistles grow instead of wheat, and cockle instead of barley." (Job 31:40)

V. KINDS OF PESTS

We speak in general terms of pests that injure our ornamentals, shrubs, trees and lawns. What are these creatures which are of vital concern to all who would cultivate plants?

A pest may be defined as an entity that attacks, invades, devours, despoils or infects our person or our property. It may be a chafer eating our prize roses, a fungus spotting a golf green, or even a small boy (our neighbor's) stamping around in our front flower bed. Control measures will presumably differ in these situations. For specificity we group pests under three headings:

A. Disease-producing organisms
B. Insects
C. Weeds and poisonous plants.

A. Disease-Producing Organisms

Although fungi are, beyond dispute, the most numerous of the plant disease-producing organisms, there are actually many fungi that are beneficial and cause no disease. However, there are over 7,000 different fungi which cause over 30,000 different plant diseases. A fungus is itself a very tiny plant. As it does not contain chlorophyll, it must obtain its food from some other living or dead organic source. The disease-producing or parasitic fungi secure their nutrient needs by attacking chlorophyll-containing plants. Fungi vary in size and stage of development from simple organisms which reproduce asexually to complicated organisms with highly refined reproductive systems. Many fungi produce one or more kinds of spores or seed, which may be blown or otherwise carried from one plant to another, thereby establishing new infections. Fungus spores have been trapped several thousand feet high in the air and have been shown to travel for hundreds of miles. Many ornamental plant and shade tree diseases are caused by fungi. Dutch elm disease (Figs. 4 and 5) is an example of a fungus-induced disease. The causal fungus, Ceratocystis ulmi, is carried by elm bark beetles, Scolytus multistriatus and Hylurgopinus refipes, which burrow beneath the bark of the tree, carry the spores of the parasitic fungus with them, and inoculate the living tissue.

Fungi have changed the eating habits of people. When the early Jamestown settlers came to the New World they brought with them the English grain, or wheat, the carbohydrate food to which they were accustomed. In the warm humid climate of Virginia, wheat rust flourished and destroyed the wheat crop. These early colonists would have quickly starved had they

FIG. 4. Dutch elm disease fungus, <u>Ceratocystis ulmi</u>, in culture.
(Photo: Davey Tree Expert Co.)

not found the Indians growing maize and grinding the grain to make flour
for bread. Today cornbread is common throughout the southern states.
On the other hand, the pilgrims in New England could grow wheat relatively
free from rust in the favorable climate, and they continued to grind wheat
grain for bread flour as in their native England. In certain areas of central
Europe, wheat fails because of rust, while rye abounds and consequently is
the main source of flour for bread. Here <u>bread</u> is neither wheat nor corn,
but rye bread.

Fungi have also caused significant migrations of populations. In Ireland, during 1845 and 1846, the potato late blight disease (<u>Phytophthora</u>
<u>infestans</u>) reduced the potato crop to masses of stinking black slime. Since
the Irish were virtually dependent upon potatoes for food, the disease un-
leashed misery, disease, and disaster on an unparalleled scale. In the
ensuing 15 years almost one-third of the Irish population, some 2 million
souls, emigrated as a direct result of this fungus-induced disease. Nearly
2 million people perished from starvation or from typhus and other diseases
which followed (34).

FIG. 5. American elm being killed by Dutch elm disease. (Photo: Illinois Natural History Survey.)

All fungi are not, however, pests. Many rot dead leaves and other organic matter, returning humus and mineral nutrients to the soil. Other fungi, such as certain species of the large fleshy mushrooms, are edible and nutritious. Some fungi provide food flavors as, for example, in Roquefort cheese. This chapter, however, is concerned with those fungi which cause diseases to our ornamental shrubs, trees, flowers, and lawns.

Bacteria, also members of the plant kingdom, are single-cell organisms which reproduce by simple binary fission. They can perhaps be considered as intermediate between the plant and animal kingdoms because, like simple animalcules, they exhibit motility through their small whips or hairlike flagellae which lash about, and propel them through liquids. Like the fungi, bacteria also lack chlorophyll and are therefore unable to produce their own food. There are some 175 different bacteria which are known to induce plant diseases. Figure 6 shows the effects of a typical phytobacterium, the fire blight disease, caused by Erwinia amylovora.

A virus is a very large, exceedingly complex molecule comprising a nucleic acid core wrapped with a protein. Occasionally a virus disease may be actually beneficial; for example, the striking variegated color in certain tulip blooms is caused by infection of the plant with a virus, and highly colored bedding plants are infected by a virus which causes the unusual coloration.

Virus diseases of shrubs and trees can induce serious damage (Fig. 7) and are generally quite difficult to control. Research has developed very precise heat treatment methods to control certain viruses that infect flowering plants; however, these methods are rather complicated and require such sophisticated equipment to apply that they are not practical for the average homeowner. Since many plant viruses are insect-vectored, insect control is often the most practical approach to the prevention of virus disease. Ornamental plant nurseries and propagators are extremely careful to use only disease-free stock, and thorough sanitary precautions are taken in all plant propagation and production workshops to discard all plants suspected of harboring viruses. The control of plant virus diseases is a wide area for future research.

Nematode parasites of plants are discussed in detail in Vol. 2, Chap. 2, so only a summary of their effect on ornamental plants is given here. The nematode is a minute, barely visible worm that lives almost exclusively in the soil. The few exceptions include the chrysanthemum stem and leaf nematode that lives on the aerial parts of the plants. Nematodes are often referred to as eelworms and belong to a distinct form of life, the Helminths. Generally, the parasitic forms on plants range in length from one-fiftieth to one-hundredth of an inch. Although over 2,000 different species of nematodes have been described, only a few hundred are associated significantly with diseases of economic plants. These parasitic worms are equipped

FIG. 6. Fire blight on Bechtel's crab. (Photo: Illinois Natural History Survey.)

FIG. 7. Phloem necrosis, a virus disease, is killing this massive American elm. (Photo: Illinois Natural History Survey.)

with a unique feeding mechanisms called a stylet, a small hollow spear located in the head which penetrates the plant root cells and ingurgitates the cell contents. When one cell is emptied the nematode moves progressively around the root, leaving behind a trail of evacuated and dead cells. Necessary root areas are thus killed, and the effective absorptive capacity of the plant is drastically reduced. Certain nematode species actually inject enzymes into root cells to facilitate dissolution of the cellular contents, and frequently these enzymes are toxic to the plant. Most nematode species dwell in the soil moisture and move freely from rootlet to rootlet. These pests are generally classified as ectoparasites, as compared to endoparasites which become embedded within the plant tissue and are therefore restricted in their movement. The rootknot nematodes are typical of the endoparasites (Fig. 8).

FIG. 8.   Rootknot nematode damage on Japanese holly.

Nematodes damage plants in various ways; the mechanical reduction of the root system produces symptoms of drought or of malnutrition, enzyme toxicity reduces growth and some nematodes vector other disease organisms. Often the nematode attack merely provides a portal of entry for other more virulent disease organisms that inhabit the soil but are incapable of attack on sound tissue.

B. Insects

The insects discussed in this chapter fall into several broad categories, chewing, piercing-sucking, internal feeders and subterranean insects.

The first group comprises those insects that damage ornamental plants by chewing off and eating the leaves, buds, stems, bark or the fruits of the plant. A typical example is the Japanese beetle, which is one of the more serious insect pests on roses (Fig. 9). It may also cause severe damage to grapes, dahlias, zinnias, linden, and elm. Grasshoppers and locusts,

FIG. 9. Japanese beetles ruin a rose bloom.

chewing pests known since antiquity, also can wreak vast devastation on forest and landscape plantings.

Some beetle larvae, called grubs, may live in the soil. Others become borers, such as the azalea borer and the bronze birch borer. The larvae of the click beetle, known as wireworms, live in the soil and can cause extensive damage to bulbs and roots of many annual and perennial ornamentals.

The larvae of moths and butterflies (caterpillars) are also classed as chewing insects. In the spring the trees planted along landscaped highways in parks or home gardens often exhibit the grayish-white weblike tents made by the tent caterpillar (Fig. 10). These tents house thousands of voracious larvae that, later in the season, leave behind them sparse, ragged foliage as mute evidence of their activity. Cutworms, leaf rollers, bagworms (Fig. 11), and the iris and peach borers, all serious pests of ornamental trees, shrubs, and flowers, are also members of the chewing group of insect pests.

FIG. 10. Tent caterpillar infestation on sassafras. (Photo: Illinois Natural History Survey.)

FIG. 11.   Bagworm defoliation of bald cypress.   (Photo: Illinois Natural History Survey.)

The second group, piercing-sucking insects, are also familiar plant pests. Aphids (or plant lice) are probably the most widespread in this group and are perhaps the most troublesome insect pests on ornamentals. They attack and feed on practically every type of living plant, both wild and cultivated. They are particularly damaging to roses, asters, dahlias, gladioli, irises and virtually all seeded annuals. These insects feed by piercing the epidermis, or outer skin, of the plant and imbibing the sap from the soft, juicy inner cells. With their slender, hollow pointed beaks, the original hypodermic needles, they pierce the plant and simply ingurgitate its sap. This group includes such familiar insect pests of ornamentals as the lace bug, which can cause considerable damage to azalea, rhododendron, and other evergreen plants. In the same way, chinch bugs cause extensive damage to home lawns. Some 2,000 species of leafhoppers belong to this group. They attack a wide variety of plants among the ornamentals, the more common of which include the maple leafhopper on Japanese and Norway maple and others which attack hawthorn, poplar, willow, birch, alder, and walnut.

Also in the piercing-sucking group are the scale insects, which are troublesome in many areas of the country and are particularly difficult to control because of the protective waxy or hard shell-like layer they form around themselves during their life cycles. This layer is extremely resistant to insecticides.

Although mites are actually classed as spiders (eight-legged) and are not true insects (six-legged), they also obtain their food through a piercing and sucking action, so for practical purposes, they can be considered within this group. They are alternatively called mites, spider mites or red spiders. They are quite minute and often form a protective webbing on the underside of leaves. Their host-feeding range is wide, and they comprise an extremely troublesome group of pests on ornamentals if not kept under control.

Mealybugs, important plant pests in the home or greenhouse, are also sucking insects and are quite difficult to control because of the white waxy material that forms a covering around their bodies. They frequently attack soft-stemmed foliage plants, such as fuchsia, coleus, cactus, ferns, gardenias, and begonias. They also may cause considerable damage to geraniums, oleanders, orchids, poinsettias, ivy, and chrysanthemums.

The third group of ornamental insect pests comprises those which feed within the plant rather than upon it. This group includes the borers, weevils, leaf miners, bark beetles and gall insects. Borers are usually the larval stage of a beetle or moth. Fruit and shade trees, together with many herbaceous plants, are particularly susceptible to borer attack (Fig. 12).

FIG. 12. Borer injury on Austrian pine. (Photo: Illinois Natural History Survey.)

Also in this group are such pests as the iris borer, the rose cane borer, the dogwood borer and the elm bark beetle, the vector of the dread Dutch elm disease.

Weevils are borers that attack fruit, nuts, and seeds; examples of these are the plum curculio and the chestnut weevil.

Leaf miners are particularly damaging to many prized ornamentals. The tiny larvae feed on the spongy tissue between the epidermal layers of leaves, hollowing out the cells, and causing ugly tunnels and malformations. Notable examples are the leaf miners on boxwood, holly, azalea, arbovitae, and other broadleaved evergreens; in the northeastern states, leaf miners regularly defoliate the picturesque white birch and often kill it. Some forms, such as the serpentine leaf miner, may damage herbaceous annuals and perennials.

Gall insects have the curious capacity to sting a plant, causing its tissue to swell up and form a home for the insect. The damage inflicted by these insects on ornamentals is not so serious to the plant as it is disfiguring and unsightly. Galls may vary in size, shape and color, ranging from the highly colored concentric spots induced by the maple leaf gall midge to the very hard, large and ugly oak twig gall. It generally takes a trained or experienced observer to determine whether the cause of a specific gall is an insect, a fungus or a bacterium. Thus, the azalea flower and leaf gall is actually incited by a fungus, although it closely resembles an insect-initiated gall. Other common galls include the goldenrod gall, which is common on the willow, and the spiny rose gall, which is caused by the sting of a small wasp. Gall wasps are also responsible for the large oak apple gall and the spiny oak gall.

The fourth group of insects important to the landscape gardener are those that spend all or most of their lives underground. The woolly apple aphid, for example, lives its entire life below ground, while the Japanese beetle has only its adult stage above ground. Such subterranean feeders cause direct injury to plants by feeding on roots, bulbs and corms; or they may cause indirect injury by opening wounds in plant tissues through which secondary fungi and bacteria may infect, producing rots and wilts.

Pests of garden and parklands also include those insects that are either annoying or harmful to man. Such insects may be an annoyance merely by their presence where people live, work, or play. The faint whine of a mosquito in a tent or bedroom can be particularly exasperating at night. Flies and cockroaches not only emit a noxious odor but also deposit an offensive taste, if not an infection, on fruits, food or tableware upon which they chance to crawl. A bug crawling on the skin is a most distracting and unpleasant sensation. Such insects as horseflies, mosquitoes, gnats, wasps, and hornets cause extreme pain and discomfort with their poisonous bite or sting. Incidentally, the giant hornet can cause severe injury to boxwood, lilac, and other plants by chewing and girdling their twigs and branches, thus causing them to die.

Finally, many insects are carriers, or vectors, of such dread human diseases as malaria, typhus, river blindness, encephalitis and dengue fever, discussed in detail in another chapter. Insects do not restrict their disease vectoring solely to diseases of humans. There are many diseases of animals, both domestic and wild, which are conveyed by blood-sucking and tissue-feeding insects. Actually, much wildlife, particularly birds, is destroyed through the ravages of insect-born disease. A number of our important and widespread diseases of ornamental plants are carried by insects. Examples are given in Table 1.

Insects are thus man's chief competitors for living space on this earth; and although beneficial insects should be protected, man must nevertheless wage a relentless battle against species that impair his health, cause him discomfort, or rob him of food, fiber, and the beauty of the landscape.

TABLE 1

Plant Diseases Carried by Insects

Host	Disease	Insect vector
Apple, pear, and quince	Fire blight	Bees, flies, aphids, leaf-hoppers
Aster	Yellows	Leafhopper, Macrosteles divisus
Azalea	Petal blight	Bees, ants, and thrips
Chestnut	Chestnut blight	Various beetles
Elm	Dutch elm disease	Elm bark beetle, Scolytus multistriatus, and Hylurgo-pinus rufipes
Norway pine	Blue stain	Beetles, Ips pini, and I. grandicollis
Oak	Wilt	Beetles
Peach	Yellows	Plum leafhopper, Macropsis trimaculata

C. Weeds and Poisonous Plants

Weeds necessarily vary in definition, for one man's weed may be another's crop or garden. In essence, a weed is a plant growing where it is not wanted. The stately maple tree, a graceful shade tree on the lawn, is a weed when its seedlings sprout in a petunia bed. Ailanthus or tree of heaven is frequently a tall, although somewhat pungent, shade tree, but Ailanthus thickets are definitely weeds.

. Many cultivated and desirable garden plants have escaped and become weeds. Typical examples are bachelor's button, morning glory, English daisy, Scotch broom, Japanese honeysuckle, spearmint, ground ivy, star-of-Bthlehem, and sweetbrier. Muenscher (23) lists over 570 weeds of general distribution and many others of only local importance.

Weed plants have several characteristics which differentiate them from useful plants. Many grow under conditions adverse to the growth of domesticated plants; they usually are able to regenerate lost plant parts, and many perennial weeds propagate by vegetative methods without seed production; most weed flowers are small and inconspicuous and may produce seed before their actual presence is suspected; some contain substances that give them disagreeable odor or taste; others are covered with spines or thorns.

The seeds of many weed species have special modifications which aid their dissemination over great distances. A typical example are the fluffy seed heads of mature dandelions which readily break up and float off in a light breeze. Common thistles and milkweed have similar seed heads on which the hairy growth facilitates airborne dissemination. Other weeds, such as beggar-ticks (Bidens bulgata) and sticktights (B. cernua), have seeds bearing sharp teeth or needles that stick to animals and clothing and are carried for long distances.

Included among the weeds are many poisonous plants; however, not all of them are weeds. Thus, several highly desirable ornamental species have plant parts which are either poisonous or irritating to man or animal. Although such decorative plants are not eradicated as weeds, it is important to recognize their toxic properties, e.g., Castor (Ricinus communis).

A poisonous plant is one containing a toxic substance which can cause discomfort, injury, or death to man or livestock. In North America over 2,000 species have been reported as poisonous (16). In New York State alone 15 species have been reported as the cause of dermatitis, and another 25 are suspected (24). A comprehensive tabulation and description of poisonous plants in North America is given by Kingsbury (16).

A few of the more important poisonous plants of import to the ornamental gardener and parkland landscaper include:

Euphorbia: or spurge, can cause dermatitis. Although there are several ornamental plants in this genus, most of them, such as E. cyparissias, cypress spurge; E. helioscopia, sun spurge; and E. peplus, petty spurge; are weeds.

Daphne species: small woody shrubs with rather showy flowers and
bright-colored berries. Many of the genus are grown as ornamentals; but unfortunately, the berries contain a glycoside and
cases of poisoning among children have been reported from eating
the berries.

Datura stramonium: jimsonweed, also known as foreign apple and
devil's apple, is a tall, branched plant with trumpet-shaped flowers yielding fruits that ripen in early fall. Severe cases of jimsonweed poisoning have been reported in the United States.

Pastinaca: wild forms of garden parsnip (P. sativa) produce symptoms
similar to those caused by poison ivy. Such wild varieties flourish
along banks of streams and near swimming beaches.

Rhus radicans (Toxicodendron radicans): poison ivy is the best known
and most widely distributed of poisonous plants. There are many
forms and varieties of this species. Although individuals may exhibit a wide variance in susceptibility to poisoning from this plant,
even the person most highly resistant to poisoning by contact is
likely to become severely affected when exposed to the oil-laden
smoke from burning poison ivy. Poison sumac also belongs to the
same genus as poison ivy; and although native from southern
Quebec to central Florida, it is found predominantly east of the
Mississippi River. It thrives in wet, swampy areas and can cause
extreme discomfort on contact.

Others of importance include the bull nettle (Jatropha stimulosa) common in the Southeast, dead nettle (Lamium amplexicaule) from eastern
North America to the Pacific Coast, horse nettle (Solanum carolinense)
common in the southern United States but less frequently in the northern
areas, stinging nettle (Urtica dioica) generally distributed, and wood nettle
(Laportea canadensis) common in deep woods in the northern and southwestern United States. Another highly poisonous plant of local occurrence
is the Manchineel tree (Hippomane mancinella), a native of Central America
and the West Indies but restricted in the United States to southern Florida.
The sap is extremely caustic, causing a severe skin irritation and even
temporary blindness in man and animals. Eating the fruit may produce
vomiting, abdominal pain and, on occasion, death. The American Indians
employed the juice from this tree to poison their arrows at the time of the
Spanish explorations.

All poisonous plants are not considered weeds, and a number of such
plants are listed in the March, 1965 issue of the Virginia Health Bulletin
(32). For example, leaves of the Christmas poinsetta contain a milky juice
that can cause blisters, loss of hair, vomiting, and diarrhea; mistletoe
berries, often used as home decorations, can be fatal when eaten. Other
examples include several common home and garden flowering plants, such
as wisteria, azalea and yew, the leaves, berries or seeds of which are
poisonous, as is also true for wild cherry, oak, elderberry, and black locust.

VI. DEVELOPMENT OF PEST CONTROL

Pest control began when man first inhabited the earth. Although no accurate records are available, paleontological and archeological records show beyond doubt that man has contested pests since the dawn of his existence; and although modern pest control practice may be more sophisticated, the battle still continues and undoubtedly will as long as man remains on earth.

Early pest control practice was largely enshrouded in superstition and mysticism, as illustrated by the early Roman annual festival to appease the rust god who, from the writings of Pliny, remained unappeased, as the ravages of black stem rust of wheat played a significant role in the decline of the empire. A few recorded instances of shrewd observations leading to pest control indicate some control was achieved without fully understanding why. As early as 1,000 B.C. Homer wrote of the "pest-averting sulfur"; it is reasonably certain, however, that he had not the slightest idea why sulfur should be effective against certain diseases and insects, especially since the causative agents of disease were not recognized for over 2,800 years thereafter.

Virtually all the early work on pest control was directed towards controlling pests which attacked food and fiber crops. Only recently have pests of ornamental shrubs, lawns, flowers and trees received special attention. The struggle for food, clothing and shelter in earlier times, understandably, left little time or energy for the common man to enjoy the beauties of a cultivated nature. It is largely due to the recent rise in general affluence and leisure time that people have placed emphasis on the beautification of home grounds, parklands, and outdoor living areas. Non-farming suburban dwellers have become critically aware that many pests jeopardize their plants and gardens.

Although orchard fruit trees are not usually considered ornamentals, they are grown on many home grounds. The early developments in disease control of domesticated plants are discussed by Lodeman (20). Interestingly, as long ago as 1629, Parkinson recommended the application of urine to control cankers on orchard trees. Although peaches and grapes have long been grown in home gardens, it was not until 1824 that Lodeman (20) suggested that flowers-of-sulfur was a specific control of powdery mildew on peach, while, in 1833, Lodeman (20) proposed that boiled lime sulfur would stop grape mildew. In both cases, although not specifically stated, presumably some insect control was also achieved.

The first reference to pest control on an ornamental plant was made in 1861 when Lodeman (20) recommended copper sulfate for the control of rose mildew. Later, in 1870, potash solutions were suggested to control scale insects on shade trees, and in 1872 Paris green was recommended to control the destructive cankerworm.

Early weed control, as indicated in biblical reference (Matthew 13:24), was by tedious hand labor; and although mechanical cultivation eased the human toil, weed control continued to be slow and costly. The chemical control of broadleaved weeds in cereal plantings with solutions of copper salts was reported as early as 1897, but a significant breakthrough occurred in 1944 when Marth and Mitchell (21) reported 2,4-D to be a selective herbicide.

Although a few references to the control of pests on ornamental plants appeared during the late 1800s, it was well into the 20th century before knowledge in this specific area accumulated rapidly. Organized pest control on ornamentals, lawns, flowers and shade trees essentially commenced with the discovery of the dithiocarbamate fungicides in 1934, with DDT as a general-purpose insecticide in 1947, and with 2,4-D as a selective herbicide in 1944.

VII. WHY CONTROL PESTS?

As stated earlier, pest problems are not new, and pests existed before man inhabited the earth. Yet, some people question the sudden present emphasis on pest control. Clearly, man's concern with pest control is not new; for the problem is as old as the pests. Early attempts at control were primitive, because knowledge of the habits and etiology of pests was either lacking or, at best, incomplete. Only recently have modern technology and pest control chemicals been available to combat those creatures that despoil the beauties of nature.

Many questions are often raised as to the necessity for controlling these pests. Generally, such attitudes arise from either lack of knowledge or vested interests. There was also similar opposition to the automobile, the electric refrigerator, central heating, paid vacations and universal suffrage.

Consideration of the overall problem of pest control leads logically to three broad categories: the economic, the public health, and the aesthetic.

A. Economic

It is expensive to develop the landscaping and beautification of home gardens, parks, industrial sites, or highways. If the plant stocks in these areas are not protected from pests, serious direct or indirect financial loss follows. Thus, holly leaf miner can ruin expensive plantings; rose black spot and mildew, together with the Japanese beetle, can reduce a rose garden to desolation; golf course maintenance will substantially increase if the greens are ruined by disease, insects, or nematodes; public tax money is lost if blights, bugs, and weeds destroy highway landscape plantings.

There are also the expense and problems besetting those who produce the plants. Nurserymen and greenhouse operators must continually contend with rising labor costs, inadequate skilled labor and, generally, narrow profit margins. These growers cannot tolerate severe losses in their crops; neither can they afford obsolete, cumbersome, and laborious procedures which fail to provide efficient production. Weed control provides an excellent example. It may cost from $50 to $600 per acre to weed ornamental plants. Although costs of herbicides may vary, as may the amount required for a particular field, there is ample evidence to show that well-planned, carefully conducted chemical weed control programs reduce weeding costs substantially. Such methods not only save the grower money, but produce better, higher-quality plants as a direct result of reduced weed competition. Similarly, intelligent use of chemical herbicides increases the efficiency of weed control in lawns and turf, and the reduction of undesirable brush along rights-of-way, including highways, railroads, and utility highlines.

The tourist business is increasing in the United States at a rapid rate. Hotel, motel and resort operators must control insect pests to ensure that their guests enjoy their vacations and will return. Such resort owners have to control insect pests: (a) in the buildings (e.g., cockroaches, bedbugs, fleas and "pantry" insects); (b) around the buildings (e.g., crickets, wasps and boxelder bugs), and (c) in outdoor areas (e.g., flies and mosquitoes).

Surveys in large metropolitan areas have clearly shown that landscaping slum areas along expressways has attracted new residents, increased apartment rentals and encouraged tenants to stay longer. Landscaping made the areas more beautiful and desirable, and property values increased from 100% to 500%. If such landscaping is unprotected from pests, however, these areas soon revert to slums, people leave and property values decline.

Peach trees are often raised in home gardens both for their abundant spring blossom and their fruit. The peach borer is an ubiquitous pest on the homegrown, as well as on the orchard-cultivated peach tree. Formerly, control was attempted by removal of the worms from the crown of the tree by hand, a very time-consuming, ineffective, and frequently injurious procedure. Now, young peach trees are protected from borer attack for two years by a simple insecticidal root dip applied before planting; and protection is extended to succeeding seasons by either topical sprays or soil-applied

systemics. The proper use of these pest control agents has virtually eliminated this severe problem on peach trees, resulting in both a saving of labor
and an increased efficiency of crop production.

Homegrown fruit trees are often prone to rodent damage, as are orchard trees, mice probably being the most destructive species. Orchard
growers list the pine, or field mouse, as the chief problem. Such rodent
damage is not necessarily restricted to fruit trees however, as indicated
in Fig. 13, where severe injury occurred on the ornamental Russian olive
(Elaeagnus), here three plants in this hedge were killed and two others so
severely damaged that they had to be replaced. Rodent control chemicals
now available on the market, when properly applied, will effectively reduce,
if not actually prevent, such mouse damage.

FIG. 13. Rodent damage on Russian olive.

B. Personal Health

Outdoor living, for recreation or health, for the young or elderly,
for the well or ailing, has become an integral part of the way of life among
the more affluent nations of the western world. Forest preserves, wild

lands, parks and beaches have been provided over vast areas to meet these open-air demands of the people. Boating, swimming, fishing, hunting, golfing, hiking, riding, camping, or simply sunbathing all require lakes, rivers, beaches, forest trails, grass swards, parklands, campsites and highways unspoiled and untrammeled by irritating insects, destructive diseases and choking or poisonous weeds. Apart from the annoyance and economic impediment of such pests in such open living areas, there are the added hazards they may present to public health. Such pests as poison ivy, oak and sumac are effective deterrents to the use of many open areas; blackflies, sand flies, chiggers, eye- and other gnats, ticks, fire ants and, of course, the Culex, Aedes and Anopheles mosquitoes, constitute not only irritating pests but, under certain conditions, may also vector serious human disease. The discovery of DDT and the other modern insecticides that quickly followed have provided whole areas of forests, parklands and other open living spaces free of these unwholesome pests and have done so without injury or significant damage to wildlife that also comprises part of the open living space. Public health literature lists numerous examples of intelligent use of modern insecticides controlling insect-carried disease that has contributed substantially to human health and to that of wildlife as well.

Apart from the public health aspects of insect-vectored crippling or fatal diseases, many insect bites and stings constitute more than mere temporary annoyance. There are many people who are highly sensitive (allergic) to mosquito, bee, wasp or ant stings and bites. These people often develop severely swollen areas that may remain inflamed and irritated for extended periods. Every year numerous fatalities are reported resulting from bee, wasp and hornet stings or ant bites. The common wood tick, which may carry the Rocky Mountain spotted fever, can also produce a painful lesion which can induce a severe paralysis in sensitive individuals.

Poisonous plants constitute a serious health hazard in outdoor living areas. In addition to plants that cause mild to severe dermatitis on contact with the individual (Fig. 14), there are a number of ornamental plants that bear brightly colored fruits that are often attractive to children. These fruits, when eaten, can cause severe internal discomfort, serious illness and even death. Although mechanical removal of such poisonous plants from home gardens may be feasible, in large outdoor areas it is completely impracticable. Only through the proper use of herbidical chemicals can such areas be made safe for occupation and enjoyment.

C. Aesthetic

No price tag or numerical evaluation can be placed on the contribution made by an aesthetic environment to man's welfare and progress. There are those who contend it is the ultimate goal in life, and all else should lead to

FIG. 14. Poison ivy climbing a backyard shade tree.

elevation of the spirit or be abandoned as worthless. Unquestionably, all civilizations from antiquity until today have recognized that man progresses more rapidly and harmoniously in an artistic and pleasant environment. As Henry Thoreau once said, "It is the marriage of the soul with nature that makes the intellect fruitful and gives birth to imagination."

Such aesthetic pleasure and stimulation can be seriously impaired by clouds of filthy flies and voracious mosquitos. The song of the night bird or the serenity of moonlight on a lake perceptibly fades in competition with blackflies or chiggers.

The increased leisure time among American working men and women becomes a social problem when no constructive outlet for its use is available. Organized outdoor recreation appears to be the major channel of release for such social pressures, and the aesthetic appeal of tastefully landscaped, clean, uninfested parklands and beaches has become essential.

The homeowner, anxious to reduce the time for routine and laborious maintenance tasks around his yard and garden, eagerly accepts simple and effective measures for fighting the pests which despoil his decorative and ornamental plantings. Thus, the intelligent use of pesticides can contribute materially to the homeowner's interest in his aesthetic surroundings.

This concept could perhaps be extended to the crowded areas of large cities. For if urban slums were improved with small parks and ornamental plantings designed to create an aesthetic environment and were protected from diseases, insects and weeds, such presently desolate areas would assume new beauty and would undoubtedly provide increased stimulation to the people in these congested areas to improve themselves, their homes, and their attitudes toward society. A program of this type could materially aid in reducing the uncertain horrors of "the long hot summers" that plague our cities today.

As greater numbers of people travel the nation's highways, emphasis on comfort and safety is increased. Today rights-of-way bounding highways are approximately equal to the area of six New England states. National, state and local road agencies have made very effort to landscape these highway borders effectively, not only to make travel more pleasant and to reduce erosion, but also to improve driving safety. Thus, state highway departments have established highway plantings to screen headlight glare from oncoming cars and to serve as traffic guides and noise abatement buffers, which are generally recognized to be exceedingly effective. It has been found that, in addition to reducing or eliminating highway glare, artistic highway plantings relieve the monotony of long-distance driving and thus reduce driver fatigue (Fig. 15). Northern states report that highway plantings not only contributed

FIG. 15. A modern highway landscaped for beauty and safety. (Photo: American Association of Nurserymen.)

aesthetic and safety benefits, but also reduced the need for snow fences. Great emphasis is placed on protecting these plantings from vandalism, but what is not generally realized is that diseases, insects and weeds cause far greater destruction than do careless and thoughtless travelers.

Finally, the basic reason for controlling pests can be seen in any flower garden. Colorful, artistic flower gardens can means many things to different people; a place where friends and neighbors congregate and admire; a quiet refuge for contemplation and creation; or a recollection of childhood. Such a garden can, however, become a source of disgust and frustration if weeds, diseases, and insects ravage it beyond control.

As Kipling put it:

Our England is a garden
But such gardens are not made
By exclaiming 'Oh! how beautiful'
And sitting in the shade.

VIII. METHODS OF PEST CONTROL

There are a number of ways to avoid, prevent, or reduce the ravages of pests in ornamental and landscape plantings. These can be divided into management procedures to escape and preclude infestation, and those which directly suppress pest invasion.

A. Sanitation

Sanitation is the first and most obvious method of pest control. Essentially, this is simply "good housekeeping" applied to outdoor areas. Covering garbage and food refuse to prevent flies, removing grass and weeds in the patio or driveway, eradicating poison ivy from flower beds and hedges are simply sanitary procedures, but many equally simple aspects in disease control are not so obvious.

Most organisms that cause diseases on ornamental plants overwinter on dead tissue, either on plants or in refuse on the ground. These disease-producing organisms follow a precise life cycle that involves an overwintering stage or, in areas where winters are mild, a dormant stage that enables the agent to pass through drought periods and other similar adverse situations. For example, the hollyhock rust fungus overwinters on old leaves and stems which have died in the fall, as does the fungus of the leafspot disease of iris and the rose blackspot agent that spends its inactive period on infected leaves and twigs. The elm bark beetle, carrier of the dread Dutch

elm disease fungus, overwinters beneath the bark on dead elm wood. Such
wood should, therefore, either be destroyed or the bark removed to eliminate shelter for the beetle.

Many sanitation procedures are effected, incidentally, merely to improve the appearance of a garden or outdoor area, without specific attention
to pest control. Thus, misshapen, blackened peony blooms are usually
removed and destroyed simply to brighten the appearance of the plant, and
most gardeners do not realize the blooms are often killed by a fungus which,
if left, will infect remaining flowers. Dead twigs pruned from evergreen
shrubs not only improve their shape but, more importantly, prevent the
spread of the disease that caused the branches to die. Such sanitary procedures are futile, however, when the debris is left to decay on the compost
pile, continuing as a source of contagion into the following growing season.
All such dead and diseased plant material should either be burned or treated
chemically to reduce the possible infection source.

Although sanitation is, in many ways, an excellent pest control method, its practical utility is largely confined to the home yard or the small
park and recreation area; though it may be effective, it requires considerable hand labor and is just impractical in extensive parks and woodlands.

B. Rotation

Crop rotation, a long-established method of pest control, is a useful
procedure for the farmer and the commercial nurserymen, but it has little
application in garden and parkland plantings. The principle of rotation is
based on the fact that wherever a number of plants of the same or related
species are grown in the same area year after year, the parasites of the
host species also increase because of the abundance of food and favorable
environment. Thus, susceptible tobacco varieties grown repeatedly in the
same field year after year generally become severely infested with the
black shank disease; since the black shank organism cannot infect pasture
grasses, a 4-year rotation of tobacco with pasture causes the organism to
die out and allows the grower to produce black shank-free tobacco every
fourth year. Such a procedure, however, is clearly not a particularly
practical pest control method for the home gardener or parkland manager.

C. Resistant Plant Stocks

Resistant varieties are exceedingly useful in pest control practice.
Resistance and immunity must be clearly understood before this method
can be effectively utilized in ornamental and landscape pest control.

Resistance is a general term referring to the inherited qualities in a plant variety that limit the damage effected by insects, disease organisms, or even adverse environmental factors, such as frost or drought. Clearly, the term resistance implies that less damage will occur to these varieties than to other susceptible varieties. Resistance may, therefore, be one of degree only. On the other hand, an immune variety is one which is not damaged at all by a specific pest under any circumstances. Actually, so-called immune varieties are generally just highly resistant rather than totally impervious to attack.

A number of economic crop plants carrying various degrees of inherited resistance to insects have been developed (25). Unfortunately very few ornamental plants have been bred with resistance to their major insect pests. An exception is a variety of Euonymus highly resistant to Euonymus scale, a serious problem on this popular shrub. Undoubtedly, as interest in protecting landscape plantings increases, more new ornamental varieties will be developed to carry insect resistance.

Disease resistance has been bred in varying degree into a number of flowering plants and shrubs. Thus, the garden seed catalogues list many annual flowering varieties which are resistant to mildew, rust, wilt and other diseases. It is obviously desirable to plant these resistant varieties whenever their form and color are compatible with the landscape design.

Wilt is a serious disease on the mimosa tree, for which, at present, there is no known cure. Fortunately, varieties of wilt-resistant Mimosa are available which are as decorative and as beautiful as the susceptible varieties but are also resistant to this devastating disease.

A great deal of research is being conducted to develop disease-resistant varieties in many species of ornamental plants. Some types of elms resistant to the Dutch elm disease have been found, but considerable breeding work remains to incorporate the desirable characters of the American elm into these resistant strains. The search for an American chestnut tree that is resistant to chestnut blight continues, as does the breeding program for resistance.

Although inherited resistance is an excellent approach to control disease, it is beset with difficulties and is exceedingly tedious. First, very few ornamental plants possess inherent resistance, at least to the more serious disease and insect pests. Second, although certain plants may possess resistance to a specific disease or insect, they are generally not immune and may succumb to infestation under conditions of severe stress. Third, resistance arises from genetic characteristics, but both insect pests and disease organisms may also modify their genetic makeup and adapt themselves to attack such resistant host varieties. A classic example is the resistance program for the control of black stem rust of wheat.

Some years ago a rust-resistant variety of hard wheat was developed and planted extensively in the western United States and Canada. A few years after its introduction, however, a race (15-B) of the wheat rust organism emerged which broke through the resistance of the new wheat variety, and the breeding program had to be renewed. This propensity for pest agents to breed out adaptive features that obviate the genetic resistance instilled by man into his domestic plant stocks requires a continuing program of breeding for pest resistance.

D. Cultural Methods

There are cultural methods of pest control which, though distinct, do at times overlap the sanitation procedures discussed above. Direct controls, such as cutting and burning infested trees and shrubs, though effective under special conditions, are quite expensive and essentially impractical for large areas. For example, it is obviously impractical to destroy an entire pine plantation because a tip weevil infestation has become established.

The familiar manual and mechanical cultural practices for weed control are cumbersome, laborious, and must be continuous to be effective.

As various plant species differ widely in their requirements for food, water, and sunlight, undesirable vegetation can often be controlled by merely taking advantage of such differences. For example, crabgrass has a high light requirement and may be kept reasonably well under control in home lawns if good turf management is employed and if the grass is not mowed below 2 inches.

Although many cultural practices will control weed and plant disease pests, only a few are effective against insects. United States Department of Agriculture scientists found that mulching squash, cucumbers, and cantaloupes with aluminum foil repels aphids. Another excellent example is the long-practiced draining of marshes and swamps to eliminate breeding places for mosquitoes; however, even though the pest population may be substantially reduced locally, it is by no means eliminated and a sufficient residue usually remains to cause injury or discomfort.

E. Eradication

Eradication of alternate host plants in disease control may also be considered as a cultural practice. The procedure is the same, although the basis for the method differs.

Several rust diseases of ornamental plants require an alternate host plant to complete their life cycles. An example is the cedar-apple rust,

caused by the fungus <u>Gymnosporangium juniperi-virginianae</u> that requires
two different hosts to complete its life cycle. The needles of red cedar or
juniper (<u>Juniperus virginiana</u>) become infected in late summer. Within 18
months the infected areas enlarge to form roundish, irregular shaped,
grayish-tan or brown galls an inch or more in diameter; these galls resemble a wad of bubble gum with a pock-marked or dimpled surface like a golf
ball. Long, orange, gelatinous spore horns grow out from these depressions during wet weather in the second spring following initial infection. As
the horns dry, spores are released and are blown to young apple and crab-
apple leaves, fruit, and shoots. These spores cannot, however, reinfect
the cedar. By late spring or early summer yellow-orange spots develop on
the upper surface of the infected apple leaves, and within a few weeks a
cluster-cup structure forms on the lower surface of the leaf in which a
second type of spore is formed. These spores cannot infect apple varieties
but must now return to cedar trees to complete the life cycle. Obviously,
elimination of either host will break the life cycle of the fungus and thus
will control the disease.

In 1914 Virginia passed a law requiring the removal of red cedars
within 1 mile of commercial apple orchards. On the other hand, ornamental
nurseries producing red cedars take care to prevent any apple trees growing in or near the nursery. Clearly, it is unwise to include both hosts in a
single landscape plan.

White pines (<u>Pinus strobus</u> and <u>P. monticola</u>) are popular trees in
many gardens and public parklands. These stately trees are highly susceptible to the blister rust fungus (<u>Cronartium ribicola</u>) to which the red and
black currant and the gooseberry (<u>Ribes</u> sp.) are alternate hosts. Ribes
eradication is by far the most practical method of blister rust control in
white pine forest areas, and such eradication is compulsory in the northeastern states where white pines are native.

Examples of other rust diseases and their alternate hosts are as follows:

Host	Disease	Alternate host
Balsam fir	Yellow witches' broom (<u>Melamsporella caryo-phyllacearum</u>)	Chickweed (<u>Stellaria</u> spp.) Mouse-ear chickweed (<u>Cerastium</u> spp.)
Virginia pine Shortleaf pine	Eastern gall rust (<u>Cronartum cerebrum</u>)	Several oak species (<u>Quercus</u> spp.)
Loblolly and other hard pines	Southern fusiform rust (<u>Cronartium fusiforme</u>)	Oak species, especially black oak (<u>Quercus</u> spp.)
Jack, lodgepole, and ponderosa pine	Sweetfern blister rust (<u>Cronartium comptoniae</u>)	Sweetfern and sweet gale (<u>Myrica gale</u>)

Host	Disease	Alternate host
Iris	Rust (Puccinia sessilis)	Canary grass (Phalaris spp.)
Rhododendron	Leaf rust (Pucciniastrum vaccinii)	Hemlock (Tsuga canadensis)

Several insects also follow the alternate host pattern, but in most cases the relationship does not actually limit their development. The spruce gall aphid, which causes swellings on the tips of blue spruce twigs, lays its eggs on the needles of the Douglas fir. Although an alternate host is not essential to this insect, the aphid population becomes markedly reduced if Douglas fir and blue spruce are not combined in the same planting design.

Again, some plants encourage insect infestations which may then spread to more desirable species. The common primrose, for example, is a favorite host for the red spider mite that readily migrates to the rose on which a heavy mite population seriously complicates control.

F. Biological

Another concept of pest control involves introducing predatory agents that automatically exercise control. Such predators must necessarily be highly specific and attack only the target organism without injuring other desirable living things. This is extremely difficult to fulfill in light of the thousands of different fungi, bacteria, viruses, nematodes, insects, and weeds that attack or interfere with the growth of our lawns, flowers, shrubs, and trees. Despite these difficulties, intensive research has achieved limited success in biological pest control. This has occurred chiefly in insect control.

1. Parasites and Predators

Biological control of insect pests is generally regarded as the introduction of parasites or predators that attack a particular insect pest. Although perhaps the most common, this approach to biological control is often misunderstood. It is widely believed that a balance prevails between destructive insects and their natural predators, and to a degree this is true; but to be effective, a sufficient population of predator must also be present. The pests are not eliminated by the predator, because both must survive in nature even though their relative numbers may fluctuate widely over a series of generations. Thus, before a predator population can build up to a dominant level, there must be an abundance of the pest insect present upon which the predator can feed and thrive. It may take many months or even years for the predator population to increase to a point where it can effectively reduce the pest numbers to a tolerable level, and during this lag period the domestic plants are ravaged and often destroyed by the pest insect. This contest between the hunter and its prey is endless and is perhaps the paramount factor in preventing pest insects from overwhelming the world.

Biological control may be accomplished by either parasites or predators. Parasites are living organisms that live in, on, and/or at the expense of other living organisms. They may be actually parasitic at only one stage of their life cycles or throughout their entire lives. Generally, parasitic insects attack larger and stronger insects and usually do not kill them immediately but may continue to live on or in their hosts for extended periods of time. Predators, on the other hand, are those insects which catch and devour smaller or more helpless insects for food. Perhaps the best known and most widely distributed insect predators are praying mantises, dragonflies, aphis lions, brown beetles, lady beetles, and flower flies.

Despite the limitations of this approach to insect control, there are certain examples of predator introduction that have proven to be quite successful. Most notable has been the introduction of the Australian lady beetle (Rodolia cardinalis Muls.) into California to control the cottony-cushion scale (Icerya purchasi Mask.) that, in the latter part of the last century, had become a very serious pest on orange trees. This species of lady beetle, a natural enemy of the cottony-cushion scale insect, had practically eliminated this serious pest from the orange groves in California within a few years following its introduction. Since that time, this lady beetle species has been released in many different countries and has never failed to bring the cottony-cushion scale rapidly under control. A few other similar predator programs have yielded effective controls but, in general, success has been neither reliable nor spectacular.

Parasitic control has been reasonably successful in certain cases, such as control of the woolly apple aphid by a tiny internal wasp parasite or the release of two small chalcidoid parasites from Australia that has effected virtually complete control of the citrus mealybug in California. Other fairly successful programs of biological control include those directed against the gypsy moth, the brown-tail moth, the European corn borer, the Japanese beetle and the oriental fruit moth.

Not all predators and parasites are beneficial, however, and so they are classed as primary, secondary, tertiary, etc. Thus, in the case of an insect that is seriously injurious to rose plants, any parasite or predator attacking it would be classed as primary and regarded as helpful to man. If such a parasite or predator is, in turn, attacked by another species of parasite or predator, then the latter becomes a secondary predator in controlling the insect pest on the rose and would be considered detrimental. Moving back another step, a tertiary parasite predatory on the secondary parasite to the primary insect pest on the rose is essentially a beneficial and helpful insect to man. There thus arises a long chain of sequential predators that are, by interpretation from the point of view of man, alternately beneficial and deleterious.

The introduction of an insect predator or parasite into a new area, where it is not native, is clearly fraught with danger and must be approached with extreme caution. For should the target species be severely reduced or totally eliminated, the introduced predator or parasite could run wild and become a pest in itself, perhaps attacking useful primary parasites or other useful species established in the area. Even when such difficulties as the discovery, collection, rearing and dissemination of the predator have been resolved, there remains the obvious limitation that no natural enemy is likely to eliminate completely those insects which provide its food. Therefore, although the pest population may be substantially reduced, the portion that remains may still be greater than man would wish to tolerate on his ornamental plantings.

Environmental factors also have a profound effect not only on the proliferation of insect pests but also upon the multiplication of its predators. Burnett (7) has shown that the predator on the greenhouse whitefly develops very poorly at temperatures of 64° F and is quite ineffective as a control measure; when, however, the temperature is raised to from 75° to 80° F, the parasite increases so rapidly that it virtually eliminates the whitefly population. Similarly, the parasite Metaphycus helvolus is quite effective at controlling the insect scale Saisstia oleae in warm areas, while in more temperate regions the predatory population may be reduced severely by frost and low temperature.

For centuries natural insect pest outbreaks have occurred over wide areas where neither man nor his chemicals could interrupt the epizootic. Before the advent of modern chemical pesticides such serious insect excursions occurred in forest lands, and many years were required before the natural predators built up to the point where the pest and its damage were reduced to normal. In early times man found it most difficult to live by the Balance of Nature exemplified in the seesaw of pest and predator; but today, under the existing conditions of such unnatural imbalances as the growth of cities and highways and intensive monocultural agricultural practice, man would find it virtually impossible to rely upon natural predators to regulate pest populations.

Biological control of certain types of undesirable vegetation has been successful in a few cases. Two rather outstanding examples are the control of the thorny shrub, Lantana camara, in the Hawaiian Islands and of the prickly pear, Opuntia spp., in Australia. Lantana was introduced into the

Islands about 1860 and promptly enjoyed a wide acceptance as an ornamental plant. It soon escaped from home gardens, however, and quickly spread over thousands of acres of fine farm and pasture land. Mechanical control was expensive and generally not successful, so other methods of control were tried. A number of insect species were known to feed on Lantana in its native Mexico, and the most voracious species were imported to the Islands with the intent that they would become established and attack and reduce the Lantana. Three species of predators were soon recognized as being particularly active on Lantana; the larvae of the tortricid moth, Crocidosema lantana, the larvae of the seed fly, Agromyza lantana, and the larvae of butterflies, Thecla echion and T. bazochi; so effective were these parasites in destroying Lantana that it has been eradicated from most of the field and crop lands in the Islands.

Prickly pear, Opuntia spp., is an interesting ornamental and hedge plant, but if it escapes it can render fields and pastures completely useless for agriculture. This plant was brought into Australia in the early 1800s and, like Lantana in Hawaii, escaped and spread over fields and pastures. Sixty million acres were covered with prickly pear by 1925 (28). The moth borer, Cactoblastis cactorum, was introduced from Argentina in 1925 and by 1935 had cleared prickly pear from 95% of the infested land in Queensland and from 75% in New South Wales. This drastic reduction in its food supply precipitated starvation for the predator, severely reducing its population, whereupon a resulting surge of regrowth of prickly pear occurred. As its food became replenished the insects multiplied and the control cycle recurred. Over succeeding cycles, however, the rate of regrowth of the prickly pear progressively diminished, and other weed species, not hosts to Cactoblastis, replaced it. Robbins et al. (28) and DeBach (3) discuss examples of biological weed control in detail.

Effective as this method has been in certain specific cases, it does have severe limitations. Success in the use of insects in a weed control program depends upon several rather critical factors. First, the weed to be controlled must not be native to the region and must not be a close relative to any economic plants cultivated in the area, or the predator may shift its attention to a desirable plant species once the target plant has been extirpated. Furthermore, the weed species, in its native habitat, must have specific predators which can be introduced free of their own parasites and must be able to become established in the control area. The weed-feeder must be a specific predator to the pest weed and unable to attack desirable plants in the area. This method of weed control is, therefore, generally limited to extensive areas where a foreign weed species has invaded and become the predominant form of wild vegetation. On this basis it has little or no value in home gardens, golf courses, parklands and similar outdoor recreation areas.

2. Diseases

Another type of biological control is the disease-producing microorganisms which infect insect pests. Although also parasites, these distinctly differ from insect parasites. Insect pathology is a relatively new scientific discipline, although it has been known since the turn of the century that insects, like all living creatures, are subject to infectious disease similar to that which infects higher animals. A good example of disease in insect control is the milky disease caused by Bacillus popillae for the control of the Japanese beetle; however, the method is rather slow in taking effect and must be applied over a large area to be really effective.

Some progress has also been made in the control of forest insect pests by biological means. Bucher (6) reports that the two bacteria Clostridium brevifaciens and C. malacosomae give appreciable control of the eastern forest tent caterpillar; however, the method is somewhat impractical. Furthermore, in view of the current public concern over human safety in the use of certain pesticides, it should benoted that these organisms which infect the forest tent caterpillar are also close relatives of Clostridium botulinum, a serious food spoilage agent that can cause illness and fatality in man.

Bacillus thuringiensis, although reported first by Berliner (2) in 1915, has recently received considerable attention (9, 10, 32) as a possible agent for economic insect control. It is quite effective against a variety of lepidopterous insect pests, including the cabbage looper, the alfalfa caterpillar, the diamond-backed moth and the tobacco hornworm. Commercial preparations of B. thuringiensis now have USDA and EPA registration and are available. It is suspected that certain useful predator insect species may also be adversely affected in addition to the pest itself.

An important advance in biological pest control was achieved in recent years by USDA Agricultural Research Service scientists in mass-production of certain viruses which attack several major insect pests of food and fiber crops. These particular viral strains do not infect insect pests of ornamental plants, but there is a distinct possibility that such strains will be found or developed that will induce disease and death in ornamental pest insects.

Production of such an insect-killing virus is initiated by finding and collecting naturally infected insects and transferring the agent to laboratory-reared insects. The instilled virus then multiplies in the insect cells and prevents these cells from functioning normally. As the virus particles cluster together within the living insect cell, they soon become covered with a protein layer and form a polyhedron. These cells containing billions of polyhedra are fed to insect larvae, where they multiply rapidly and eventually kill the larvae. The larvae are then processed to extract the virus, which is then applied against susceptible insects in the field.

Field efficacy tests indicate that these polyhedral viruses only infect members of the Heliothis species of insects. These insects are pests on several important food crops. This viral research appears to be an encouraging step forward in biological control; but before final approval can be obtained for their commercial field use, extensive toxicity tests must be completed to establish that the virus will have no adverse effect on man.

Although it is generally considered that these microbial and viral agents that can control insect pests have no effect on man, animals or plant life, a critical evaluation of their long-term chronic effects is mandatory before they can be released safely into the biosphere on an extensive basis.

3. Chemosterilants

A quite new approach to biological control is by means of chemosterilants. The sustained release of large numbers of sterilized males of specific insect species has yielded remarkable control of certain insect pests of domestic animals and crop plants. Such a program is both difficult and costly and must be applied over a large area to obtain satisfactory results. Its success depends upon: (a) Efficient, productive and specific mass rearing techniques; (b) Sterility induction in the male without destroying its sexual competitiveness (libido); (c) Pest population reduction to a low initial level; (d) No damage from the sterile insects themselves to crops, livestock or man. Although this method of insect control has promise for extensive areas such as forests, national parks and whole states, it will probably never be particularly practicable in home grounds, small parks, cemeteries, golf courses and rights-of-way.

Sterilization can be effected in two ways; physically, by means of radiation, x-rays, or gamma rays (8, 9, 19) or chemically, by feeding compounds of known chemosterilant activity, such as the folic acid antagonist aminopterin, fluorouracil derivatives or the aziridinyl compounds tepa and metepa (18) or the sulfuraziridines (4). Exposure can be carried out at the rearing station or in male traps suitably baited with irresistible female sex hormones which can be strategically placed in the flyways. The lured males will come into the traps, where they are exposed to radiation or a diet laced with the chemosterilant chemical. Upon release they promptly mate with fertile females who then lay only sterile eggs. In species where the female is monogamous and mates but once, it is obvious that the population will quickly decline when the sterile male count exceeds that of the wild unscathed males. Knipling (17) has computed that the rate of decline of the entire population increases exponentially as the ratio of sterile to nonsterile males exceeds 1.00. This is inverse to the customary decline rate of an insect population under the impact of chemical insecticidal attrition. Here the decline rate decreases exponentially after the population has been reduced 50%. Therefore, a combination of the two methods, chemical insecticides initially and male sterility ultimately, should rapidly reduce a large invasive population.

A vast amount of long, painstaking, and expensive research remains to be done on the possible effect of these chemicals on beneficial insect.

plants, mammalian wildlife, and on man before their widespread use can be contemplated.

Thus, nonchemical methods of pest control are both old and new; although in certain circumstances such methods are sufficient in themselves, more often than not their most effective use is in combination with the chemical pesticides.

G. Pesticides

Pesticides, by definition, are chemicals which kill pests. The word is derived from the Latin caedo, to kill; thus, a fungicide kills fungi, an insecticide insects, a herbicide plants, a molluscocide mollusks (snails, slugs) and a rodenticide rodents.

There is a question as to whether a particular chemical actually kills the pest or merely renders it harmless or static. Specifically, for example, does a fungicide actually kill the fungus Diplocarpon rosae (blackspot of rose), or does it elicit subtle changes in the fungal metabolism so that it can no longer infect rose leaves? Is it then a fungicide or a fungistat? The distinction is essentially academic rather than practical, the objective is to control the pest regardless of the mechanism or mode of action by which the chemical accomplishes this purpose. For basic understanding of pesticidal activity, however, studies on the mechanics of action of pesticides are exceedingly important, increased knowledge in this field will facilitate the development of more effective, efficient, and presumably safer pesticides.

Pesticides are classified chemically in many ways, as discussed in detail in chapter 1, vol. 1 of this treatise. In summary, the major groupings of importance in the control of ornamental pests are:

1. Fungicides

   a. Inorganic, including:
      (1) Sulfur
          The oldest fungicide.
          Used as dust or wettable formulation.
          Lime sulfur, or calcium polysulfide, most widely applied
             compound.
      (2) Copper
          As copper sulfate (26).
          Other inorganic compounds, as copper oxides, the carbonate
             and the oxychloride or as complex organic compounds (30).
      (3) Mercury
          Long known for its fungicidal properties, largely restricted
             to seed protectants due to its high toxicity to plant foliage.
          Newer organic mercury compounds—less phytotoxic.
          Useful as fruit and foliage sprays.
          Have extended residual activity.

(4) Tin

>Inorganic tin compounds are inactive as fungicides. Organo-tin compounds, the alkyl- and aryl-tins, e.g., tributyltin hydroxide and triphenyltin hydroxide and triphenyltin chloride, hydroxide TPTH (DU-TER) and the acetate (BRESTAN). The alkyl tins are generally phytotoxic, the aryl tins are not toxic at fungicidal levels. The triphenyltin chlorides and sulfates tend to be more toxic, the sulfides and disulfides less toxic.

>Triphenyltin acetate or fentin acetate (BRESTAN) controls several important <u>Cercospora</u> spp. and <u>Phytophthora</u> spp. diseases of plants.

(5) Cadmium

>Several inorganic and a few organic cadmium compounds are active fungicides particularly against turf grass diseases. These include cadmium chloride (CADDY), a cadmium-calcium copper zinc chromate complex (CRAGTURF FUN-GICIDE 531), cadmium succinate (CADMINATE), cadmium sebacate (+ potassium chromate + malachite green + thiram) (KROMAD), and PURETURF—a complex of organo-cadmium compounds.

b. <u>Organic</u>

>Highly versatile compounds.
>Now widely employed in plant disease control.
>Include:

(1) Carbamate fungicides

>Derivatives of dithiocarbamic acid.
>The most widely used organics:

>>Thiram (tetramethylthiuram disulfide) (NOMERSAM, ARASAN, POMASOL or THYLATE)
>>>Formulated as a seed or foliar protectant and as a turf fungicide

>>Metallic dithiocarbamates
>>>Chiefly the iron (ferbam—ferric dimethyldithiocarbamate), zinc (ziram—zinc dimethyldithiocarbamate) dimethyl derivatives.
>>>Ferbam generally applied for fruit and ornamental diseases and ziram for vegetable diseases.

>>Ethylene bisdithiocarbamates
>>>Three important metallic derivates.
>>>Sodium (nabam—disodium ethylene bisdithiocarbamate) (DITHANE-D14 or PARZATE), zinc (zineb—zinc ethylene bisdithiocarbamate) (DITHANE-Z78, PARZATE-C), and manganese (maneb—manganese ethylene bisdithiocarbamate) (DITHANE-M22 or MANZATE).
>>>All have broad-spectrum activity.

(2) Glyoxalidine (imidazoline) derivatives:

Glyodin (2-heptadecyl-2-imidazoline acetate) (CRAG 341 or GLYOXIDE).

An important apple and cherry fungicide for control of apple scab (<u>Venturia inaequalis</u>) and cherry leaf spot (<u>Coccomyces hiemalis</u>).

Also effective against chrysanthemum leaf spot (<u>Septoria chrysanthemi</u>), rose blackspot (<u>Diplocarpon rosae</u>) and snapdragon rust (<u>Puccinia antirrhini</u>).

Benomyl (methyl 1-(butylcarbamoyl)-2-benzimidazole-carbamate)

Broad spectrum on turf and ornamentals. Some success in control of Dutch elm disease.

(3) Quinone compounds

Chloranil (tetrachloro-p-benzoquinone) (SPERGON)

Seed protectant and bulb and tuber dip treatment.

Also used against brown patch of turf.

Dichlone (2,3,-dichloro-1,4-naphthoquinone) (PHYGON)

Seed protectant and foliage fungicide.

Effective against certain diseases of apple, azalea, cherry, peach and rose.

(4) Phthalimide derivatives

Captan (N-trichloromethylthio-4-cyclohexene-1,2-dicar-boximide) (ORTHOCIDE) first of series.

Folpet (N-trichloromethylthiophthalimide) (PHALTAN)

Both broad-spectrum and effective against many diseases of ornamentals, fruits and vegetables.

Folpet particularly effective against powdery mildew diseases.

(5) Guanidine compounds

Dodine (n-dodecylguanidine acetate) (CYPREX)

Particularly effective against diseases of apples, cherries, pears and of some ornamentals.

(6) Chlorinated triazines

DYRENE (2,4-dichloro-6-(o-chloroanilino)-s-triazine)

Effective against several diseases of ornamentals and turf.

c. <u>Soil Fungicides</u>. Soil fungicides are a separate category because the pests involved frequently differ from foliar diseases, and both inorganic and organic compounds are employed in their control. Some of the foliar chemicals described above are also effective in the control of soilborne pathogens; others are used almost exclusively as soil fungicides and include:

(1) Methyl bromide

Tasteless, odorless, colorless but very toxic to humans.

Must be handled with CAUTION.

Frequently combined with 1% to 2% chloropicrin (tear gas) as warning agent.

Has great penetrating ability

Gives excellent control of soil fungi, insects, nematodes and weeds.

Formulations available containing methyl bromide, chloropicrin and propargyl bromide for improved disease control.

Soil fumigation a preplant method only.

Treatment area covered with plastic tarp to confine gas.

(2) Chloropicrin (trichloronitromethane) (LARVACIDE)

Soil fumigant.

Highly toxic to plants and animals.

Must be used with CAUTION.

Usually injected to depth of 6 to 8 in. below soil surface.

Plastic cover enhances treatment.

(3) SMDC—Metham—(sodium N-methyldithiocarbamate) (Vapam)

Liquid soil fumigant.

Toxic to plants.

Recommended as preplant treatment for control of soilborne pests.

Not used in greenhouses containing plants.

Irritating to eyes and skin but relatively safe.

(4) DMTT (3, 5-dimethyl-1, 3, 5, 2H-tetrahydrothiadiazine-2-thione) (MYLONE)

Usually wettable powder mixed with fertilizer or suspended in water and applied as a drench.

Aeration period of 14 to 21 days suggested between treating and planting.

Soilborne fungi, nematodes and weeds primary targets.

(5) Allyl alcohol

Used alone or in combination with ethylene dibromide.

Generally not too effective against soilborne fungi.

(6) Formaldehyde

Soil fumigant largely replaced by newer materials.

(7) PCNB (pentachloronitrobenzene)

Controls Sclerotinia and Rhizoctonia root-rot diseases on wide range of ornamental shrubs and flowering plants.

Not effective against root-rot diseases caused by Pythium or Phytophthora.

Formulations applied as dry soil mix, surface application, drench or seed treatment.

(8) DEXON (p-dimethylaminobenzene diazo sodium sulfonate)

Application method similar to PCNB but specific activity complementary.

Mixtures of PCNB and DEXON effective for control of broad-spectrum of root-rot fungi.

d. _Antibiotics._ Recently antibiotics have entered the field of plant disease control. Antibiotics can reduce certain bacterial plant disease organisms, previously difficult to control. Some exhibit systemic action within plant tissues which increases their efficacy. Diseases controlled by antibiotics include: fire blight of apple and pear, peach bacterial spot, crown gall on many hosts, bacterial wilt of chrysanthemum and walnut blight.

Three antibiotics now recommended either alone or combined for plant disease control are cycloheximide, streptomycin, and terramycin. Cyclo-heximide is also a fungicide recommended for control of powdery mildew, cherry leaf spot and rust diseases, including white pine blister rust.

2. Nematicides

Nematodes, discussed earlier, are small eelworms which cause seri-ous damage to many ornamental trees, shrubs, flowering plants, and lawns. Nematicides are now available which eliminate, or vastly reduce, such nematode injury.

Some of the soil fungicides have nematicidal activity (e.g., methyl bromide, chloropicrin, SMDC, and DMTT), and alternatively other chemi-cals, primarily nematicides, have some fungicidal activity.

The principal nematicidal fumigants currently employed are the following:

(1) 1,2-dibromo-3-chloropropane (DBCP), dichloropropane—dichloro-propene (DD) and ethylene dibromide (EDB). These are volatile liquids which, when injected into the soil, exert control through fumigation. DBCP is diluted in water and merely poured or sprayed on the soil surface. Additional water is then applied to carry the chemical down to the depth required. It is also available in a granular formulation.

(2) VORLEX, or MORTON EP-162, is a mixture of methyl isothiocya-nate and chlorinated C_3 hydrocarbons. It is used for preplant soil injection only. Some weed control has also been noted.

(3) V-C 13 (O-(2,4-dichlorophenyl)O,O-diethyl phosphorothioate), a liquid formulation drenched on the soil surface or injected and, like DBCP, may be used in the vicinity of established plants.

(4) Thionzain (O,O-diethyl O-2-pyrazinyl phosphorothioate) (ZINOPHOS) is a broad-spectrum nematicide and soil insecticide applied by injection or broadcast as granules. Its high mammalian toxicity, however, limits its use by homeowners and greenhouse operators.

This brief list of nematicides refers only to the more popular materi-als employed on ornamentals. There are, however, many other chemicals possessing nematicidal activity, and research continues for the control of plant parasitic nematodes.

3. Insecticides

There may be single active compounds or mixtures of chemicals ap-plied to kill insects which attack plants or animals or infest premises.

Mites, such as red spider mites, through not true insects, are often destroyed by insecticides. There are, however, specific materials designed for mite control, called miticides or acaricides, which are here included for convenience under the general heading of insecticides.

Insecticides can be classified, on the basis of their mode of action, as stomach poisons, contact poisons, long residuals, short residuals, fumigants, repellants or systemics, but for practical purposes it is convenient to group them according to their source or composition.

a. Inorganics
   (1) Arsenates
          Used for over a century.
          Calcium arsenate cheap, compatible with lime sulfur but
             not with nicotine sulfate.
          Unstable.
          May cause foliage injury.
          Relatively highly toxic to mammals.
          Used widely to control the cotton boll weevil.
       Lead arsenate
          Developed to control the gypsy moth.
          More stable and less phytotoxic than other arsenicals.
       Paris green (copper acetoarsenite)
          Today of secondary importance.
          Actually a metal organic compound.
          Formulated into poison baits for armyworms, cutworms
             and grasshoppers.
          Also used as a dust or a spray.
   (2) Sulfur
          Both elemental and compound forms employed.
          Flowers-of-sulfur—an effective miticide for rhododendron,
             azaleas, boxwoods, etc.
          Lime sulfur—also a fungicide, effective in control of scale
             insect pests on fruit trees and woody ornamentals.
   (3) Fluorine compounds
          Cryolite (sodium fluoaluminate)
             A stomach poison.
             Largely replaced by newer compounds, still in limited use
                for control of fruit insects.

b. Organics. A large group of insecticides divided conveniently into botanicals, the halogenated (usually chlorinated) hydrocarbons, the carbamates and the organophosphates. Examples of each group of importance in control of insects on ornamentals are listed:

(1) Botanicals—insecticides of plant origin.

    Pyrethrum, derived from genus <u>Chrysanthemum</u>, one of
the oldest and safest insecticides known.

    Low toxicity to warm-blooded animals, making it a de-
sirable spray for pets, household sprays and in picnic
areas, parks, and home grounds.

    Nicotine, a product of tobacco

    Fumigant and contact insecticide useful in greenhouses.

    Nicotine and its sulfate very effective against aphids on
flowers and shrubs.

    Modern insecticides having reduced the demand for nico-
tine, no longer produced in the United States; some imported from England.

    Rotenone

    Obtained from roots of derris, cube or timbo plants.

    Most commercial rotenone coming from Peru.

    Moderately toxic to warm-blooded animals and very toxic
to fish.

    Avoid pond and stream contamination.

    Useful, however, to eliminate undesirable fish species
which may dominate farm ponds, enabling desirable
species to be reestablished.

    Citronella (oil of citronella)

    An insect repellant often formulated into creams, lotions,
sprays and into repellant wax candles.

    Extracted from the fragrant grass, <u>Cymbopogon nardus</u>,
from southern Asia.

    Ryania

    From roots and stems of <u>Ryania speciosa</u>, grown in
Trinidad.

    Relatively safe.

    Useful for control of insect pests on homegrown apple trees.

    Sabadilla

    From seeds of a lily-like plant of the genus <u>Schoenocaulon,</u>
native in Venezuela.

    Use dates back to the 16th century.

    Very limited use.

(2) Halogenated hydrocarbons

    Most members of group contain chlorine.

    One useful member, methyl bromide, contains bromine.

    Group characterized by broad-spectrum insecticidal activity,
marked toxicity to warm-blooded animals and long persistence
in the environment.

    Methyl bromide

    A non-persistent gas used to fumigate grain ele-
vators, ships, greenhouses and other enclosed
areas.

Methyl bromide (Continued)

    Hazardous unless applied with adequate control by trained operator—not recommended for home use by layman.

    Effective as soil fumigant to control soil insect pests.

    Also controls nematodes, weeds and soil disease organisms.

DDT (dichlorodiphenyl trichloroethane)

    First described in 1874 but insecticidal properties not discovered until 1939.

    The first major breakthrough in modern insecticides.

    Has excellent residual properties and broad spectrum.

    Widely used for control of insect pests in outdoor living areas—e.g., for flies and mosquitoes.

    Insect resistance a major problem.

    Currently banned for general use in the United States and some other countries

DDD or TDE (dichlorodiphenyl dichloroethane) (RHOTHANE)

    Closely resembles DDT in efficacy.

    Toxicity to warm-blooded animals about 10% of DDT.

Lindane (GAMMEXANE)

    Gamma isomer of benzene hexachloride (BHC); when 99% + pure, has greater insecticidal activity with less taste and odor than commercial BHC.

    A broad-spectrum insecticide with low toxicity to warm-blooded animals.

    Particularly effective against azalea lace bug, holly and boxwood leaf miner, pine aphids and stem borers on various woody or herbaceous ornamentals.

Methoxychlor (2,2-bis(p-methoxyphenyl)-1,1,1-trichloro-ethane) (DMDT, MARLATE)

    The methoxy analog of DDT.

    Is much less toxic to warm-blooded animals.

    Has long residual activity.

    Widely used to protect ornamentals from insect damage.

    Particularly effective against cankerworms and Japanese beetle.

Chlordane (1,2,4,5,6,7,8,8-octachloro-2,3,3a,4,7,7a-hexahydro-4,7-methanoindene) (OCTACHLOR)

    Persistent water-insoluble compound.

    Very useful for control of insects in lawns, turf and household.

    Important pests controlled include: grubs, ants, webworms, wireworms, armyworms, cutworms, chiggers, ticks, fleas and leafhoppers.

    Many cockroach strains are now resistant to chlordane and require other materials for control.

    Recently banned in U.S. except for special permit uses.

Aldrin (1,2,3,4,10,10-hexachloro-1,4,4a,5,8,8a-hexa-
hydro-1,4-endo-exo-5,8-dimethanonaphthalene) (OCTA-
LENE) and related dieldrin (1,2,3,4,10,10-hexachloro-
6,7-epoxy-1,4-4a,5,6,7,8,8a-octahydro-1,4-endo-exo-
5,8-dimethanonaphthalene) (OCTALOX)
Aldrin gives a quick initial kill but dissipates in a few
days.
Dieldrin, slower acting, retains its activity in soil up to
several years.
Both are safe for plants and do not affect soil microor-
ganisms.
Recently banned in U.S. except for special permit uses.
Toxaphene (octachlorocamphene) (PHENACIDE or
PHENATOX)
Chlorinated camphene.
Acts as both stomach and contact poison.
Useful in outdoor living areas for control of grasshoppers,
armyworms and cutworms.
Also controls insect pests of pets and livestock.
(3) Carbamates
Carbaryl (1-naphthyl-N-methylcarbamate) (SEVIN) is chief
member of this group.
Has broad spectrum.
Is free of handling hazards.
Particularly effective against Japanese beetle.
(4) Organophosphates
First reported in late 1940s; an important breakthrough in
insect control.
A wide field of compounds.
A few that apply to ornamental pest control include:
TEPP (tetraethyl pyrophosphate) (TETRON)
One of the earliest members.
Very toxic to warm-blooded animals.
A short residual life; breaks down within 48 hr to harm-
less compounds.
Very active against mites and soft-bodied insects.
Parathion (O,O-diethyl-O-p-nitrophenyl phosphorothioate)
(FOLIDOL, NIRAN, RHODIATOX, THIOPHOS, ALKRON)
An early development in group.
Has wide efficacy on many field, fruit, ornamental and
greenhouse plants.
CAUTION—has extremely high toxicity to warm-blooded
animals.
Must be handled with extreme care.
Not recommended for nonprofessional use.

Methyl parathion (O, O-dimethyl O-p-nitrophenyl phosphorothioate) (DALF, NITRIXO)
Closely related to parathion.
Has same general spectrum of activity.
Slightly less toxic.
CAUTION—also not recommended for nonprofessional use.

Malathion (O, O-dimethyl S-(1, 2-dicarbethoxyethyl) phosphorodithioate)
Highly versatile, safe insecticide.
Controls a broad array of insect pests, including aphids, mites, scales, horseflies, mosquitoes and many other insect pests of ornamentals, fruits, vegetables and animals.

Carbophenothion (S-(p-chlorophenylthio)methyl)O, O-diethyl phosphorodithioate) (TRITHION)
More toxic than malathion.
Is much safer than parathion.
Is a particularly effective miticide.

DDVP-dichlorovos (2, 2-dichlorovinyl O, O-dimethyl phosphate) (VAPONA)
Used extensively for control of household and public health pests.
Is a stomach and contact poison and a fumigant.
Formulated as baits, emulsifiable concentrates, space sprays or aerosols.

The systemic action of certain insecticides offers an approach to insect control with the advantages of long residual action and a high degree of safety to the user, to pets and to wildlife. Several systemic insecticides are now commercially available and are effective in the control of many insect pests of trees, shrubs and flowering plants. CAUTION—the systemic organophosphates now on the market are highly toxic in their formulated state and should be employed by home gardeners with strict attention to label instructions and warnings.

Demeton (mixture of O, O-diethyl O-[2-(ethylthio) ethyl] phosphorothioate-demeton O and O, O-diethyl S-[2-(ethylthio) ethyl] phosphorothioate-demeton S (ISOSYSTOX), disulfoton (O, O-diethyl S-[2-(ethylthio) ethyl] phosphorodithioate) (DI-SYSTON, DITHIOSYSTOX, FRUMIN-AL, SOLVIREX) and methyl demeton (METASYSTOX)
Similar chemicals.
Are effective systemics against a variety of insect plant pests.

Demeton (Continued)
  Applied to the soil, they are absorbed by plants which
  become toxic to insects.
  Demeton also used as a foliage spray.
  Disulfoton for seed treatment.
Phorate (O, O-diethyl S-[(ethylthio) methyl] phosphorodithioate) (THIMET)
  Effective in granular form; worked into the soil around
  a variety of ornamentals—e.g., rose aphids, various
  leaf miners, borers, etc.; controlled for extended
  periods following a single soil application.

The systemic organophosphate insecticides are spectacularly effective
against insect pests that bore deep into the host plant, thus becoming virtually impervious to topical pesticides that remain on the surface of the plant.
Thus, such penetrating pests as the leaf miners of boxwood, holly and birch
and the borers like those which attack irises can be readily reduced by
early spring applications of granular systemics well worked or watered into
to the surrounding soil. Protection will endure for an entire season if applications are sufficient and correctly made. It is essential to follow label
instructions on the container in detail.

4. Herbicides

Over the past 20 years advances have been achieved in the use of
chemicals to regulate plant growth and to control weeds. Effective economic chemical weed control has opened up new areas of vegetation control.
Thus, noxious plant growth in lakes, ponds, streams and irrigation ditches
need be tolerated no longer; highway, railroad, and power line rights-of-
way traversing rough terrain are also areas where chemical brush control
can enhance the natural beauty of the landscape at a cost far below that of
laborious hand methods. Because herbicides are chemicals that can kill
plants, and many herbicides can kill desirable plants as readily as they can
weeds, it is essential that only the correct herbicide be applied for a particular plant species or group of species and that the instructions for its
use be followed precisely.

Of the many herbicides now commercially available, only a few will
be mentioned here, with specific reference to maintenance of gardens,
parklands, and rights-of-way.

A practical classification of herbicide chemicals can be based on their
mode of action. Three groups are clearly defined: the contact (either selective or nonselective), the systemic, and the soil sterilants. A few herbicides fall into more than one group, depending on how they are applied.

For example, atrazine applied at a low rate will kill weeds in corn selec-
tively, yet it does not harm the corn. Applied at a high rate, however,
atrazine is a soil sterilant, killing all plant life present.

The following list includes representative herbicides found in each
group:

a. Contact. These herbicides are toxic only to plant cells with which
they come into contact because their movement within the plant is limited.
Two types of contact herbicides are employed; the selective and the nonselective.

Selective contact herbicides, as implied, kill some plant species but
not others. Thus, DNBP (dinitrobutylphenol) can control weeds among
small grains without injury to the crop, and VARSOL, an aromatic oil, sup-
presses weeds in pine tree nurseries without injury to the pine seedlings.

Nonselective contact herbicides, in contrast, destroy all of the vegetation with which they come into contact. DNBP, regarded as a selective,
becomes a nonselective when dispersed in fuel oil. Other examples are
paraquat [1,1'-dimethyl-4,4'-bipyridinium di(methylsulfate)] and mixtures
of PCP (pentachlorophenol) and fuel oil.

b. Systemics. Systemic herbicides are absorbed by roots or above-
ground parts and are then translocated throughout the plant. They do not
kill rapidly, as do the contact chemicals, because their mode of action lies
in their ability to disturb the plant's metabolism or growth processes. Most
herbicides in this group tend to be selective as to the plants they can destroy.

2,4-D(2,4-dichlorophenoxyacetic acid) is the earliest and most widely
known member of this group. Commercially available, 2,4-D formulations
are generally salts, amines or esters of the parent acid. 2,4-D preparations are employed widely for controlling broadleaf weeds in lawn and turf
areas.

Dalapon (2,2-dichloropropionic acid), conversely, is effective for the
control of undesirable grasses but is quite safe on most broadleaf plants.
Another selective systemic herbicide is 2,4,5-T (2,4,5-trichlorophenoxy-
acetic acid), very effective against woody shrubs. An exception is amitrol
(3-amino-1,2,4 triazole) which, although systemic, is selective on only a
few species (e.g., grapevines).

c. Soil Sterilants. These kill virtually all green plants and prevent
regrowth for extended periods. The length of time such treated soil will
remain toxic to plants depends upon the particular herbicide used, its rate
and method of application, the type of soil, the soil moisture at time of
application, and the subsequent temperature, rainfall, and aeration. These
factors, which affect the duration of herbicide activity in the soil, express

their effects through the rate of chemical breakdown, the susceptibility of the compound to degradation by microorganisms, its volatility and its sensitivity to light.

Soil sterilization with minimal residual toxicity can be achieved with methyl bromide. This colorless, tasteless, odorless, and extremely poisonous compound is a liquid at low temperatures and high pressures but readily converts to a very active gas under normal atmospheric conditions. Sold in pressurized cans or cylinders, it must be used with great caution. The area treated must be sealed under a gasproof cover, beneath which the methyl bromide vaporizes, and where it is confined for 24 to 36 hr. It becomes safe to plant in such treated soil following an aeration period of from 48 to 72 hr after the cover is removed. Although this method is exceedingly efficient for sterilizing home gardens, lawn and turf areas prior to renovation, greenhouse soil and potting mix, seed beds, nursery propagation, and liner beds, and so forth, the high toxicity of the gas to man necessitates its application by qualified experts only.

Such short residual toxicity is very useful in many situations, but in others it is completely undesirable. Thus, weed and brush control in industrial areas, marshalling yards, along railroad and power line rights-of-way, under highway guard rails and similar areas needs to be as complete and permanent as possible. In practice, sterilization is regarded as permanent when there is no regrowth for at least two consecutive growing seasons.

Boron (as sodium borate), an essential plant nutrient in minute amounts, is an effective soil sterilant when applied at 1,000 lb/acre or more. Three thousand pounds of borax or 1,000 lb of arsenic per acre will completely prevent weed growth for 2 years. Rates of this magnitude are, however, neither practical nor economic in most situations. They would certainly not be advisable for home driveways or parking yards.

Simazine (2-chloro-4,6-bis(ethylamino)-s-triazine), effective at low rates as a selective herbicide for controlling a number of weeds in ornamentals, is a useful soil sterilant when applied at 40 lb/acre, providing sterilization for as long as 2 years. Other examples of long-residual soil sterilants are diuron [3-(3,4-dichlorophenyl)-1,1-dimethyl urea] (KARMEX or MARMER), fenuron (3-phenyl-1,1-dimethyl urea) (DYBAR), monuron [3-(p-chlorophenyl)-1,1-dimethyl urea] (TELVAR), and erbon [2-(2,4,5-trichlorophenoxy)ethyl-2,2-dichloro propionate] (BARON). In all cases it is imperative that the details of the label recommendations be adhered to scrupulously to avoid disastrous spillover or washdown onto prized ornamentals.

Advances in chemical weed control continue to progress rapidly, and although the major thrust of the research is directed at economic crop and pasture weed control, many of the achievements are singularly useful to the home gardener and parkland manager.

IX. SELECTING A PEST CONTROL PROGRAM

Pest control is not an end in itself but merely a means to an end.
Control methods should be adjusted to the problems that need to be solved.
This section considers how control methods should relate to the specific
pest problem involved.

Prior to establishing a pest control program two basic questions need
to be considered: (a) Is the pest sufficiently serious to warrant control?
(b) If control is justified, then what method(s) should be employed in relation to the severity of the problem, the area involved, the degree of control
desired, its cost and its safety?

For example, though a lone mosquito in the bedroom at night may be
both annoying and irritating, obviously it hardly justifies the fumigation of
the entire house. Effective control can more practicably be achieved with
the primitive, though efficient, flyswatter. Where clouds of mosquitoes
infest a parkland area however, a control program is clearly necessary.

Because mosquito eggs only hatch in water, the first step is to eliminate or to treat areas of standing water wherever possible and practical.
Hatched larvae are readily killed by temophos, ABATE, lindane, malathion,
chlordane, methyoxychlor, or similar insecticides when applied according
to label instructions. Where such chemicals are used, however, the water
so treated must not be drunk by humans, livestock or pets, or used in bird-
baths or aquariums, at least until the treatment chemical has become diluted to the point of insignificance.

Adult mosquitoes are readily controlled with sprays containing malathion, lindane, ronnel or fenthion. Wettable powders are generally safe
when applied to vegetation but emulsions and oil solutions may cause some
foliage burning. Aerosols, though useful around the porch and patio, are
quite impractical in large open areas.

The azalea leaf and flower gall disease provides a good example of
how the severity and the area of infection influence the control procedure
selected. A few galls on azaleas call for simple sanitation by the homeowner, picking the galls by hand and burning them, which usually provides
satisfactory control. Where the infection is severe, however, and a large
number of plants are threatened, then it is clear that an organized fungicidal program is required.

Absolute, complete, and permanent eradication of all pest problems
in our outdoor living areas is obviously impossible. It becomes necessary,
therefore, to decide on the degree of control desired. A lawn disease and
weed control program can be exceedingly expensive, time-consuming and
mandatory if a high degree of control is demanded. The home gardener
must decide whether he wishes to pay for a lawn that looks like an unblemished

velvet carpet or whether he will tolerate an occasional patch of dollar spot disease and a few weeds that may survive a more modest program. Similarly, the disease and weed control programs applied to golf greens are more intensive and costly than those applied to the fairways simply because control of higher degree is demanded.

Cost is usually the major factor determining the method of control employed. Ornamental nursery stock growers find that a preplant soil fumigation with a methyl bromide formulation (BROZONE or TRIZONE) repays itself in reduced weeding costs. In addition, soilborne diseases, insects, and nematodes are also controlled, resulting in better quality and more salable stock. Around landscaped shrubs at home, however, mechanical weeding is generally more economical and practical than selective herbicide treatment.

Safety to humans, pets, and wildlife is paramount in all pest control programs. It is also important to avoid injury to desirable plants in the treated area. Parathion and its closely related phosphates, though highly effective insecticides, should be avoided in all home garden and most parkland programs due to their high mammalian toxicity. They can be used with safety by skilled, trained custom operators who exercise the necessary precautions; they should not be applied at all by homeowners or amateurs. On the other hand, 2,4-D is a relatively harmless chemical to humans and animals and is generally quite safe to use, but an improper formulation or its careless application can induce serious damage to nearby desirable plants. Although many insecticides will kill fleas, some can also kill the pet and, obviously, must not be used.

All registered pesticidal chemicals, regardless of their form, package, formulation or method of application, are required by law to state the necessary precautions, attendant on their correct use, clearly on the label. Where risk of hazard may occur, WARNING must be plainly indicated. The homeowner and landscape gardener should never buy a pesticide without first reading the label carefully and thoroughly, and the compound should never be applied until the label has been read each and every time the material is used to ensure its application is correct.

Experience has shown that home accidents and regrettable fatalities from pesticides invariably arise from mishandling, misuse, and misunderstanding, primarily because of failure to read the label.

X. OBSERVATIONS ON PEST CONTROL

Although it is often implied that nature is kind and particularly favorable to man, tilting the so-called balance of nature to his side, the fact remains that no species is favored beyond its own ability to fend for itself. Thus, within their own inherent power, pests will always continue to threaten and encroach upon man's health, food supply, and enjoyment, as their own needs dictate.

It is man, therefore, who through his intelligence, tilts the balance in his own favor in myriad ways. This is, however, an uncertain advantage that, at times, man himself impairs presumably through his own ignorance, avarice, indifference, and misuse of his intelligence. The contest between man and other living creatures for the exploitation of the environment is never static but always dynamic. Every year the cycle is repeated in field and orchard, in house and garden, in forest and plain. Each weapon man invents to protect his food, his health, his habitation, and his living space is eventually circumvented by pests and he must renew his effort afresh. Unless man continues to wage unrelenting battle against the competition of nature, he will cease to be the dominant form of life on this planet, because he will subside into biological obscurity and other more capable creatures will inherit his place on earth.

Paradoxically, man would like to enjoy the thrill and beauty of life in the wild, and many argue furiously for maintenance of the crude, untamed beauties of nature, yet virtually without exception man insists that such pleasure in the wild be untrammeled by disease, serious hardship, insecurity, injury, and sudden death. Yet these are the very hallmarks that characterize nature in the raw, and from which early man fled as he began his long ascent to modern civilization.

Clearly, in spite of the vociferous clamor of the "nature lovers," man cannot have it both ways. Either he will rest his case on the balance of nature, accepting "the fang and claw" with the "call of the wild" or he will tilt the scales in his own favor and forego some of the wonders of the wilderness.

This basic choice appears continually, whether it be the eradication of mosquitoes, leaf spot disease, or poison ivy from a flower garden so that it can be enjoyed; or whether it be the suppression of locusts in grain fields, gypsy moths in forests, disease-vectoring insects in swamps or ravenous rodents in orchards. Each case must be weighed on its own merits in terms of the greatest good for the many. Today the needs of man come first, whether it be justified or not, and the choice made generally is to suppress the competition.

The wild beauty of the open lands is being continually and inexorably contracted by the pressures of an exploding population. This enlarging pressure will not be relieved by resigning our environment to the ravage of insect, infection and weed; if anything, the injury to all wildlife would increase. If mankind can be fed, clothed, sheltered and educated to loftier pursuits there is a better chance he will appreciate the beauties of nature more and will protect them. Since most people would like to enjoy the

beauties of nature as well as be safe and secure, albeit by the intelligent
use of pest control practice, it will become fully feasible for man to achieve
the best of both worlds and at once provide for his health, needs, and shelter while large areas of wildlands and their creatures are preserved.

To combat the incessant threat to the living beauty in man's environment, a wide array of pesticides and combined formulations are today pro-
ferred on the market. These products are the fruit of considerable investment and vast research, which is finally distilled into the simple instructions and recommendations on the label. For the convenience of the home
gardener, many of these formulations are designed to possess a broad
spectrum in activity in order to control several insects and diseases. Lindane or malathion are usually recommended for the control of a wide variety of insect pests on trees, shrubs and many herbaceous annuals and
perennials, while dicofol, dimethoate, tetradifol, and malathion are often
recommended for the control of mites on flowers and ornamental shrubs,
and carbaryl (SEVIN) is preferred for the control of Japanese beetle.
Ants can be controlled easily with chlordane applied in accordance with
label instructions.

There are several good all-purpose fungicides available for general
home garden use. For particular purposes, however, there are ferbam,
specific for hemlock rust and cedar apple rust, and cycloheximide (ACTI-
SPRAY), also effective for controlling cedar apple rust but of limited spectrum. For the practical convenience of the home gardener there are several home fruit tree spray combinations available that can control both the
disease and insect pests on his fruit trees with both maximum safety and
minimum effort and expense.

Weeds are a continual problem and annoyance in home lawns, particularly if poisonous plants are involved; e.g., poison ivy. There are several
herbicidal formulations designed for the homeowner that are relatively
nonhazardous to children, pets, wildlife, and desirable plants, provided
the label instructions are strictly followed. There is efficient, precise,
and low-priced application equipment available to meet the specific needs of
the home gardener. Spot treatments can be made conveniently with aerosols, weed canes, or swabs, or by a directed stream from a plastic squirt
bottle, while handdrawn sprayers and granular spreaders are available for
larger areas. A relatively new device for home garden application of herbicides incorporates the active compound into a semisoft wax which has
been formed into bars of various sizes that can be rolled or drawn over the
weedy areas. The wax bearing the herbicide brushes off lightly onto the
lawn plants, killing those for which it was selected. Dalapon, formulated
in wax and shaped into a small bar with a hand grip, is useful for edging
walks, patios, driveways, and so forth, for the control of unwanted grass
species. A longer, larger wax bar containing a 2,4-D amine controls
broadleaf weeds in lawn areas. These bars have a cord attached to them so

that they can be drawn by hand or behind a lawn mower or small tractor. Virtually any wax-soluble compound can be formulated in this simple method to combat specific weeds.

Pest problems in public parklands are generally extensions of those of the backyard. However, pest control programs in these areas necessarily require more careful evaluation to decide upon a specific program. Many of the same pesticides used on private home grounds are equally applicable to public areas, but the formulations, method, and timing of application may differ because of the larger areas involved and because the presence of the public in these areas necessitates absolute safety precautions at all times.

The enlarging public use of our parks, campsites, and other outdoor recreation areas creates, or at least intensifies, certain pest problems that formerly were not considered to be problems at all. Thus, the occasional camper usually cleaned up his campsite and burned or buried his garbage, but today the thousands of campers and picnickers using public parks and campsites often leave their garbage and refuse to accumulate and encourage flies and other insect populations. Moreover, this human encroachment drives the birds to migrate to less crowded habitats, and as these natural insect predators withdraw, insect numbers increase. Sanitation, though important, is not enough. A properly planned and executed pest control program is necessary to protect the health and comfort of the vacationing public. A pesticide program should be designed to cause minimal injury to wildlife (5, 11, 29).

Wildlife, as well as humans, can benefit from the proper use of pesticides. It is well recognized that insecticidal sprays, dust, and soaps effectively control insect pests that plague household pets. Reduction of burgeoning aquatic weeds permits food plants needed by fish and other aquatic life to flourish. Fish also may be protected from predators, as demonstrated vividly in the lamprey eel control program in the Great Lakes. The lamprey, with a suction mouth ringed by tiny teeth, attaches itself to fish and literally drains out their life. Fish populations in Lakes Ontario, Erie, Michigan and Huron dropped at alarming rates as the lamprey inexorably advanced and increased. A chemical pesticide was developed that destroyed the lamprey without harm to fish. Following careful treatment of the affected Great Lakes, fish populations are being restored.

Herbicides have effectively cleared rangelands of poisonous plants and inedible brush and the infested areas have been reseeded to forage crops suitable for grazing by antelope, deer, elk, and other wildlife species as well as by cattle. The poisonous whorled milkweed (Asclepias verticillata) was found to be responsible for 19 of 23 cattle deaths in 1961 and 38 or 41 in 1962, on an experimental range near Las Cruces, New Mexico. When these plants were sprayed with the herbicide, 2,4,5-T

dispersed in diesel oil and water, cattle losses from plant poisoning declined to but one certain and one suspect case in 1963, and no losses were
reported in 1964 or 1965.

Pesticides are classed as economic poisons; if they were not so, they
would be completely useless. However, proper use is the touchstone applicable to the use of all chemicals, whether they be medicinal, industrial,
household, or agricultural. Thus, common household aspirin has caused
many fatalities, principally among children, when the tablets were mistaken
for candy—an improper use. Unfortunately, criteria for the safe use of
many modern technical products are neither standardized nor understood
and are too often just learned by our own experiences or through those of
others. Misuse of ethical drugs and medicines is largely prevented by the
prescription system, and, similarly, many safeguards surround the use of
pesticides.

As the use of pesticides has increased so have the laws that control
them. Significant early legislation includes the Federal Food, Drug and
Cosmetic Act of 1938. Attention was focused more directly on pesticides
in 1947 by passage of the Federal Insecticide, Fungicide and Rodenticide
Act (FIFRA). This act became effective in 1948 and, in brief, included six
important requirements:

1. Registration of economic poisons or chemical pesticides prior to
   their sale or movement in interstate or foreign commerce.
2. Prominent display of poison warnings on labels of highly toxic
   pesticides.
3. Coloring or discoloring of dangerous white powdered pesticides
   to prevent them being mistaken for foodstuffs.
4. Inclusion of warning statements on the label to prevent injury to
   people, animals, and plants.
5. Inclusion of precise instructions for use.
6. Information be furnished the administrator of the act with respect
   to the delivery, movement, or holding of pesticides.

Administration of the act became the responsibility of the U.S. Department of Agriculture with respect to efficacy, utility and application and of
the Food and Drug Administration with respect to safety and residue tolerances upon treated edible products.

In 1970 a significant change in these administrative responsibilities
was initiated with the creation of the Environmental Protection Agency
whose administrator reported directly to the President and not to the Secretaries of Agriculture; of Health, Education and Welfare; or of the Interior.
Responsibility for final registration of all new pesticides and for new uses
of already registered compounds came under the new agency, although both
the U.S. Department of Agriculture and the FDA continued to exercise

decision over efficacy and safety. This sweeping and initially uncertain change was swiftly followed by Congressional passage in 1972 of the Federal Environmental Pesticide Control Act (FEPCA), which substantially revises FIFRA, particularly with respect to the application of pesticides. FEPCA will become fully operative by 1977.

The public is protected further by the 1954 Miller Pesticide Chemical Amendment to the Federal Food, Drug, and Cosmetic Act of 1938, which provides that the FDA should set tolerance levels for pesticide residues found on or in food or feed products. These tolerances are derived from long-term toxicology data and probable hazards of use. The USDA and FDA work in concert with the land grant colleges, state experiment stations, and industry and other scientific institutions to establish "proof of safety" and efficacy before the pesticide can be registered for interstate commerce. The two chief objectives are: (a) the health and safety of those who use pesticides and those who consume the treated products, and (b) the protection of fish and wildlife by prevention of undue contamination of air, soil or water from misuse of pesticides. Further monitoring is maintained by the Federal Committee on Pest Control, which reviews all government-sponsored pest control programs and coordinates efforts in related research and education.

The ultimate responsibility for safety lies, however, with the user. Improper and irresponsible use is the major cause of all pesticide accidents. A national program conducted by the Cooperative Extension Service and supported by the National Agricultural Chemicals Association, the Manufacturing Chemists Association and other related groups is designed to educate users as to the most effective way of applying pesticides, to the importance of following label instructions and to the dangers inherent in the mishandling of these powerful and valuable tools of modern crop production and landscape maintenance.

The label on a pesticide container is probably one of the most costly pieces of literature in existence. Chemical companies estimate a total cost of from $3 to $5 million to research and develop a new pesticide and bring it through U.S. Federal and State registration and label approval for sale. This high cost is due to the extensive and exhaustive studies, often extending beyond eight years, necessary to accumulate the data required by government agencies to establish proof of efficacy and safety to man and animals. This procedure, required by law, assures that no pesticide can move in interstate commerce without a label approved and registered by the U.S. Environmental Protection Agency, and where residues might remain on the edible product, concurred on by the FDA. The label is both valuable and useful to the pesticide user, for it is a "Doctor's prescription" for the correct use of the pesticide and should be read carefully each time before the product is used. It represents the concentrated distillation of several million dollars and many thousands of man-hours of highly sophisticated technical work.

The following information must appear on every approved label:

1. The brand or trade name of the product.
2. A statement of intended use, including target pests, host plants or animals, dosage levels, times and frequency of application, and precautions for safety of user, treated product and consumer.
3. Correct chemical name of the active ingredients and the percent of each present in the product. Generic or common names are given when available. Percent inert ingredients.

EXAMPLE

Active Ingredients
  Zineb (Zinc ethylenebisdithiocarbamate) . . . . X%
Inert Ingredients . . . . Y%
  Directions for use
    Pests to be controlled
    Crops, animals, or sites that can be treated.
    Dosage, time, and method of application.
    Prohibited uses.
  Warnings
    To protect the user and the general public.
    To protect the consumer of treated plants or animals.
    To protect beneficial plants and animals.
  Antidotes and first aid treatments.
  Net contents.
  Manufacturer's name and address.
  EPA registration number.
  EPA establishment number.

Pesticides can be used invariably with safety and with benefit, if the user will but read the label, follow all instructions carefully and obey all warnings.

Control measures vary with problem severity, desired extent of control, area involved, cost and safety, all of these call for a careful analysis of the control program. Experience has abundantly demonstrated that, when clearly indicated, pesticides, properly used, provide an effective, economical, and safe solution to most pest problems.

XI. PEST CONTROL IN THE FUTURE

Pest control is necessarily exceedingly dynamic. Predictions today as to its specific course in the future are, therefore, subject to substantial modification, but present trends do allow for a measure of speculation.

Research to develop pest-resistant plants has intensified in recent years and will continue on an enlarging scale. Understandably the emphasis has been primarily on pest-resistant food, feed, and fiber crop plants, with but limited and essentially incidental interest in ornamentals. However, some success has been achieved with resistant ornamentals as discussed earlier, and as people become more aware of the need to control pests in decorative plantings, increased emphasis will occur on the development of pest-resistant flowering plants, shrubs, and trees. Research in this field is quite slow, and it takes many years before its effects are recognized by the general public.

Biological pest control provides many fascinating possibilities, and at present a substantial portion of the research on insects is directed at the development of new biological controls for the major pests, and at the discovery of basic information concerning insects. Research into new ways for the control of weeds, plant diseases, and nematodes also has intensified. Here again the emphasis is placed on food and fiber crop protection, but there will no doubt be developments applicable to ornamental plants.

Undoubtedly the search for predators and parasites of pests will continue. A few excellent results in both weed and insect control have been obtained already, and hopefully increased success will be realized in the future.

Diseases of pests are now recognized as possible useful tools in certain pest control programs. Examples were cited earlier. Research in this field is also expanding and some useful new developments are anticipated. Nonchemical control of insects may be applied through the Insect Population Suppression (IPS) technique. The classic case of the sterilized male screwworm flies, discussed earlier, that effectively reduced native populations to the vanishing point, and eliminated the species from the test area, will unquestionably be extended to other equally susceptible species. The practicality of this method must await further research to prove the efficacy of the IPS technique in controlling other major insect pests.

Biological control methods, though of definite value in many specific pest control programs, are by no means a panacea. Insects and disease organisms constantly mutate to strains or races which can break through the inbred resistance in plants; insect predators may become pests themselves; diseases of insect pests may mutate and become a hazard to beneficial insects, pets, wildlife, or even man.

Biological methods of insect and weed control are characteristically slow in action and are generally effective only over large areas. Thus, several years are required for the milky spore disease to reduce a Japanese beetle population to acceptable levels, and then only if large areas are treated. Certainly treating a single home lawn would be quite futile unless simultaneously the entire neighborhood or community was treated also.

As pointed out earlier, the object of control is to reduce the pest level to that considered acceptable, but what is an acceptable level? Undoubtedly, as the human population expands, acceptable levels of pests will recede as they have in the past. Although exceptions may be found, most biological control methods merely reduce, but seldom eliminate, a pest population. It seems highly probable that as standards of food quality advance, biological controls of insects and disease will be unable to provide the freedom from damage and blemish the consumer will demand.

In terms of public health, the control of insects that vector human and livestock disease will have to be absolute in order to assure eradication of such plagues as malaria, typhus, and encephalitis.

Improved cultural and mechanical methods of pest control are receiving more attention. For example, black plastic film is now used successfully as a mulch both to conserve soil moisture and to control weeds. This method, because of its cost, is confined essentially to home gardens, high-priced ornamental plantings, and to commercial row crop and nursery stock culture that provide a high return per acre. Flame control of weeds and certain insects promises effective results. A piece of experimental equipment is shown in Figure 16. There still remains a number of technical problems that must be resolved before this method becomes practicable to control weeds and insect pests in garden and parkland areas. It has been applied for several years, now, however, to driveways, parking lots, rights-of-way and industrial sites.

FIG. 16. Experimental equipment for flame control of weeds and certain insects.

Chemical pest control, despite its great advances and benefits over the past two decades, is still largely in its infancy. The miracles of modern chemistry now contribute to plant and livestock protection in many ways, and in the future still further contributions can be anticipated to the health, comfort, prosperity and happiness of mankind. Chemical pest control offers an area of virtually unlimited possibilities.

Research concerning safety in Federal, State, and industrial laboratories and experiment stations will continue to receive increased emphasis in chemical pest control. Such studies will be directed not only to pesticide chemicals themselves but even more to the method of application. Certain significant breakthroughs have already been achieved. For example, the new low-volume (LV) methods of spraying, both aerial and ground, provide more effective control with considerably less drift and therefore reduced hazard to contiguous plantings and wildlife.

Extensive research is in progress to develop and synthesize pesticidal compounds with wider margins of safety between vertebrates and invertebrates, to reduce the hazard to man and wildlife. Simazine and atrazine, for example, are two weed killers that are essentially nonpoisonous to humans and animals, are noncorrosive to equipment and do not destroy desirable soil bacteria. They are particularly useful to control unwanted herbage along highway fences, in parking lots, and along parkland pathways.

Plant growth regulators such as gibberellic acid, CCC (chlorocholine-chloride) and maleic hydrazide comprise another advance in the development of safer chemicals for crop production. Maleic hydrazide is especially valuable in slowing the rate of growth of grasses and is now applied to many thousands of acres of roadside grass as well as to golf courses, cemeteries, and lawns around industrial plants. It is a safe and effective maintenance tool, nonirritating, and not harmful to humans, animals, or birds when applied at recommended rates.

Weed control was radically changed in 1942 with the discovery of 2,4-D, but its use is limited by the fact that it is extremely active and can damage many desirable broadleaf plants. When, however, the carboxyl group on the 2,4-D molecule is replaced with a hydroxyl, a 2,4-D alcohol is formed which is relatively inactive as a herbicide, and this alcohol may then be converted to an inert material by forming the sulfate ester. In contact with the soil bacteria this compound is hydrolyzed, the sulfate group is detached and the alcohol oxidized back to the active 2,4-D. When applied as the sulfate ester, the compound will not harm plants, but in the soil the restored active 2,4-D destroys undesirable weed seedlings; it does not accumulate because it too becomes hydrolyzed by other soil bacteria into acetic acid and dichlorophenol which in turn is degraded by various soil microflora into completely harmless products. Such clever use of natural biochemical properties in the environment can be applied in many ways to effect control of target pests without endangering desirable plants, animals and microorganisms and will undoubtedly comprise a major course of future research in the pesticide field.

Research chemists continually search for analogues and derivatives of known compounds that combine high pesticidal activity with low toxicity to man and animals. Research with the organic phosphate insecticides is a good example.

Parathion (O,O-diethyl-O-p-nitrophenyl phosphorothioate), a very effective insecticide, is also extremely toxic to vertebrate animals and therefore should not be used around homes or in populated public areas. Chemists modified the thiophosphate molecule and derived malathion (O,O-dimethyl phosphorodithioate of diethyl mercaptosuccinate). This compound, approximately as toxic to insects as parathion, is about 1/300 as toxic to man and animals, and as it decomposes rapidly following application, the residue problem is negligible. This molecular manipulation promises to lead to the development of many safer yet highly effective insecticides in the future.

Research has revealed that the plant enzyme systems of certain crop plants differ radically from those present in many weed plants, which offers the opportunity to develop herbicidal chemicals that are detoxified by desirable plants and yet remain lethal to those weeds unable to degrade the compound. Such chemicals will provide precise control of the tares among commercial food crops, and undoubtedly many of them will find useful application in home garden, parkland, and right-of-way ornamental plantings.

Another approach to the safe use of pesticides concerns the regulation of the rate of pesticide absorption by man and animal. DDT, for example, is lipid soluble and tends, therefore, to accumulate in animal fat. Thus, detectable amounts of DDT are often found in milk from cows sprayed with the pesticide to control flies or fed on hays or pasture that have been treated with the compound. This has resulted in severe economic losses from condemnation of dairy products that have been found to contain illegal residues. The problem of lipid solubility was resolved by substituting two methoxyl groups for two of the chlorine atoms in the DDT molecule. The resulting compound, methoxychlor, is one-quarter as toxic as DDT to mammals, and accumulation in animal tissue is sharply reduced. The compound does not pass into the milk when sprayed onto the cows. This principle of absorption regulation can be extended to many other pesticides to enhance their safety.

The inherent safety and efficacy of systemic pest control chemicals will undoubtedly encourage the search for new systemic pesticides, particularly fungicides, in the future.

For the home gardener, the parkland manager, and the highway, right-of-way, and industrial site superintendent biological and mechanical control of pests will continue to provide enlarging, though probably limited, methods; meanwhile the application of chemical pesticides specifically designed to provide both safety and efficacy will continue into the future to comprise the major instrument of control of insects, diseases, weeds, nematodes, and rodents among ornamental plantings.

XII. PESTS OF ORNAMENTALS
AND THEIR CONTROL

A comprehensive listing of pests and recommended control procedures would extend beyond the scope of this chapter and would not apply to all areas of the U.S. nor to all situations extant within any one area. Likewise, pesticide use patterns on nonfood crops, including ornamentals, have come under closer scrutiny and recommendations are changing rapidly.

The average homeowner, beset by many pests among his ornamental plantings, may become discouraged and conclude that effective control is impossible. Fortunately, this is not usually the case. Abundant control information is available for the asking. The homeowner, gardener, or parkland manager should consult Cooperative Extension Service specialists, county agents, or his State College of Agriculture. In addition, local garden centers are usually qualified to offer advice for solving local pest problems.

Safety is imperative in the use of any pesticide; it is therefore essential to use the right material, in the right way, and at the proper time. The label on the pesticide container provides the information necessary for its safe and proper use. All those who apply pesticides, whether they be homeowners, gardeners, parkland managers, or others, are urged to read the label each time a pesticide is used, follow all directions, and heed all warnings and precautions.

When properly used, pesticides can help preserve the beauty of our outdoor environment, but their misuse can cause harm to humans, wildlife, and the beauties of nature that we are trying to protect.

XIII. SUMMARY

Decorative and ornamental planting of flowers, lawns, shrubs, and trees constitutes an integral part of the life of modern man. Such plantings in home gardens, institutional and industrial sites, highways, canals, rights-of-way, parklands, and recreational areas are all subject to invasion and destruction by insects, diseases, and vermin, and are thus readily despoiled. This damage is both costly and unsightly, vitiating, in most cases, the purpose of the landscaping.

Such pests are not new, for their ravage is recorded in ancient history, but the habitat and social custom of modern life frequently encourage their proliferation. Man has countered these pests through both manual and procedural methods with limited and varied success in the past. With the advent of agricultural science, particularly chemistry, in the production of food and fiber crops, many pesticides effective for pest control in the field have been adapted to the protection of ornamental plantings with marked success.

Fungicides and nematicides that reduce diseases, insecticides that restrain insects, and herbicides that eliminate weeds and unwanted vegetation and that are effective, safe, and economical are now available to the home gardener, nurseryman, lawn and greens keeper, and parkland manager.

The touchstone of utility of these useful chemicals is their remarkable efficacy, versatility, practicality and their safety to humans, livestock and wildlife when applied in strict accordance with the recommendations and the instructions on the label.

It is imperative to read and follow the label on each package of pesticide purchased and applied. This label summarizes in simple, unmistakable language an investment in research and development by the manufacturer and Federal and State workers of over $3 million and many thousands of hours of highly technical labor. Failure to read it is tantamount to wasting the cost of the product and jeopardizing not only the treated plants but human and animal safety alike.

When employed properly with intelligence and discrimination modern pesticides are extraordinarily useful instruments for the maintenance of ornamental plantings. They will doubtless be improved in the future as mounting pressures for increased food demand more effective pesticides for general crop and livestock production.

REFERENCES

1. M. J. Berkeley, Vegetable Pathology, Gardener's Chronicle, 1854–1857.

2. E. Berliner, Angew. Entomol., 2:29–56 (1915).

3. Biological Control of Insect Pests and Weeds, P. DeBach (ed.), New York, Reinhold, 1964, 844 pp.

4. A. B. Borkovec and C. W. Woods, Adv. Chem. Ser., 41:47–55 (1962).

5. A. W. A. Brown, Can. J. Pub. Health, 44:1–8 (1953).

6. G. E. Bucher, J. Insect Patho., 3:439–445 (1961).

7. T. Burnett, Ecology, 30:113–134 (1949).

8. R. C. Bushland and D. E. Hopkins, J. Econ. Entomol., 44:725–731 (1953).

9. R. C. Bushland and D. E. Hopkins, J. Econ. Entomol., 46:648–656 (1953).

10. G. E. Cantrell, S. R. Duthy, J. C. Keller, and C. C. Thompson, J. Insect Pathol., 3:143–147 (1961).

11. C. Cottam, Mass. Wildlife, 14:2 (1963).

12. A. DeBary, Untersuchungen uber die Brandpilze und die durch sie verursachten Krankheiten der Pflanzen mit Rucksicht auf das Getreide und ander Nutzpflanzen, Berlin, Muller, 1853.

13. S. A. Forbes, The Insect, the Farmer, the Teacher, the Citizen, and the State, Bull. Ill. State Lib. Mat. Hist., 1915.

14. The Greek Herbal of Dioscorides, R. T. Gunther (ed.), New York, Hafner, 1959.

15. J. G. Horsfall, Fungicides and Their Action, Waltham, Massachusetts, Chronica Botanica, 1945.

16. J. M. Kingsbury, Poisonous Plants of the United States and Canada, Englewood Cliffs, Prentice-Hall, 1964.

17. E. F. Knipling, J. Econ. Entomol., 48:459–462 (1955).

18. G. C. La Brecque, Adv. Chem. Ser., 41:42–46 (1963).

19. A. W. Lindquist, J. Econ. Entomol., 48:467–469 (1955).

20. E. G. Lodeman, The Spraying of Plants, New York, MacMillan, 1896.

21. P. C. Marth and J. W. Mitchell, through G. C. Klingman, Weed Control as a Science, New York, Wiley, 1961.

22. H. N. Moldenke and A. L. Moldenke, Plants of the Bible, New York, The Ronald Press, 1952.

23. W. C. Muenscher, Weeds, Ed. 2, New York, MacMillan, 1955.

24. W. C. Muenscher, Some Plants Poisonous to Touch, New York, State Coll. Agric. Cornell Ext. Bull. 441, 1959.

25. R. H. Painter, Insect Resistance in Crop Plants, New York, Mac-Millan, 1951.

26. B. M. Prevost, Memoire sur la Cause Immediate de la Carie ou Charbon des Bles et de Plusieures Autres Maladies des Plantes, et sur les Preservatifs de la Carie, Impremeur-Libraire, Chez Bernard, 1807.

27. V. Quinn, Shrubs in the Garden, New York, Frederick A. Stokes, 1940.

28. W. W. Robbins, A. S. Crafts, and R. H. Raynor, Weed Control, New York, McGraw-Hill, 1942.

29. R. L. Rudd and R. E. Genelly, Ca. Dep. Fish and Game Fish Bull. 7, 1956.

30. E. G. Sharvelle, The Nature and Uses of Modern Fungicides, Minneapolis, Miss., Burgess Publishing Co., 1961.

31. State Department of Health, Va. Health Bull., 17:11, Ser. 2.

32. E. A. Steinhaus, Principles of Insect Pathology, New York, McGraw-Hill, 1949, 757 pp.

33. F. Unger, Untersuchungen uber die Brandpilze und die durch sie verursachtan Krankheiten der Pflanzen mit Rucksicht auf das Getreide und ander Nutzpflanzen, Berlin, Muller, 1853.

34. C. Woodham-Smith, The Great Hunger, New York, Harper and Row, 1964.

35. W. Woodville, Medical Botany, Vol. 3 and Supplement, London, James Phillips, 1790-1794, through J. M. Kingsbury, Poisonous Plants of the United States and Canada, Englewood Cliffs, New Jersey, Prentice-Hall, 1964.

36. J. B. Zallinger, De Morbis Plantarum Cognoscendis et Curandis Dissertatio ex Phaenomenis Deducta, Oemponti, 1773, through H. H. Wetzel, An Outline of the History of Phytopathology, Philadelphia, Saunders, 1918.

Chapter 4

THE ROLE OF PESTICIDES IN THE CONTROL OF ECTOPARASITES IN DOMESTIC LIVESTOCK AND PETS

Arthur J. Hackett
Department of Animal Sciences
Rutgers University
New Brunswick, New Jersey

I. INTRODUCTION

A parasite is an animal that lives within or on a different kind of animal or plant, called the host, drawing benefit from it and, at the same time, doing it harm (88). Thus parasitism is one of several ways in which animals may live together in association. Ectoparasites are parasitic species associated with the exterior of the body of the host for, at least, part of their life cycle. Members of the ectoparasitic species, which attack man and domestic livestock and pets, belong to the Phylum Arthropoda, which includes lobsters, shrimp, crabs, and their relatives (Crustacea) the centipedes and millipedes (Myriapoda), the insects (Insecta), and the scorpions, ticks, and mites (Arachnida) (88). Many of the above species do not harm man or his domesticated animals, whereas others, particularly the insects, ticks, and mites are true ectoparasites.

Probably the most devastating role ectoparasites play is that of disease transmission. For example, the parasite that causes malaria is transmitted by the anopheline mosquito, and has caused more human deaths, than all other diseases combined, for millenia. This is a classic example by which an ectoparasite transmits a disease from man to man. It is significant that tens of millions of human lives have been saved by eradication of this vector through the use of DDT (75). Unconscionably, many irresponsible adversaries of DDT have succeeded in having it banned in the United States and several other countries.

In Vietnam alone at least 19 arthropod-borne diseases are transmitted by ectoparasites and are known to affect man: 11 are mosquito-borne; 2 are flea-borne; 1 is mite-borne; 3 are tick-borne, and 2 are sandfly-borne (152). Other protozoan diseases transmitted by ectoparasites in domestic animals include anaplasmosis (19, 49, 132, 146, 149, 186, 187, 188), piroplasmosis (81, 98, 156, 164, 170), filariasis (118), and dirofilariasis (176). There are numerous arthropod-borne microorganisms which cause diseases in animals. In 1968 there were 1617 equine animals that suffered from arthropod-borne encephalitides (177). Deaths occurred in 317 cases from viruses of western equine encephalitis, eastern equine encephalitis, and St. Louis encephalitis.

The tropical bont tick, <u>Amblyomma variegatum,</u> is the vector of <u>Rickettsia ruminatum,</u> which causes heartwater in cattle, sheep, and goats and a condition called Nairobi sheep disease (67). Transmission of Q-fever, a rickettsial infection, by this tick to man and animal has also been reported. Included in its broad host range are cattle, sheep, goats, asses, horses, antelopes, giraffe, zebra, elephant, buffalo, wart hog, ant bear, cheeta, lion, and rhinoceros. The first reported incidence of the tick <u>A. variegatum</u> in the United States actually occurred in the Virgin Islands. Immediate steps were taken to eradicate this species, using carbaryl as a spray and as a dip. Another report described the bont tick, <u>Amblyomma hebraeum,</u> as present on a newly arrived rhinoceros (26). Fortunately, the tick was dead, as was another bont tick found on a rhinoceros being shipped from New York to California. Rocky Mountain spotted fever, a rickettsial disease is transmitted from dogs to humans by the American dog tick (<u>Dermacentor variabilis</u>) (16).

A more direct type of infection by ticks is that of tick paralysis, which includes native and imported steers and heifers and mature cows and Shetland ponies (92). Thus ticks can induce diseases in livestock by transmitting toxins, as well as microorganisms and they therefore act both as reservoirs and as vectors of disease (112). The toxic principle of tick paralysis has not been identified, but it has been reported in man, domestic animals, and birds (92, 113). In East, Central, and South Africa, another toxin of the tick <u>Hyalomma transiens</u> causes a condition known as sweating sickness in cattle, sheep, pigs, and goats (112).

In addition to vectoring diseases, ectoparasites can also cause disease conditions directly, which may be manifested at various stages of their life cycle, and often give rise to severe economic losses. For example, damage by demodectic mange and cattle grubs (Fig. 1) have greatly lowered the quality of hides and primal cuts, thereby reducing their economic value to the farmer. Among household pets, ectoparasites can be a constant source of discomfort, irritation, and unesthetic appearance. Hookworm larvae or

FIG. 1. <u>Hypoderma bovis</u> (cattle grub). Larval stage emerging from the epidermis of the back. (Credit drawing to USDA Bureau of Entomology and Plant Quarantine.)

nematodes, <u>Rhabditis strongyloides</u>, nonarthropods, can also cause unsightly dermatitides similar to those produced by flea bites (13, 78, 151).

Each species of domestic livestock and pets can be afflicted with a host of ectoparasites, some of which are host specific and others less so. Ectoparasites are members of the classes Insecta and Arachnida. Included in Class Insecta is the Order Diptera, true flies. There are two suborders, Nematocera—gnats, mosquitoes and midges—of which the mosquitoes affect livestock and pets, and Cyclorrapha, which include the flies. The other two important orders of the Class Insecta are Phthiraptera (lice) and Siphonaptera (fleas). In the Class Arachnida are the mites and ticks. For discussion purposes, the important ectoparasites will be discussed under the animal host they attack.

II. ECTOPARASITES OF BOVINES

A. Flies

For over half a century the cattle grubs, or ox warbles, <u>Hypoderma lineatum</u> (De Villers) and <u>Hypoderma bovis</u> (De Greer) (Fig. 1), members of the family Oestridae, have plagued the livestock industry with estimated annual losses of up to $300 million. The presence of parasites or lesions caused by larval migration may render the organs in the carcass unwholesome, and some meat inspectors feel the affected organ or carcass, in whole or part, should be judged unfit for food (127). The adult heel fly elicits much anxiety among cattle, causing them to run frantically (36). Thus, when cattle are harassed by these flies, they lose weight and weight-gain due to their increased activity and reduced foraging time.

To eradicate or control any pest effectively, its life cycle must be known so that it may be attacked at its weakest link, while at the same time employing the most economical and practical approach. <u>H. bovis</u> (Fig. 1) lays its eggs (approximately 100 per fly) singly, during the warm weather of June and July, only on the legs and rumps of cattle (88). On the other hand, <u>H. lineatum</u> lays its eggs (200-300) in rows, only in the heel region; some of the eggs hatch after 3 to 6 days, and in the case of <u>H. bovis,</u> the larvae penetrate the skin and migrate through the subcutaneous tissue, finally reaching the back of the animal (Fig. 1) or, in the case of <u>H. lineatum,</u> the connective tissue of the esophagus. Occasionally, however, aberrant <u>H. bovis</u> larvae enter the spinal canal, brain, renal capsule, lungs, abdominal organs, and lymph nodes (3, 24, 55, 66).

Early in the spring the larvae of both these species migrate to the back skin on either side of the spinal canal, and, as third-stage larvae, pierce the skin for respiratory purposes. Some 30 days later, they force themselves out through the skin, drop to the ground, and pupate for another 35 days.   They then emerge as adults to repeat the cycle.   Because little can be done to exercise control over the breeding of the adult flies or the laying of their eggs, control measures must necessarily be established to eliminate the first or second instar larvae as they migrate through the sub-cutaneous tissue or as the third-stage instar larvae emerge from the backs of cattle.   There are several effective topical and systemic insecticides currently available commercially to control this parasite, for example ronnel, famphur, and coumaphos.

Some members of the superfamily Calypteratae attack and affect cattle in various ways.

Stomoxys calcitrans, the stable fly (Fig. 2), a bloodsucking fly, bites the lower parts of the body of cattle, usually during sunny mornings.   For most of the day this pest rests on posts and in the stable.   It can cause lowered milk production by irritating dairy cows.

Horn (Fig. 3) or buffalo flies, members of the genus Lyperosia (or Haematobia), spend most of their life cycle on the host, rather like lice, and leave only long enough to lay their eggs.   They are blood suckers and

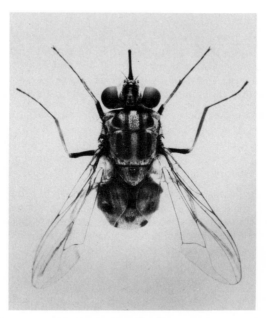

FIG. 2.   Stomoxys calcitrans (stable fly).   Dorsal view of adult female. (Photo: Photography Division, U.S. Department of Agriculture.)

FIG. 3. Lyperosia (Haematobia) species (horn fly or buffalo fly).
Dorsal view of adult female. (Photo: Photography Division, U.S. Department of Agriculture.)

cause such irritation that the animal repeatedly rubs itself against trees, fence posts etc., thereby opening wounds for secondary invaders. There may be over 1000 flies on a single animal, around its head, horns, withers, shoulders, and flanks.

Since 1953 Musca autumnalis, the face or black bush fly, has spread throughout the Middle West and particularly prefers the white faces of cattle (153). They are most active between 10:30 AM and 4:00 PM (17). In 13 widely separated pastures on a 4600-acre farm, there seemed to be a close relationship between the number of face flies and the incidence of pink eye.

Although it is not of too great importance to cattle, the common housefly, Musca domestica, can be a passive or incidental transmitter of diseases between animals, between humans, or from animal to man or from man to animal. It does, however, irritate cattle, depressing their growth and milk production, and readily breeds in their dung.

The tsetse fly, Glossina species, is also closely related to the stable and housefly. Tsetse flies do not lay eggs (88). The larva is nourished in the body of the female fly in a pouch of the uterus, which receives only one egg at a time. Milk glands enter the dorsum of the uterus to provide nourishment. Fortunately they are confined to Africa where they transmit trypanosomes to man (African sleeping sickness) and domestic livestock (nagana) and pets.

The life cycles of the above members of the superfamily Calypteratae are almost identical. Eggs are laid in fresh manure and require moisture; they hatch in 12 to 24 hr and the entire life cycle of each species of three instar larvae and a pupal stage takes about 12 days. Control measures include proper disposal or chemical treatment of manure and garbage, and insecticides, such as topical sprays of 0.125% to 0.25% CIODRIN or the systemic ronnel, fed in a mineral block formulation of 5.5% (17, 18, 27, 39, 63, 80).

In the southwestern part of the United States, several members of the subfamily Calliphorinae [blowflies (Fig. 4) and screwworm flies (Fig. 5)] attack cattle and cause extensive damage (Fig. 6). These flies are large and exhibit a variety of metallic colors. The adults do not bite but merely lay their eggs on and in wounds where the hatching larvae inflict the damage. Any elective surgery, such as dehorning or castration, should never be performed during the season of these flies.

The most important species in this group of ectoparasites is Cochliomyia americana (Fig. 5) (Callitroga hominivorax), a bluish-green fly with a yellowish-red face, three blue dorsal stripes, and dark hair on its abdomen (153). It is most dangerous in the maggot stage and is called a screwworm. Screwworms do not crawl about on the surface but become embedded in and devour the live flesh. If the invaded host animal is left untreated, death often ensues. This parasite lives only on the living flesh of warm-blooded animals and cannot penetrate through the unbroken skin of healthy animals.

Other members Lucilia caesar (green bottle fly), Caleiphora erythrocephala (blue bottle fly), and Cochliomyia macellaria are termed secondary flies because before they can attack an animal or a carcass, it must have been prepared by the action of lytic substances previously emitted from the maggots of preceding primary fly pests, such as Lucilia sericata one of the green bottle flies. Then, in turn, the maggots of the secondary flies,

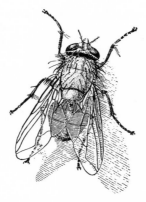

FIG. 4. Phormia species (black blowfly). Dorsal view of an adult blowfly. (Photo: Photography Division, U.S. Department of Agriculture.)

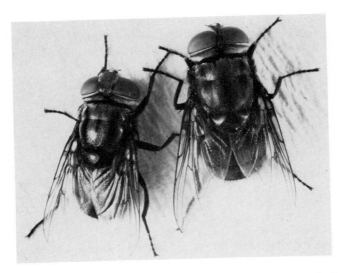

FIG. 5. Cochliomyia americana (screwworm fly). Dorsal view of adult screwworm flies—female on right and male on left. (Photo: Photography Division, U.S. Department of Agriculture.)

having hatched from eggs deposited by their adults, sequentially prepare the animal or carcass for the tertiary flies of the genus Phormia (the black blowflies) (Fig. 4). The life cycle includes the attraction of the flies to putrefactive areas where the female lays clusters of eggs (Fig. 7), each containing 20 to 250 eggs. During her adult life of about 1 month, she may lay up to 3000 eggs. The eggs hatch into larvae (Fig. 8) from between 8 hr and 3 days, depending upon the moisture content of the deposition site. These larvae mature in from 2 to 7 days, depending upon the availability of food and the temperature, and then drop off from their host and pupate (Fig. 9) for from 10 days to 3 weeks. Thus the entire life cycle may be as short as 2 weeks or less, but may extend up to 4 weeks if the weather is cold or dry. Because the adults mate only once in their lifetime, this weak link in the life cycle leads to sterilization of the males by irradiation as a control measure. Sterilized males are released in prodigious numbers to mate with indigenous females so that sterile eggs are laid (1, 169, 178). Other methods of control include burying the contaminated carcasses to prevent further proliferation of the pest or the use of insecticides, such as a dip or spray of 0.1% fenthion or 0.25% coumaphos (34). Comparatively little can be done by the farmer to control these flies under practical conditions.

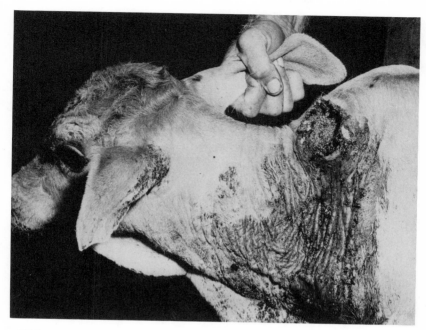

FIG. 6.   Larvae in wound of Braham calf.   (Photo: Photography Division, U.S. Department of Agriculture.)

FIG. 7.   Female screwworm fly laying eggs.   (Photo: Photography Division, U.S. Department of Agriculture.)

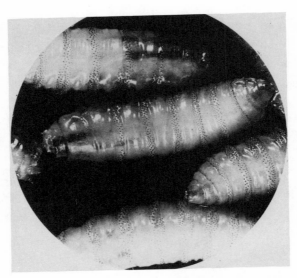

FIG. 8. Screwworm larvae. (Photo: Photography Division, U.S. Department of Agriculture.)

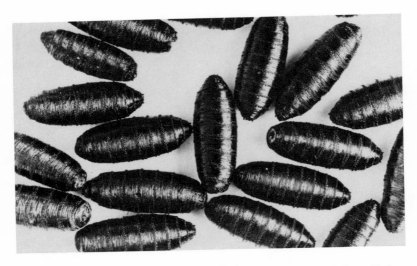

FIG. 9. Screwworm pupae. (Photo: Photography Division, U.S. Department of Agriculture.)

B. Lice

Lice cause domestic livestock losses of $30 to $52 million annually (20). In cases of heavy infestation, adult cattle suffer severely from anemia, which may be fatal because the packed cell volume (PCV) of blood may drop to as low as 22.2%, when the normal range is 31% to 47%, with a mean of 38.4%. Thus, moderate infestation approximately doubled winter weight loss in Hereford females. For heavily infected animals, treatment with insecticides is economically justified.

Lice are permanent parasites, highly host-specific, usually dying within a few days of leaving their host. In the first suborder, Anoplura, are the sucking lice Haematopinus eurysternus or the short-nosed cattle louse, and Linognathus vituli, the long-nosed cattle louse, both of which cause irritation, elicit anemia, induce loss of hair and reduced weight in cattle, and lowered milk production among lactating cows when heavy infestations occur. They attack the head, neck, shoulders, withers, back, and rump. In some cases, the anemia may become so severe that the animal may die. The population of lice on steers is significantly affected by ration (37).

Biting lice (suborder Mallophaga) are not quite as serious as are the sucking lice, living as they do on epithelial debris. One species that affects cattle is Damalina bovis (Fig. 10), the biting louse of cattle. Heavy infestations of this louse may also be related to nutrition (179). This louse is more active than sucking lice; it moves about more, causes more irritation, and the host-cattle, in trying to rid themselves of this pest, may bite themselves or rub against solid objects, thereby abrading their skin and removing their hair. There is no metamorphosis; the egg (or nit) hatches and undergoes

FIG. 10. Damalina (Bovicola) bovis (biting louse of cattle). (Photo: U.S. Department of Agriculture.)

three ecdysis and four nymphal instars before becoming an adult. The adult lives out its life on the host and lays its eggs attached to the hairs of the host; development from egg to adult varies from 9 to 29 days. The best control is accomplished by good management of the cattle but residual insecticides, as sprays or powders, can be used. These include sprays of 0.03% diazinon, 0.1% coumaphos, 0.1% carbaryl, 0.1 to 0.5% CIODRIN, 0.5% malathion, pour-on powders of ronnel, coumaphos, famphur, and rotenone (50, 60, 79, 97, 137).

C. Fleas

The Siphonaptera (fleas) is an order of the Class Insecta and there are no specific fleas associated with cattle. Occasionally, fleas may attack cattle, taking a blood meal, but usually they are host selective. Any effect on cattle would be irritation and, possibly, disease transmittance.

D. Acarids—Mites and Ticks

Throughout the world, virtually all animals suffer from infestations of mites and ticks, which are members of the Order Acarina of the Class Arachnida.

In the tanning industry, it has been estimated that demodectic mange in cattle results in losses of from $5 to $10 million annually (41). When examined by transmitted light, 92% of the hides inspected had nodules associated with demodectic mites at various levels within the corium. Approximately 30% of the hides so examined had nodules present in sufficient numbers to cause significant damage to leather. The actual incidence of Demodex bovis in the cattle was not, however, clinically diagnosed. In another study, shoulder lesions were detected in 17 of 55 live cattle, and small firm nodules could be detected within the skin (134). In 10 of these cows the lesions contained caseous pus and were microscopically positive for mites. There are two major families of mites that cause mange, the Sarcoptidae and the Psoroptidae, the most important species of which are Sarcoptes scabei and Chorioptes bovis. There are also reports about psorergatic acariasis caused by Psorergates bos mites in cattle (140). Occasionally, species of the genus Psoroptes cause scabies or mange in cattle (143, 165).

Generically, mange is any of a group of contagious animal diseases characterized by dandruff accumulation on the skin. In cattle there are four types of mange or cattle scab; tail mange caused by Chorioptes species, common scab caused by Psoroptes communis bovis, follicular cattle mange caused by Demodex folliculorum var. bovis, and barn itch caused by

Sarcoptes scabei bovis. Psorergates bos mites have been identified on cattle in New Mexico and Texas (140). Grain mites (Tyroglyphus) and chigger mites (jigger, red bug, harvest, or Trombicula irritans) may, on occasion, infest cattle. Most mites are selective but not specific, and the lesions include irritation, follicular papules, hypersensitivity, exudation of serum, encrustation and thickening of skin; secondary invasion of bacteria, and emaciation are common sequelae.

The life cycle of mites varies as to species. The Sarcoptic female burrows into the horny layer of skin where she remains for the rest of her life. As she burrows, she lays 2 to 3 eggs daily in the burrow over a period of about 2 months. When hatched 3 to 5 days later, the larvae stay in the burrows or make new ones and reach a hair follicle where they become first-stage nymphs after about 4 to 6 days. After about 2 days, the nymph moults and becomes a male or pubescent female. The Psoroptic and Chorioptic mites are surface dwellers and lay their eggs on the skin, where they hatch, in from 4 to 5 days, into larvae. In 2 to 3 days the larvae moult to the nymph stage, which, in turn, lasts for from 3 to 4 days. The smaller nymphs become males in about 6 days, while the females become pubescent earlier. The pubescent female of all species copulates with a male, moults 2 days later, and commences to lay eggs. The entire life cycle runs from 8 to 30 days. The adult female can survive up to 20 days away from its host and, therefore, precautions must be taken to avoid reinfestation after treatment. The life history of the Demodex species is not well-known but probably is similar to that of the other Acarina. Tyroglyphus and Trombicula species hatch from the egg into larvae that pass through two and one nymphal instars, respectively, to become adults. For control of mite infestations, reduction of contact between infested and non-infested animals is essential; this can be materially aided by accurately diagnosing the condition rapidly. CIODRIN (0.25%) or CIODRIN (1%) with dichlorvos (0.25%) has been used effectively in spray form (97, 158). An older treatment is lindane (88).

Closely related to mites are ticks, comprising over 300 species, some of which attack cattle, cause tick paralysis, transmit disease, suck blood, and result in milk losses of up to 20% (153). Those affecting cattle include members of both the Ixodidae (hard ticks) and Argasidae (soft ticks) (19, 32, 33, 35, 88, 142, 153).

Family Argasidae
   Ornithodorus megnini (spinose ear tick) (Fig. 11)
   Otobius megnini (ear tick)
Family Ixodidae
   Amblyomma americana (lone star tick) (Fig. 12)
   Amblyomma hebraeum (bont tick)
   Amblyomma maculatum (Gulf Coast tick)

Boophilus annulatus (cattle fever tick)	
(Figs. 13 to 15)	a one-host tick
Boophilus microplus	a one-host tick
Dermacentor albipectus (winter tick)	a three-host tick
Dermacentor andersoni (Rocky Mountain spotted fever tick)	a three-host tick
Dermacentor occidentalis (Pacific Coast tick)	a three-host tick
Dermacentor variabilis (wood tick)	a three-host tick
Ixodes cookei	a three-host tick
Ixodes pacificus	a three-host tick
Ixodes ricinus (castor bean tick)	a three-host tick
Ixodes scapularis (shoulder tick)	a three-host tick
Rhipicephalus appendiculatus	a two-host tick

The life cycles of the soft ticks differ markedly according to species. Those of Genus Ornithodorus lay eggs in sand in batches of about 100 eggs (88). The larvae do not hatch out, but develop and moult inside the eggs; thus the larvae are not parasitic and the nymph is the stage that leaves the egg. There are several nymphal instars and both nymphs and adults are parasitic on man and on domesticated and wild animals. Those of the Genus Otobius lay eggs in crevices and cracks in areas where the hosts are kept. Cattle kept in the open are not attacked by the ticks of this species. The larvae are parasitic and attach themselves deep in the ear, below the hair, feed for 5 to 10 days, and moult while on the host's ears. As nymphs, they remain on the host's ear until fully grown to 1/5- to 2/3-in. in length. They then drop off, climb up fences, trees, or walls, and hide in available crevices, becoming nonparasitic nonfeeding adults. Intermittently, the females may lay eggs for as long as 6 months and then die. Unfertilized females may actually live longer than a year.

The habits and life history of the hard ticks differ from those of the soft ticks. They are parasitic blood suckers during part of their life and infest most mammals and some birds. The life stages exhibit a distinct host preference; the larval forms prefer small animals, while the adults are attracted to larger animals. They also spend a considerable proportion of their life cycle away from and between hosts, depending upon the season, temperature, and humidity. During their relatively long life span of up to 2 to 3 years, they successively pass from egg to larva, nymph to adult, moulting and hibernating until the right temperature and environment exist.

Depending on the species, there are one-host, two-host, and three-host ticks. A one-host tick spends all its life stages on a single host; the newly hatched larvae promptly ingest a blood meal, moult into a nymph, eat another blood meal, and moult into an adult. After taking another blood

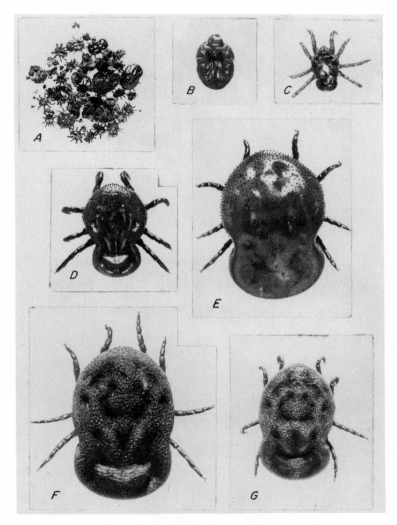

FIG. 11. <u>Ornithodorus megnini</u> (spinose ear tick). (A) Ear ticks and debris from ear of animal. (B) Engorged larvae. (C) Young tick. (D) Partly engorged young tick. (E) Fully engorged young tick. (F) Adult female. (G) Adult male. (Photo: U.S. Department of Agriculture.)

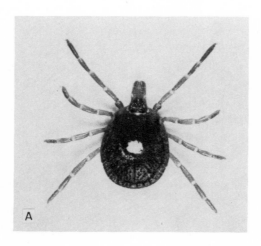

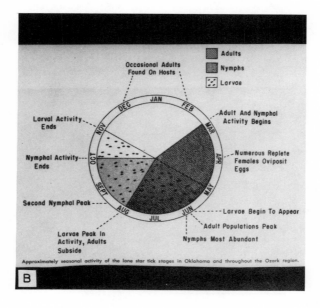

FIG. 12. <u>Amblyomma americana</u> (lone star tick). (A) Lone star tick adult. (B) Diagram of life cycle of the lone star tick. (Photo: U.S. Department of Agriculture.)

FIG. 13. <u>Boophilus annulatus</u> (cattle fever tick). Adult ticks on a cow. (Photo: U.S. Department of Agriculture.)

meal, the adult female drops off and lays its eggs, copulation having occurred either on or off the host. The larvae of the two-host tick attaches itself to a host animal, takes a blood meal, and then moults on the host animal to become a nymph. It then drops off and moults into an adult. The adult soon attaches itself again to a second host animal, ingests a blood meal, and then finally drops off the host animal to lay its eggs. In the triple phases of its life cycle, the larvae of a three-host tick feed on one host, the nymphs on another, and the adults upon a third. The adult lays its eggs on the ground and the hatched larvae remain dormant until the right environmental conditions prevail before migrating to an appropriate host animal where they ingest a blood meal over the succeeding 5 to 6 days. They then drop off and moult into nymphs. Again when conditions are suitable, they will migrate to a second host and have a blood meal for another 5 to 6 days. The fed nymphs again drop off, moult into adults, and promptly reach for a third host, upon whom they feed on blood for about a further 2 weeks, drop off, and the females lay their eggs. Most of the hard tick species follow a life cycle of this type.

FIG. 14. <u>Boophilus annulatus</u>. Cattle fever tick laying eggs. (Photo: U.S. Department of Agriculture.

FIG. 15. <u>Boophilus annulatus.</u> (A) Female. (B) Male. (C) Nymph. (D) Engorged female. (Photo: U.S. Department of Agriculture.)

The control of ticks is both difficult and expensive. Burning of rough pastures, cultivation, and drainage of the land help but have not been particularly successful. Spraying known infested land with 1 lb of DDT per acre has been tried in some areas. Treating the infested host animals with sprays, dips, or emulsions of residual insecticides to kill the tick larvae, nymphs, and adults generally gives the most effective results. Where the infestation is severe, repeating the contact insecticide treatment every 3 weeks is suggested. For ticks that lodge in the pinna of the ear, hand application of the acaricide may be required (154). Sprays of up to 0.75% of chlorfenvinphos, dioxathion or toxaphene, and BANOL (0.25%), bromophos (1.0%), CIODRIN (0.3%), carbaryl (5%), trichlorfon (1%), chlorpyrifos (0.1%), and fenthion (0.1%) are effective (32, 33, 144, 172, 184). Some strains of ticks may become resistant to organophosphorus ixodicides (56, 155, 162). Conversely some cattle can also become resistant to cattle ticks (61, 119, 144).

III. ECTOPARASITES OF OVINES AND CAPRINES

Sheep suffer afflictions from many of the same ectoparasites that attack cattle. Stable flies (Fig. 2), horn flies, and flesh flies (Sarcophaga) (Fig. 3) also attack sheep, invading wounds, laying eggs, and producing flesh-eating maggots. The housefly annoys sheep by crawling about and feeding on eye secretions.

A. Flies

Probably one of the most obnoxious flies to sheep is Oestrus ovis, also referred to as sheep nasal fly, head-maggot fly, sheep botfly, or sheep gadfly (153). Somewhat larger than the common housefly, it is active during the warm part of the day and rests during the cooler periods. These flies bear living young (viviparous), depositing their larvae around the sheep's nostrils. These minute larvae migrate into the nasal passages, remaining for sometime, then they migrate to the frontal sinuses, where they complete their development. After maturation, they return to the nasal passages, drop to the ground, and pupate for from 3 weeks to 2 months, depending upon the temperature and moisture content of the soil. The adults then emerge, become active, and annoy the sheep. As larvae, they irritate the mucosa with their oral hooks and spines. The sheep react by sneezing, lose their appetite, and become vulnerable to secondary infections. Systemic insecticides, such as dimethoate, are used to treat sheep with nose bots (106). Other measures that aid in control are fly repellants and shelter during hot sunny weather.

Strike is a condition in sheep caused by the larvae of several species of flies, namely the black blowfly (Fig. 4), the blue bottle fly, and the green bottle fly. These flies are attracted to the wool of sheep's rump and thighs, particularly when it is soiled by feces and urine. The flies lay eggs and the larvae or maggots hatch and feed, initially, on the soiled wool, subsequently attacking the skin, infesting any wound, and devouring the flesh. This condition is somewhat similar to that caused by blowflies or screwworm larvae in cattle. Control can best be obtained by using residual insecticides, such as dieldrin, which give up to 12 weeks of protection, or several of the organophosphorus compounds, such as ronnel, chlorfenvinphos, and dichlofenthion (8, 94, 189). Any condition that tends to soil the wool should be prevented, for example, Mule's operation, tail docking, or clipping the wool around the tail and rear quarters. Chemosterilization of male blowflies with apholate offers some hope for eradication (191).

The sheep ked (Mallophagus ovinus), also called the sheep tick, is a bloodsucker that infests the wool and, particuarly, the skin of the neck, breast, shoulders, belly, and thighs of sheep (53). It is not a true tick but a fly with poorly developed wings. The egg of the ked is retained for about 7 days in the body of the female until it develops into a pupa. The ked then attaches the pupa to the wool of the sheep with a gluey secretion. After about 12 hr, the pupa, which initially is covered with a soft white membrane, becomes brown and hard. Some 19 to 23 days later, the adult emerges and rapidly matures in from 3 to 4 days. The adults seldom leave their host except by direct contact with other sheep. Signs of ked affliction are revealed by the sheep biting, scratching, or rubbing and evidence of a ragged fleece. The keds themselves can readily be seen by parting the wool. Other effects include loss of blood, irritation, and damaged wool by ked feces. For direct control, dips, sprays, or powders of residual insecticides are effective. Nutrition also is important in sheep ked infestation. Thus, low protein and vitamin A deficiency delay the development of resistance to sheep keds in lambs (117).

B. Lice

Four important species of lice attack sheep. These include Linognathus ovillus (sheep-body louse) and Linognathus pedalis (sheep-foot louse) and two bloodsucking lice, Trichodectes ovis (round-headed sheep louse) and Damalinia ovis (biting louse). They cause little problem in summer but annoy sheep in the winter months. Various organophosphorus compounds, such as carbophenothion (0.021%), coumaphos, diazinon, and fenchlorphos (5 to 50 ppm), used for dipping, successfully control the lice on sheep (120, 173).

C. Fleas, Mites, and Ticks

Fortunately, fleas do not cause much of a problem for sheep and, if present, can be readily controlled by the use of insecticides, as is done with cattle. Sheep, however, like cattle, are hosts for both mites and ticks. Presently, the United States is declared to be free of sheep scabies. Common scab on sheep is caused by the mite Psoroptes communis ovis, while foot scab is caused by the mite Chorioptes ovis. The sarcoptic mite, Sarcoptes scabei bovis, when found on sheep, causes head mange, and Demodex folliculorum ovis causes follicular sheep mange; psorergatic acariasis, caused by the itch mite, Psorergates ovis, has also been reported (141). Dips, using residual insecticides, are used to treat mite infestation. Treatment may have to be repeated in heavy infestations (105, 165, 166). The ticks parasitizing sheep are the Gulf Coast tick, the lone star tick (Fig. 12), the Pacific Coast tick, the shoulder tick, and the spinose ear tick (Fig. 11). The life cycles and effects on sheep are similar to those of the cattle ticks and they can be controlled in the same manner.

Of the domesticated ruminants, goats also host numerous ectoparasites. In general, some of the pests that affect both cattle and sheep will also attack goats, and the treatments and controls recommended are essentially the same.

IV. ECTOPARASITES OF EQUINES

A. Flies

About 57 species of parasites that attack the horse have been listed (28). Of the three most injurious ectoparasites, bots rank high. There are two common species in North America, Gasterophilus intestinalis (common botfly) and Gasterophilus nasalis (throat botfly), while on occasion Gasterophilus haemorrhoidalis (nose botfly) also attacks horses (Fig. 16). While laying eggs, the adult botflies (Order Diptera), cause considerable annoyance to horses. The eggs of all three species are attached to hairs on the host animal and, after a few days, the maggots emerge; curiously, eggs of G. intestinalis do not hatch unless licked or rubbed or exposed to friction. The first larvae cannot develop unless conveyed to the mouth, where they burrow for a few weeks, pass on to the stomach or intestine, attach to its mucosa or lining, cause inflammation, and interfere with digestion (Fig. 17). Later they pass out through the anus of the horse and fall to the ground. There the skin hardens and the bot changes into a botfly. This pupal stage may last from 20 to 70 days. Control of this pest includes fly repellants and systemic insecticides, butonate, carbon disulfide, and dimethoate

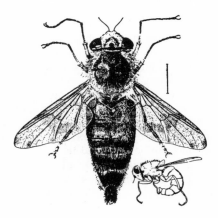

FIG. 16. <u>Gasterophilus</u>. Horse botfly with sketch of side view. (Photo: U.S. Department of Agriculture.)

FIG. 17. <u>Gasterophilus</u>. Horse stomach with complete infestation. (Photo: U.S. Department of Agriculture.)

(23, 25, 29, 107, 147, 153, 159). Another method is to remove the eggs by clipping the hair or by hatching them in warm water (28).

Various other flies also attack horses, either annoying them or, more significantly, transmitting disease. The common horsefly (also called the gadfly, earfly, or black horsefly) is a vicious bloodsucker belonging to the genus Tabanus. A close relative is the greenhead horsefly, which is also a tabanid. These flies bite, cause considerable pain, and remove rather large quantities of blood. They lay their eggs on plants, growing in or over water or on stones in streams. After hatching, the larvae drop off, burrow into mud or gravel and feed upon soft-bellied microfauna. After nearly a year, they pupate in the moist soil near the water and soon emerge as an adult. These flies are extraordinarily strong and fast fliers. Because of their peculiar breeding habits, these flies are difficult to control in the pre-adult stage; however, drainage of swamps and the careful control of irrigation ditches can materially help. Insecticides and repellent treatments effectively protect horses from tabanids; as tabanid flies are particularly sensitive to DDT, it is most effective when sprayed (5%) around stables and corrals where horses and ponies are kept.

Stable flies (Fig. 2) are the intermediate hosts for the small-mouthed stomach worm of horses (153). Face flies feed on tears around the eyes of horses. Houseflies may occasionally bite and annoy horses. Sand flies, in large numbers, seriously irritate horses, especially nervous and highstrung animals. All of the above flies can be controlled by repellants or insecti-cides, for example, CIODRIN, dichlorvos, or pyrethrin (108, 153).

B. Mosquitoes and Fleas

Mosquitoes attack horses, suck blood, and, more importantly, trans-mit several serious equine diseases caused by protozoa, microorganisms, and viruses (153). The human flea is occasionally found on horses. The only effective control of these pests is pesticides.

C. Lice

Species of lice living only on horses, mules, and asses include Haematopinus asini (bloodsucking horse louse) and Trichodectes pilosus and Trichodectes parunpilosus (biting lice) (153). All of these species congre-gate in colonies on the host animals, causing them to rub and bite them-selves, with the result their coats become rough and patchy. Usually lice tend to locate on the sides of the neck, around the flank, and under the jaws, but they may also be found all over the body and legs of the host animal. Lice are readily carried between animals on currycombs, brushes, blankets, harness, saddle, or other interchangeable equipment. All hair clippings should be burned and the ground where the animals stand or lie should be treated with a residual insecticide. There are a various number of registered insecticides, such as lindane, which are available and effectively control lice infestations among equines.

D. Acarids—Mites and Ticks

In the equine, mange is caused by Sarcoptes, Chorioptes, Psoroptes, and Demodex species (83, 153). The condition, also called scabies, scab or itch, exists as common horse mange, foot mange, psoroptic mange and demodectic mange, respectively. The life cycle for these mites is essentially the same as for mites that infest cattle. Their treatment and control are effectively accomplished by the use of approved acaricides.

Ticks feeding on animals, leave open wounds that may become infected with organisms, or they may transmit diseases from an infected animal to a clean animal, or more significantly from animal to man. One horse reportedly dropped a burden of 13,000 ticks with a total weight of 14 lb in 3 days. Young, infested animals become stunted, while a single mature animal is reported to have lost up to 200 lb of blood in one season (153). Allergy induced by the tick Amblyomma americanum has been reported (174). In the United States, Dermacentor nitens (the tropical horse tick) has been found to transmit equine piroplasmosis in Texas and Florida (81, 135, 148, 164). Other tick species have also been incriminated in the transmission of this disease, including members of the genera Dermacentor, Ixodes, Rhipicephalus and Hyalomma, which include the Gulf and Pacific Coast ticks, the shoulder tick, the winter tick, the wood tick, and the Rocky Mountain spotted fever tick. Occasionally, horses are parasitized by the brown dog-tick (Rhipicephalus sanguineus), which in the immature state is called seed tick, and by the cayenne tick (Amblyomma cajennense), which is found in southern Texas and the Panama Canal Zone (153). Infesting thoroughbred race horses, the ear tick (Otobius megnini), causes a high degree of irritability when the larval and nymphal ticks engulf the ear (168). For control of all species of ticks, various insecticides are used. Some of those recommended include toxaphene (0.5%), lindane (1%) in vegetable oil, naled (2%), CIODRIN (0.3%), and coumaphos (0.25%) (12, 30, 31, 168, 176).

V. ECTOPARASITES OF SWINE

There are fortunately relatively few ectoparasites that are economically significant in swine husbandry. Lice and mange mites are the most important pests. The use of the chlorinated hydrocarbon and the organo-phosphate insecticides, has facilitated control of the external parasites in swine.

A. Flies and Fleas

The common housefly (Musca domestica) and the stable fly (Stomoxys calcitrans) are frequent pests to swine. The green screwworm fly (Coch-liomyia americana) will swiftly attack swine that have any open wounds. Occasionally, the human flea (Pulex irritans) and the jigger flea (Tunga penetrans) infest swine. Control of the flies and fleas on swine is with recommended insecticides and also by strict sanitation of the breeding grounds, particularly manure deposits.

B. Lice

The only louse infesting swine in the United States is the bloodsucking louse (Fig. 18) (Haematopinus suis) (153). A severe infestation results in poor growth in young pigs and causes shrinkage in the weight gains of fat-tening hogs. The affected animals rub against any object that both destroys the hair and injures the skin. These lice may also transmit swine fever

FIG. 18. Haematopinus suis. Female hog louse. (Photo: U.S. Department of Agriculture.

and they may vector swine pox virus. Ronnel, dimethoate, and toxaphene are recommended to control lice in swine (71, 145).

C. Acarids—Mites and Ticks

The sarcoptic mite (Sarcoptes scabei suis) and the demodectic mite (Demodex phylloides) are the only causes of mange in swine (9, 83, 104, 153). Both species are contagious but generally do not affect swine to any serious degree. Various acaricidal preparations are used to control the disease on the animal host. Burning used litter and spraying the facilities with acaricides will substantially reduce reinfestation.

Ticks that infest swine include the brown dog tick, Rocky Mountain spotted fever tick, shoulder tick, wood tick, and spinose ear tick. These ticks can be readily treated with a number of approved acaricides.

VI. ECTOPARASITES OF CANINES

Even with the new and improved external parasiticides, infestations by such common organisms as mosquitoes, flies, fleas, mites, and ticks remain a major concern for dog owners and for practitioners of canine medicine. In a heavily contaminated environment, parasites of dogs and other pets constitute health hazards to humans because of the close contact. For example, canine scabies or mange due to the canine sarcoptic mite, Sarcoptes scabei var. canis can infect humans (43). Fortunately it is not a serious conditions, is self-limiting, and is amenable to treatment. Another dog mite, Cheyletiela parasitovorax can also cause a skin condition in man (87, 183). The close contact of humans with dogs in houses allows for the ready transfer of fleas, ticks, and mites to man (160). Incidentally, the traffic is not just one way, for example a dog was found harbouring a number of Phthirus pubis, a species of lice which infests the pubic hairs of humans.

A. Mosquitoes

In the southern United States particularly, numerous species of mosquitoes transmit the injurious parasite Dirofilaria immitis, the heartworm of dogs (93). This vectored disease constitutes the major harm mosquitoes convey to dogs. Mosquitoes are slender pubescent flies with a long slim proboscis ideally adapted for piercing. Only the female bites or feeds upon the host. She lays 40 to 100 eggs singly, or on rafts, usually in quiet water.

Within 24 hr to a few days, depending upon temperature, the larvae emerge and moult several times to reach a fourth instar. They then enter pupation and within a couple of weeks become adults. The females live from 3 weeks to as long as 4 months and require a blood meal for reproduction. Control includes both the elimination of favorable breeding sites and residual surface sprays to destroy the adults. Unfortunately, some strains of mosquitoes have become resistant to certain insecticides that formerly were very effective. The larvae can be destroyed readily by putting oil or an emulsified larvicide on the water surface.

B. Flies

Generally, flies are not a serious problem to dogs although stable flies may bite them around their ears. Sandflies (Phlebotomus spp.) however do transmit leishmaniasis to man and certain animals and such a case has been reported in dog in the United States (171). This dog's particular origin was traced to Greece from whence it had come some 18 months previously. Thus, it is important that care be taken when animals are imported from foreign countries where insect vectors or diseases occur that are exotic to the North American continent.

Miasis by larvae of the fly Dermatobia hominis occurs only occasionally in dogs in North America, but the larvae of the genus Cuterebra frequently affect dogs in this country, causing paralysis or subcutaneous abscess-like lesions in the neck (59, 64). The adult fly lays her eggs in the habitat of the host, such as in the burrows of rabbits or on nearby grass. After hatching, the larvae attach themselves by their hooks to the skin of any passing dog, penetrate it, and continue to develop. Usually only two or three larvae will attack one dog simultaneously, and the treatment and prognosis are good. Gently remove the larvae and flush the wound with an antiseptic solution to prevent further attack by screwworm flies. The adult flies may be controlled by area application of several approved insecticides.

C. Lice

There is one bloodsucking louse (Linognathus piliferus) and two biting lice, Trichodectus canis and T. subrostratus, that infest dogs (153). These lice generally cause little, if any, injury to healthy dogs. The human crab louse (Phthirus pubis) is reported to have infested a dog that had shared a bed with its owner, who was presumably infested, but the pest was removed with a flea-and-tick spray and an insecticidal shampoo (48).

D. Fleas

Misunderstanding the flea problem on dogs causes much concern among their owners. Persistence of fleas on household or kenneled dogs results from the ignorance of many owners of the flea's life history, their endurance by the host dog, and the failure to treat the dog adequately with a suitable insecticide (72). An initial flea burden may be tolerated by a dog and actually remains unknown to the owner until an eczematous dermatitis occurs. Once the dermatitis occurs, the dog's tolerance to the flea burden diminishes. The host dog will bite and scratch the infested areas to reduce the flea population, but the skin damage promptly intensifies and the degree of skin irritation is directly proportional to the patient's hypersensitivity (78). About 15% of all dogs are sensitive to flea bites and frantically chase buzzing insects, especially in the summer months when the fleas are more active (51, 77). Flea eradication is the main treatment, usually with approved insecticides, for example, malathion (0.5%), lindane (1%), ronnel 500mg/ 10 lb body weight, or trichlorphon, although injections of a flea antigen have been used effectively (146-150). The use of flea collars (impregnated with dichlorvos) on both dogs and cats has raised some controversy, but their judicious use, accompanied by frequent checking of the neck for any adverse reaction, effectively reduces the incidence of fleas on the animals (44, 65, 76, 95, 110, 125, 139). The dog flea Ctenocephalus canis is the vector of a dog tapeworm Dipylidium caninum. It can transmit brucellosis and can also attack both man and cats. The human flea Pulex irritans is also often found on dogs and cats. All species of fleas that infest household pets and often, coincidentally, attack humans can be readily controlled by treating their breeding sites with sprays or dusts of approved residual insecticides.

Adult fleas lay their eggs on the host animal or in dust and the larvae hatch in 2 to 16 days. The larvae have no legs and require very little food. The larvae moult twice reaching the third larval stage in 9 to 15 days, depending on environmental conditions, then it spins a cocoon, inside which it pupates (88). This stage lasts from 7 days to a year, thus the entire cycle will vary from 18 days to months.

E. Acarids—Mites and Ticks

Demodectic mange is a disease primarily of young shorthaired dogs and affects the head, the limbs, and occasionally the entire body (2). On occasion the condition may develop into a pustular form, which involves almost the entire body. The mite responsible for this disease can complete its entire life cycle within a hair follicle, while those mites which migrate to other tissues may have been transported via the blood or lymph (2, 82). The follicular mite (Demodex canis) may be present without causing skin lesions. This suggests that the nutritional level and resulting skin metabolism of the host may be involved, to some extent, in the pathogenesis of follicular mange (54, 83, 85, 167). In fact, a dietary regimen of raw meat may alleviate so-called red mange (101). All the life stages of Demodex have been found in blood lymph nodes and other tissues of infested dogs with skin lesions, and this has a bearing on the results obtained with topical treatment (45). The skin lesions are actually a severe dermatitis, presumably caused by the mites dilating the hair follicles and sweat glands, thereby encouraging bacteria to invade the skin and elicit inflammation (85). These lesions may vary from small squamous areas with denuded hair to extensive pustular lesions covering the entire body. The early type usually occur around the head, while the later pustular lesions occur over the entire body, with purulent material oozing from the reddened, hypertrophied wrinkled skin. Treatment with either oral or topical organophosphate acaricides are effective even in advanced cases. These include cythioate, benzyl benzoate, rotenone, ronnel, ruelene, and trichlorphon (57, 85, 100, 102, 111, 123, 124, 180, 182, 192). Trypan blue has also been used with other therapeutic medicaments (68, 90, 121). To control bacterial infection antibiotic and sulfa drug ointments are applied. Feeding a high grade protein diet and supplementing it with a high level multivitamin therapy is also indicated.

There are other mites which affect dogs, namely Otodectes cynotis, the ear mite, and Cheyletiella parasitivorax, a mite most often associated with rabbits (4, 40, 46, 47, 87, 133, 161, 183). The ear mite causes a mild to severe irritation of the ear canal with the formation of fetid black wax. Treatment includes thorough irrigation of the auditory canal and the introduction of a mild topical acaricide. For the free, living mange mite Cheyletiella, whose life cycle, as yet, is unknown, treatment includes thoroughly washing the infected site and application of an acaricide. As these mites may also affect both humans and cats, they constitute a public health problem.

Because canine scabies, Sarcoptes scabei var. canis, also causes pustular lessions on man, care must be taken by people to preclude transfer when handling infected dogs (43, 157). Fortunately the condition in man is not actually severe because the parasite is rather host specific and furthermore it can be readily treated with gamma benzene hexachloride (lindane) applied in a cream base. Treatment of dogs is more difficult, but with thorough washing of the affected areas the use of approved acaricides and improvement of the general health of the animal will rid it of this parasite. VAPONA (dichlorvos) bar treatment in the bedding has also been used effectively as a control (185).

Tick infestations on canines pose considerable distress, especially during warmer weather. In all areas of the country the brown dog tick, Rhipicephalus sanguineus (Fig. 19) has become harder to control because it has developed resistance to acaricides, particularly the chlorinated

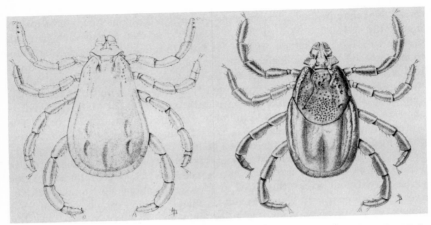

FIG. 19. <u>Rhipicephalus sanguineus</u>. The brown dog tick. (Left) Male.
(Right) Female. (Drawing by H. B. Bradford. Photo by courtesy of Pratt.)

hydrocarbons (51). Infestation of the home can readily occur if dog owners
fail to treat their premises adequately, and human parasitism by ticks has
been frequently reported (115). An engorged gravid female tick can lay up
to 4000 eggs, which hatch into larvae in about 30 days (128). After ingesting
host blood for from 2 to 6 days, these larvae drop off, moult in from 5 to
23 days to become sexually immature nymphs. They can live for months
without blood, but after attaching to a host again and feeding for from 4 to 9
days, they again drop off the host animal and mature to adults in from 11 to
73 days. The male impregnates the female and the cycle is repeated.

Because some tick species and strains have become resistant to the
chlorinated hydrocarbon acaricides, several organophosphorus compounds
(viz., malathion, diazinon, dicapthon, dioxathion, dichlorvos, ronnel,
and coumaphos) which are very effective are now employed to control ticks
on dogs and other domestic animals (5, 73, 89, 116, 128).

VII. ECTOPARASITES OF FELINES

Among the ectoparasites of the cat, the ear mite (<u>Otodectes cynotis</u>),
the common mange mite (<u>Notoedres cati</u>), and the flea (<u>Ctenocephalus felis</u>)
are encountered most frequently. The latter can reside on human hosts
also. Oral systemic insecticide therapy and topical applications have been
successful in the treatment of ticks, fleas, lice, demodectic mange,

sarcoptic mange, and ear mites of the cat (14, 15, 38, 161). Flea collars (impregnated with dichlorvos) are also used for flea control in cats (42). Occasionally Cuterebra americana or C. horripilum larvae cause lesions in cats (58, 115). Curiously the clover mite (Bryobia praetiosa) has been reported to cause acariasis in the cat (91). This mite thrives in a warm environment and will often enter homes in the fall and sequester in cracks and crevices. Sarcoptic mites (Felicola subrostratus), the only cat louse, and some ticks may, on occasion, parasitize cats and can be treated by the use of approved acaricides.

VIII. ECTOPARASITES OF SMALL ANIMALS AND AVIANS

With the increasing demand for and the importance of laboratory animals, the care and disease control of rabbits, rats, mice, guinea pigs, and hamsters are being thoroughly investigated. These species can be hosts for a wide variety of fleas, lice, mites, and ticks, of which only a few will be mentioned here.

Rabbits can become infested with the ear mites (Psoroptes communis cuniculi and Chorioptes cuniculi), causing ear mange or canker. Other mites include Dermanyssus and Lyonyssus; the common red mites of the domestic hen; Lyponyssus bacoti, the tropical rat mite which can transmit endemic typhus in man; sarcoptic and notoedric mites, which cause scabby irritations; demodectic mites and Cheyletiella parasitivorax (3, 5, 7, 88, 163, 193); Ctenocephalides cuniculi the common flea; and Hemodidsus ventricosus, the rabbit louse, have also been found on rabbits. The tick, Haemaphysalis leporis paulstris, is a reservoir for Rocky Mountain spotted fever and is found on wild rabbits. All of the above ectoparasites can be readily controlled with approved insecticides.

Lice commonly infest mouse (Polyplax serrata) and rat (Polyplax muris) colonies; both can transmit protozoan blood parasites (6). Often severe pruritus on the face and head occur primarily from the mites Myobia musculi (mice) and Myobia ratti (rats); there is no damage to hair follicles. And Psorergates simplex, the follicle inhabiting mite, produces small white nodules on the visceral skin surface; intracutaneous, thick-walled brown eggs are commonly found. Rat fleas can vector bacterial diseases to man from both mice and rats. Little can be done to prevent the breeding of rat fleas other than that all such ectoparasites can be swiftly controlled by insecticidal powders or sprays.

Though hamsters are generally free of spontaneous disease, they can become infected with scabies, usually due to improper care (62). In the guinea pig, two common lice species (Gryopus ovalis and Gliricola porcelli) may cause emaciation and rough hair coat (10). As with other species, these parasites can be controlled with appropriate residual insecticides, such as DDT or lindane.

Both commercial poultry and pet birds are subject to flea, lice, and mite infestations, with most of those species that affect poultry being cross-communicable also to pet birds. Mites that suck blood from the surface of the body are the red mite (chicken mite or Dermanyssus gallinae) and the feather mite (northern fowl mite or Lyponyssus sylviarum). The life cycles of these mites vary somewhat. The red mite lays its eggs in crevices and under dry crusts of manure near or on perches, while the northern fowl mite cements its eggs on to the soft part of the host's feathers (153). Under favorable conditions, the eggs hatch in about 48 hr and the young mites promptly grow to maturity. To eradicate either of these species of mites, the poultry house should be thoroughly cleaned and sprayed with a coal tar type disinfectant or acaricide. Painting the roosts with nicotine sulfate has also been recommended but is rather ephemeral in effect. The scaly-leg mite (Cnemidocoptes mutans) is an itch mite that will attack chickens, turkeys, pheasants, and caged birds, usually starting between the toes and spreading up the leg. Dipping the feet and shanks in a crude oil or kerosene materially aids in control but is obviously tedious in large flocks. Effective pesticides are more practical. Other mites that attack birds include Cytodites nudus (air-sac mite), Trombicula irritans (chigger mite), Cnemidocoptes gallinae (depluming mite), Syringophilus bipectinatus (quill mite), Laminosioptes cysticola (subcutaneous mite), and Lyponyssus bursa (tropical fowl mite). There have been reports of Sternostoma tracheacolum being associated with respiratory ailments in the budgerigar—a parakeet from Australia (96).

Two important species of fleas infest poultry (153). The western hen-flea (Ceratophyllus niger) is a voracious feeder and that also occasionally attacks man. Infested birds become emaciated, produce less eggs, and, in severe cases, actually die. In the Southern states, the sticktight flea (Echidnophaga gallinacea) frequently infests poultry and will also attack the ears of dogs and cats. This parasite occasionally infests horses, mules, asses, swine, cattle, and humans. On poultry, this flea clusters on the comb, wattles, and around the eyes, and when heavily infested, young fowls will often quickly die. Thorough steam cleaning of the poultry houses followed by appropriate insecticide spraying is required to eradicate these two flea species from poultry.

At least seven different species of biting lice attack poultry (153). They spend their entire life cycle on the bird, living on various parts of their host. The constant irritation of such infestations cause poultry to become droopy and to develop ruffled feathers, low vitality, and low resistance to disease and to elicit a decrease in egg production. These lice feed on feathers, dry scales, and scabs on the skin. Some of the species include Lipeurus heterographus (head louse), Menacanthus stamineus (body louse) (Fig. 20), Menopon

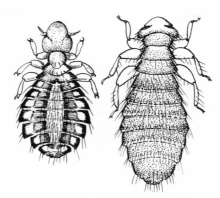

FIG. 20. Poultry lice <u>Lipeurus heterographus</u>—head louse (left). <u>Menacanthus stamineus</u>—body louse (right). (Photo: U.S. Department of Agriculture.)

<u>gallinae</u> (shaft louse), <u>Lipeurus caponis</u> (wing louse), and <u>Eomenacanthus stamineus</u> (yellow body louse). One control of chicken lice, is to paint the top of the perches with undiluted nicotine sulfate just before the flock roosts. Residual insecticides, applied either as powders or sprays, are generally more effective.

In some parts of the south, chickens and sometimes both pigeons and turkeys become infested with the powerful bloodsucking chicken tick (<u>Argas persicus</u>). This tick is very persistent and all hiding places should be removed prior to applying an acaricide. A suitable residual disinfecting compound, such as carbolineum, should be applied to all roosts, roost supports, and nests. If only a few birds are infested, they may be dipped in a 2% solution of creolin.

IX. SYSTEMIC PESTICIDES IN ECTOPARASITE CONTROL OF LIVESTOCK AND PETS

After World War II, there was a great increase in the use of synthetic chemical insecticides (84). Some of these newer compounds are systemic parasiticides, so named because they are absorbed from internal or external body surfaces or from subcutaneous or muscle tissue, and are transported by the blood vascular system to remote tissues, where they destroy the target parasites (129). Such systemic parasiticides are generally organophosphate or carbamate compounds (see Vol. 1, Part 1, Chap. 1). For example, the first instar grub of <u>Hypoderma lineatum</u> and <u>Hypoderma bovis</u> can be killed before they migrate to the back of cattle, where they cause economic damage, by the oral administration of a single bolus of ronnel at the rate of 100 mg/kg body weight (22, 74, 103, 109, 129, 131). Ronnel will also coincidentally destroy several other ectoparasites when so administered, but the action is so fleeting it offers no great advantage over more commonly available compounds (22). Applied as a spray, a technical grade of ronnel is recommended as a topical parasiticide to control horn-flies, lice, screwworms, fleeceworms, and sheep keds (94, 129). Used as a "pour-on," ronnel is also effective in controlling cattle grubs and lice, while eliciting no harmful effects to pregnant cows or to young calves nursing such treated cows (70, 138). Coumaphos (CO-RAL, BAYER 21/199), another organophosphate insecticide, controls first-instar cattle grubs when applied to cattle as a spray at a concentration of 0.5% of the active chemical (11, 52, 129). It will also kill screwworms, hornflies, lice, and ticks when so applied. Both also show some anthelmintic activity when orally administered. Other systemic insecticides, some of which may also have anthelmintic properties, include dimethoate, dioxathion, reulene, fenthion, famphur, trichlorophan, dichlorvos, phenothiozine, naphthalophos, and are effective against biting flies, mosquitoes, fleas, lice, ticks, and mites (21, 106, 114, 122, 126, 129, 136, 142, 153, 181). Dust bags, back rubbers, sprays, dips, low volume applications of ready-to-use oil base sprays, ultra low volume aerial applications of highly con-centrated formulations, and impregnated mineral blocks are the various methods available to administer such systemic insecticides.

The ideal parasiticide should rid the host of all its parasitic burden with a single administration (129). It should be safe for all species of animals, including wildlife, and it should leave no residue in the edible tissues of the treated host that could create food hazards. The term residue might be defined as persistent small quantities of a chemical or its active metabolites in a living organism, as opposed to deposits of such chemicals on the surface of a treated host or that of an inaminate object (106). Of the many substances that are registered as drugs or pesticides, only some 17 are permitted tolerances in meat, none are permitted in milk, and only two are permitted in eggs (190). This "zero tolerance" or "no residue" in milk poses awkward problems for the farmer because with the use of ex-ceedingly refined and sophisticated modern assay methods, the limits of detection of pesticide residues reach down to the order of parts per trillion, where previously they were at a tenth part per million (1, 130). The physiology and metabolism of the active compound must be known. The active compound must be readily absorbed through the surface to which it is applied and then transported throughout the animal's body to be an effective systemic parasiticide. Often the derived metabolites may be more toxic to the host than is the parent compound. Organic phosphorus insecticides that are approved for use as animal systemics are usually readily metabolized by the treated animals. They destroy the parasite

either by the action of the original compound or by one or more degradation
products produced by the host or by the parasite itself (129). The route of
administration of the pesticide, its compatibility with other drugs applied
simultaneously or sequentially, and its toxicity must be carefully considered. Also, there is the important effect which massive quantities of
dead parasites might exert upon the host. Most of the organophosphate
insecticides, including some systemics, are used only on crop plants,
for at insecticidal levels they are quite toxic to livestock (99).

X. CONTACT PESTICIDES IN ECTOPARASITE CONTROL
OF LIVESTOCK AND PETS

The earlier group of insecticides applied to livestock, and still in
current use, is the chlorinated hydrocarbons (169). These compounds,
however, tend to accumulate in the fat and persist and they therefore create a more serious residue problem than do the organophosphates. They
act on contact as neutral and general protoplasmic toxins primarily because
of their affinity for lipids. Examples are chlordane, DDT, TDE, hepta-
chlor, toxaphene, aldrin, dieldrin, endrin, benzene hexachloride (its
gamma isomer-lindane), o-dichlorobenzene, dilan, methoxychlor, and
strobane (153). Many of the above are restricted to use only as crop
insecticides, but their indiscriminate use in the field has induced toxicity
in livestock. Resistance by some livestock insects to the organochlorine
insecticides has been reported (175). Another group of livestock insecticides includes the synthetic organic botanicals (e.g., allethrin, pyrethrin
for control of lice, houseflies, and mosquitoes); the ill-defined fossil
hydrocarbons (e.g., coal-tar, creosate dip, crankcase oil, carbolineum—for mites and ticks); the inorganics (e.g., silica aerogel and arseni-
cals for dips); the carbamates (e.g., carbaryl—for fleas, lice, ticks,
and mites, houseflies, and mosquitoes) (153).

XI. CONCLUSIONS

Without the use of pesticides for livestock animals, meat, milk,
egg, and wool production would seriously decline, and food and fiber prices
would sharply increase (69). As a nation, the United States can be proud

of its role in the research, development, and application of these chemical compounds. Recently, however, there has been a great deal of public concern, and perhaps rightly so, concerning the indiscriminate use of pesticides. Unfortunately, much of this agitation is overplayed by the press, radio, and TV, which has led to unnecessarily restrictive legislation. Before any new pesticidal compound can be released for general use in the environment, industry and government must spend millions of dollars to prove both the efficacy and the safety of each new product. The essence of this extensive and expensive research is distilled into the registered label, which too often some people fail to read and therefore fail to follow explicit directions, with the result that accidents, some of which end in tragedy, too often occur. Indiscriminate use of pesticides by farmers (who in general tend to be exceedingly cautious), and more frequently by urban and suburban homeowners, has also contributed to environmental pollution, albeit to a relatively minor extent. The rational use of pesticides must, however, be continued to sustain our present dietary standards, at least until biological methods for the control of ectoparasites affecting livestock, pets, and birds can be achieved. With the careful and discriminate use of some of the appropriate compounds discussed, it will be possible to control, both practically and efficiently and, in some instances, perhaps eradicate numerous species of flies, fleas, mosquitoes, lice, mites, and ticks that currently impair and harass our domestic livestock and pets.

APPENDIX

Ectoparasite Control of Livestock—Read Labels

Pest	Recommended pesticide	Formulation	Application Method	Application Rate	Permissible tolerance	Remarks
Cattle-beef barn and feed lots						
Housefly (Musca domestica)	Fenthion	BAYTEX E.C. (46%)	Spray	1 gal, 1.5%/500 ft²	0.1 ppm in meat and fat	Remove animals to spray
	Dimethoate	CYGON 2E (23.4%)	Spray	1 gal, 1%/100 ft²	0.02 ppm in meat and fat	Remove animals to spray
	Dichlorvos	DDVP	Mist spray	5 gal, 0.5%/acre	0.02 ppm in meat and fat	Remove animals, avoid feed and water
Horn fly (Haematobia irritans)	Methoxychor	E.C. (25%)	Backrubber	5% oil, 1 gal/20 ft	3 ppm in fat	Avoid feed and water
		Dust (10%)	Self-treatment	10 lb/bag	3 ppm in fat	Avoid feed and water
		Wettable powder (50%)	Hand dust individuals	10 lb/bag	3 ppm in fat	Avoid feed and water
Horn fly, lice sucking 'short nose', Haematopinus eurysternus 'long nose'	Toxaphene	Dust (5%)	Self-treatment	10 lb/bag	7 ppm in fat	Avoid feed and water
		E.C. (60%) Wettable powder (40%)	Spray or dip	0.5%	7 ppm in fat	Avoid feed and water

APPENDIX (Continued)

Cattle—beef barn and feed lots (Continued)

Pest	Recommended pesticide	Formulation	Application Method	Rate	Permissible tolerance	Remarks
Lignognathus vituli, lice biting Damalina bovis, ticks, various Amblyomma sp.	Ronnel	E.C. (60%)	Backrubber	5% oil, 1 gal/20 ft	7 ppm in fat	Avoid feed and water
		KORLAN	Spray or dip	0.5%	10 ppm in fat	Apply at 2-wk intervals, wait 7 days
		E.C. (24%)	Backrubber	0.25%	4 ppm in fat	No wait[a] period
			Backrubber	1% in oil, 1 gal/20 ft	10 ppm in fat 4 ppm in meat	No wait[a] period
Boophilus sp., Dermacentor sp., Ixodes sp. and other biting flies	Coumaphos	CO-RAL	Spray	0.5%	1 ppm in fat or meat	Not less than 90 days of age
		E.C. (11.6%)		0.25%		
		Wettable powder (25%)	Dip			
		E.C. (11.6%)	Backrubber	1% in oil, 1 gal/20 ft	1 ppm in fat or meat	Avoid feed and water
		Dust (1%)	Dust individuals	25 lb/bag	1 ppm in fat or meat	Avoid feed and water

	Chemical	Product	Application	Dosage	Tolerance	Remarks
Grub (Hyperderma bovis) and lice (see above)	Crufomate	RUELENE E.C. (25%)	Spray, pour on	0.5%, 1 qt. – 1 gal/hd 8.3%, 1 fl oz/cwt	1 ppm in fat and meat	Only once each 30 days, wait 7 days
		RUELENE (9.4%)	Pour on as is	1 fl oz/cwt	1 ppm in fat and meat	Only once each 30 days, wait 7 days
		RUELENE E.C. (35%)	Dip	0.25%		Follow label instructions
	Famphur	WARBEX (13.2%)	Pour on as is	0.5 fl oz/cwt	0.1 ppm in fat and meat	Once per season only, wait 35 days
		FAMIX medicated premix	In feed	0.199%, 12 lb premix/T feed 0.25 lb/cwt daily to 100 0.125 lb/cwt daily to 300	0.1 ppm in fat and meat	Must not be used coincident with other insecticides, wait 4 days Do not treat animals under stress
Grubs, lice, and hornfly	Phosmet	PROLATE E.C. (11.5%)	Pour on	4%, 1 fl oz/cwt	0.2 ppm in fat and meat	Maximum 8 fl oz/hd, wait 21 days
			Spray to wet	0.25%	0.2 ppm in fat and meat	Apply at 7-day intervals only, wait 21 days
	Trichlorfon	NEGUVON W.P. (80%)	Spray to wet	1.0%	0.1 ppm in fat and meat	Do not apply coincident with other medicaments or under stress, wait 14 days

APPENDIX (Continued)

Pest	Recommended pesticide	Formulation	Application		Permissible tolerance	Remarks
			Method	Rate		
Cattle–beef barn and feed lots (Continued)						
Hornfly, lice, ticks, and other biting flies	Crotoxyphos	CIODRIN	Spray	1%, 1 qt/hd	0.02 ppm in fat and meat	
		E.C. (14.4%)	Pour on	1% in oil, 1 gal/20 ft		
			Backrubber			
		Dust (3%)	Self-treatment	8 lb/bag	0.02 ppm in fat and meat	
	Crotoxyphos and dichlorvos	CIODRIN (1%) plus DDVP	Oil mist spray	1 fl oz/hd	0.02 ppm in fat and meat	Apply daily
	Dioxathion	DELNAV	Spray	1 qt, 0.15%/hd	1 ppm in fat	Only once each 30 days
		E.C. (21%)	Backrubber	1.5% in oil, 1 gal/20 ft		
	Malathion	Tech (95%)	ULV aerial spray	6-8 oz/acre	4 ppm in fat and meat	
		E.C. (57%)	Spray individuals	0.5%		
			Backrubber	2% in oil, 1 gal/20 ft	4 ppm in fat	
		Dust (5%)	Dust (5%) individuals	5%		Apply as required to cover

Grubs only	Trichlorfon	NEGUVON (8%)	Pour on as is	0.5 fl oz/cwt maximum 4 fl oz/hd	0.1 ppm in fat and milk	See above, wait 21 days
	Rommel	TROLENE FM-premix	In feed (0.6%)	0.3 lb premix/cwt daily to 7 days	10 ppm in fat, 4 ppm in meat	Do not apply coincident with othermedicaments or under stress, wait 10 days
			In mineral supplement, 3%–6%	0.6 lb, 3% or 0.3 lb, 6%/10 cwt daily to 7 days		
	Fenthion	TIGUVON (3%)	Pour on as is	0.5 fl oz/cwt	0.1 ppm in fat and meat	Apply at 2–wk intervals, wait 35 days. Do not apply coincident with other medicaments or under stress
	Rommel	STEER-KLEER 5.5% in mineral block	Feed as is	1 block/15 hd or 0.25 lb/cwt to 75 days	10 ppm in fat, 4 ppm in meat	Repeat at 2–wk intervals
Face fly (Musca autumnalis)	Crotoxyphos	CIODRIN (3%) dust	Dust individuals as is		0.02 ppm in fat and meat	Repeat at 2–wk intervals
	Crotoxyphos and dichlorvos	CIOVAP (1%) CIODRIN + (0.025%) DDVP	Mist spray as is	2 fl oz/hd	0.02 ppm on fat and meat	Apply at not less than 180 days age; not to Brahmans

APPENDIX (Continued)

Pest	Recommended pesticide	Formulation	Application		Permissible tolerance	Remarks
			Method	Rate		
Cattle–beef barn and feed lots (Continued)						
Dichlorvos		DDVP 1% oil	Mist spray	2 fl oz/hd	0.02 ppm in fat and meat	Apply at not less than 180 days of age; not to Brahmans
		E.C. (21.8%)	Spray	1.8%, 1–2 fl oz/ hd daily	0.02 ppm in fat	Apply at not less than 180 days of age; not to Brahmans
			Face paint	Add 1 lb sugar/ gal 1% spray	0.02 ppm in fat and meat	Apply at not less than 180 days of age; not to Brahmans
Maggots in wounds, e.g., screwworm, Callitroga hominivorex	Coumaphos	CO–RAL (1%–5%) dust	Hand dust individual wounds	1%–5% as needed	1 ppm in fat and meat, 0.5 ppm in milk	Repeat to complete control
		Spray (3%) pressur- ized	Spot spray wounds	3% as needed	1 ppm in fat and meat, 0.5 ppm in milk	Wait[a] 14 days
	Romnel	KORLAN spray (0.25%) pressur- ized	Spot spray wounds	2.5% as needed	10 ppm in fat, 4 ppm in meat	Repeat to complete control

Cattle–Dairy barns

Housefly	Dimethoate	CYGON 2E (23.4%)	Spray inside and outside barns	1%	0.002 ppm in milk; 0.02 ppm in fat and milk	Remove animals to spray; not in milkrooms
	Fenthion	BAYTEX E.C. (46%)	Spray	1 gal, 1.5%/ 500 ft²	0.01 ppm in milk; 0.1 ppm in fat and milk	Remove animals to spray

Dairy Cattle and Milkrooms

Housefly	Rommel	KORLAN E.C. (24%)	Spray	1%	1.25 ppm in milk and fat	Remove animals to spray; avoid milking equipment
	Dichlorvos	DDVP E.C. (21.8%)	Spray	1 qt, 0.5%/ 1000 ft²	0.02 ppm in milk, fat, and meat	Avoid feed, water, milking equipment
			Mist spray	1 pt 1%/8000 ft³		Do not expose animals treated with other medicaments within 24 hr
		DDVP (1% oil)	Aerosol mist as is	1 pt/8000 ft³	0.02 ppm in milk and fat	Avoid feed, water and milking equipment
						Do not expose animals treated with other medicaments within 24 hr
		DDVP (20%) pest strip	Suspended as is	1 strip/ 1000 ft³	0.02 ppm in milk	No restrictions

APPENDIX (Continued)

Pest	Recommended pesticide	Formulation	Application		Permissible tolerance	Remarks
			Method	Rate		
Dairy cattle and milkrooms (Continued)						
Housefly	Pyrethrins and piperonyl butoxide	0.1%–0.6%	Aerosol mist as is	Fog area completely	0.5 ppm in milk, fat; 0.1 ppm in fat and meat	Very short-term action, repeat as needed with reinfestation
			As synergist to pyrethrins	Fog area completely	0.1 ppm in milk, 0.25 ppm in fat and meat	
Cattle—dairy, milking						
Horn fly, biting flies, lice	Coumaphos	CO-RAL (1%–5%) dust	Self dusting	Apply as is in bag and suspended in exits	1 ppm in fat and meat, 0.5 ppm in milk fat	Avoid feed and water
		CO-RAL E. C. (11.6%)	Backrubber	1 gal 1% oil/ 20 ft	1 ppm in fat and meat, 0.5 ppm in milk fat	Avoid feed and water
	Romnel	KORLAN 4E (41.2%)	Backrubber	1 gal 1% oil/ 20 ft	10 ppm in fat, 4 ppm in meat, 1.25 ppm in milk	Avoid feed and water
	Cryotoxyphos	CIODRIN dust (3%)	Hand or self treatment,	As is in dusting bags	0.02 ppm in fat, meat, and milk	Hand dust at 2-wk intervals, self dusting daily

Pest	Common name	Formulation	Application	Rate	Tolerance	Remarks
Horn fly, face fly, biting flies, lice		CIODRIN E.C. (14.4%)	suspended bags at exits; Spray or pour on Backrubber	1 qt 1%/hd, 1 gal 1% oil/20 ft	0.02 ppm in fat, meat, and milk	Apply at weekly intervals ad libitum
	Dichlorvos	CIOVAP (1%)	Mist spray	1–2 fl oz oil/hd	0.02 ppm in fat, meat and milk	Apply daily, do not wet animals
	Malathion	Dust (4%–5%)	Hand dust	3–4 tbsp/hd	4 ppm in fat, 0.5 ppm in milk and fat	Wait 5 hr before milking
	Methoxychor	Wettable powder (50%)	Hand dust	5 g/hd	0.05 ppm in milk	Apply at 3-wk intervals to back and neck, rub into coat
	Dichlorvos	DDVP (2.18%)	Spray daily, face paint daily	1–2 fl oz, 1% hd; add sugar lb/gal	0.02 ppm in fat, milk, and meat	Do not apply below 180 day age or to Brahmans
		DDVP (1%) oil	Spray as is daily	1–2 fl oz/hd	0.02 ppm in fat, milk and meat	Do not apply below 180 days age or to Brahmans
Maggots in wounds, e.g., screwworm	Coumaphos	CO-RAL (1%–5%) dust	Hand dust to wound	As is	1 ppm in fat and meat, 0.5 ppm in milk	Repeat as needed to control
	Rommel	KORLAN (2.5%) pressure can	Aerosol spray to wound area	As is	1.25 ppm in milk and fat	Repeat as needed to control

APPENDIX (Continued)

Pest	Recommended pesticide	Formulation	Application		Permissible tolerance	Remarks
			Method	Rate		
Swine						
Houseflies and biting flies (see also poultry below)	Pyrethrins plus piperonyl synergist	0.1%–0.6%, 0.6%–6%	Mist spray pen quarters	As is	0.1 ppm in fat and meat, 0.1 ppm in fat and meat	Very short-term action; repeat as needed with reinfestation
	Dichlorvos	DDVP (21.8%) E.C.	Dry sugar bait	1 lb/500 ft^2	0.1 ppm in fat and meat	Wear rubber gloves to prepare bait or spray, avoid skin contact, do not spray animals
	DIBROM	NALED E.C. (37%)	Spray pens	0.5%	Presently extended, may be restricted—check	Avoid skin contact, do not spray animals
			Sugar bait	1 lb/500 ft^2		
		(1%) oil	Spray pens	3%	Presently extended, may be restricted—check	Do not apply coincident with other medicaments or stress
			Mist spray daily	1 fl oz/3000 ft^3		
	Dimethoate	CYGON 2E (23.4%)	Spray pens and litter as larvacide	1 gal 1%/150 ft^2	0.02 ppm in fat and meat	Avoid spraying animals, feed or water

Parasite	Pesticide	Formulation	Method	Rate	Tolerance	Remarks
Lice	Tetrachlor-vinphos	RABON (50%) wettable powder	Spray pens and litter as larvacide	1 gal 1%/100 ft^2	0.1 ppm in meat; 0.75 ppm in fat	Avoid animals, feed, and water
	Rommel	KORLAN granules (5%)	Broadcast by hand in pens	0.5 lb/100 ft^2	3 ppm in fat, 2 ppm in meat	Not more than 14-day intervals
		KORLAN (24%)	Spray individuals	Up to 1 qt, 0.5%/hd	3 ppm in fat, 2 ppm in meat	Only spray at 30 days age and over
	Malathion	E. C. (57%)	Spray individuals	Up to 1 qt, 0.5%/hd	4 ppm in fat and meat	Repeat to control as needed
		Dust (4%-5%)	Hand dust individuals	Sufficient to cover	4 ppm in fat and meat	
	Dioxathion	DELNAV E. C. (21%)	Spray individuals	Up to 1 qt, 0.15%/hd	1 ppm in fat and meat	Apply at 90 days age or over only
	Coumaphos	CO-RAL E. C. (11.6%)	Spray individuals	Up to 1 qt, 0.125%/hd	1 ppm in fat and meat	Apply at 90 days age or over only
	Crotoxyphos	CIODRIN E. C. (14.4%)	Spray individuals	Up to 1 qt, 0.25%/hd	0.02 ppm in fat and meat	Apply at 30 days age or over only
Lice and mange	Lindane	E. C. (20%) wettable powder	Spray individuals	Up to 1 qt, 0.05%/hd	4 ppm in fat	Apply at 30 days of age or over only, wait periods: spray = 30 days; dip = 60 days

APPENDIX (Continued)

Pest	Recommended pesticide	Formulation	Application Method	Application Rate	Permissible tolerance	Remarks
Swine (Continued)						
	Toxaphene	E.C. (60%)	Spray individuals	Up to 1 qt, 0.5%/hd	7 ppm in fat	Apply at 90 days age or over only
Maggots in wounds	Lindane	(3%) oil	Apply locally in wound area	As is	4 ppm in fat	Repeat to complete control
Sheep and goats						
Lice, keds, ticks, mites (mange), wool maggots	Coumaphos	CO-RAL E.C. (11.6%)	Spray or dip each animal	0.125%	1 ppm in fat	Wet or immerse animals thoroughly; do not treat at lactation or within 14 days of freshening or before 90 days of age or sick or medicated animals; wait[a] 15 days.
		Wettable powder (25%)			0.5 ppm in milk	
		CO-RAL (0.5%) dust	Rub in individually	1-2 oz/hd		
	Dioxathion	DELNAV E.C. (21%)	Spray or dip each animal	0.15%	1 ppm in fat	Wet or immerse animals thoroughly; do not treat milking goats at less than 14-day intervals or animals less than 90 days old

Lindane	E.C. (20%)	Spray or dip each animal	0.05%	7 ppm in fat	Wet or immerse animals thoroughly; wait[a] – spray 30 days, 60 days; none before 90 days age
	Wettable powder (25%)				
Malathion	E.C. (57%)	Spray or dip each animal	0.5%	4 ppm in fat	Wet or immerse animals thoroughly; do not apply to lactating does; no wait[a] period
	Dust (4%–5%)	Hand dust each animal	As is		
Ronnel	KORLAN E.C. (24%)	Spray or dip each animal	0.5%	10 ppm in fat, 4 ppm in meat	Wet or immerse animals thoroughly; wait[a] 28 days; apply at 14-day intervals; do not apply to lactating does or within 7 days of freshening
Toxaphene	E.C. (60%)	Spray or dip	0.5%–0.25%	7 ppm in fat	Wet or immerse animals thoroughly; wait[a] 28 days; do not apply to lactating does
Maggots in wounds	Coumaphos CO-RAL dust (5%)	Apply locally to wound	As is	1 ppm in fat and meat; 0.5 ppm in milk	Repeat to complete control

APPENDIX (Continued)

Pest	Recommended pesticide	Formulation	Application Method	Rate	Permissible tolerance	Remarks
Sheet and goats (Continued)						
	Lindane	E. Q. 335 Smear (3%)	Apply locally to wound	As is	7 ppm in fat	Repeat to complete control
	Ronnel	KORLAN Smear (5%) Aerosol (2.5%)	Apply locally to wound	As is	10 ppm in fat, 4 ppm in meat	Repeat to complete control
Horses						
Bots	Trichlorfon	ANTHON (90%) powder	Mix in feed	5 gm/250 lb body weight		Do not treat more than at 30-day intervals, not before 120 days of age or within 10 days of other insecticides, or when sick or debilitated or mares within 30 days prepartum, or animals intended for human consumption.

Parasite	Pesticide	Formulation	Application	Concentration	Remarks
Horsefly, deer fly, other biting flies	Pyrethrins plus piperonyl butoxide synergist	0.05% 0.5%	Mist spray or wipe on	As is	Sprays often frighten horses, hand wipe-on is simpler, dose must be accurate because horses are very sensitive to toxicants; action very ephemeral; repeat as necessary
	Dichlorvos	DDVP 1% in oil	Mist spray or fog	As is	Do not apply to horses within 8 hr of other insecticide treatments
Lice, ticks, horn fly	Coumaphos	CO-RAL E.C. (11.6%) Wettable powder (25%)	Spray to good coverage	0.125%	Do not treat before 90 days age or if sick or distressed or coincident with other medicaments
	Crotoxyphos	CIODRIN E.C. (14.4%)	Spray to good coverage	0.5%	Not more than 7-day intervals
	Dioxathion	DELNAV E.C. (21%)	Spray to good coverage	0.15%	Not more than 30-days interval
	Malathion	E.C. (57%)	Spray to good coverage	As is	As needed at up to 2-day intervals

APPENDIX (Continued)

Pest	Recommended pesticide	Formulation	Application Method	Application Rate	Permissible tolerance	Remarks
Horses (Continued)						
Ear ticks	Rommel	KORLAN (5%) dust	Dust ears lightly	As is		Repeat as needed
Poultry—caged, layers, and others on wires						
Housefly	Dichlorvos	DDVP E.C. (21.8%)	Bait sugar	1 lb/500 ft^2	0.05 ppm in meat, fat, or eggs	Wear rubber gloves to mix; keep away from animals
			Spray roosting area	0.5%		Do not spray birds; avoid feed and water
	NALED	DIBROM E.C. (37%)	Bait sugar (dry)	1 lb/500 ft^2	Extended, may become restricted	Wear rubber gloves to mix; avoid skin contact; keep away from animals
			Sugar	3%		Do not spray birds; avoid feed and water
		(1%) oil	Mist spray or fog	1 fl oz/3000 ft^3	Extended, may become restricted	Use daily or as needed for control; do not use coincident with other medicaments or under stress

	Pyrethrin plus piperonyl butoxide synergist	(0.1%) (0.5%)	Mist spray or fog	As is	Extended, may become restricted	Action very ephemeral; repeat as necessary
	Dimethoate	CYGON 2E (23.4%)	Spray as larvacide on pens and droppings	1 gal 1%/150 ft²	0.02 ppm in fat, meat or eggs	Avoid spraying birds, feed, or water
	Tetrachlorvinphos plus dichlorvos	RABON wettable powder	Spray as larvacide on pens and droppings	1 gal 1%/100 ft²	0.1 ppm in eggs and meat, 0.75 ppm in fat	Avoid spraying birds, feed, or water
		RA–VAP E.C. (4 lb RABON + 0.25 lb DDVP/gal)	Spray as larvacide on pens and droppings	1 gal 1%/100 ft²	Dichlorvos 0.05 ppm in fat, meat, and eggs	Avoid spraying birds, feed or water

Poultry—caged or floor birds

Lice and mites	Coumaphos	CO–RAL wettable powder (25%)	Spray—high pressure	1 gal 0.25%/100 birds	1 ppm in fat and meat, 0.1 ppm in eggs	Avoid spraying food or water; use at 7-day intervals; do not use within 10 days of other insecticides

APPENDIX (Continued)

Pest	Recommended pesticide	Formulation	Application Method	Application Rate	Permissible tolerance	Remarks
Poultry—caged or floor birds (Continued)						
	Coumaphos	CO-RAL 0.5% dust	Dust birds and litter	1 lb/100 birds or 1 lb/100 ft^2	1 ppm in fat and meat, 0.1 ppm in eggs	Avoid spraying food or water; use at 7-day intervals; do not use within 10 days of other insecticides
	Carbaryl	SEVIN wettable powder (80%)	Spray—high pressure	1 gal 0.5%/100 birds	5 ppm in fat and meat, none in eggs; under extension, may become restricted	Apply at 30-day intervals; avoid food and water; do not apply within 10 days of other insecticide; wait[a] 7 days
		SEVIN dust (5%)	Dust birds and litter	1 lb/100 birds or 1 lb/40 ft^2	5 ppm in fat and meat, none in eggs, under extension, may become restricted	Apply at 30-day intervals; avoid food and water; do not apply within 10 days of other insecticides; wait 7 days
	Tetrachlor-vinphos	RABON wettable powder (50%)	Dust litter Self-treatment dust boxes	2.5 oz/100 ft^2 2.5 oz/50 birds	0.1 ppm in eggs and meat	Avoid feed and water

	Formulation	Application	Dosage	Tolerance	Remarks
Malathion		Spray—high pressure	1 gal 0.5%/100 birds		At 14-day intervals
	E.C. (57%) or dust (4.5%)	Spray—high pressure	1 gal 0.5%/100 birds	4 ppm in fat and meat	At 7-day intervals; avoid feed and water; do not apply within 10 days of other insecticide treatment or under stress
		Dust litter	2.5 oz/100 ft²	0.1 ppm in eggs	
NALED	DIBROM E.C. (37%)	Spray pens	0.3%	Extended, may become restricted	Avoid birds, feed, and water
Turkeys					
Chiggers					
Malathion	Wettable powder (25%)	Spray range	1 lb/acre	4 ppm in fat and meat, 0.1 ppm in eggs	Treat range 1 day before setting out birds; avoid water and feed
	E.C. (57%)	Individual dip	0.5% solution		
	Dust	Range broadcast	20 lb/acre		

[a]Days between treatment and slaughter.

Key: E.C., emulsifiable concentrate; W.P., wettable powders; (%) percent active ingredient. Agricultural Chemicals Handbook, Ext. Serv., Clemson University, Clemson, South Carolina, and USDA.

REFERENCES

1. R. J. Anderson, J. Am. Vet. Med. Assoc., 147:1584 (1965).

2. K. P. Baker, Vet. Rec., 86:90 (1970).

3. W. N. Beesley, Res. Vet. Sci., 3:203 (1962).

4. P. Berg and R. R. Shomer, J. Am. Vet. Med. Assoc., 143:1224 (1963).

5. R. E. Berns, Mod. Vet. Pract., 41:35 (1960).

6. G. Bjotvedt, Vet. Scope, 9:14 (1964).

7. W. P. Blount, Natl. Rabbit Raiser (1960).

8. P. R. M. Brown, D. A. M. Ferguson, K. W. Page, M. S. Smith, and J. C. Wood, Vet. Rec., 77:920 (1965).

9. W. M. Brownlie and I. R. Harrison, Vet. Rec., 72:1022 (1960).

10. N. Brewer, All Pets, 32:53 (1961).

11. H. M. Brundrett, W. S. McGregor, and R. C. Bushland, Agric. Chem., 12:36 (1957).

12. J. E. Bryant, J. B. Anderson, and K. H. Willers, J. Am. Vet. Med. Assoc., 154:1034 (1969).

13. D. L. Buelke, J. Am. Vet. Med. Assoc., 158:735 (1971).

14. G. R. Burch, Allied Vet., 31:69 (1960).

15. G. R. Burch and D. C. Brinkman, Small Anim. Clin., 2:21 (1962).

16. E. P. Cawley and C. E. Wheeler, J. Am. Med. Assoc., 163:1003 (1957).

17. T. H. Cheng, J. Econ. Entomol., 60:598 (1967).

18. T. H. Cheng, A. A. Hower, and R. K. Sprenkel, J. Econ. Entomol., 58:910 (1965).

19. J. F. Christensen and J. A. Howarth, Am. J. Vet. Res., 27:1473 (1966).

20. R. C. Collins and L. W. Dewhirst, J. Am. Vet. Med. Assoc., 146:129 (1965).

21. D. D. Cox, M. T. Mullee, and A. D. Allen, Am. J. Vet. Res., 30:1933 (1969).

22. G. L. Crenshaw, Down to Earth, 12:4 (1956).

23. G. Danelius, Cornell Vet., 52:16 (1962).

24. C. L. Davis and W. A. Leadbetter, North Am. Vet., 33:703 (1952).

25. M. Delak and I. Mijatovic, Vet. Med. Rev., 2:124 (1968).

26. G. Diamant, Vet. Med./Small Anim. Clin., 60:847 (1965).

27. C. K. Dorsey, J. O. Heishmann, and C. J. Cunningham, J. Econ. Entomol., 59:726 (1966).

28. J. H. Drudge, Vet. Scope, 6:2 (1961).

29. J. H. Drudge, S. E. Leland, Z. N. Wyant, G. W. Elam, and E. T. Lyons, Am. J. Vet. Res., 22:1106 (1961).

30. R. O. Drummond and J. M. Ossorio, J. Econ. Entomol., 59:107 (1966).

31. R. O. Drummond, S. E. Ernst, C. C. Bartlett, and O. H. Graham, J. Econ. Entomol., 59:395 (1966).

32. R. O. Drummond, T. M. Whetstone, and S. E. Ernst, J. Econ. Entomol., 60:1021 (1967).

33. R. O. Drummond, T. M. Whetstone, and S. E. Ernst, J. Econ. Entomol., 60:1735 (1967).

34. R. O. Drummond, S. E. Ernst, J. L. Trevino, and O. H. Graham, J. Econ. Entomol., 60:199 (1967).

35. R. O. Drummond, S. E. Ernst, J. L. Trevino, and O. H. Graham, J. Econ. Entomol., 61:467 (1968).

36. Editorial, J. Am. Vet. Med. Assoc., 131:431 (1957).

37. D. G. Ely and T. L. Harvey, J. Econ. Entomol., 62:341 (1969).

38. B. P. English, Aust. Vet. J., 36:85 (1960).

39. H. W. Essig and W. A. Pund, Feedstuffs, 37:34 (1965).

40. S. A. Ewing, J. E. Mosier, and T. S. Foxx, J. Am. Vet. Med. Assoc., 151:64 (1967).

41. W. F. Fisher, J. Am. Leather Chem. A., 65:547 (1970).

42. I. Fox, I. G. Bayona, and J. L. Armstrong, J. Am. Vet. Med. Assoc., 155:1621 (1969).

43. A. Freeman, J. Am. Vet. Med. Assoc., 150:529 (1967).

44. A. Freeman, J. Am. Vet. Med. Assoc., 156:6 (1970).

45. F. E. French, Cornell Vet., 54:272 (1964).

46. R. C. Frost and W. P. Beresford-Jones, Vet. Rec., 72:375 (1960).

47. R. C. Frost, J. Sm. Anim. Pract., 2:253 (1961).

48. F. L. Frye, J. Am. Vet. Med. Assoc., 152:1113 (1968).

49. N. L. Garlick, J. Am. Vet. Med. Assoc., 147:1576 (1965).

50. W. J. Gibbons, Mod. Vet. Pract., 45:78 (1964).

51. L. S. Goyings, Mich. St. Univ. Vet., 22:16 (1961).

52. O. H. Graham, J. Econ. Entomol., 51:359 (1958).

53. J. H. Greve, Mich. St. Univ. Vet., 19:84 (1959).

54. J. H. Greve and S. M. Gaafer, Am. J. Vet. Res., 25:520 (1964).

55. J. H. Greve and D. R. Cassidy, J. Am. Vet. Med. Assoc., 150:627 (1967).

56. R. J. Hart and P. Batham, J. So. Afr. Vet. Med. Assoc., 40:284 (1969).

57. J. W. Harrison, Sm. Anim. Clin., 1:362 (1961).

58. B. C. Hatziolos, J. Am. Vet. Med. Assoc., 148:787 (1966).

59. B. C. Hatziolos, Cornell Vet., 57:129 (1967).

60. W. O. Haufe, Can. J. Comp. Med. Vet. Sci., 29:102 (1965).

61. R. W. Hewetson and J. Nolan, Aust. J. Agri. Res., 19:323 (1968).

62. E. Hindle and H. Magalhaes, U.F.A.W. Handbook, 324 (1957).

63. R. A. Hoffman, I. L. Berry and O. H. Graham, J. Econ. Entomol., 58:815 (1965).

64. B. E. Hooper, Missouri Vet., 10:19 (1961).

65. R. T. Hopwood and W. Migden, J. Am. Vet. Med. Assoc., 146:1115 (1965).

66. H. Horne, Zeitschr. Fleisch und Milch Hyg., 5:126 (1895).

67. J. L. Hourrigan, R. K. Strickland, O. L. Kelsey, B. E. Knisley, C. C. Crago, S. Whittaker and O. J. Gilhooly, J. Am. Vet. Med. Assoc., 154:540 (1969).

68. L. Hurov, Mod. Vet. Pract., 42:62 (1961).

69. L. G. K. Iverson, J. Am. Vet. Med. Assoc., 151:1806 (1967).

70. J. B. Jackson, J. Russell, and S. F. Rosner, J. Am. Med. Vet. Assoc., 144:1127 (1964).

71. W. T. Johnson, J. Econ. Entomol., 54:821 (1961).

72. D. W. Jolly, Mod. Vet. Pract., 40:33 (1959).

73. D. W. Jolly, Mod. Vet. Pract., 41:36 (1960).

74. R. H. Jones, H. M. Brundrett, and R. D. Radeleff, Agric. Chem., 12:45 (1957).

75. T. H. Jukes, Am. Scientist, 51:355 (1963).

76. R. M. Kibble, Aust. Vet. J., 44:456 (1968).

77. A. Kissileff, Sm. Anim. Clin., 2:132 (1962).

78. A. Kissileff, Vet. Med./Sm. Anim. Clin., 64:580 (1969).

79. F. W. Knapp, J. Econ. Entomol., 58:585 (1965).

80. F. W. Knapp, J. Econ. Entomol., 58:836 (1965).

81. R. C. Knowles, R. M. Mathis, J. E. Bryant, and K. H. Willers, J. Am. Vet. Med. Assoc., 148:407 (1966).

82. F. R. Koutz, J. Am. Vet. Med. Assoc., 131:45 (1957).

83. F. R. Koutz, Speculum, 17:20 (1963).

84. F. R. Koutz, J. Am. Vet. Med. Assoc., 146:35 (1965).

85. F. R. Koutz, H. F. Groves, and C. M. Gee, Vet. Med., 55:52 (1960).

86. F. Kral, Mod. Vet. Pract., 44:34 (1963).

87. F. Kral and J. P. Uscavage, J. Small Anim. Pract., 1:277 (1961).

88. G. Lapage, Veterinary Parasitology, Edinburgh, Oliver & Boyd, 1956.

89. W. A. Lawrence, Mod. Vet. Pract., 41:70 (1960).

90. P. H. LeRoux, Mod. Vet. Pract., 43:66 (1962).

91. D. E. Lindo and H. H. Grenn, Can. Vet. J., 9:254 (1968).

92. E. C. Loomis and R. B. Bushnell, Am. J. Vet. Res., 29:1089 (1968).

93. K. W. Ludlam, L. A. Jackowski, and G. F. Otto, J. Am. Vet. Med. Assoc., 157:1354 (1970).

94. W. C. Marquardt and W. W. Hawkins, J. Am. Vet. Med. Assoc., 132:429 (1958).

95. P. S. K. Mason, J. So. Afr. Vet. Med. Assoc., 32:381 (1961).

96. W. J. Mathey, J. Am. Vet. Med. Assoc., 150:777 (1967).

97. J. G. Matthysse, R. F. Pendleton, A. Padula, and G. R. Nielson, J. Econ. Entomol., 60:1615 (1967).

98. F. D. Mauer, Proc. 1962 Am. Assoc. Equine Pract. Conv., p. 241 (1963).

99. R. T. McCarty, M. Haufler, M. G. Osborn, and C. A. McBeth, Am. J. Vet. Res., 30:1149 (1969).

100. V. D. McCreary, Allied Vet., 31:165 (1960).

101. W. R. McCuistion, Vet. Med./Sm. Anim. Clin., 63:683 (1968).

102. J. K. McGregor and L. H. Lord, Vet. Med., 56:166 (1961).

103. W. S. McGregor and R. C. Bushland, J. Econ. Entomol., 50:246 (1957).

104. E. A. McPherson, Vet. Rec., 72:869 (1960).

105. W. P. Meleney and I. H. Roberts, J. Am. Vet. Med. Assoc., 151: 725 (1967).

106. W. P. Meleney and S. A. Apodaca, J. Am. Vet. Med. Assoc., 155: 136 (1969).

107. M. Mimioglu and F. Sayin, Vet. Med. Rev., 2:130 (1968).

108. G. A. Mount, J. B. Gahan and C. S. Lofgren, J. Econ. Entomol., 58:685 (1965).

109. J. O. Mozier, J. Am. Vet. Med. Assoc., 154:1206 (1969).

110. G. Muller, Sm. Anim. Clin., 1: 185 (1961).

111. G. Muller, Vet. Med. /Sm. Anim. Clin., 59:846 (1964).

112. W. O. Neitz, Onderst. J. Vet. Res., 27:115 (1956).

113. W. O. Neitz, Onderst. J. Vet. Res., 27:197 (1956).

114. D. L. Nelson, R. G. White, J. O. Mozier, and A. D. Allen, Am. J. Vet. Res., 31:199 (1970).

115. V. A. Nelson, J. Econ. Entomol., 62:710 (1969).

116. V. A. Nelson, J. Econ. Entomol., 62:719 (1969).

117. W. A. Nelson and R. Hironaka, Exp. Parasitol., 18:274 (1966).

118. L. Niilo, Can. Vet. J., 9:132 (1968).

119. J. C. O'Kelley and G. W. Seifert, Aust. J. Biol. Sci., 22:1497 (1969).

120. D. K. O'Neill and S. P. Hebden, Aust. Vet. J., 42:207 (1966).

121. A. Pailet, Mod. Vet. Pract., 43:66 (1962).

122. J. S. Palmer, J. Am. Vet. Med. Assoc., 146:221 (1965).

123. Panel Mod. Vet. Pract., 40:46 (1959).

124. Panel Mod. Vet. Pract., 51:50 (1970).

125. W. M. Pang, Vet. Med., 57:704 (1962).

126. S. Partosoedjono, J. H. Drudge, E. T. Lyons, and F. W. Knapp, Am. J. Vet. Res., 30:81 (1969).

127. C. J. Prchal, J. Am. Vet. Med. Assoc., 146:39 (1965).

128. M. A. Price and J. D. McCrady, Southwestern Vet., 14:287 (1961).

129. R. D. Radeleff, J. Am. Vet. Med. Assoc., 136:529 (1960).

130. R. D. Radeleff, J. Am. Vet. Med. Assoc., 143:248 (1963).

131. E. S. Raun and J. B. Herrick, J. Am. Vet. Med. Assoc., 131:421 (1957).

132. G. B. Rea, J. Am. Vet. Med. Assoc., 147:1567 (1965).

133. C. M. Reed, J. Am. Vet. Med. Assoc., 138:306 (1961).

134. J. F. S. Reid and I. M. Lauder, Vet. Rec., 79:482 (1966).

135. G. P. Retief, J. Am. Vet. Med. Assoc., 145:912 (1964).

136. G. B. Rich, Can. J. Comp. Med. Vet. Sci., 29:30 (1965).

137. G. B. Rich, Can. J. Anim. Sci., 46:125 (1966).

138. L. A. Riehl, H. W. Lembright, and P. D. Ludwig, Allied Vet., 36:137 (1964).

139. C. Ritter, R. Hughes, G. Snyder, and L. Weaver, Am. J. Vet. Res., 31:2025 (1970).

140. I. H. Roberts and W. P. Meleney, J. Am. Vet. Med. Assoc., 146:17 (1965).

141. I. H. Roberts and W. P. Meleney, and H. P. Colbenson, J. Am. Vet. Med. Assoc., 146:24 (1965).

142. I. H. Roberts, W. P. Meleney, and S. A. Apodaca, J. Am. Vet. Med. Assoc., 155:504 (1969).

143. I. H. Roberts and W. P. Meleney, J. Am. Vet. Med. Assoc., 158:372 (1971).

144. J. A. Roberts, J. Parasitol., 54:657 (1968).

145. R. H. Roberts and R. D. Radeleff, J. Econ. Entomol., 52:322 (1960).

146. R. H. Roberts, W. A. Pund, H. F. McCrory, J. W. Scales, and J. C. Collins, J. Am. Vet. Med. Assoc., 153:205 (1968).

147. S. J. Roberts and J. Bentinck-Smith, Cornell Vet., 52:596 (1962).

148. T. O. Roby and D. W. Anthony, J. Am. Vet. Med. Assoc., 142:768 (1963).

149. J. W. Safford, J. Am. Vet. Med. Assoc., 147:1570 (1965).

150. G. B. Schnelle, J. Am. Vet. Med. Assoc., 156:393 (1970).

151. R. M. Schwartzman, J. Am. Vet. Med. Assoc., 145:25 (1964).

152. H. G. Scott, J. Am. Vet. Med. Assoc., 151:1741 (1967).

153. R. Seiden, Insect Pests of Livestock, Poultry and Pets, and their Control, New York, Springer-Verlag, 1964.

154. R. D. Shaw and J. A. F. Baker, Vet. Rec., 78:864 (1966).

155. R. D. Shaw, M. Cook, and R. E. Carson, J. Econ. Entomol., 61:1590 (1968).

156. W. L. Sippel, D. E. Cooperrider, J. H. Gainer, R. W. Allen, J. E. B. Mouw, and H. B. Teigland, J. Am. Vet. Med. Assoc., 141:694 (1962).

157. E. B. Smith and T. F. Claypoole, J. Am. Med. Assoc., 199:59 (1967).

158. H. J. Smith, Can. Vet. J., 8:88 (1967).

159. J. P. Smith and R. R. Bell, Southwestern Vet., 21:293 (1968).

160. E. J. L. Soulsby, Pennsylvania Vet., 8:11 (1966).

161. T. W. Staggs, Southeastern Vet., 12:152 (1961).

162. B. F. Stone, Aust. J. Biol. Sci., 21:309 (1968).

163. H. Strasser, Kleintier Praxis, 8:212 (1963).

164. R. K. Strickland and R. R. Gerrish, J. Am. Vet. Med. Assoc., 144:875 (1964).

165. R. K. Strickland and R. R. Gerrish, J. Am. Vet. Med. Assoc., 148:553 (1966).

166. R. K. Strickland, R. R. Gerrish, J. L. Hourrigan, and F. P. Czech, Am. J. Vet. Res., 31:2135 (1970).

167. J. V. Tacal and F. L. Maledda, Phillipine J. Vet. Med., 1:65 (1962).

168. I. B. Tarshis and W. D. Ommert, J. Am. Vet. Med. Assoc., 138:665 (1961).

169. K. E. Taylor, J. Am. Vet. Med. Assoc., 143:245 (1963).

170. W. M. Taylor, J. E. Bryant, J. B. Anderson, and K. H. Willers, J. Am. Vet. Med. Assoc., 155:915 (1969).

171. P. Theran and G. V. Ling, J. Am. Vet. Med. Assoc., 150:82 (1967).

172. G. E. Thompson and J. A. F. Baker, J. So. Afr. Vet. Med. Assoc., 39:61 (1968).

173. P. J. Treeby, Vet. Rec., 70:326 (1966).

174. L. G. Tritschler, Vet. Med./Sm. Anim. Glin., 60:219 (1965).

175. U.S.D.A., J. Am. Vet. Med. Assoc., 132:283 (1958).

176. U.S.D.A.-A.R.S., J. Am. Vet. Med. Assoc., 150:825 (1967).

177. U.S.D.A.-A.R.S., J. Am. Vet. Med. Assoc., 156:638 (1970).

178. U.S.L.S.A., J. Am. Vet. Med. Assoc., 150:73 (1967).

179. K. B. W. Utech, R. H. Wharton, and L. W. Wooderson, Aust. Vet. J., 45:414 (1969).

180. G. S. Walton, Vet. Rec., 75:355 (1963).

181. F. P. Ward and C. L. Glicksberg, J. Am. Vet. Med. Assoc., 158:457 (1971).

182. J. B. Webb, Allied Vet., 34:75 (1963).

183. R. W. Weitkamp, J. Am. Vet. Med. Assoc., 144:597 (1964).

184. R. H. Wharton, K. L. S. Harley, P. R. Wilkinson, K. B. W. Utech, and B. M. Kelley, Aust. J. Agri. Res., 20:783 (1969).

185. L. F. Whitney, Vet. Med./Sm. Anim. Clin., 64:993 (1969).

186. E. H. Willers, J. Am. Vet. Med. Assoc., 147:1573 (1965).

187. B. H. Wilson and R. B. Meyer, Am. J. Vet. Res., 27:367 (1966).

188. B. H. Wilson, J. Am. Vet. Med. Assoc., 153:203 (1968).

189. J. E. Wood, K. W. Page, P. R. M. Brown, M. S. Smith, and D. A. M. Ferguson, Vet. Rec., 77:896 (1965).

190. R. A. Yeary, J. Am. Vet. Med. Assoc., 149:145 (1966).

191. G. H. Yeoman and B. C. Warren, Vet. Rec., 77:922 (1965).

192. B. C. Youmans and A. K. Robinson, Sm. Anim. Clin., 1:281 (1961).

193. P. E. Zollman, Cornell Vet., 54:191 (1964).

AUTHOR INDEX

Numbers in brackets are reference numbers and indicate that an author's work is referred to although his name is not cited in the text. Underlined numbers give the page on which the complete reference is listed.

A

Abbott, E. V., 125(142), 126(108, 142), 137(108), $\underline{179}$, $\underline{180}$
Adams, L. E., 84(1), 85(1), $\underline{174}$
Ahlgren, G. H., 121(3), 122(3), $\underline{174}$
Albert, J. A., 63(4), 162(4), $\underline{174}$
Alexander, L. J., 135(163), 136(163), 137(163), 138(163), 139(163), $\underline{181}$
Alexopoulos, C. J., 14, 17(5), 22(5), 28(5), 31(5), $\underline{174}$
Allen, A. D., 358(21, 114), $\underline{380}$, $\underline{384}$
Allen, R. W., 326(156), $\underline{386}$
Anderson, H. W., 9(181), 62, 73(6), 74(6), 84(6), 163(116), $\underline{174}$, $\underline{179}$
Anderson, J. B., 326(170), 348(12), $\underline{380}$, $\underline{386}$
Anderson, R. J., 332(1), 358(1), $\underline{380}$
Anthony, D. W., 348(148), 352(148), $\underline{385}$
Apodaca, S. A., 337(142), 343(106), 358(106, 142), $\underline{384}$, $\underline{385}$
Ark, P. A., 9(7), $\underline{174}$
Armstrong, J. L., 355(42), $\underline{381}$
Arneson, P. A., 79(22), 80(22), 85(23), 86(24), $\underline{175}$
Arthur, J. G., 4, $\underline{174}$
Atkins, J. G., 103(9), $\underline{174}$
Atkinson, G. F., 190, $\underline{252}$

B

Baker, J. A. F., 343(154, 172), $\underline{386}$, $\underline{387}$
Baker, K. P., 352(2), $\underline{380}$
Bardin, R., 154(84), 155(84), 157(84), $\underline{178}$
Barnes, G. L., 91(10), $\underline{175}$
Barnes, M. M., 92(11), $\underline{175}$
Bartlett, C. C., 348(31), $\underline{381}$
Bates, J. D., 63(115), $\underline{179}$
Batham, P., 343(56), $\underline{382}$
Bauer, A. H., 84(12), $\underline{175}$
Bayona, I. G., 355(42), $\underline{381}$
Beattie, J. H., 159(13), $\underline{175}$
Beattie, W. R., 159(13), $\underline{175}$
Beesley, W. N., 328(3), 355(3), $\underline{380}$
Bell, D. K., 108(111), 110(111), $\underline{179}$
Bell, R. R., 347(159), $\underline{386}$
Bentinck-Smith, J., 347(147), 352(147), $\underline{385}$
Benton, D. A., 113(156), $\underline{181}$
Bereford-Jones, W. P., 353(46), $\underline{381}$
Berg, P., 353(4), $\underline{380}$
Berkeley, M. J., 189, $\underline{252}$, 262, $\underline{323}$
Berliner, E., 294, $\underline{323}$
Berns, R. E., 354(5), 355(5), $\underline{380}$
Berry, I. L., 331(63), $\underline{382}$
Bessey, E. A., 6(16), 22(16), $\underline{175}$, 190, $\underline{252}$

389

390

Garlick, N. L., 326(49), 382
Garrett, W. N., 91(73), 177
Garriss, H. R., 108(74), 116(74),
    132(75, 76), 133(76), 164(76),
    165(76), 177
Gee, C. M., 353(85), 383
Genelly, R. E., 313(29), 324
Gerrish, R. R., 336(165), 345(165,
    166), 348(164), 386
Gibbons, W. J., 336(50), 382
Gilgut, C. J., 129(78), 130(78), 178
Gilhooly, O. J., 327(67), 382
Glicksberg, C. L., 358(181), 387
Goenaga, A., 7(28), 175
Goldsworthy, M. C., 8, 122(79),
    178
Good, J. M., 252
Goodey, J. B., 252
Goodman, R. N., 9(80), 10,
    64(128), 178, 180
Goyings, L. S., 352(51), 354(51),
    382
Graham, O. H., 331(63), 332(34),
    337(35), 348(31), 358(52), 381,
    382
Gray, R. A., 9, 178
Green, E. L., 8(79), 122(79), 178
Grenn, H. H., 355(91), 383
Greve, J. H., 328(55), 344(53),
    353(54), 382
Grogan, R. G., 154(84), 155(84),
    157(84), 178
Groves, H. F., 353(85), 383
Guba, E. F., 160(85), 178
Gubb, G. L., Jr., 84(1), 85(1), 174

H

Haesler, C. M., 84(1), 85(1), 174
Hafen, L., 146(122), 147(122),
    150(122), 151(122), 153(122),
    154(122), 157(122), 158(122),
    159(122), 160(122), 164(122),
    167(122), 180
Hains, A. F., 9(43), 176
Halbert, J. R., 117(124), 180
Hall, D. H., 75(151), 80(151),
    83(151), 84(151), 92(151),
    128(151), 181
Hamilton, J. M., 9, 178
Hammer, C. L., 9(66, 69), 177
Hardenburg, E. V., 128(89), 178
Hardenburg, R. E., 132(126), 180
Harlan, J. R., 121(90), 178
Harley, K. L. S., 343(184), 387

Harrar, J. G., 40(201), 151(201),
    183
Harris, M. R., 129(91), 130(91),
    178
Harrison, I. R., 350(9), 380
Harrison, J. W., 353(57), 382
Harrison, R., 147(170), 182
Harry, J. B., 8(92), 178
Hart, R. J., 343(56), 382
Harvey, T. L., 335(37), 381
Hatziolos, B. C., 351(59), 355(58),
    382
Haufe, W. O., 336(60), 382
Haufler, M., 359(99), 384
Hawkins, W. W., 344(94), 358(94),
    383
Heald, F. D., 46, 178
Heath, M. E., 121(106), 122(106),
    124(106), 179
Hebden, S. P., 344(120), 384
Heiser, C. B., Jr., 103(94), 178
Heishmann, J. O., 331(27), 381
Henneberry, T.J., 154(209),
    155(209), 183
Hepler, P. R., 163(116), 179
Herrick, J. B., 358(131), 385
Heuberger, J. W., 7, 8(59), 9(96),
    63(4, 95, 115), 162(4), 174,
    177, 178, 179
Hewetson, R. W., 343(61), 382
Higgins, D. J., 9(43), 176
Hildebrand, E. M., 78(100),
    132(99), 133(99), 178, 179
Hills, O. A., 154(229), 155(229),
    184
Hindle, E., 355(62), 382
Hironaka, R., 344(117), 384
Hislop, E. C., 68(101), 179
Hodgin, B., 147(170), 182
Hoffman, R. A., 331(63), 382
Hooper, B. E., 351(64), 382
Hopkins, D. E., 294(9), 295(8, 9),
    323
Hopwood, R. T., 352(65), 382
Horn, N. L., 61(102), 83(102),
    179
Horne, H., 328(66), 382
Horsfall, J. G., 6, 7(103), 8(59,
    103), 9(103), 63, 64(103),
    65(103), 177, 178, 179,
    278(15), 323
Hough, W. S., 7(104), 64(104),
    179
Hourrigan, J. L., 327(67),
    345(166), 382, 386

392

393

394

A

ABATE, 309
Acanthorhyncus vaccinii, 89
Acariasis, cat, 355
Acaricides
  cat, 355
  horse, 348
  ornamentals, 301
  poultry, 356, 357
  swine, 350
Acarids
  cattle, 336
  dog, 352-354
  horse, 348
  swine, 350
Acarina, order, 336, 337
Acervuli, definition, 15
Acervulus, 20, 28, 46
Achenes, 81
Acorn, 90
Actidione, 9
Actinomyces scabies, 128
ACTISPRAY, 312
Active ingredient, 316
Adenophorea class, nematodes, 198
Aeciospore, 26, 27, 28, 30
Aecium, 26, 27
Aedes sp., 282
Aerial pathogenic nematodes, 199, 201, 203
Aerial spraying
  cereals, 95
  rice, 103
Aerosol herbicides, 312
Aerosol insecticide
  cattle, 368, 369
  sheep, 374
Aerosol mist insecticide, cattle, 367
Aesthetics, pests, 282-285
Aesthetic values, 256

African marigold, nematodes, 223
African sleeping sickness, 330
Aggregate fruit, 81, 86
Agricultural Research Service, 57
Agricultural science, 322
Agrobacterium, 35
Agromyza lantana, 293
Agropyron cristatum, 121
Agrosol S, sorghum, 119
AGSCO DB GREEN, cereals, 100
AGSCO DB YELLOW, cereals, 100
AGSCO-ZINEB, potato, 129
Ailanthus, 276
Air pollution losses, 58
Air sac mite, poultry, 356
Alabama Polytechnic Inst., 190
Alae, nematodes, 195
Albugo occidentalis, 165
Albugo ipomoeaepanduratae, 133
Alcohol disinfection, sugarcane, 126
Alcohol, potato, 128
Alderleaf hopper, 272
Aldicarb, nematicide, 234
Aldrin, 305
  livestock, 259
Alfalfa
  caterpillar, 294
  crown rot, 56
  diseases, 111
  losses, 58
  stem, nematode, 201, 208, 219, 224, 225
ALKRON, 304
  nematicide, 235
Alkyl tins, 297
Allergies, pests, 282
Allergy tick, horse, 348
Allethrin, livestock, 359
Allium spp., 151
Allium ascalonicum, 152
Allium cepa, 152

Asparagus
  postharvest fungicides, 172
  U.S. production, 163
Aspergillus niger, 153
Aspergillus spp.
  carrot, 159
  corn, 117
  cotton, 105
  peanut, 109
Ass, ectoparasites, 347
Aster aphid, 272
Aster yellows, 275
  carrot, 48, 158, 159
  lettuce, 155, 156
  parsnip, 160
Atrazine, 307, 318
Atrophy, 50
Auburn University, 190
Australian lady beetle, 291
  parakeet, 356
Austrian pine borer, 273
Avena byzantine, 94
Avena sativa, 94
Avaian ectoparasite, 355-357
Awl, nematodes, 199
Azalea, 277
  borer, 271
  fungicides, 298
  history, 260
    lace bug control, 303
  lacewing, 272
  leaf disease, 309
    history, 260
    mites, 301
    petal blight, 275
  leaf gall, 274
  leaf miner, 274
Aziridinyl, 295

B

Bachelor's button, 276
Bacillus, bacteria, 37
Bacillus popillae, 294
Bacillus thuringiensis, 294
Backrubber, insecticide
  cattle, 361, 362, 364, 368, 369
Backrubbers, livestock, 358
Bacteria, 33
  bacillus, 37
  coccus, 37
  flagella, 35
  plant pathogens, 35
  spirillum, 37

Bactericide, definition, 62
Bacterial black chaff, cereal, 97
Bacterial blight
  beans, 165
  carrot, 158
  cereal, 97
  cotton, 105
  filbert, 90
  forage grass, 122
  walnut, 91, 92
Bacterial cane blight, blueberry, 88
Bacterial canker
  apricot, 77
  peach, 78
  stone fruits, 76
  tomato, 44, 136
Bacterial diseases, 267
  citrus, 93
Bacterial dwarfing, 50
Bacterial fruit spot, tomato, 40
Bacterial infection, 47
Bacterial leaf blight, corn, 59
Bacterial leaf spot, 47, 57, 64
  cauliflower, 150
  celery, 161, 162
  plum, 81
  stone fruits, 76
  tomato, 40
Bacterial nematode synergism, 214
Bacterial pathogens, nematodes, 227
Bacterial ring rot, 128
Bacterial rods, 33
Bacterial rot, escarole, 155
Bacterial soft rot, 41
  celery, 163
  escarole, 155
  lettuce, 156
  potato, 129
Bacterial speck, tomato, 136, 137
Bacterial spherical, 33
Bacterial spiral, 33
Bacterial spot
  peach, 10, 78, 299
  pepper, 35, 140, 141
  sorghum, 120
  stone fruit, 35
  tomato, 35, 134, 136-138
Bacterial stripe
  cereal, 97
  sorghum, 120
Bacterial wilt
  chrysanthemum, 299
  cucumber, 42, 144
  cucurbits, 143
  muskmelon, 145

Boophilus annulatus, 338, 341, 342
Boophilus microplus, 337
Boophilus spp.
  cattle control, 362
Bordeaux mixture, 7, 63, 66
  almonds, 75
  bramble, 86
  grape, 85
  pear, 73
  pecans, 91
  plum, 80
  stone fruit, 77
  vegetables, 172
  walnut, 92
Borer control, 306
Borers, ornamentals, 273
Boron
  deficiency, celery, 161, 162
  weed control, 308
Boston Ivy, 260
Botanical insecticides, 301, 302
  livestock, 359
BOTRAN, 9
  carrot, 159
  escarole, 155
  onion, 154
  peanut, 109
  pear, 74
  stone fruit, 77
  sweetpotato, 132
  tomato, 139
  uses, 66
  vegetables, 172
Botfly, horse, 345, 346
Bot rot, apple, 69, 70, 72
Botryosphaeria corticis, 88
Botryosphaeria dothidea, 88
Botryosphaeria ribis, 69, 87
Botryosphaeria spp., apple, 9
Botryotinia ricini, 125
Botrytis, 61
  blast, onion, 153
  blossom blight, cherry, 79
  greenrot, stone fruit, 77
  stem rot, tomato, 136, 139
Botrytis cinerea, 56, 74, 82, 84, 86, 88,
    114, 137, 142, 155, 159, 160, 163
Botrytis spp., 9
  castor bean, 125
  lettuce, 155, 156
  onion, 152, 156
Bots horse, 345
  control, 374
Bottom rot, lettuce, 155, 156
Bovicola bovis, 335, 336

Bovine ectoparasite, 328-343
Boxelder bugs, 280
Boxwood
  history, 259
  hornet, 275
  leaf miner, 274, 306
    control, 303
  mites, 301
Bramble, 30
  diseases, 86
  losses, disease, 86
  spray schedule, 86
Branch wilt walnut, 92
Brassica campestris var. napobrassica,
    148
Brassica chinesis, 148
Brassica Japonica, 148
Brassica juncea, 148
Brassica juncea var. crispifolia, 148
Brassica napobrassica, 148
Brassica nigra, 148
Brassica oleracea var. acephaja, 148
Brassica oleracea var. botrytis, 148
Brassica oleracea var. capitata, 148
Brassica oleracea var. gemmifera, 148
Brassica oleracea var. gongyloides, 148
Brassica oleracea var. virides, 148
Brassica pekinensiss, 148
Brassica rapa, 148
BRASSICOL, vegetables, 173
BRAVO
  crucifer, 151
  cucurbits, 146
  onions, 153
  peanut, 111
  potato, 130
  salad crops, 157
Bremia lactucae, 154, 156
BRESTAN, 297
  granular insecticide, swine, 371
Broadleaf weed control, 307
Broccoli
  diseases, 148
  fungicides, 168
  U.S. production, 148
Brome grass, 121
  losses, 123
BROMOFUME, nematicide, 230, 232
Bromomethane, nematicide, 231
Bromophos, ticks, 343
Bromus inermis, 121
Bronze birch borer, 271
Brooks spot, apple, 69
Broomcorn, 118
  disease, 118

Brown beetles, 291
Brown blossom rot
  apricot, 77
  nectarine, 77
  peach, 77
Brown dog tick, 353, 354
  horse, 348
  swine, 350
Brown leaf spot
  pecan, 91
  rice, 103, 104
  tobacco, 115, 116
Brown rot
  almond, 75
  apricot, 77
  cherry, 79
  citrus, 93
  peach, 78
  plum, 80, 81
  stone fruit, 46, 65, 76
Brown stripe, sugarcane, 126
Brown tail moth, 291
BROZONE, 310
  nematicide, 230
  tobacco, 115
  tomato, 136
Brucellosis, dog, 352
Brush control, 306
Brussel sprouts
  fungicides, 168
  U.S. production, 148
Bryobia praetiosa, 355
Buccal capsule, nematode, 196
Buchloe dactyloides, 122
Buckeye rot, tomato, 135-137
Budgerigar, 356
Buffalo fly, 329, 330
Buffalo grass, 122
Bulb crop
  diseases, 151-154
  losses, disease, 152
Bulb protectants, 298
Bulb rot, onion, 153
Bull nettle, 277
Bunt
  cereals, 101, 102
  wheat, 3, 97
Burning bush, 260
BUSAN, cotton, 106
Bush fly, 330
Butonate horses, 345
Butternut, 90
Butyric acid, nematodes, 218
Buxus, 259

C

Cabbage
  disease, losses from, 149
  diseases, 149
  fungicide, 168
  U.S. production, 148
Cabbage looper, 294
Cactoblastis cactorum, 293
Cactus mealybug, 273
CADDY, 297
CADMINATE, 297
Cadmium calcium-copper-zinc-chromate-
    complex, 297
Cadmium chloride, 297
Cadmium fungicide, 297
Cadmium sebacate, 297
Cadmium succinate, 297
California
  almonds, 75
  citrus losses, 61
  pears, 73
  plums, 80
  prunes, 80
  stone fruit, 77
  strawberry, 83
Calcium arsenate, 301
Calcium deficiency
  celery, 161, 162
  tomato, 12, 135
Calcium polysulfide, 7, 297
Calliphora erythrocephala, 331
Calliphorinae, subfamily, 331
Callitroga hominivorax, 331, 366
Calomel, 8
Calypteratae superfamily, 329, 330
Canary grass, 290
Caneblight
  bramble, 86
  currant, 87
  gooseberry, 87
Canine
  ectoparasites, 350-354
  mange, 350
  sarcoptic mite, 350
  scabies, 350
    dog, 353
Canker
  chestnut, 17
  orchard trees, 278
  parsnip, 160
  plum, 81
  rabbit, 355
  tomato, 37

0,0,Diethyl-0-p(methylsulfinyl)phenyl
   phosphorothioate, nematicide, 234
0,0,Diethyl-0-p-nitrophenyl phosphorothioate, 304, 320
   nematicides, 235
0,0,Diethyl-0-2-pyrazinyl phosphorothioate,
   300
   nematicide, 234, 237
DIFOLATAN
   cherry, 79
   cucurbits, 146
   onions, 154
   potato, 130
   uses, 66
   vegetables, 160-170
Digestive system, nematodes, 196, 197
2,3-Dihydro-2,2-dimethyl-7-benzenofur-
   anyl methyl carbamate, 236
DIKAR
   apple, 70, 72
   pear, 74
   uses, 66
DILAN, livestock, 359
Dimethoate, 312
   cattle, 361, 367
   horses, 345
   livestock, 358
   poultry, 377
   sheep, 343
   swine, 350, 370
Dimethyl fungicides, 297
p-Dimethylaminobenzene diazo sodium sulfonate, 299
1,1-Dimethyl-4'-bipyridinium di(methyl
   sulfate), 307
0,0,Dimethyl S-(1,2-dicarbenthoxymethyl)
   phosphorodithioate, 305
0,0,Dimethyl-0-p-nitro-phenyl-phosphoro-
   thioate, 305
0,0,Dimethylphosphorodithioate of diethyl
   mercaptosuccinate, 320
3,5-Dimethyl-1,3,5-2H tetrahydrothiadi-
   azine-2-thione, 299
   tobacco, 115
Dinitrobutylphenol, 307
   grape, 84
Dinocap, 9, 63, 65
   apple, 70
   bean, 164
   cherry, 79
   cucurbits, 147
   currant, 87
   pear, 74
   stone fruit, 77
   strawberry, 82

(Dinocap)
   uses, 67
   vegetables, 172
Dinoseb, grape, 84, 85
Dioecious, nematodes, 197
Diorchic, nematodes, 197
Dioscorides, 263
Dioxathion
   cattle, 364
   dog, 354
   horses, 375
   livestock, 358
   sheep, 372
   swine, 371
   ticks, 343
Dip insecticides
   cattle, 361, 362, 363
   livestock, 358
   sheep, 372, 373
Diplocarpon earliana, 65, 82
Diplocarpon rosae, 296, 297
Diplodia fruit rot
   cucurbits, 143
   pumpkin, 147
Diplodia gossypina, 105, 143, 144
Diplodia theobrome, 132
Diplodia tubericola, 132
Diplodia zeae, 12, 117
Diplodina lycopersici, 142
Diplogaster spp., morphology, 250, 251
Diptera, order, 328, 345
Dipylidium caninum, 352
Directed anteriad nematode, 253
Directed posteriad nematode, 253
Diofilaria immitis, 350
Disease
   alfalfa, 111
   almonds, 75
   asparagus, 163, 164
   barley, 94
   beans, 164, 165
   beets, see diseases sugarbeets
   birdsfoot trefoil, 111
   blueberry, 88
   bramble, 86
   broccoli, 148
   broomcorn, 118
   bulb crop, 151-154
   cabbage, 149
   cantaloupe, 144, 145
   carrots, 158, 159
   castor bean, 124, 125
   cauliflower, 150
   celery, 161, 163
   cereals, 94-98

Elective surgery, ectoparasites, 331
Elk, 313
Elm bark beetle, 264, 274, 275, 285
Elm, resistant, 287
Elsinoi ampelina, 84
Elsinoi fawcettii, 93
Elsinoi veneta, 86
Emulsified larvacides, 351
Encephalitides, 326
Endemic typhus, man, 355
Endive, fungicides, 169
Endocarp, 74
Endoconidiophora paradoxa, 126
Endoparasitic nematodes, 199, 203, 214,
    218, 269
Endrin, livestock, 359
English daisy, 276
English gardens, 261
English walnut, 91
Entyloma spp., forage grass, 122
Environment, nematodes, 215, 241, 242
Environmental Protection Agency, 166, 314
Enzymes, nematode, 205, 269
Eomenacanthus stamineus, 357
EPA, 314
EPA establishment number, 315
Epidermis, nematodes, 195
Epiphytotic, potato, 40
Epithelium, nematode, 194
Equine ectoparasite, 345-348
Equine piroplasmosis, 348
Eradication, pest control, 288
Erbon, 308
Ergot
    forage grass, 122
    poisoning, 122
Ericaceae, 88
Erosion, nematodes, 217
Erwinia amylovora, 37, 43, 69, 267
Erwinia aroidea, 129
Erwinia atroseptica, 128
Erwinia carotovora, 37, 41, 114, 129, 155,
    156, 159, 160, 163
Erwinia phytophthora, 128
Erwinia spp., 37
    potato, 129
Erwinia tracheiphyla, 42, 143, 144
Erysiphe cichoracearum, 21, 143, 144,
    155, 164
Erysiphe graminis, 17, 122
Erysiphe polygoni, 150, 164
Escaped cultivars, 276
Escarole
    fungicides, 169
    losses, disease, 155, 156

Esophageal intestinal valve, nematode,
    197
Esophagus, nematode, 194
Esters, 2, 4-D, 307
Ethazol
    cotton, 107
    peanut, 109
    sorghum, 119
    vegetables, 172
Ethmolaimus spp., morphology, 250, 251
Ethoprop, nematicide, 233
0-Ethyl S,S-dipropyl phosphorodithioate,
    nematicide, 233, 238
Ethylene bisdithiocarbamates, 297
Ethylene bis thiuram monosulfide diffusion,
    68
Ethylene dibromide, 299, 300
    nematodes, 237, 230, 232
Ethyl parathion, nematicide, 235
Etiology, nematodes, 201, 202
Eubacteriales, 35
Euonymus scale, 287
Euphorbia cyparissias, 276
Euphorbia helioscopia, 276
Euphorbia peplus, 276
Euphorbia spp., 276
Euphoriaceae, 124
Europe, famine, 40
European corn borer, 291
European grape, 83
European rye, 95
Evergreen lacewing, 272
EVERSHIELD CM, legumes, 112
Excretory canal, nematode, 198
Excretory pore, nematode, 194, 195, 198
Excretory system, nematode, 198, 199
Exocarp, 74
Exodus, 262
EXOTHERM, tomato, 138
External parasiticides, canines, 350
Eye gnats, 282
Eyespot disease, sugarcane, 126

F

Fabrea maculata, 73
Face fly, 330
    cattle control, 365, 369
    horse, 347
Face paint insecticide, cattle, 366
Facultative definition, 13, 14
Fallowing nematodes, 216, 217
Fall potatoes, nematodes, 219
Famphur cattle, 336, 363
    livestock, 329, 358

421

(Fungicides)
cereals, wettable powder, 96
concentrations
apple, 72
corn, 118
contact, 65
cotton, 107
cotton belt, 106
cotton in furrow, 106
definition, 62
grape, dust, 83
history, 6-9
labels, 103
legumes, 112
mobility, 68
mode of action, 68
organic, 8
ornamentals, 297, 299
peanuts, 108-111
performance, 63
potato, 129
program, 68
properties, 66, 67
residual, 63, 65
residues, 166
restrictions, 171-173
blueberry, 88
bramble, 87
cranberry, 90
peach, 79
plum, 81
strawberry, 83
salad crops, 157
seed treatment, cereals, 100
seed treatment, vegetables, 166
spray schedule, apple, 70
spreader-sticker, 111
systemic cereals, 96
tomato, 136-139
tolerances, 166, 171-173
vegetables, 168-173
waterfilm, 65
Fungi
classification, 14
morphology, 13
taxonomy, 14
toxicant, 65
FUNGICLOR, vegetables, 173
Fungi imperfecti, 22
nematode parasites, 225
Fungistat, 296
definition, 64
Fungus
definition, 10

(Fungus)
disease
citrus, 93
ornamentals, 264
races, 40, 44
FURADAN, 236
Furrow treatment, onion, 153, 154, 173
Fusarium moniligorme, 12, 105, 117, 126
Fusarium nivale, 122
Fusarium oxysporum f. apii, 161
Fusarium oxysporum f. asparagi, 163
Fusarium oxysporum f. batatas, 131
Fusarium oxysporum f. conglutinans, 149
Fusarium oxysporum lini, 120
Fusarium oxysporum f. lycopersici, 134
Fusarium oxysporum var. nicotianae, 113
Fusarium oxysporum var. vasinfectum, 140
Fusarium root rot, onions, 152-154
Fusarium spp., 22
bulbs, 152
carrot, 159
castor bean, 125
cereal, 96, 97
cotton, 105, 106
cucumber, 144
cucurbits, 143
flax, 120
forage grass, 122
peanuts, 109
potato, 128, 129
rice, 104
stalk wilt, asparagus, 163
sugarcane, 126
tobacco, 114
wilt, 64
celery, 161
cotton, 214, 216
cucumber, 144
cucurbits, 143, 146
flax, 120
pepper, 140
potato, 128
sweetpotato, 131, 132
tobacco, 113
tomato, 134, 136, 214, 216
watermelon, 145
Fusarium yellows
cabbage, 149
celery, 161, 162
kale, 150
Fusicladium dendriticum, 22
Fusiform nematode, 194

Greasy spot, citrus, 93
GREAT LAKES BROMO-GAS, 115
Great Lakes Fish, 313
Greek gardens, 261
Greenbottle fly
  livestock, 331
  sheep, 344
Greenhead horsefly, 347
Greenhouse
  cucumbers, 144
    fungicides, 172
  disease
    cucurbits, 143
    fumigation,
    fungicides, 172
    lettuce, 154
    rhubarb, fungicides, 172
    tomato, 135
    tomato, fungicides, 172
    treatment, tomato, 137
    whitefly, 292
  escarole, 155
Green screwworm fly, swine, 349
Greens keeper, 322
Gray leaf spot, tomato, 133–138
Gray mold fruit rot
  eggplant, 142
  strawberry, 82
Grey mold rot
  celery, 163
  escarole, 155
  grape, 84, 85
  lettuce, 155, 156
  onion, 152
  strawberry, 83
  tomato, 137, 138
Grey wall, tomato, 135
Ground ivy, 276
Grub
  cattle control, 363, 364
  control, 303
  ornamentals, 271
Gryopus ovalis, 356
Guanidine fungicides, 298
Gubernaculum, nematode, 197
Guignardia bidwellii, 84
Guignardia vaccinii, 89
Guinea pig ectoparasites, 355, 356
Gulf Coast citrus, losses, 61
Gulf Coast tick
  cattle, 337
  horse, 348
  sheep, 345
Gumbo, disease, 164

Gummy stem blight
  cucurbits, 143, 146
  pumpkin, 147
Gymnoconia interstitialis, 30
Gymnosperm, 31
Gymnosporangium clavipes, 69
Gymnosporangium globosum, 69
Gymnosporangium juniperi virginianae,
  30, 32, 34, 69, 289
Gypsy moth, 291, 301, 311

H

Haemaphysalis leporis paulstris, 355
Haematobia, genus, livestock, 329
Haematobia irritans, 361
Haematopinus asini, 347
Haematopinus eurysternus, 335, 361
Haematopinus suis, 349
Haggai, 262
Hail injury, apple, 72
Halo blight
  bean, 35
  oats, 97
Halogenated hydrocarbons, 301–304
Hamster ectoparasites, 355
Hard ticks, cattle, 337
Harvest mite, cattle, 337
Hawthorn leaf hopper, 272
Hawthorne rust, apple, 69
Hay losses, 58
Hazel nut, 90
HCB
  cereals, 100
  forage grass, 122
Head blight, cereal, 96, 97
Head louse, poultry, 356, 357
Head maggot fly, sheep, 343
Head mange, sheep, 345
Head smut, forage grass, 122
Headlight glare, 284
Heartwater, ruminants, 327
Heartworm, dog, 350
Heat control, nematodes, 226
Heat treatment, viruses, 267
Heel fly, 328
Hegari, 118
Helicobasidium purpureum, 127
Helicopter spraying
  cereals, 95
  rice, 103
Heliothus spp., disease, 295
Helminthosporium avenae, 98

(Human flea)
  swine, 349
Humulus lupulus, 124
Hyalomma transiens, 327
Hyalomma spp., horse, 348
Hydrocooling, 83
  peaches, 79
Hydrogen sulfide, nematodes, 218
Hylurgopinus rufipes, 264
Hyperplasia, 50
Hypertrophy, 50, 55
  nematode, 205
Hypoplasma, 50
Hypha, 33
  definition, 13
  haploid, 36
  receptive, 26, 27
  septate, 15
Hypoderma bovis, 327, 328, 329, 257, 263
Hypoderma lineatum, 328, 329, 357
Hypodermis, nematode, 194, 195

I

Icerya purchasi, 291
Idaho prunes, 80
Ilex aquifolium, 260
Illinois corn loss, 59
Imidazoline, 297
Immune cultivars, 287
Impregnated mineral block livestock, 358
Incubation period, 44
Industrial plant grass, 319
Inert ingredient, 316
Injury, livestock
  ectoparasites, 329
  flies, 328
Inoculum, source, 46
Inorganic fungicides, 296, 297
Inorganic insecticides, 301
  livestock, 259
Insecta, class, 326, 328, 336
Insecticidal shampoo, 351
Insecticide, 322
  cattle, 336
  contact, 301
  ornamentals, 300, 306
  residual, 301
  stomach, 301
Insect
  control, 317
  ornamentals, 270, 275
  population suppression, 317
  vectored disease, 282, 318

(Insect)
  vectors, 275
    tomato, 135
Insect vectored viruses, 267
Interacting pathogens, nematodes, 216
Inter-current infection, 270, 274
Internal browning, tomato, 135
Internal cork disease, sweetpotato, 131
Intestine, nematode, 197
Isthmus region, nematode, 198
Ivy mealybugs, 273
Ipomoea batatas, 131
IPS, 317
Ips gradicollis, 275
Ips pini, 275
Ireland famine, 40
Iris
  aphid, 272
  borer, 271, 274
    control, 306
  leafspot, 285
  rust, 290
  stem nematode, 226
Irish potato, 128
  famine, 265
Iron fungicide, 297
Irradiation, males, 332
Irrigation, nematodes, 223
ISOSYSTOX, 305
Itch mite
  horse, 348
  poultry, 356
  sheep, 345
Itersonilia peplexans, 160
Ithyphallus rubicundis, 126
Ixodes cookei, 338
Ixodes pacificus, 338
Ixodes ricinus, 338
Ixodes scapularis, 338
Ixodes spp.
  cattle control, 362
  horse, 348
Ixodicide, cattle, 343
Ixodidae, family, 337

J

Jack pine, 289
Jamestown settlers, 264
Japanese beetle, 270, 274, 280, 291, 294, 303, 317
  control, 304, 312
  gardens, 261
  holly, 269
  honeysuckle, 276

Japanese maple leaf hopper, 272
Jatropha stimulosa, 277
Java black rot, sweetpotato, 132
Jigger flea, swine, 349
Jiggers, cattle, 337
Jimson weed, 277
Job, 263
Johnson grass disease, 118
Johnson, Pres. Lyndon B., 256
Juglans nigra, 91
Juglandaceae, 90
Juniper, 289
Juniperus virginiana, 289

K

Kafir, 118
Kafir ant, sorghum, 119
Kale, 148
  disease, 150
  fungicides, 169
  losses, disease, 150
Kaoliang, 118
KARATHANE, 9
  bean, 164
  cucurbits, 147
  pear, 74
  stone fruit, 77
  uses, 67
  vegetables, 172
KARMEX, 308
Karyogamy, 36
Keds
  goats, control, 372
  sheep, control, 372
Kentucky bluegrass, 122
  loss, 123
Keratin, nematodes, 195
Kernel blight, rice, 104
Kernel smut, rice, 104
Kerosene foot dip, poultry, 356
KOBAN, vegetables, 172
Koch's postulate, 4
KOCIDE
  beans, 165
  carrot, 159
  crucifer, 151
  nuts, 90
  peanut, 110
  walnut, 92
Kohlrabi, 148
  fungicides, 169
KOLKERMETHYL BROMIDE, tomato,
  136

KORLAN
  cattle, 362, 367-369
  horse, 376
  sheep, 373, 374
  swine, 371
KROMAD, 297
Kudzu disease, 111

L

Label, pesticides, 322
Labial muscles, nematodes, 196
Lace bug, ornamentals, 272
Lactuca sativa, 154
Lady beetles, 291
Lake Erie, 313
  grapes, 84
Lake Huron, 313
Lake Michigan, 313
Lake Ontario, 313
Laminosioptes cysticola, 356
Lamium ampleyicaule, 277
Lamprey eel, 313
Land grant, colleges, 315
Land management, nematicides, 216
Landscape
  architecture, 257
  values, 256
Lantana camara, 292
Laportea canadensis, 277
Large oak apple gall, 274
LARVACIDE, 299
  nematicide, 230, 231
  poultry, 377
  tobacco, 115
  tomato, 136
Late blight
  celery, 162
  potato, 40, 128, 130, 265
  tomato, 51, 133-137
Late leaf blight, celery, 161
Latent life, 203
Lawn
  growth control, 319
  insect control, 303
  keeper, 322
  nematodes, 238
  weed control, 307, 312
Lead arsenate, 301
Leaf blight
  almond, 75
  celery, 161
  corn, 118
  forage grass, 122

431

Mercury
  inorganic, uses, <u>67</u>
  methyl, <u>8</u>
  organic, <u>8</u>
    fungicides, 297
  phenyl, <u>8</u>
MERTECT
  sugarbeet, 127
  vegetables, 173
MERTECT 340 F, sweetpotato, 132
Mesocarp, 74
Metal organic fungicides, <u>297</u>
Metal organic insecticides, 301
Metallic dithiocarbamates, 297
Metallic fungicides, 297
<u>Metaphycus helvolus</u>, 292
METASYSTOX, 305
Metatepa, 295
Metham, 299
  nematicide, <u>233</u>, 237
METHOCELL, onion, 153
Methoxychlor, 303, 309, 320
  cattle, <u>361</u>, <u>369</u>
  livestock, 359
  seed treatment, cereals, <u>102</u>
  sorghum, 119
2,2-bis(p-Methoxyphenyl)1,1,1-trichloro-
    ethane, 303
Methyl bromide, 298, 300, 302, 303, 308,
    310
  almonds, 75
  nematicides, 230, <u>231</u>, 226
  pepper, 140
  sweetpotato, 132
  tobacco, 114, 115
  tomato, 136
Methyl-1-(butylcarbamoyl-2-benzimada-
    zole) carbamate, 298
Methyl cellulose, onion, 153
Methyldemeton, 305
N-methyldithiocarbamate, tobacco, 115
Methylisothiocyanate, 300
  nematicide, 230, 232, 237
  tomato, 137
2-2-Methyl-2-(methylthio)propionaldehyde-
    0-(methylcarbamoyl) oxime,
    nematicide, <u>234</u>
Methyl parathion, 305
Metiram
  asparagus, 164
  cucurbits, 146
  peanut, <u>110</u>, <u>111</u>
  potato, <u>129</u>, <u>130</u>
  salad crops, 157

(Metiram)
  sorghum, 119
  tobacco, 116
  vegetables, 172
Michigan plums, 80
Microorganisms, unicellular, 33
MICRO-SPERSE, peanut, 111
Middle lamella, nematodes, 205
Midwest, apricots, 77
Mildew
  aerolate, cotton, 106
  downy, 15
    grape, <u>52</u>
  grape, 278
  powdery, 9, 16, 17
    cucumber, 21
Mildew rose, 278, 280
Milk insecticide tolerances
  cattle, <u>365-369</u>
  goats, <u>372</u>, <u>373</u>
Milkroom insect, control, <u>367</u>, <u>368</u>
Milkweed, 276
Milky disease, <u>294</u>, 317
MILLERS CAPTAN, sorghum, 119
MILLER MICROFUME 25D, tobacco, 115
Miller Pesticide, Chemical Amendment,
    315
MILLER 658Z, celery, 162
Milk, 118
Mimosa, wilt resistant, 287
Mineral block insecticide, cattle, <u>365</u>
Mineral oil
  nuts, 90
  walnut, 92
Minor element deficiencies, citrus, <u>93</u>
Mistletoe, 277
  history, 260
Mist spray insecticide
  cattle, <u>365-369</u>
  horse, <u>375</u>
  poultry, <u>376</u>, <u>377</u>
  swine, <u>370</u>
MIT
  pepper, 141
  tomato, 137
Mite
  borne diseases, 326
  cattle, <u>336</u>, <u>337</u>
  control, 305, 358
    livestock, 359
  dogs, 350, <u>352</u>, <u>353</u>
  goats, control, <u>372</u>
  horse, 348
  mice, 355

MYLONE, 299
  tobacco, 115
Myobia musculi, 355
Myobia ratti, 355
Myriapoda, 326
Myrica gale, 289

N

Nabam, 8, 297
  diffusion, 68
  onion, 153
  uses, 67
Nagana, 330
Nailhead spot, tomato, 136, 137
Nairobi sheep disease, 327
Naled
  horse, 348
  poultry, 376, 379
  swine, 370
Naphthalophos, livestock, 358
1-Naphthyl-N-methyl carbamate, 304
National Agricultural Chemicals Assoc.,
  315
National beautification, 256
National forests, 257
Neck rot, onions, 152
Necrosis, nematode, 205
Nectarine diseases, 74, 76, 77
Needle nematode, 215
NEGUVON, cattle, 363, 365
NEMACIDE VC 13, nematicide, 233
NEMAGON
  peach, 78
  nematicide, 230, 232, 234
NEMAFOS, nematicide, 234
Nematicidal compounds, 218
Nematicides, 216, 226-240, 332
  application, 229, 230
  dosage, 229
  economics, 238
  emulsion, 230
  formulations, 230
  granular, 230
  ornamentals, 300
  phytotoxicity, 230
  soil cover, 230
  solution, 230
  vapor pressure, 23
  volatility, 230
Nematocera, suborder, 328
Nematodal dwarfing, 50
Nematode, 188
  alfalfa, 208
  crucifers, 150

(Nematode)
  cucurbits, 146
  control, 216-240
  destroyers, 220
  disinfestation, tobacco, 116
  distribution, 201, 202
  economics, 239
  environment, 241
  etiology, 201, 214-216
  female reproductive system, 253
  free stock, 216, 224
  inter-relations, 214
  kinds, 199
  livestock, 328
  losses from, 193
  male reproductive system, 253
  morphology, 194-200, 246-253
  ornamentals, 267, 270
  onions, 153
  parasitism, 200
  pathogenesis, 202, 202-204
  pathology, 204-205
    potato, 211
    sugarbeet, 212
    synergism, 214
    tomato, 213
  root parasites, 200
  second stage larvae, 222
  survival, 216, 217
  sweetpotato, 132
  symptoms, 202, 203
  synergism, 215, 216
  trapping fungi, 225
  virus vectors, 215
Nematologists, 240, 241
Nematology, history, 189, 189-192
Nematologica, 192
NEMASTER, tobacco, 115
NEMEX, tobacco, 115
Neofabrae malicortus, 69
Neotylenchus spp., morphology, 250,
  251
NEPHIS, nematicide, 230, 232
NEPO, viruses, 215
Nerve ring, nematodes, 198, 198
Nervous system, nematodes, 198
Net blotch barley, 98
Nettles, 277
NETU viruses, 215
Neural system, nematodes, 198
NIACIDE, uses, 67
Nicotiana glauca, 113
Nicotiana rustica, 113
Nicotiana tabacum, 113
Nicotine, 302

Nicotine sulfate, 301
  poultry, 356, 357
Nigrospora oryzea, 12, 117
NIRAN, 304
  nematicide, 235
NITROX, 305
Nits, cattle, 335
New England pilgrims, 265
Nodule inoculant legume, 111
NOMERSAN, 297
Non-chemical control, 61
North America, rye, 95
Northwestern anthracnose, apple, 69, 72
Northern fowl mite, poultry, 356
Norway maple, leaf hopper, 272
Norway Pine, blue stain, 275
Nose botfly, horse, 345
Notoedres cati, 354
Notoedric mite, rat, 355
Nucleus
  diploid, 33
  haploid, 33
Nurseryman, 322
Nut
  diseases, 90
  losses, 58
    disease, 60
  weevils, 274
Nutrient deficiency, nematodes, 223
Nutrition, sheep-keds, 344
Nymphal instars, 336, 337

O

Oak, 277
Oak wilt, 275
Oat blight, 57
Oat
  diseases, 94
  losses, disease, 59, 94
  rust, 40
  U.S. production, 95
Oatstem nematode, 225
Obligate parasitic nematodes, 217
Obligate pathogen, 13, 14, 30
OCTACHLOR, 303
OCTALENE, 304
OCTALOX, 304
Octochlorocamphene, 304
1,2,4,5,6,7,8,8-Octachloro-2,3,3a,4,7,
    7a-hexahydro-4,7-methanoindene,
    303
Odontostylet, nematode, 196

Oestridae, 328
Oestrus ovis, 343
Oil base spray, livestock, 358
Oil glands, citrus, 92
Oil mist spray insecticide, cattle, 364
Okra
  diseases, 164
  seed treatments, 167
Oleander mealybug, 273
Olivaceous lesion, 46
Olive diseases, 74
Olpidium brassicae, 114
Olpidium seedling blight, tobacco, 114
One host ticks, cattle, 338
Onion
  diseases, 151, 154
  fungicides, 169, 172
  losses, disease, 152, 153
  U.S. production, 152
Ontario, grapes, 84
Ophiobolus graminis, 122
Opuntia spp., 292, 293
Oral bolus, 358
Oral organophosphate insecticides, dog,
    353
Orange
  diseases, 92
  losses, disease, 60
Orchard grass, 121
  losses, 123
Orchid mealybug, 273
Oregon pears, 73
  prunes, 80
Organ replacement, 50
  transformation, 50
Organic fungicides, 297
Organic insecticides, 301
Organo-cadmium fungicides, 297
Organophosphate insecticides, 301,
    304-306
  livestock, 358, 359
  swine, 349
Organophosphate nematicides, 237, 238
Organophosphate systemics, livestock, 357
Organophosphorus acaricides
  dogs, 354
  sheep, 344
Organophosphorus ixodicides, 343
Organotin fungicides, 297
Oriental fruit moth, 291
Oriental gardens, 261
Ornamental
  diseases, 264-267
    control, 297

Phenyl mercury acetate, 8
Phlebotomus spp., dog, 351
Phleum pratense, 121
Phloem-necrosis, elm, 268
Phoma betae, 127
Phoma destructiva, 136
Phoma fruit rot, tomato, 136
Phoma lingam, 149
Phoma rot, tomato, 137
Phoma spp.
  crucifer, 150
  parsnip, 160
  pepper, 141
Phomopsis blight, eggplant, 142
Phomopsis diachenii, 160
Phomopsis vexans, 141, 142
Phorate, 306
  nematicide, 235, 238
  livestock, 331, 332
Phosmet, cattle, 363
Phthalimide fungicides, 298
Phthiraptera, order, 328
Phthirus pubis, 350, 351
Phycomycetes
  morphology, 14, 15
  nematode parasites, 225
PHYGON, 298
  potato, 129
  sorghum, 119
Phyllosticta batatas, 133
Phyllosticta solitaria, 69
Phylum arthropoda, 326
Phymatotrichum omnivorum, 127
Physalis sp., 133
Physalospora obtusa, 69
Physalospora rhodina, 105, 126
Physalospora tucumanensis, 126
Physical control, nematodes, 226
Physical sterilants, 295
Physiogenic disease, tomato, 134
Physiological disease
  citrus, 93
  pecan, 91
Physiological races, 44
  disease, 81
Physiology, nematodes, 241
Physoderma brown spot, corn, 59
Phytopathogenic
  bacteria, 215
  fungi, 215
  viruses, 215
Phytophthora blight, pepper, 140, 141
Phytophthora cactorum, 69
Phytophthora capsici, 135, 140
Phytophthora drechslere, 127, 135

Phytophthora erythroseptica, 126
Phytophthora fragariae, 40
Phytophthora infestans, 40, 51, 128, 133,
  134, 265
Phytophthora megasperma, 126
Phytophthora nicotianae, 114
Phytophthora nicotianae var. parisitica,
  113, 192, 214
Phytophthora parasitica, 114, 135
Phytophthora phaseoli, 165
Phytophthora spp., 15, 36, 297
  carrot, 159
  citrus, 93
  legumes, 112
  root rots, 299
  tobacco, 114
Phytotoxic
  copper sulfate, 7
  cycloheximide, 9
  fungicides, 297, 299
  grape copper, 85
  grape-folpet, 85
  grape sulfur, 85
  insecticides, 301
  lime sulfur, 7
  nematicides, 237, 238
PICFUME
  nematicide, 230, 231
  tobacco, 115
  tomato, 136
Pickling cucumbers, 143, 144
PICRIDE, tobacco, 115
Piercing, sucking insects, 272, 273
Pierce's disease, grape, 83
Pin nematodes, 199
Pinckster, 260
Pine aphid control, 303
  mouse, 281
  seedling weed control, 307
Pineapple nematicides, 238
Pink root, onions, 152-153
Pink spray, apple, 70
Pinkeye vectors, 330
Pinna ear, 343
Pinus monticola, 289
Pinus strobus, 289
Piperonyl butoxide
  cattle, 368
  horse, 375
  poultry, 377
  swine, 370
Piricularia grisea, 103
Piricularia oryzale, 122
Piscicides, 302
Plantago aristata, 208

(Potato)
 fungicides, 169, 172
 losses, disease, 128
 root eelworm, 191
 root nematode, 189, 211
 U.S. production, 128
Potato mozaic, pepper, 140
Poultry ectoparasites, 357
 control, 376-379
Pour-on insecticide
 cattle, 363, 364, 365, 369
 livestock, 358
 powders, cattle, 336
Powdery mildew, 64, 65, 84, 298
 apple, 69-72
 beans, 164
 bramble, 87
 cherry, 79
 crucifer, 150
 cucumber, 144
 cucurbits, 143
 currant, 87
 gooseberry, 87, 88
 forage grass, 122
 grape, 83, 85
 lettuce, 155, 157
 muskmelons, 145
 okra, 164
 ornamentals, 300
 pea, 164
 peach, 278
 pumpkin, 147
 spinach, 165
 strawberry, 82
 vegetables, 164, 172
Pox, sweetpotato, 131
Pratylenchus brachyurus, 204
Pratylenchus spp., nematodes, 199, 205
Pratylenchus penetrans, 214
Praying mantis, 291
Predaceous fungi, nematodes, 219, 226
Predaceous nematodes, 218
Predatory insects, 290-292
Prepink spray, apple, 70
Preservative, wood, 8
Prickly pear, 292, 293
Prickly pear moth borer, 293
Priest, physicians, 261
Primary parasite, 292
Primary predators, 291
Primrose, 290
Procedural pest control, 322
Prodelphic nematode, 253
PROFUME
 tobacco, 115

(PROFUME)
 tomato, 136
PROLATE, cattle, 363
Promycelium, 25, 27, 28, 31, 33, 36
Propargul bromide, 230, 298
Prophos, nematodes, 238
Prophylaxis potato, 129, 131
Propionic acid, nematodes, 218
Protein sheep keds, 344
Protein diet, dog, 353
Protophyta, 35
Protorhabdions, nematodes, 196
Protostom, nematode, 196
Protozoa
 diseases, 326
 horse, 347
Protozan, parasites, 355
Protrusible spear, nematode, 196
Prune
 diseases, 74, 76, 80
 losses, disease, 60
Pseudocel, nematodes, 195
Pseudocoelomate, nematode, 194
Pseudomonas adropogoni, 120
Pseudomonas angulata, 114, 116
Pseudomonas apii, 161, 162
Pseudomonas coronafaciens, 97
Pseudomonas holci, 120
Pseudomonas lachrymans, 143, 144
Pseudomonas maculicola, 150
Pseudomonas pastinaceae, 160
Pseudomonas phaseolicola, 165
Pseudomonas solanacearum, 113, 114,
    134, 192, 214
Pseudomonas spp., 35
 blueberry, 88
 escarole, 155
 forage grass, 122
Pseudomonas striafaciens, 97
Pseudomonas syringae, 76, 120, 165
Pseudomonas tabaci, 113, 114
Pseudomonas tomato, 136
Pseudonadales, 35
Pseudoperonospora cubensis, 143, 144
Pseudoperonospora humuli, 124
Pseudopeziza ribis, 87
Psorergate bos, 336, 337
Psorergate ovis, 345
Psorergate simplex, 355
Psorergatic acariasis
 cattle, 336
 sheep, 345
Psoroptes communis bovis, 336
Psoroptes communis coniculi, 355
Psoroptes communis ovis, 345

Psoroptes sp.
  cattle, 336
  horse, 348
Psoroptic mange, horse, 348
Psoroptic mites, cattle, 337
Psoroptidae family, 336
Psylla, pear, 73
Public health pests, 281, 282
  control, 305
Puccinia antirrhini, 297
Puccinia arachidis, 108, 109
Puccinia asparagi, 163
Puccinia graminis, 27, 28, 40
Puccinia gramiinis var. tritici, 29, 44
Puccinia sessilis, 290
Puccinia spp., forage grass, 122
Puccinia substriata, 142
Pucciniastrum vaccinii, 290
Puffballs, 22
Pulex irritans, 349, 352
Pumpkin
  diseases, 142, 146-148
  fungicides, 169, 172
  U.S. production, 146
Puratized agricultural spray, 8
PURETURF, 297
Purple blotch, onions, 152, 153
Pycnia, 25, 27, 30
Pycnidium, definition, 15, 20
Pycnium, 26
Pycniospores, 25, 26, 28, 30
Pyracantha, history, 260
Pyrenochaeta terrestris, 152
Pyrenophora avenae, 98
Pyrenophora teres, 98
Pyrethrin, 302
  cattle, 368
  horse, 347, 375
  livestock, 359
  poultry, 377
  swine, 370
Pyricularia oryzae, 104
Pythium spp., 15
  carrot, 158, 159
  castor bean, 125
  celery, 161
  cereal, 96
  corn, 117
  cotton, 105, 106
  crucifer, 150
  cucumber, 144
  cucurbits, 143, 147
  eggplant, 141
  flax, 120
  forage grass, 122

(Pythium spp.)
  legumes, 112
  lettuce, 155, 156
  peanuts, 109
  pepper, 139
  potato, 129
  pumpkin, 147
  rice, 104
  root rots, 299
  sugarbeet, 127
  sugarcane, 126
  tobacco, 114
  tomato, 136

Q

Quercus spp., 289
Q-Fever, 327
Quill mite poultry, 356
Quince
  diseases, 69
  fire blight, 37, 275
Quinone fungicides, 298

R

Rabbit ectoparasites, 355
RABON
  poultry, 377, 378
  swine, 371
Races, pathogenic fungi, 60
Radish, 148
  fungicides, 170
Radopholus similis, 192
Ramularia leaf spot, sugarbeet, 127
Ramularia pastinacae, 160
Rangeland, management, 313
Range loss, 58
Raphanus sativus, 148
Raspberry
  diseases, 86
  losses, disease, 60
Rat ectoparasites, 355
Ratoon stunting, sugarcane, 126
RA-VAP, poultry, 377
Recreation areas, 257
Rectal glands, nematode, 197
Rectal muscles, nematodes, 196
Rectum, nematode, 194, 197
Redbug, cattle, 337
Red cedar, 289
Red currant, 289
Red mange, dog, 353

450

(Wintertick)
  horse, 347
Wirestem, crucifer, 150
Wireworm control, 303
  ornamentals, 271
  sorghum, 119
Wisteria, 277
Woodrot fungi, 22
Wood nettle, 277
Wood tick
  cattle, 338
  horse, 348
  swine, 350
Woody shrub control, 307
Wool clipping, 344
Wool maggots, sheep, control, 372
Woolly apple aphid, 272, 291
Wounds, ectoparasites, 331
Wrappers, citrus fruits, 93

X

Xanthomonas campestris, 149
Xanthomonas carotae, 159
Xanthomonas corylina, 90
Xanthomonas holcicola, 120
Xanthomonas juglandis, 91
Xanthomonas malvacearum, 105
Xanthomonas phaseoli, 165
Xanthomonas pruni, 38-39, 86
Xanthomonas solanacearum, 113
Xanthomonas spp., 35
  forage grass, 122
Xanthomonas stewartii, 38
Xanthomonas translucens, 97
Xanthomonas vesicatoria, 40, 134, 135,
  140
Xiphinema spp., nematodes, 191, 199,
  215
X-ray sex sterilization, 295

Y

Yellow body louse, poultry, 357
Yellow dwarf virus, onion, 152, 153
Yellow dwarf disease, sweetpotato, 131
Yellow witches broom, 289
Yellow disease
  eggplant, 142
  sugarbeet, 127
Yellow virus
  carrot, 158
  parsnip, 160
Yew, 277

Z

Zinc deficiency
  citrus, 93
  pecan, 91
Zinc dimethyldithiocarbamate, 297
Zinc ethylenebisdithiocarbamate, 297
Zinc fungicide, 297
Zinc
  pecans, 91
  seed treatment, cereals, 101
  sulfate, lime, peach, 78
Zineb, 8, 9, 63, 297
  apple, 70
  asparagus, 164
  beans, 165
  blueberry, 89
  carrot, 158
  celery, 162
  cereals, 96
  cherry, 79
  citrus, 93
  corn, 118
  cotton, 105, 106
  crucifer, 151
  cucurbits, 146-147
  diffusion, 68
  eggplant, 142
  grape, 85
  hops, 124
  lettuce, 157
  nuts, 90
  onions, 153
  parsnip, 160
  peach, 78
  pepper, 141
  plum, 80
  potato, 129
  strawberry, 81
  tobacco, 114, 115, 116
  tomato, 137, 138
  vegetables, 168-170
  uses, 67
ZINOPHOS, 300
  nematicides, 234
Ziram, 8, 297
  almonds, 75
  beans, 165
  carrots, 158
  celery, 162
  cucurbits, 147
  eggplant, 141
  salad crops, 157
  stone fruit, 77
  uses, 67

On War

The Best Military Histories

From the Pritzker Military Museum & Library Award–winning authors

On War

The Best Military Histories

From the Pritzker Military Museum & Library Award–winning authors

Rick Atkinson ▪ Carlo D'Este ▪ Max Hastings

James M. McPherson ▪ Allan R. Millett

Tim O'Brien ▪ Gerhard L. Weinberg

PRITZKER
MILITARY
MUSEUM & LIBRARY

2013

All excerpts reprinted with permission:

Battle Cry of Freedom, by James M. McPherson, copyright © 1988 by Oxford University Press, published by Oxford University Press.

Military Innovation in the Interwar Period, edited by Williamson R. Murray and Allan R. Millett, copyright © 1996 by Cambridge University Press, published by Cambridge University Press.

Visions of Victory: The Hopes of Eight World War II Leaders, by Gerhard L. Weinberg, copyright © 2005 by Gerhard L. Weinberg, published by Cambridge University Press.

An Army at Dawn: The War in North Africa, 1942–1943, by Rick Atkinson, copyright © 2002 by Rick Atkinson, published by Henry Holt and Company, LLC.

Patton: A Genius for War, by Carlo D'Este, copyright © 1995 by Carlo D'Este, published by HarperCollins Publishers, Inc.

Catastrophe 1914, by Max Hastings, copyright © 2013 by Max Hastings, published in arrangement with Alfred A. Knopf, an imprint of the Knopf Doubleday Publishing Group, a division of Random House, Inc.

The Things They Carried, by Tim O'Brien, copyright © 1990 by Tim O'Brien, published by Houghton Mifflin Harcourt.

Executive Editor: Kenneth Clarke
Editor: Michael W. Robbins
Designer: Wendy Palitz

ISBN: 978-0-9897928-1-3
Limited Edition ISBN: 978-0-9897928-2-0

The trade edition of *On War: The Best Military Histories* was preceded by a limited leather-bound edition of 26 lettered copies signed by each of the authors, as well as 100 numbered copies, each housed in a clamshell case, accompanied by a sterling silver commemorative coin and signed by each of the authors. The trade edition will be followed by a magazine edition.

First Edition

To the citizen soldier

BATTLE OF THE BULGE: GIS AT AID STATION
NEAR OURTHE RIVER, BELGIUM, 1944

CREDITS

The Pritzker Military Museum & Library Archives
Cover: Army Signal Corps; v: Leslie Bigelow Collection;
viii: United States Government Printing Office;
5; 17; 80, 104: Karduck Collection;
155: Bernard Perlin/U.S. Government Printing Office;
185: Army Signal Corps; 197; 202; 244: Franz
Altschuler Collection

The Library of Congress
Endsheets: detail of Gettysburg battlefield, ca. 1863,
John B. Bachelder; 2: Timothy H. Sullivan; 11: Jacob F.
Coonley; 21; 41: Bain News Service; 113: Harris & Ewing;
128: U.S. Army; 133: Harris & Ewing; 166: Office
of War Information; 186: Bain News Service;
193: Harris & Ewing; 229

The National Archives and Records Administration
28: Chief Photographer's Mate Robert F. Sargent/U.S.
Coast Guard; 35: Office of War Information; 50: U.S. Army;
59: Photographer's Mate 1st Class Harry R. Watson/U.S.
Coast Guard; 70: U.S. Army; 83: Office of Alien Property;
91, 99: Office of War Information; 119: Clifford
Berryman/Records of the U.S. Senate; 144, 149, 160, 168,
176, 183, 220, 233: U.S. Army

U.S. Army Center of Military History: 173

Weider History Group Archives: 207; 215

CONTENTS

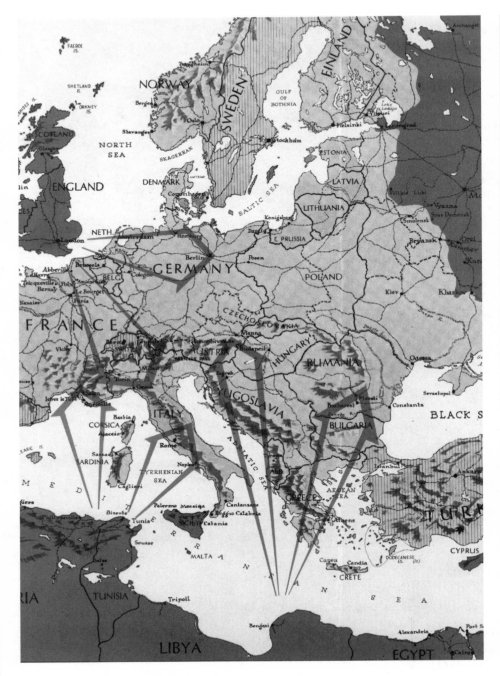

ALLIED AIR OFFENSIVE ON OCCUPIED EUROPE, AUGUST 1943

TO HONOR THE CITIZEN SOLDIER

THE SEVEN AUTHORS IN THIS ANTHOLOGY, all recipients of the Pritzker Military Museum & Library Literature Award for Lifetime Achievement in Military Writing, have achieved recognition akin to being a household name for the work they have done in military history and fiction. As the name of this award suggests, it took a lifetime for these authors to achieve their successes, making their words in these seven chapters, which they have selected for us, all the more meaningful.

I was born in 1950, five years after the end of World War II and in the year the United States intervened in Korea. Most of the men in my father's generation had some military service, and many of my ancestors served in World War I. Military terms and references were common in my childhood and youth. My career in the U.S. Army and the Illinois National Guard gave me the opportunity to become professionally qualified at something universally recognized, and I met many great and interesting people in the military. My experience is not out of the ordinary for those times, but today fewer Americans are serving in the military, and perhaps more important, fewer civilians are paying attention to what their military is doing.

A democratic form of government means "rule of the people," and an essential function of such a government is to provide for the defense of the nation. That means that under a democratic system, all citizens are potential soldiers. The concept of citizens having the right to defend themselves and their communities, as opposed to subjects of a hereditary monarchy that has a monopoly over the means of defense, is one of the reasons for the Second Amendment of the U.S. Constitution.

Given the vast sums of money this country spends on defense each year, it is essential for the general public to have access to ample sources of information on military affairs. How can we have a democracy with civilian control of the military if civilians do not have knowledge of and contact with the military? A society that divides its military and civilian communities into "us" and "them" is going to have significant problems. Accordingly, I founded the Pritzker Military Museum & Library as a place where civilian and military communities can come together to learn about each other through a not-for-profit nongovernment organization—and this anthology is one tool to support that mission.

I have been collecting books since before I could read them. My parents, grandparents, and great-grandfather all enjoyed books. The study of history enables one to connect the past with the present to make more accurate projections of the future. The authors in this anthology are among the best military history writers of all time and they actively help us remember what happened yesterday, so we don't make the same mistakes today and tomorrow.

The range of perspectives offered by the authors in this book is staggering. James M. McPherson, who began publishing on the Civil War in 1964, is the world's leading scholar on the topic. Allan R. Millett, the world's leading scholar on the Korean War, is a towering military historian. Gerhard L. Weinberg began publishing works about World War II in 1953 and his book *A World at Arms* is one of the most important publications about World War II. Embedded journalist and military historian Rick Atkinson is drawing everyday readers into his meticulously researched and compelling histories about Iraq, World War II, and other topics. Carlo D'Este's biographies of George S. Patton, Winston Churchill, and Dwight D. Eisenhower are the definitive works on these important leaders. The works of embedded journalist and military historian Max Hastings on the Falklands War, World War I, and World War II have brought him a large, discerning international readership. Tim O'Brien's novels about the Vietnam War are considered the world's best fiction on the topic.

It is my hope that this anthology will lead both military and civilian readers to discover—or rediscover—these fine authors.

—Colonel (IL) J. N. Pritzker, IL ARNG, (Retired)

Principles and rules are intended to provide a
thinking man with a frame of reference.

—*Count Carl von Clausewitz*

SLAVE FAMILY, FIVE GENERATIONS ON SMITH'S PLANTATION, BEAUFORT, SOUTH CAROLINA, 1862

CHAPTER ONE

JAMES M. McPHERSON

"We Are All Americans"
from *Battle Cry of Freedom*

IN HIS LANDMARK HISTORY OF THE AMERICAN CIVIL WAR era, James McPherson portrays the complex events of four "lifetime" years of fighting, suffering, and national upheaval, plus the social, economic, and political landscape upon which the military events exploded. He begins with the Mexican War and ends with "What now?" questions about the fates and futures of the freed slaves. In so doing, he brilliantly defines an era of growing conflict between competing national visions and cultures, and competing definitions of "freedom," that culminates tragically in the Civil War. That epic bloodletting did not settle every conflict, but was the sharp point of what he later characterizes as America's own hundred-years war.

"We Are All Americans," the finale of *Battle Cry*, covers the closing days of the war in early 1865. McPherson highlights the Confederacy's struggle with the conundrum of whether to deploy its slaves as soldiers defending slavery; the fall of Richmond; the face-to-face conversation at the hour of surrender and victory—and the universal sense that an uncertain new American order looms.

I.

THE CONFEDERACY HAD ONE LAST STRING TO ITS BOW—a black string. Early in the war a few voices had urged the arming of slaves to fight for their masters. But to most Southerners such a proposal seemed at best ludicrous and at worst treasonable. With a president who denounced the North's emancipation and recruitment of slaves as "the most execrable measure recorded in the history of guilty man," it required rash courage to suggest that the Confederacy itself put arms in the hands of slaves.[1]

After the fall of Vicksburg and the defeat at Gettysburg, however, the voices suggesting such a thing had become less lonely. Several newspaper editors in Mississippi and Alabama began speaking out in extraordinary fashion. "We are forced by the necessity of our condition," they declared, "to take a step which is revolting to every sentiment of pride, and to every principle that governed our institutions before the war." The enemy was "stealing our slaves and converting them into soldiers....It is better for us to use the negroes for our defense than that the Yankees should use them against us." Indeed, "we can make them fight better than the Yankees are able to do. Masters and overseers can marshal them for battle by the same authority and habit of obedience with which they are marshalled to labor." It was true, admitted the *Jackson Mississippian*, that "such a step would revolutionize our whole industrial system" and perhaps lead to universal emancipation, "a dire calamity to both the negro and the white race." But if we lose the war we lose slavery anyway, for "Yankee success is death to the institution...so that it is a question of necessity—a question of a choice of evils....We must...save ourselves from the rapacious North, WHATEVER THE COST."[2]

General Patrick Cleburne had been thinking along similar lines. He wrote down his ideas and presented them to division and corps commanders in the Army of Tennessee in January 1864. The South was losing the war, said Cleburne, because it lacked the North's manpower and because "slavery, from being one of our chief sources of strength at the commencement of the war, has now become, in a military point of view, one of our chief sources of weakness." The Emancipation Proclamation had given the enemy a moral cause to justify his drive for conquest, Cleburne continued, had made the slaves his allies, undermined the South's domestic security, and turned European nations against the Confederacy. Hence we are threatened with "the loss of all we now hold most sacred—slaves and all other personal property, lands, homesteads, liberty, justice, safety, pride, manhood." To save the rest of these cherished possessions we must sacrifice the first. Let us recruit an army of slaves, concluded Cleburne, and "guarantee freedom within a reasonable time to every slave in the South who shall remain true to the Confederacy."[3]

Twelve brigade and regimental commanders in Cleburne's division endorsed his proposal. This was a potentially explosive matter, for these were not just editors

STIRRING APPEAL.

CHIVALRIC SOUTHERNER. "Here! you mean, inferior, degraded Chattel, jest kitch holt of one of them 'ere muskits, and *conquer my freedom for me!*"

CHATTEL. "Well, dunno, Massa; guess you'd better not be free: you know, Massa, *slave folks is deal happier than free folks.*"

CARTOON LAMPOONING SOUTHERN ATTITUDES TOWARD FREEDOM AND SLAVERY expressing an opinion, but fighting men on whom the hopes for Confederate survival rested. Cleburne's arguments cut to the heart of a fundamental ambiguity in the Confederacy's raison d'être. Had secession been a means to the end of preserving slavery? Or was slavery one of the means for preserving the Confederacy, to be sacrificed if it no longer served that purpose? Few Southerners in 1861 would have recognized any dilemma: slavery and independence were each a means as well as an end in symbiotic relationship with the other, each essential for the survival of both. By 1864, however, Southerners in growing numbers were beginning to wonder if they might have to make a choice between them. "Let not slavery prove a barrier to our independence," intoned the *Jackson Mississippian.* "Although slavery is one of the principles that we started to fight for...if it proves an insurmountable obstacle to the achievement of our liberty and separate nationality, away with it!"[4]

At the time of Cleburne's proposal, however, such opinions still seemed dangerous. Most generals in the Army of Tennessee disapproved of Cleburne's action, some of them vehemently. This "monstrous proposition," wrote a division commander, was "revolting to Southern sentiment, Southern pride, and Southern honor." A corps commander abhorred it as "at war with my social, moral, and political principles." A shocked and angry brigadier insisted that "we are not whipped, & cannot be whipped. Our situation requires resort to no such remedy....Its propositions contravene the principles upon which we fight."[5]

Convinced that the "promulgation of such opinions" would cause "discouragements, distraction, and dissension" in the army, Confederate president Jefferson Davis ordered the generals to stop discussing the matter.[6] So complete was their compliance that the affair remained unknown outside this small circle of Southern officers until the U. S. government published the war's *Official Records* a generation later. The only consequence of Cleburne's action seemed to be denial of promotion to this ablest of the army's division commanders, who was killed ten months later at the battle of Franklin.

By then the South's dire prospects had revived the notion of arming blacks. In September 1864 the governor of Louisiana declared that "the time has come for us to put into the army every able-bodied negro man as a soldier." A month later the governors of six more states, meeting in conference, enigmatically urged the impressment of slaves for "the public service as may be required." When challenged, all but two of the governors (those of Virginia and Louisiana) hastened to deny that they meant the *arming* of slaves. On November 7, Jefferson Davis urged

Congress to purchase 40,000 slaves for work as teamsters, pioneers, and laborers with the promise of freedom after "service faithfully rendered." But this cautious proposal proved much too radical for most of the press and Congress. It would crack the door of abolition, declared the *Richmond Whig*. The idea of freeing slaves who performed faithfully was based on the false assumption "that the condition of freedom is so much better for the slave than servitude, that it may be bestowed upon him as a reward." This was "a repudiation of the opinion held by the whole South...that servitude is a divinely appointed condition for the highest good of the slave."[7]

Congress did not act on the president's request. But the issue would not go away. Although Davis in his November 7 message had opposed the notion of arming blacks *at that time*, he added ominously: "Should the alternative ever be presented of subjugation or the employment of the slave as a soldier, there seems no reason to doubt what should then be our decision." Within three months the alternative stared the South starkly in the face. The president and his cabinet made their choice. "We are reduced," said Davis in February 1865, "to choosing whether the negroes shall fight for or against us."[8] And if they fought for us, echoed some newspapers, this would not necessarily produce wholesale abolition. Perhaps those who fought must be offered freedom, but that would only "affect units of the race and not the whole institution." By enabling the South to whip the Yankees, it was the only way to save slavery. "If the emancipation of a part is the means of saving the rest, then this partial emancipation is eminently a pro-slavery measure." Some advocates went even further and said that discipline rather than the motive of freedom was sufficient to make slaves fight. "It is not true," declared General Francis Shoup, "that to make good soldiers of these people, we must either give or promise them freedom....As well might one promise to free one's cook...with the expectation of thereby securing good dinners."[9]

Such talk prompted one exasperated editor to comment that "our Southern people have not gotten over the vicious habit of not believing what they don't wish to believe."[10] Most participants in this debate recognized that if slaves became soldiers, they and probably their families must be promised freedom or they might desert to the enemy at first opportunity. If one or two hundred thousand slaves were armed (the figures most often mentioned), this would free at least half a million. Added to the million or so already liberated by the Yankees, how could the institution survive? asked opponents of the proposal.

These opponents remained in the majority until February 1865. Yet with the Yankees thundering at the gates, their arguments took flight into an aura of unreality. We can win without black help, they said, if only the absentees and stragglers return to the ranks and the whole people rededicate themselves to the Cause. "The freemen of the Confederate States must work out their own redemption, or they must be the slaves of their own slaves," proclaimed the *Charleston Mercury*, edited by those original secessionists the Robert Barnwell Rhetts, father and son. "The day that the army of Virginia allows a negro regiment to enter their lines as soldiers they will be degraded, ruined, and disgraced," roared Robert Toombs. His fellow Georgian Howell Cobb agreed that "the moment you resort to negro soldiers your white soldiers will be lost to you....The day you make soldiers of them is the beginning of the end of the revolution. If slaves will make good soldiers our whole theory of slavery is wrong."[11]

And was not that the theory the South fought for? "It would be the most extraordinary instance of self-stultification the world ever saw" to arm and emancipate slaves, declared the Rhetts. "It is abolition doctrine...the very doctrine which the war was commenced to put down," maintained a North Carolina newspaper. It would "surrender the essential and distinctive principle of Southern civilization," agreed the *Richmond Examiner*.[12] Many Southerners apparently preferred to lose the war than to win it with the help of black men. "Victory itself would be robbed of its glory if shared with slaves," said a Mississippi congressman. It would mean "the poor man...reduced to the level of a nigger," insisted the *Charleston Mercury*. "His wife and daughter are to be hustled on the street by black wenches, their equals. Swaggering buck niggers are to ogle them and elbow them." Senator Louis Wigfall of Texas "wanted to live in no country in which the man who blacked his boots and curried his horse was his equal." "If such a terrible calamity is to befall us," declared the *Lynchburg Republican*, "we infinitely prefer that Lincoln shall be the instrument of our disaster and degradation, than that we ourselves should strike the cowardly and suicidal blow."[13]

But the shock effect of Lincoln's insistence at Hampton Roads on unconditional surrender helped the Davis administration make headway against these arguments. During February many petitions and letters from soldiers in the Petersburg trenches poured into Richmond to challenge the belief that white soldiers would refuse to fight alongside blacks. While "slavery is the normal condition of the negro...as indispensable to [his] prosperity and happiness...as is liberty to the whites," declared the 56th Virginia, nevertheless "if the public exigencies require that any number of our male slaves be enlisted in the military service in order to [maintain] our Government, we are willing to make concessions to their false and unenlightened notions of the blessings of liberty."[14]

Robert E. Lee's opinion would have a decisive influence. For months rumors had circulated that he favored arming the slaves. Lee had indeed expressed his private opinion that "we should employ them without delay [even] at the risk which may be produced upon our social institutions." On February 18 he broke his public silence with a letter to the congressional sponsor of a Negro soldier bill. This measure was "not only expedient but necessary," wrote Lee. "The negroes, under proper circumstances, will make efficient soldiers. I think we could at least do as well with them as the enemy....Those who are employed should be freed. It would be neither just nor wise...to require them to serve as slaves."[15]

Lee's great prestige carried the day—but just barely. Although the powerful *Richmond Examiner* dared to express a doubt whether Lee was "a 'good Southerner'; that is, whether he is thoroughly satisfied of the justice and beneficence of negro slavery," even this anti-administration newspaper recognized that "the country will not venture to deny to General Lee...*anything* he may ask for."[16] By a vote of 40 to 37 the House passed a bill authorizing the president to requisition a quota of black soldiers from each state. In deference to state's rights, the bill did not mandate freedom for slave soldiers. The Senate nevertheless defeated the measure by a single vote, with both senators from Lee's own state voting No. The Virginia legislature meanwhile enacted its own law for the enlistment of black soldiers—without, however, requiring the emancipation of those who were slaves—and instructed its senators to vote for the congressional bill. They did so, enabling it to pass by 9 to 8 (with several abstentions) and become law on March 13. In the few weeks of life left to the Confederacy no other state followed Virginia's lead. The two companies of black soldiers hastily organized in Richmond never saw action. Nor did most of these men obtain freedom until the Yankees—headed by a black cavalry regiment—marched into the Confederate capital on April 3.[17]

A last-minute diplomatic initiative to secure British and French recognition in return for emancipation also proved barren of results. The impetus for this effort came from Duncan F. Kenner of Louisiana, a prominent member of the Confederate Congress and one of the South's largest slaveholders. Convinced since 1862 that slavery was a foreign-policy millstone around the Confederacy's neck,

Kenner had long urged an emancipationist diplomacy. His proposals got nowhere until December 1864, when Jefferson Davis called Kenner in and conceded that the time had come to play this last card. Kenner traveled to Paris and London as a special envoy to offer abolition for recognition. Davis of course could not commit his Congress on this matter, and these lawmakers could not in turn commit the states, which had constitutional authority over the institution. But perhaps European governments would overlook these complications.

Kenner's difficulties in getting out of the Confederacy foretokened the fate of his mission. The fall of Fort Fisher prevented his departure on a blockade-runner. He had to travel in disguise to New York and take ship from this Yankee port for France. Louis Napoleon as usual refused to act without Britain. So James Mason accompanied Kenner to London, where on March 14 they presented the proposition to Lord Palmerston. Once again the Confederates learned the hard lesson of diplomacy: nothing succeeds like military success. "On the question of recognition," Mason reported to Secretary of State Benjamin, "the British Government had not been satisfied at any period of the war that our independence was achieved beyond peradventure, and did not feel authorized so to declare [now] when the events of a few weeks might prove it a failure....As affairs now stood, our seaports given up, the comparatively unobstructed march of Sherman, etc., rather increased than diminished previous objections."[18]

II.

WHILE THE SOUTH DEBATED THE RELATIONSHIP of slavery to its Cause, the North acted. Lincoln interpreted his reelection as a mandate for passage of the Thirteenth Amendment to end slavery forever. The voters had retired a large number of Democratic congressmen. But until the 38th Congress expired on March 4, 1865, they retained their seats and could block House passage of the amendment by the necessary two-thirds majority. In the next Congress the Republicans would have a three-quarters House majority and could easily pass it. Lincoln intended to call a special session in March if necessary to do the job. But he preferred to accomplish it sooner, by a bipartisan majority, as a gesture of wartime unity in favor of this measure that Lincoln considered essential to Union victory. "In a great national crisis, like ours," he told Congress in his message of December 6, 1864, "unanimity of action among those seeking a common end is very desirable—almost indispensable."

TENNESSEE COLORED BATTERY, CAMP AT JOHNSONVILLE, TENNESSEE, 1864

This was an expression of an ideal rather than reality, since most war measures, especially those concerning slavery, had been passed by a strictly Republican vote. For the historic achievement of terminating the institution, however, Lincoln appealed to Democrats to recognize the "will of the majority" as expressed by the election.[19]

But most Democrats preferred to stand on principle in defense of the past. Even if the war had killed slavery, they refused to help bury it. The party remained officially opposed to the Thirteenth Amendment as "unwise, impolitic, cruel, and unworthy of the support of civilized people." A few Democratic congressmen believed otherwise, however. The party had suffered disaster in the 1864 election, said one, "because we [would] not venture to cut loose from the dead carcass of negro slavery." Another declared that to persist in opposition to the amendment "will be to simply announce ourselves a set of impracticables no more fit to deal with practical affairs than the old gentleman in Copperfield."[20] Encouraged by such sentiments, the Lincoln administration targeted a dozen or so lame-duck Democratic congressmen and subjected them to a barrage of blandishments. Secretary of State William H. Seward oversaw this lobbying effort. Some congressmen were promised government jobs for themselves or relatives; others received administration favors of one sort or another.[21]

This arm twisting and log rolling paid off, though until the House voted on January 31, 1865, no one could predict which way it would go. As a few Democrats early in the roll call voted Aye, Republican faces brightened. Sixteen of the eighty Democrats finally voted for the amendment; fourteen of them were lame ducks. Eight other Democrats had absented themselves. This enabled the amendment to pass with two votes to spare, 119 to 56. When the result was announced, Republicans on the floor and spectators in the gallery broke into prolonged—and unprecedented—cheering, while in the streets of Washington cannons boomed a hundred-gun salute. The scene "beggared description," wrote a Republican congressman in his diary. "Members joined in the shouting and kept it up for some minutes. Some embraced one another, others wept like children. I have felt, ever since the vote, as if I were in a new country." By acclamation the House voted to adjourn for the rest of the day "in honor of this immortal and sublime event."[22]

The Thirteenth Amendment sped quickly through Republican state legislatures that were in session; within a week eight states had ratified it and during the next two months another eleven did so. Ratification by five additional Northern states was certain as soon as their legislatures met. Of the Union states only those carried by McClellan in the presidential election—New Jersey, Kentucky, and Delaware—held out.[23] The "reconstructed" states of Louisiana, Arkansas, and Tennessee ratified readily. Since the Lincoln administration had fought the war on the theory that states could not secede, it considered ratification by three-quarters of *all* the states including those in the Confederacy to be necessary. One of the first tasks of reconstruction would be to obtain the ratification of at least three more ex-Confederate states to place the amendment in the Constitution.

Among the spectators who cheered and wept for joy when the House passed the Thirteenth Amendment were many black people. Their presence was a visible symbol of the revolutionary changes signified by the amendment, for until 1864 Negroes had not been allowed in congressional galleries. Blacks were also admitted to White House social functions for the first time in 1865, and Lincoln went out of his way to welcome Frederick Douglass to the inaugural reception on March 4. Congress and Northern states enacted legislation that began to break down the pattern of second-class citizenship for Northern Negroes: admission of black witnesses to federal courts; repeal of an old law that barred blacks from carrying mail; prohibition of segregation on streetcars in the District of Columbia; repeal of black laws in several Northern states that had imposed certain kinds of discrimination against Negroes or barred their entry into the state; and steps to submit referendums to the voters of several states to grant the ballot to blacks (none of these referendums passed until 1868).

Perhaps the most dramatic symbol of change occurred on February 1, the day after House passage of the Thirteenth Amendment. On that day Senator Charles Sumner presented Boston lawyer John Rock for admission to practice before the Supreme Court, and Chief Justice Salmon P. Chase swore him in. There was nothing unusual in this except that Rock was a black man, the first Negro accredited to the highest court which eight years earlier had denied U.S. citizenship to his race. The court had been virtually reconstructed by Lincoln's appointment of five new justices including Chase. The transition from Roger Taney to Chase as leader of the court was itself the most sweeping judicial metamorphosis in American history.[24]

Important questions concerning emancipation and reconstruction were sure to come before this new court. Two such questions might well grow out of actions taken in the area of freedmen's affairs during the winter of 1864–65. Thousands of contrabands had attached themselves to Sherman's army on its march from Atlanta to the sea. Reports filtered northward of Sherman's indifference to their welfare and ill treatment of them by some officers and men. To sort out the rumors and problems, Secretary of War Edwin M. Stanton journeyed to Savannah in January for a talk with Sherman and an interview with black leaders, most of them former slaves. Among the questions Stanton asked these men was how best they could support their families in freedom. "The way we can best take care of ourselves," they answered, "is to have land, and turn in and till it by our labor....We want to be placed on land until we are able to buy it, and make it our own."[25]

Stanton and Sherman thought this a good idea, so the conservative general prepared the most radical field order of the war. Issued January 16, Sherman's "Special Field Orders, No. 15" designated the sea islands and the rich plantation areas bordering rivers for thirty miles inland from Charleston down to Jacksonville for settlement by freedmen. Each head of family could be granted forty acres of land, to which he would be given a "possessory title" until Congress "shall regulate their title."[26] This land had of course belonged to slaveholders. Their dispossession of it by Sherman's order, like Lincoln's dispossession of their slaves by the Emancipation Proclamation, was a military measure carried out under "war powers." The Thirteenth Amendment confirmed Lincoln's action; it remained to be seen how court, Congress, and executive would deal with the consequences of Sherman's Order No. 15. The army did not wait, however. During the next several months General Rufus Saxton, an abolitionist who commanded the Union occupation forces on the South Carolina sea islands, supervised the settlement of 40,000 freedmen on land designated in Sherman's order.

The wartime experience of Union army officers who governed occupied territory, of treasury agents who had charge of abandoned plantations, and of freedmen's aid societies that sent missionaries and teachers to the freed slaves made clear the need for a government agency to coordinate their efforts. All too often these various groups pulled in different directions—and the freedmen themselves in still another direction. In 1863 Congress first considered legislation to create a Freedmen's Bureau. Disagreement whether this agency should be part of the War or the Treasury Department prevented final enactment of a bill until March 3, 1865. By then Chase was no longer secretary of the treasury, so radical Republicans who had wanted him in charge of the bureau were now willing to place it in the War Department. The function of the bureau (formally called the Bureau of Refugees, Freedmen, and Abandoned Lands) would be to dispense rations and relief to the hundreds of thousands of white as well as black refugees uprooted by the war and to assist the freedmen during the difficult transition from slavery to freedom. Congress also gave the bureau control of "abandoned" land with the provision that individual freedmen "shall be assigned not more than forty acres" of such land at rental for three years and an option to buy at the end of that time with "such title thereto as the United States can convey."[27] Here was Sherman's Order No. 15 writ large. Whether Congress would be able to convey any title was a troublesome question, given the constitutional ban on bills of attainder and the presidential power of pardon and restoration of property. In any event, the Freedmen's Bureau represented an unprecedented extension of the federal government into matters of social welfare and labor relations—to meet unprecedented problems produced by the emancipation of four million slaves and the building of a new society on the ashes of the old.

The success or failure of the Freedmen's Bureau would be partly determined by the political terms of reconstruction. This matter occupied much of Congress's time during the winter of 1864–65. Prospects seemed good for a compromise with the president. The afterglow of a Republican electoral sweep had dissolved the bitterness that had crested with Lincoln's pocket veto of the Wade-Davis bill. Chase's elevation to the Supreme Court was another step toward rapprochement between Lincoln and radical Republicans. The president's reference to reconstruction in his message to Congress on December 6 hinted at the likelihood of "more rigorous measures than heretofore."[28] This willingness to meet Congress halfway set the stage for an attempt to enact a new reconstruction bill. The outlines of such legislation soon emerged in negotiations between Lincoln and congressional leaders: Congress would accept the already reconstructed regimes of Louisiana and Arkansas (soon to be joined by Tennessee) in return for a presidential promise to sign legislation similar to the Wade-Davis bill for the other Confederate states.

As first introduced in the House this new bill enfranchised "all male citizens"—including blacks. Lincoln persuaded the committee chairman in charge of the bill to modify it to limit this provision to black soldiers. During the next two months the measure went through a bewildering series of revisions and amendments in committee and on the House floor. At one stage it enfranchised literate blacks as well as soldiers; at another it removed racial qualifications altogether; at still another it applied these provisions to Louisiana and Arkansas as well as to the other states. Democrats voted with moderate Republicans to defeat the more radical versions of the bill and joined radicals to defeat the more conservative versions. Consequently no bill could be passed. Not wishing to create a precedent for executive reconstruction by seating Louisiana's senators and representatives in Congress, radicals teamed with Democrats to deny them admission. Thus the Thirty-Eighth Congress expired without any further action on reconstruction. Radical Republicans thought this just as well. The next Congress would have more radicals and fewer Democrats, noted one of them, and "in the meantime I hope the nation may be educated up to our demand for universal suffrage."[29]

The prospect of "educating" Lincoln up to this demand seemed promising. He had moved steadily leftward during the war, from no emancipation to limited emancipation with colonization and then to universal emancipation with limited suffrage. This trajectory might well carry him to a broader platform of equal suffrage by the time the war ended. The entreaty in Lincoln's second inaugural address for "malice toward none" and "charity for all" provided few clues on this question, though it seemed to endorse generous treatment of ex-rebels. At the same time this address left no doubt of Lincoln's intention to fight on until slavery was crushed forever. "Fondly do we hope—fervently do we pray—that this mighty scourge of war may speedily pass away," said the nation's sixteenth president at the beginning of his second term. "Yet if God wills that it continue, until all the wealth piled up by the bondman's two hundred and fifty years of unrequited toil shall be sunk, and until every drop of blood drawn with the lash, shall be paid by another drawn with the sword, as was said three thousand years ago, so it must be said 'the judgments of the Lord, are true and righteous altogether.' "[30]

III.

ULYSSES S. GRANT WAS DETERMINED THAT it should not take that long. During the winter, Union forces at Petersburg had fought their way westward to cut off the last road into town from the south and threaten the last open railroad. With Lee's army of 55,000 melting away by desertions while the oncoming spring dried roads after an exceptionally raw, wet winter, the final success of Grant's 120,000 seemed only a matter of time. Sherman could be expected on Lee's rear by late April, but Grant wanted the Army of the Potomac to "vanquish their old enemy" without help that might produce future gloating by Sherman's veterans. "I mean to end the business here," the general in chief told Phil Sheridan. Grant's main concern now was that he might wake one morning to find Lee gone to join John-

SAMUEL PRAYER, AFRICAN-AMERICAN CIVIL WAR VETERAN, CA. 1900

ston's 20,000 for an attack on Sherman.[31] Lee had precisely that in mind. In his effort to accomplish it, however, he gave Grant a long-sought opportunity to drive the ragged rebels from their trenches into the open.

By March, Lee had become convinced that he must soon abandon the Petersburg lines to save his army from encirclement. This would mean the fall of Richmond, but better that than loss of the army, which was the only thread holding the Confederacy together. To force Grant to contract his lines and loosen the stranglehold blocking a rebel escape, Lee planned a surprise attack on the enemy

Among the troops who marched into Richmond as firemen and policemen were units from the all-black 25th Corps.

position just east of Petersburg. Southern corps commander John B. Gordon sent false deserters to fraternize with Yankee pickets in front of Fort Stedman on the night of March 24–25. The "deserters" suddenly seized the dumbfounded pickets, and Gordon's divisions swarmed into sleepy Fort Stedman. Capturing several batteries and a half-mile of trenches, the Confederates seemed to have achieved a smashing breakthrough. But a Northern counterattack recaptured all lost ground plus the forward trenches of the Confederate line, trapping many of the rebels and forcing them to surrender. Lee lost nearly 5,000 men; Grant only 2,000. Instead of compelling Grant to shorten his lines, Lee had to thin his own to the breaking point. And Grant lost little time in breaking them.

On March 29 he ordered an infantry corps and Sheridan's cavalry, recently returned from the Shenandoah Valley, to turn the Confederate right ten miles southwest of Petersburg. Lee sent George Pickett with two divisions of infantry through a drenching downpour to help the worn-down rebel cavalry counter this move. Hard fighting across a sodden landscape on the last day of March stopped the Federals temporarily. But next afternoon they launched an enveloping attack against Pickett's isolated force at the road junction of Five Forks. Sheridan's rapid-firing troopers, fighting on foot, attacked head-on, while Gouverneur K. Warren's 5th Corps moved sluggishly against Pickett's flank. Storming up and down the line cajoling and god-damning the infantry to move faster and hit harder, Sheridan finally coordinated an assault that achieved the most one-sided Union victory since the long campaign began eleven months earlier in the Wilderness. Pickett's divisions collapsed, half of their men surrendering to the whooping Yankees and the other half running rearward in rout. When the news reached Grant that evening, he ordered an assault all along the line next morning, April 2.

At dawn it came, with more élan and power than the Army of the Potomac had shown for a long time. And the Army of Northern Virginia—weary, hungry, shorn of more than one-fifth of its strength by the fighting on March 25 and April 1—could no longer hold the Yankees off. Sheridan got astride the last railroad into Petersburg, and the blue infantry punched through Confederate lines at several places southwest of the city. The rebels fought desperately as they fell back, but it was only to hold on to the inner defenses until dark in order to get away.

For Lee knew that he must pull out. As Jefferson Davis worshipped at St. Paul's Church in Richmond this balmy Sunday, a messenger tiptoed down the aisle and gave him a telegram. It was from Lee: Richmond must be given up. Turning pale, the president left the church without a word. But parishioners read the message on his face, and the news spread quickly through the city. Everyone who could beg, borrow, or steal a conveyance left town. Government officials crowded aboard ramshackle trains headed for Danville with the Treasury's remaining gold and as much of the archives as they could carry, the rest being put to the torch. So was everything of military and industrial value in Richmond. As night came and the army departed, mobs took over and the flames spread. Southerners burned more of their own capital than the enemy had burned of Atlanta or Columbia. When the Yankees arrived next morning, their first tasks were to restore order and put out the flames. Among the troops who marched into Richmond as firemen and policemen were units from the all-black 25th Corps.

Following the Northern soldiers into Richmond came a civilian—the number one civilian, in fact, Abraham Lincoln. The president had taken a short vacation from Washington to visit the Army of the Potomac, arriving just before it broke up the Confederate attack on Fort Stedman. Wanting to be there for the end, which now seemed imminent, Lincoln had stayed on as Grant's guest. Commander-in-chief and general-in-chief entered Petersburg on April 3 only hours after the Army of Northern Virginia had left. Grant soon rode west on the chase to head off Lee. Lincoln returned to the Union base on the James River and told

Admiral David D. Porter: "Thank God I have lived to see this. It seems to me that I have been dreaming a horrid dream for four years, and now the nightmare is gone. I want to see Richmond."[32] Porter took Lincoln upriver to the enemy capital where the president of the United States sat down in the study of the president of the Confederate States forty hours after Davis had left it.

Lincoln's visit to Richmond produced the most unforgettable scenes of this unforgettable war. With an escort of only ten sailors, the president walked the streets while Porter peered nervously at every window for would-be assassins. But the Emancipator was soon surrounded by an impenetrable cordon of black people shouting "Glory to God! Glory! Glory! Glory!" "Bless the Lord! The great Messiah! I knowed him as soon as I seed him. He's been in my heart four long years. Come to free his children from bondage. Glory, Hallelujah!" Several freed slaves touched Lincoln to make sure he was real. "I know I am free," shouted an old woman, "for I have seen Father Abraham and felt him." Overwhelmed by rare emotions, Lincoln said to one black man who fell on his knees in front of him: "Don't kneel to me. That is not right. You must kneel to God only, and thank Him for the liberty you will enjoy hereafter."[33] Among the reporters from Northern newspapers who described these events was one whose presence was a potent symbol of the revolution. He was T. Morris Chester, who sat at a desk in the Confederate Capitol drafting his dispatch to the *Philadelphia Press*. "Richmond has never before presented such a spectacle of jubilee," he wrote. "What a wonderful change has come over the spirit of Southern dreams."[34] Chester was a black man.

FOR ROBERT E. LEE AND HIS ARMY THE DREAMS HAD TURNED into a nightmare. Reduced to 35,000 men, the scattered divisions from Petersburg and Richmond rendezvoused at Amelia Courthouse thirty-five miles to the west, where the starving men expected to find a trainload of rations. Because of a mixup they found ammunition instead, the last thing they needed since the worn-out horses could scarcely pull the ordnance the army was carrying. A delay to forage the countryside for food proved fatal. Lee had intended to follow the railroad down to Danville, where he could link up with Johnston and where Jefferson Davis on April 4 issued a rallying cry to his people: "Relieved from the necessity of guarding cities...with our army free to move from point to point...and where the foe will be far removed from his own base...nothing is now needed to render our triumph certain, but...our own unquenchable resolve."[35] But the foe was closer to

BURNED DISTRICT, RICHMOND, VIRGINIA, APRIL 1865

Danville than Lee's army was. Racing alongside the retreating rebels a few miles to the south were Sheridan's cavalry and three infantry corps. On April 5 they cut the Danville railroad, forcing Lee to change direction toward Lynchburg and the Blue Ridge passes beyond.

But this goal too was frustrated by the weariness of Lee's despondent men and the speed of Union pursuers who sniffed victory and the end of the war. Stabbing attacks by blue cavalry garnered scores of prisoners, while hundreds of other Southerners collapsed in exhaustion by the roadside and waited for the Yankees to pick them up. Along an obscure stream named Sayler's Creek on April 6, three Union corps cut off a quarter of Lee's army, captured 6,000 of them, and destroyed much of their wagon train. "My God!" exclaimed Lee when he learned of this action. "Has the army been dissolved?"[36]

Not yet, but it soon would be. As the remaining rebels trudged westward on April 7, Grant sent Lee a note under flag of truce calling on him to surrender. Lee responded with a feeler about Grant's terms. The Northern commander offered the same terms as at Vicksburg: parole until exchanged. Since Lee's surrender would virtually end the war, the part about exchange was a mere formality. As the tension mounted on April 8—Grant had a splitting headache and Meade suffered from nausea—Lee parried with a vague proposal to discuss a general "restoration of peace," a political matter on which Grant had no authority to negotiate. Grant shook his aching head and commented: "It looks as if Lee meant to fight."[37]

Lee did have that notion, intending to try a breakout attack against Sheridan's troopers blocking the road near Appomattox Courthouse on the morning of April 9. For the last time rebel yells shattered the Palm Sunday stillness as the gray scarecrows drove back Union horsemen—only to reveal two Yankee infantry corps coming into line behind them. Two other Union corps were closing in on Lee's rear. Almost surrounded, outnumbered by five or six to one in effective troops, Lee faced up to the inevitable. One of his subordinates suggested an alternative to surrender: the men could take to the woods and become guerrillas. No, said Lee, who did not want all of Virginia devastated as the Shenandoah Valley had been; the guerrillas "would become mere bands of marauders, and the enemy's cavalry would pursue them and overrun many sections they may never [otherwise] have occasion to visit. We would bring on a state of affairs it would take the country years to recover from." With a heavy heart Lee decided that "there is nothing left for me to do but go and see General Grant, and I would rather die a thousand deaths."[38] Lee sent a note through the lines offering to surrender. Grant's headache and Meade's illness vanished. The bleeding and dying were over; they had won. To the home of Wilmer McLean went Lee and Grant for the surrender formalities. In 1861, McLean had lived near Manassas, where his house was a Confederate headquarters and a Yankee shell had crashed into his dining room. He moved to this remote village in southside Virginia to escape the contending armies only to find the final drama of the war played out in his living room. The vanquished commander, six feet tall and erect in bearing, arrived in full-dress uniform with sash and jeweled sword; the victor, five feet eight with stooped shoulders, appeared in his usual private's blouse with mud-spattered trousers tucked into muddy boots—because his headquarters wagon had fallen behind in the race to cut off the enemy. There in McLean's parlor the son of an Ohio tanner dictated surrender terms to the scion of a First Family of Virginia.

The terms were generous: officers and men could go home "not to be disturbed by U.S. authority so long as they observe their paroles and the laws in force where they may reside." This clause had great significance. Serving as a model for the subsequent surrender of other Confederate armies, it guaranteed Southern soldiers immunity from prosecution for treason. Lee asked another favor. In the Confederate army, he explained, enlisted men in the cavalry and artillery owned their horses; could they keep them? Yes, said Grant; privates as well as officers who claimed to own horses could take them home "to put in a crop to carry themselves and their families through the next winter." "This will have the best possible effect upon the men," said Lee, and "will do much toward conciliating our people." After signing the papers, Grant introduced Lee to his staff. As he shook hands with Grant's military secretary Ely Parker, a Seneca Indian, Lee stared a moment at Parker's dark features and said, "I am glad to see one real American here." Parker responded, "We are all Americans."[39]

The surrender completed, the two generals saluted somberly and parted. "This will live in history," said one of Grant's aides. But the Union commander seemed distracted. Having given birth to a reunited nation, he experienced a post-partum melancholy. "I felt...sad and depressed," Grant wrote, "at the downfall of a foe who had fought so long and valiantly, and had suffered so much for a cause, though that cause was, I believe, one of the worst for which a people ever fought." As news of the surrender spread through Union camps, batteries began firing joyful salutes until Grant ordered them stopped. "The war is over," he said; "the rebels are our countrymen again, and the best sign of rejoicing after the victory will be to abstain from all demonstrations."[40] To help bring those former rebels back into the Union, Grant sent three days' rations for 25,000 men across the lines. This perhaps did something to ease the psychological as well as physical pain of Lee's soldiers.

So did an important symbolic gesture at a formal ceremony three days later when Confederate troops marched up to stack arms and surrender their flags. As they came, many among them shared the sentiments of one officer: "Was this to be the end of all our marching and fighting for the past four years? I could not keep back the tears." The Union officer in charge of the surrender ceremony was Joshua L. Chamberlain, the fighting professor from Bowdoin who won a medal of honor for Little Round Top, had been twice wounded since then, and was now a major general. Leading the Southerners as they marched toward two of Chamberlain's

The news of Lee's surrender traveled through a North barely recovered from boisterous celebrations of Richmond's capture.

brigades standing at attention was John B. Gordon, one of Lee's hardest fighters who now commanded Stonewall Jackson's old corps. First in line of march behind him was the Stonewall Brigade, five regiments containing 210 ragged survivors of four years of war. As Gordon approached at the head of these men with "his chin drooped to his breast, downhearted and dejected in appearance," Chamberlain gave a brief order, and a bugle call rang out. Instantly the Union soldiers shifted from order arms to carry arms, the salute of honor. Hearing the sound General Gordon looked up in surprise, and with sudden realization turned smartly to Chamberlain, dipped his sword in salute, and ordered his own men to carry arms. These enemies in many a bloody battle ended the war not with shame on one side and exultation on the other but with a soldier's "mutual salutation and farewell."[41]

The news of Lee's surrender traveled through a North barely recovered from boisterous celebrations of Richmond's capture. The fall of the rebel capital had merited a *nine-hundred* gun salute in Washington; the surrender of Lee produced another five hundred. "From one end of Pennsylvania Avenue to the other," wrote a reporter, "the air seemed to burn with the bright hues of the flag....Almost by magic the streets were crowded with hosts of people, talking, laughing, hurrahing and shouting in the fullness of their joy. Men embraced one another, 'treated' one another, made up old quarrels, renewed old friendships, marched arm-in-arm singing." The scene was the same on Wall Street in New York, where "men embraced and hugged each other, kissed each other, retreated into doorways to dry their eyes and came out again to flourish their hats and hurrah," according to an eyewitness. "They sang 'Old Hundred,' the Doxology, 'John Brown,' and 'The Star-Spangled Banner'...over and over, with a massive roar from the crowd and a unanimous wave of hats at the end of each repetition." "My only experience of a people stirred up to like intensity of feeling," wrote a diarist, "was the great Union meeting at Union Square in April 1861." But this time the feeling was even more intense because "founded on memories of years of failure, all but hopeless, and the consciousness that national victory was at last secured."[42]

Lincoln shared this joyous release of pent-up tension, but he was already thinking more of the future than of the past. While in Richmond he had met with John A. Campbell, one of the Confederate commissioners at the earlier Hampton Roads conference. Campbell was now ready to return to the Union on Lincoln's terms. He suggested an apparent way to undermine what was left of the Southern war effort: allow the Virginia legislature to meet so it could withdraw the state's troops from the Confederacy. The president thought this a good idea and on April 6 gave the necessary permission. But Campbell misconstrued Lincoln's position to be one that recognized the legislature as the legitimate government of the state. Lincoln had no such purpose. He had authorized a meeting of "the gentlemen who had *acted* as the Legislature of Virginia...having power *de facto* to do a specific thing," but did not intend to recognize them as "the rightful Legislature." Lee's surrender, which included nearly all of Virginia's soldiers, made the whole matter academic, so Lincoln revoked his permission for the legislature to meet. And on April 11 he delivered from a White House balcony a carefully prepared speech on peace and reconstruction to a crowd celebrating Union victory. "There is no authorized organ for us to treat with," he said—thereby disposing of state governments as well as Jefferson Davis's fugitive government. "We must simply begin with, and mould from, disorganized and discordant elements." This he had done in Louisiana, Arkansas, and Tennessee. Defending the government of Louisiana, Lincoln conceded that he would prefer it to have enfranchised literate Negroes and black veterans. He hoped that it would soon do so; as for the unreconstructed states, Lincoln promised an announcement soon of a new policy for their restoration to the Union.[43]

At least one listener interpreted this speech as moving Lincoln closer to the radical Republicans. "That means nigger citizenship," snarled John Wilkes Booth to a companion. "Now, by God, I'll put him through. That is the last speech he will ever make."[44]

Notes

1. Davis quoted in Robert F. Durden, ed., *The Gray and the Black: The Confederate Debate on Emancipation* (Baton Rouge: Louisiana State University Press, 1972), 24.

2. These quotations are from editorials in the *Jackson Mississippian* reprinted in *Montgomery Mail*, Sept. 9, 1863; *Montgomery Weekly Mail*, Sept. 2, 1863; and *Mobile Register*, Nov. 26, 1863, all reprinted in Durden, *Gray and Black*, 30–35, 42–44.

3. *War of Rebellion, A Compilation of the Official Records of the Union and Confederate Armies* (hereafter *OR*), ser. 1, vol. 52, pt. 2, 586–92.

4. As reprinted in *Montgomery Weekly Mail*, Sept. 9, 1863, in Durden, *Gray and Black*, 31–32.

5. Patton Anderson in *OR*, ser. 1, vol. 52, pt. 2, 598–99; Alexander P. Stewart to William H. T. Walker, Jan. 9, 1864, William B. Bate to Walker, Jan. 9, 1864, Civil War Collection, Henry E. Huntington Library.

6. *OR*, ser. 1, vol. 52, pt. 2, 608.

7. *OR*, ser. 1, vol. 41, pt. 3, 774; Dunbar Rowland, ed., *Jefferson Davis, Constitutionalist: His Letters, Papers, and Speeches* (Jackson: Mississippi Department of Archives and History, 1923), vol. 4, 394–97; *Richmond Whig*, Nov. 9, 1864, in Durden, *Gray and Black*, 110.

8. Rowland, *Davis*, vol. 6, 396; *OR*, ser. 4, vol. 3, 1110.

9. *Lynchburg Virginian*, Nov. 3, 1864; *Richmond Sentinel*, Nov. 24, 1864; article by Shoup in *Richmond Whig*, Feb. 20, 1865, all in Durden, *Gray and Black*, 79, 121, 214.

10. *Charlottesville Chronicle*, reprinted in *Richmond Sentinel*, Dec. 21, 1864, in Durden, *Gray and Black*, 147.

11. *Charleston Mercury*, Nov. 3, 1864, in Durden, *Gray and Black*, 99; Toombs quoted in Shelby Foote, *The Civil War: A Narrative* (New York: Random House, 1974), vol. 3, 860; Cobb in *OR*, ser. 4, vol. 3, 1009–10.

12. *Charleston Mercury*, Nov. 3, 19, 1864, *North Carolina Standard*, Jan. 17, 1865, in Durden, *Gray and Black*, 99, 114, 177; *Richmond Examiner*, Jan. 14, 1865, quoted in Paul D. Escott, *After Secession: Jefferson Davis and the Failure of Confederate Nationalism* (Baton Rouge: Louisiana State University Press, 1978), 154.

13. Mississippi congressman quoted in Durden, *Gray and Black*, 140; *Charleston Mercury*, Jan. 26, 1865, quoted in Bell Irvin Wiley, *Southern Negroes 1861–1865* (New Haven, CT: Yale University Press, 1938), 156–57; Louis Wigfall quoted in E. Merton Coulter, *The Confederate States of America 1861–1865* (Baton Rouge: Louisiana State University Press, 1950), 268; *Lynchburg Republican*, Nov. 2, 1864, in Durden, *The Gray and the Black*, 94.

14. Published in *Richmond Whig*, Feb. 23, 1865, reprinted in Durden, *Gray and Black*, 222–23.

15. Lee to Andrew Hunter, Jan. 11, 1865, in *OR*, ser. 4, vol. 3, 1012–13; Lee to Ethelbert Barksdale, Feb. 18, 1865, in Durden, *Gray and Black*, 206.

16. *Richmond Examiner*, Feb. 16, 25, in Durden, *Gray and Black*, 199, 226.

17. War Department regulations governing the recruitment of slave soldiers bootlegged a quasi-freedom into the process by stipulating that a slave could be enlisted only with his own consent and that of his master, who was required to grant the slave in writing, "as far as he may, the rights of a freedman." Whether this ambiguous language actually conferred freedom, as several historians maintain, must remain forever moot. See Durden, *Gray and Black*, 268–70; Escort, *After Secession*, 252; and Emory Thomas, *The Confederate Nation 1861–1865* (New York: Harper Torchbooks, 1979), 296–97.

18. Frank Lawrence Owsley, *King Cotton Diplomacy: Foreign Relations of the Confederate States of America* (Chicago: University of Chicago Press, 1931), 550–61; quotation from 560.

19. Roy P. Basler, *The Collected Works of Abraham Lincoln* (hereafter *CWL*) (New Brunswick, NJ: Rutgers University Press, 1953–55), vol. 8, 149.

20. Christopher Dell, *Lincoln and the War Democrats* (Rutherford, NJ: Fairleigh Dickinson University Press, 1975), 290; Anson Herrick of New York in *Congressional Globe* (hereafter *CG*), 38th Cong., 2d Sess., 525—25; Samuel S. Cox quoted in Joel H. Silbey, *A Respectable Minority: The Democratic Party in the Civil War Era, 1860-1868* (New York: Norton, 1977), 183.

21. James G. Randall and Richard N. Current, *Lincoln the President: Last Full Measure* (New York: Dodd, Meade, and Co., 1955), 307–13; La Wanda Cox and John H. Cox, *Politics, Principle, and Prejudice 1865-1866: Dilemma of Reconstruction America* (New York: Free Press, 1963), 1–30.

22. "George W. Julian's Journal," *Indiana Magazine of History*, 11 (1915), 327; *CG*, 38th Cong., 2d Sess., 531.

23. New Jersey ratified the amendment in 1866 after Republicans gained control of the legislature.

24. In June 1864, Lincoln had finally accepted Chase's third offer to resign from the cabinet. In October, Taney died, and two months later the president appointed Chase chief justice in part as a gesture of conciliation to the radical wing of his party.

25. *The Liberator*, Feb. 24, 1865.

26. *OR*, ser. 4, vol. 47, pt. 2, 60–62.

27. U.S. Statutes at Large, vol. 13, 507–9.

28. *CWL*, vol. 8, 152.

29. Congressman James M. Ashley in *Boston Commonwealth*, March 4, 1865. For a skillful analysis of these confusing debates and votes on reconstruction see Herman Belz, *Reconstructing the Union: Theory and Policy during the Civil War* (Ithaca, NY: Cornell University Press, 1969), 244–76.

30. *CWL*, vol. 8, 333.

31. Bruce Catton, *Grant Takes Command* (Boston: Little, Brown, 1969), 437; *Personal Memoirs of U. S. Grant*, 2 vols. (New York: Charles L. Webster, 1886), vol. 2, 424–25, 430–31, 459–61.

32. Foote, *Civil War*, vol. 3, 896.

33. Burke Davis, *To Appomattox: Nine April Days, 1865* (New York: Rhinehart and Co., 1959), 184; Foote, *Civil War*, vol. 3, 897; Charles Carleton Coffin, *The Boys of '61* (Boston: Estes and Lauriat, 1896), 538–42.

34. *Philadelphia Press*, April 11, 12, 1865.

35. Rowland, *Davis*, vol. 6, 529–31.

36. Douglas Southall Freeman, *R. E. Lee: A Biography*, 4 vols. (New York: Charles Scribner's Sons, 1934–35), vol. 4, 84.

37. Catton, *Grant Takes Command*, 460.

38. Freeman, *Lee*, vol. 4, 120–23.

39. *OR*, ser. 1, vol. 46, pt. 1, 57–58; Horace Porter, "The Surrender at Appomattox Courthouse," *Battles and Leaders of the Civil War*, eds. Robert U. Johnson and Clarence C. Buell (New York: The Century Co., 1888; reprint ed., Seacaucus, NJ, 1982), vol. 4, 739–40; Davis, *To Appomattox*, 386.

40. Davis, *To Appomattox*, 387; *Personal Memoirs of Grant*, vol. 2, 489; Porter, "The Surrender at Appomattox Courthouse," 743.

41. Davis, *To Appomattox*, 362; Joshua L. Chamberlain, "The Last Salute of the Army of Northern Virginia," in *Southern Historical Society Papers*, 32 (1904), 362.

42. Foote, *Civil War*, vol. 3, 900; *The Diary of George Templeton Strong: The Civil War 1860-1865*, ed. Alan Nevins and Milton Halsey Thomas (New York: Macmillan, 1952), 574–75.

43. *CWL*, vol. 8, 406–7, 399–405.

44. William Hanchett, *The Lincoln Murder Conspiracies* (Urbana: University of Illinois Press, 1983), 37.

U.S. infantrymen disembark from LCVP at 0630 on June 6, 1944, at Omaha Beach.

CHAPTER TWO

ALLAN R. MILLETT

"Assault from the Sea"
from *Military Innovation in the Interwar Period*

HOW DO ARMED FORCES CHANGE? That deceptively simple question is posed by Allan R. Millett and coauthor Williamson Murray at the outset of their study of military innovation in the years between World War I and World War II. They sought to identify the driving forces for change—technological, strategic, managerial, political, financial. To answer the crucial question of why some innovations succeed, that is, become effective operational doctrine, while many other innovations are discarded, they focused on seven areas of warfare that changed dramatically during the 1920s and 1930s: armored warfare, amphibious warfare, strategic bombing, close air support, carrier-borne aviation, submarine warfare, and electronic communication.

In "Assault from the Sea," Millett analyzes the approaches of three nations' military services to the challenges of amphibious warfare: the American, British, and Japanese experiences. He examines the strategic contexts in which the different doctrines and forces were developed, and compares their successes and shortcomings in World War II.

On the morning of November 12, 1921, Charles Evans Hughes, secretary of state of the United States, surveyed the thousand expectant spectators with a coolness matched only by the autumn winds. Hughes came to the podium of Continental Hall in Washington, D.C., to begin the International Conference on Limitation of Armament. Much to the surprise of the delegations and audience that crowded the hall, Hughes proposed a comprehensive plan to limit naval forces and to create a regime of non-aggression in Asia, the most potent remaining arena of international rivalry. By the time the delegates had completed their work in February 1922, they had cobbled together three major, interrelated treaties that shaped the naval policies of Britain, the United States, and Japan through most of the interwar period.[1]

One of the three major treaties—the Treaty for the Limitation of Armament—carried important implications for amphibious operations, a form of naval warfare not addressed in the negotiations. Although the conferees discussed the development of battleships, naval aviation, and submarines, they failed to discuss the ability of their navies to conduct assaults from the sea. Ironically, the Japanese Army was the world's foremost amphibious force in 1921. Within twenty years Britain, the United States, and Japan all identified a need for amphibious capability, but Japan alone possessed the doctrine, tactical concepts, and forces for such operations in 1939.

The demand for such forces could be inferred from Article XIX of the "Five Power Treaty," which limited the size and modernity of the battle fleets of all three great navies. The signatories of the Five-Power Treaty (which also applied to France and Italy) agreed to establish no new bases in Asia and the western Pacific outside of carefully defined "home lands" and to hold the military capability (coastal defenses and fleet repair and maintenance facilities) of existing bases within the treaty area to 1921 levels. This provision limited any further development of the American naval bases in the Philippines and Guam, the British base at Hong Kong, and the Japanese bases on the island of Formosa and in the central Pacific islands held under a League of Nations mandate. If any of the three navies fought one another, they would have to occupy and develop extemporized bases or capture the bases that survived the Five-Power Treaty.[2]

The British, American, and Japanese had a rich heritage of amphibious operations before World War I but had made no special effort to develop amphibious forces under naval control or as joint (army-navy) forces. By World War I each of these three naval forces had demonstrated on several occasions a similar basic concept for successful landing: land where there is no opposition from ground forces. The British had done so during the Egyptian campaign (1882) and the Punitive Expedition to China (1900); the Americans had invested Santiago, Cuba, and Manila, the Philippines (1898) after unopposed landings; and the Japanese had twice besieged Port Arthur (1894 and 1904–1905) after transporting armies across the Yellow Sea. The principal danger in such an operation was an attack upon the invasion force by the opposing navy. In the Japanese and American experiences, mishandled and inferior Russian and Spanish squadrons had perished in attempts to disrupt naval operations, and the entire history of the nineteenth-century British Empire demonstrated the naval impotence of Britain's conquests.[3]

The post–World War I treaty regime lasted well into the 1930s because of the burden of wartime governmental debt and the collapse of revenues during the Great Depression. Even after Japan repudiated the Washington Conference treaties in 1936, hope and bankruptcy combined to keep the Royal Navy and United States Navy under the treaty limits in terms of warship construction. Neither could afford a major base improvement program, especially far from their national shores. Japan, however, started to improve and fortify its naval facilities in the Mandates in 1937 with the most ambitious effort focused on Yap in the western Carolines, "the Gibraltar" of the central Pacific. The Japanese tested Western resolve in 1931 by establishing a puppet regime in Manchuria and in 1932 by launching an air and land punitive expedition against the Chinese of Shanghai for abusing Japanese property and lives. In both "incidents" Western nations, collectively or singly, responded with sanctions insufficient to persuade the Japanese that imperialism did not pay. By 1937, when Japan started its full-scale war to dominate China, the three principal signatories of the Five-Power Treaty had embarked upon major naval rearmament programs. Yet the problem of defending and creating advanced naval bases for extended naval campaigns still remained a low priority for all three navies. This inattention had roots in World War I and in the tumult of the Russian, Greek, and Turkish civil wars that followed.[4]

World War I provided ample examples of the feasibility of amphibious operations, as well as evidence that some kinds of amphibious operations could result in prohibitive losses. The most ambitious operation of the war, the Allied expedition to the Dardanelles in 1915, was also its most celebrated failure. The controversies that clustered about the Dardanelles campaign made rational investigation difficult, for at issue were the political career and credibility of Winston Churchill, the expedition's sponsor as First Lord of the Admiralty; command relations between the Royal Navy and the British Army; the distrust between British commanders and the Australian–New Zealand expeditionary corps; the timid conduct of fleet operations before the landing, which might have made an amphibious assault unnecessary; and disagreements with the French forces assigned to the campaign. Tactical pundits claimed that the slaughter of the British 29th Division on "V" and "W" beaches proved that amphibious assaults could not prevail against prepared defenses armed with machine guns and artillery. Virtually every phase of the campaign received searching investigation after the war. The weight of the evidence seemed to give the Dardanelles all the elements of a fated Attic disaster, not a laboratory for identifying and correcting the operational flaws that might have swung the campaign to the Allies.[5]

World War I, however, also provided evidence that amphibious landings were not necessarily a forlorn hope. The Germans conducted successful assaults against the Russians on the Baltic islands of Courland and Oesel in the Gulf of Riga. Despite the Dardanelles failure, the Admiralty continued to give serious attention to amphibious operations. In 1918 it mounted several raids against German submarine bases in Belgium, the most dramatic being a night assault on Zeebrugge in April. The same year the Royal Navy examined the feasibility of amphibious landings along the Adriatic with the enthusiastic support of Admiral William S. Sims, commander of U.S. naval forces in European waters. After the war, seaborne expeditionary forces went to Murmansk, Vladivostok, the newly independent Baltic states, and cities along the Black Sea to try to confront the Red armies during the Russian civil war. Expeditionary forces also arrived by ship to provide safe havens (and some diplomatic leverage) during the war between Greece and Turkey. In a sense these expeditions repeated the nineteenth-century pattern of avoiding defended ports or beaches, or ensuring through prior diplomacy that the disembarkation site would be in friendly hands. The British armed forces, in fact, viewed amphibious operations as a type of "combined operations," a campaign in which the navy, the army, and the newly independent Royal Air Force each might have an important role. This definition tended to burden any consideration of amphibious operations with all the interservice conflicts that beset the British armed forces.[6]

Of all the belligerents of World War I the Japanese showed they had not forgotten the amphibious concepts they had used against the Chinese in 1894 and against the Russians in 1904–1905. In September 1914 they sent an expeditionary force by sea to the Shantung peninsula to invest the German leasehold of Tsingtao. Greatly outnumbered, the German garrison could do little, and after a brief siege it surrendered. A similar fate befell German colonial forces on the Pacific islands of Saipan, Palau, and Truk as well as on some atolls. In the Pacific islands the landing forces came from the ranks of the Imperial Japanese Navy rather than from the army. Four years after its whirlwind campaign of 1914 Japan showed it had lost none of its expertise in moving land forces by sea when it sent an army of 70,000 to Vladivostok as the leading edge of Allied intervention in Siberia.[7]

The Strategic Context of Amphibious Operations

Britain, Japan, and the United States faced different strategic dilemmas in the 1920s and 1930s, all of which contributed to their varied development of amphibious forces. The British Army and Royal Navy had a minor interest in "combined operations," and the Royal Air Force had virtually none. For much of the interwar period the British armed forces struggled to justify their estimates under the "Ten Year Rule," a judgment by civilian leaders that the armed forces would not face a major war for at least ten years. When the possibility of war increased in the 1930s, the British had no choice but to view Germany as the principal enemy and Europe as the most important potential theater of operations. As the British Chiefs of Staff Committee and service staffs examined force requirements, structure, and service roles and missions, they envisioned a European war primarily of motorized armies and of air forces with wide capabilities for air defense, strategic bombardment, and ground force support.

Any modernization of the Royal Navy was distinctly secondary, let alone an investment in amphibious forces. The navy did not view amphibious capability as important as strengthening the battle line, carrier aviation, the destroyer force, or even the submarine force. A naval war would be fought from fixed bases under Allied control, sited either in the United Kingdom or along the route from Britain to India. Although the Japanese invasion of China in 1937 sent clear signals that a Far Eastern crisis had come, the Chiefs of Staff and their civilian ministers could not justify a higher priority for amphibious forces. They recognized that such forces might be essential to any naval campaign east of Suez, but even the defense of Hong Kong and Singapore did not look promising if Japan decided to extend its interests south from China. The fate of the British Empire in East Asia especially Malaya, Burma, and India might rest in the hands of two uncertain allies, the United States and the Soviet Union. Nevertheless, the British had great trouble believing that the Royal Navy and Royal Air Force, backed by the armies of the Commonwealth, could not stop the Japanese armed forces along the Malay Barrier.[8]

The Japanese political and military elites believed that their nation's destiny, the very survival of the Japanese people, would require a major war for the domination of Asia. Such a contest should subjugate China and eliminate the Russians, the Americans, and the Europeans as regional economic and military powers. Some Japanese diplomats and civilian leaders believed that war was unnecessary for Japan's continued economic growth and regional hegemony, but their voices failed in the 1930s under extreme domestic pressure. More optimistic Japanese military leaders believed that a growing world crisis centered in Europe would allow Japan to reap further territorial gains.

The real barriers to expansionism were not the European colonial powers but the United States and the Soviet Union. Both played an unwelcome role in supporting an independent China, but Japan could not fight every potential foe at once, although the most emotional and naive officers thought otherwise. Essentially, Japanese planners had to join one of two geopolitical schools and a wrong guess not only endangered careers but lives. One school might have been called the "continentalists," mostly army officers, who believed that Japan's future depended on its domination of mainland Asia, especially China, Manchuria, the Russian Maritime Province, and even Siberia. Any influence south of China (French Indochina, Siam, Burma) would be welcome, but not essential. The other school, the "imperialists," was most influenced by the British experience and envisioned an empire (the "co-prosperity sphere") that might include mainland enclaves in China, but which drew its strength from "the South Seas resource area." The prime targets were Malaya and the Dutch East Indies with French Indochina and the Philippines added principally for strategic security. These objectives implied a commitment to eliminate American bases west of Hawaii and to conquer the South Pacific islands under Australian mandate. Thereby the Japanese would establish a strategic outpost line that could confound a U.S. Navy counteroffensive through an "interception-attrition" strategy.

Japanese strategic planning, however, hinged not only on geopolitical considerations but also on timing. The Japanese decision to conduct an "imperial-navalist" campaign in 1941 stemmed from assessments impossible before that

JAPANESE SOLDIERS STAND GUARD ON THE COAST OF CHINA, JUNE 1939.

year. The planners realized that the Soviet Union could do little to oppose Japan while it fought for survival against Germany, and that the United States had decided to rebuild its armed forces—and deploy some of them—in anticipation of eventual involvement in the European war. (In 1941, for example, the U.S. Marine Corps deployed a brigade from California to Iceland so that it could keep a marine division on the East Coast intact for contingency missions in the Atlantic and Caribbean.) The coalition of Japanese army and navy leaders who eventually drove their government to war tested the Russians in the Nomonhan Incident of 1939 and found their own forces wanting. A campaign that began with surprise naval operations, a characteristic of earlier Japanese wars, promised better results. The Japanese planned for the "southward advance" in full confidence that they could move and land sufficient expeditionary forces.[9]

For the United States in the interwar period, Japan represented the most likely enemy and one that only an extended naval campaign across the vast spaces of the central Pacific Ocean could defeat. While political, business, educational, and religious leaders discounted the likelihood of war with Japan, senior officers of the American armed forces, especially those of the U.S. Navy, believed even before World War I that war with Japan had become inevitable. Part of this assumption rested, no doubt, on racism, but most of it stemmed from a close examination of Asian politics and the obligation to defend the Philippines and the mid-Pacific islands of Guam, Wake, and Midway, all especially vulnerable after the Japanese gained control of Germany's Pacific colonies in World War I.

No responsible American military planner believed that a European power represented a danger until the late 1930s, as only Britain had the military capability to threaten the United States for many years after World War I. Japan, in contrast, had both a fleet and a motive to challenge the United States, champion of Chinese nationalism (within limits) and governor of the Philippines. With the navy taking the lead, the military planning committees attached to the Joint Army-Navy Board focused on one contingency plan after 1919, Plan ORANGE, the strategic conception for a conflict with Japan. Although it went through predictable alterations and controversies—for example, could the Philippines be held?—Plan ORANGE shaped the navy's force structure in the interwar period and also influenced organization and deployment of the army, especially those ground and air units assigned to China, the Philippines, and Hawaii.[10]

The senior officers of the U.S. Navy and Army differed from their contemporaries in Britain and Japan with their focus upon a single major war plan with a single potential foe. Not until 1938 did the military planning bureaucracy of the army and navy consider the possibility of a German threat to the Western Hemisphere; earlier contingency plans for operations against Canada, Mexico, and Cuba had faded into irrelevance in the 1930s. Planning a war with Japan gave both services ample roles and missions to justify budget requests to the White House and Congress, even if their requirements were never met in full. The army, for example, argued that the Japanese threat to the Philippines, Hawaii, Alaska, and the Canal Zone—not to mention the west coast of the United States—justified continued investment in long-range bombers, heavy coast artillery, tactical aviation, and mobile defense forces of combined arms. Horse cavalry regiments were about the only element that the army could not easily relate to War Plan ORANGE. Those units the army deployed to protect the Mexican border from raiders, still a threat in the 1930s. Experimental armored forces also appeared to have limited relevance to a war with Japan, but they were so small (one brigade in the mid-1930s) that they were not a budgetary burden.

For the navy a possible war with Japan required development of forces capable of operating across 7,000 miles of ocean. The navy assumed that only a major fleet engagement in the western Pacific and an eventual sea-air siege of the Japanese home islands could determine the outcome of such a war. The campaign would require battleships, heavy and light cruisers, destroyers, and submarines, all capable of operating for weeks at sea. The surface fleet would need protection from island-based air attack, which meant development of fleet carriers and embarked air groups; the major debate about naval aviation was not its general utility, but its offensive capability away from surface combatants. The navy also required long-range amphibian patrol aircraft for reconnaissance duties, a function shared with submarines, and needed a mobile service force capable of fueling, arming, repairing, and reprovisioning the fleet from extemporized bases and anchorages.

After the Japanese seized control of the German central Pacific islands in 1914, navy planners assumed that the fleet would have to wrest bases from a Japanese defense force. Although various revisions of ORANGE differed in the priority given to the use of advanced bases, planners focused on several potential sites in the western Marshalls (with emphasis on Eniwetok) and Truk island group, whose special advantages for anchorages and airfields had not escaped the Japanese. Even if the fleet reached the Philippines without a major engagement, it would still have to establish a major base on the southern coast of Mindanao before reconquering Luzon, because a direct thrust into Manila Bay or Lingayen Gulf would give the Japanese unacceptable operational advantages for air and naval action. In addition to this consideration, army planners added another objective in the 1930s, the seizure of islands (especially the Marianas) suitable for long-range bombers.[11]

In sum, Britain, Japan, and the United States all had some strategic rationale for creating amphibious forces, or at least studying amphibious operations in the conceptual sense, but only the United States had a distinct requirement to create advanced naval bases and, if necessary, to seize such bases from a determined defender. All three nations, therefore, required amphibious capability, but the United States alone identified the need to conduct an *opposed landing*. Of course, a force that could capture a defended base could also land at an unopposed site because it had already mastered the intricacies of ship-to-shore movement under fire. The different paths of the three great naval powers in developing amphibious forces, however, rested not just in the geostrategic context in which they functioned but reflected as well a complex interaction of organizational politics, funding, operational doctrine, and technological challenges.

The Organization of Amphibious Forces

BRITAIN

Given its history of joint expeditions, naval campaigns, imperial defense, and disengagement from continental alliances, the British had every reason to maintain some sort of amphibious assault force. Although they recognized the likely wartime demand for such a capability, they did not have such a force. The shock and frustration of failure in the Dardanelles campaign provided some justification for the neglect, but this military trauma was not the only, nor the most significant, reason for the neglect of amphibious warfare. Britain's interwar military and naval staff committee system, along with its staff college adjuncts, did a good deal of thinking about joint operations. But the British failed to create a modest amphibious capability in the sense that the Royal Navy and army did not conduct exercises with amphibious shipping and landing forces on a regular basis, even to test some of the theories produced by the planning staffs. Instead the British military establishment made it difficult for operational development of any sort to receive a fair hearing, and the parliamentary form of government itself simply reinforced innate military conservatism. Budgetary restraints provided a convenient excuse for maintaining the status quo, but the British armed forces might have done more within the prison of fiscal restraint, as the American experience suggests. The Royal Navy and army simply assumed that amphibious operations presented too many problems for them to reach a solution in peacetime.[12]

British interest in amphibious warfare after World War I began with the general consideration of interservice cooperation—complicated by the birth of the independent Royal Air Force and its capture of naval aviation—and the specific issue of naval base defense. In 1920 the Board of Admiralty, the navy's highest policy-making committee, ordered the naval staff to prepare a mobile base plan, with special consideration of air operations, because fixed bases might not be adequate in another war. This Admiralty interest reflected similar work by a special all-services syndicate of 169 officers, drawn from the three service staff colleges and convened for a week at the army's staff college in October 1919. The syndicate examined the defense of Singapore and Hong Kong and reported to their respective ministries that the existing guidance of joint operations, the *Manual of Combined Naval and Military Operations* (1913), needed thorough review and revision in light of the recent world war. The Admiralty, working through the Royal Naval Staff College, convinced the army and the Royal Air Force to participate in a joint committee that would review the staff colleges' recommended changes and publish a revised manual on joint operations. The Inter-Department Committee on Combined Operations met in June 1920 and ended its deliberations with a staunch defense of the status quo, service independence, and rejection of the importance of preparing for opposed landings.

Dissatisfied with the first committee effort, the Admiralty called for a second review in 1921, which produced a new publication, the *Manual of Combined Naval Military and Air Force Operations-Provisional*, 1922. The new manual reflected sound study at the Royal Naval Staff College about the inherent problems of amphibious operations but did not demonstrate any broad institutional commitment, only guidance for the staff colleges in planning their biannual academic exercises on joint operations. A 1925 revision of the *Manual* incorporated the "lessons learned" from these problems but had no force outside the school system.

The operating forces contributed little to the limited operational debate in the staff colleges, and the two actual field exercises in the 1920s demonstrated little more than the initiative of local commanders. In 1924, Rear Admiral Sir Herbert Richmond, naval commander in East Indian waters, staged a landing in the Bay of Bengal with a scenario drawn from the plans to defend Singapore. The landing force, a reinforced battalion, operated under the supervision of the students of the staff college at Quetta, and Richmond, a champion of amphibious operations, took this experience back to England when he became director of the senior officers' course at the Royal Naval Staff College. A 1928 landing in the Moray Firth, Scotland, originated with a Scots battalion commander who wanted to give his troops some novel training, in this case a trip to sea on three destroyers, an unopposed landing in ships' boats, and a dramatic assault on the Black Watch's own barracks at Fort George. The next exercise did not occur until 1934, but it benefitted from the interest of Admiral Lord Cork and Orrery, commander in chief, Home Fleet, who used forty-one of his own ships to land an army brigade group at the mouth of the Humber River against an opposing army brigade. This landing, however, provided little useful training, since its principal purpose was to serve as a communications test.[13]

Amphibious warfare development in Britain lacked any organizational commitment and institutional memory, both of which the Royal Marines might have provided. The Royal Marines, however, did not seize the limited opportunity to become the champions of amphibious operations. From a wartime high of 40,000 officers and other ranks, the peacetime Royal Marines shrank to 9,000 men. The predictable turmoil of force contraction grew more complicated after a 1922 decision to merge the Royal Marine Artillery and Royal Marine Light Infantry, a distinction that dated from 1855. This decision meant a merger of messes (always traumatic in British forces) and a reorganization of the marine barracks and training establishment.

Moreover, in 1923 the Admiralty called for a thorough study of the wartime functions of the Royal Marines and its training for wartime missions. An Admiralty committee, chaired by Admiral Sir Charles Madden and including ranking marine officers, studied the issue for six months and issued its report in August 1924. The Madden Committee reasserted the need for marines aboard warships for landing party and gun crew service—traditional duties—but urged that marines serving ashore (in "divisions" identified with each marine barracks) train to seize and defend temporary bases and to conduct amphibious raids upon enemy naval bases and coastal positions. To provide such a peacetime "striking force," the Royal Marines should be increased to approximately 16,000 officers and men. Without such an increase in strength, only about one-third of the marines would be available for amphibious warfare training.[14]

The Madden Committee report brought no substantial change to the Royal

BRITISH LANDING PARTY AT KUM KALEH AT THE DARDANELLES, TURKEY, 1915

Marines despite the fact that the Admiralty and senior marine officers agreed with its conclusions. The Admiralty decided that it could not ask for the money to increase the Royal Marines, except to provide detachments for warships. It faced no resistance from senior officers of the Royal Marines, who accepted their lot, happy or not. Overwhelmed with the requirement to provide elite detachments for ships and ad hoc battalions for colonial service, the leaders of the Royal Marines discouraged dissent among their own officers, some of whom urged a real commitment to the amphibious assault mission despite the limitations of manpower and funding. Moreover, so the admirals and marine generals argued, the Royal Marines had another mission: the preparation of their component of the Mobile Naval Base Defense Organization. This organization, conceived in 1920 but undeveloped until the 1930s, offered the Royal Marines a more attractive wartime function than the amphibious assault and built upon skills already present in the by then defunct

Royal Marine Artillery because the operational core of the Royal Marines Group, Mobile Naval Base Defense Organization, was an antiaircraft artillery brigade and a coast defense artillery brigade. A single Royal Marines infantry battalion was the only mobile defense component of the group in a total force of over 7,000 officers and men. To the degree that the Royal Marines interested themselves in amphibious operations into the late 1930s, they organized annual exercises to test landing craft, vehicles, and equipment needed to disembark a heavy artillery force over an

The Japanese used fourteen ships to do a job that British planners believed would require forty vessels.

undefended beach or at a friendly port. Although some of these experiments had applicability to offensive amphibious operations, their scope and scale provided little useful experience for an amphibious assault.[15]

In the meantime, the Admiralty remained restive about its lack of knowledge and experience in amphibious operations, a disquiet fed by reports from the staff college exercises and reports on parallel developments in the United States and Japan. The Japanese, for example, demonstrated at Shanghai in 1932 that they needed far fewer transports and landing craft than the British had estimated were necessary to put ashore a combined arms force of 12,000. (The Japanese used fourteen ships to do a job that British planners believed would require forty vessels.) In 1936 the Royal Naval Staff College, again preparing revisions for the *Manual of Combined Operations*, provided a persuasive report, organized by Captain Bertram Watson, RN, that allowed the Admiralty to persuade the War Office and Air Ministry to take action on studying amphibious operations.

The Watson memorandum could not have come at a more propitious time, for the Deputy Chiefs of Staff Committee working on the *Manual* had concluded that joint operations needed a permanent institutional advocate in the armed forces. In 1938, as recommended by the Watson memorandum and after much bureaucratic memo-passing, the Chiefs of Staff established the Inter-Service Training and Development Sub-Committee, which it made directly responsible to the

Deputy Chiefs of Staff, and established the Inter-Service Training and Development Centre at Eastney Barracks, Portsmouth, a Royal Marines base. Captain L. E. H. Maund, RN, a Richmond protégé, with an interservice staff of three (army, RAF, Royal Marines) and some clerks, opened the center in July 1938 with £30,000 to conduct research and development on amphibious operations. The Sino-Japanese war was already almost a year old, and the Munich crisis came only two months later. In the meantime, the Deputy Chiefs of Staff issued the 1938 edition of the *Manual of Combined Operations*, which still provided more wisdom on the problems of joint operations than decisions on their solution.[16]

In the short year before the outbreak of war, the Inter-Service Training and Development Centre struggled to coordinate the theoretical studies of amphibious operations (still the province of the staff colleges and the Admiralty staff) and the development of prototype amphibious ships and landing craft. It also studied the relationship of airborne operations to landings and the special engineering requirements of defending and overcoming shore defenses. Perhaps its most important contribution was its educational effort within the highest circles of British defense planning, the Chiefs of Staff, and the Committee of Imperial Defense. Maund, who had seen Japanese amphibious operations in Shanghai in 1937, proved to be a forceful and persuasive advocate of amphibious development. He won an important convert, Major General Hastings Ismay, secretary to both the Deputy Chiefs of Staff and Committee of Imperial Defense, who reflected after one report that "as our strong suit was command of the sea, we should be making poor use of our hand if we relegated combined operations to the background of our war plans. Indeed it seemed that we should be laying ourselves open to justifiable criticism if we made no effort in time of peace to train on modern lines for a possible combined operation in time of war."[17]

As Ismay observed, the British armed forces had allowed themselves to fall behind Japan and the United States in amphibious development, and the situation could not be reversed overnight. By the time war came, Brigadier Bernard Law Montgomery was the only senior army officer who had commanded an amphibious landing force, his own 9th Infantry Brigade, which had made an opera-bouffe assault in July 1938, at Slapton Sands, Devonshire. The weight of British military opinion bore down the arguments of amphibious warfare enthusiasts. Only the navy had senior spokesmen in favor of the concept such as Admirals Sir Roger Keyes, Richmond, and Lord Cork and Orrery.

More importantly, the British political elite provided no external connection within Parliament for amphibious warfare, and the discipline of the party system and civilian control made it difficult for military missionaries to sidestep the bureaucracy. Winston Churchill remained in isolation, and no one took his place. In the world of military journalism and punditry, Captain Basil Liddell Hart flirted with the concept of the amphibious assault as part of his advocacy of the "indirect approach," but he failed to embrace the cause, probably because of his inherently inconsistent advocacy of British isolationism and mechanized warfare. With the exception of a few isolated and specialized shipbuilders, no important industrial interests embraced amphibious warfare; the Board of Trade, which exercised governmental power over the merchant marine, endorsed military views on transport requirements without much interest. The British armed forces could have emphasized amphibious operations but chose to direct little attention toward this type of complex military capability until 1940.[18]

JAPAN

Although the Japanese military establishment drew inspiration during the Meiji Restoration from the German army and the Royal Navy, it surpassed its European models in attention to amphibious warfare after World War I. In terms of usable military capability Japan entered World War II as well prepared as the United States, both in terms of operational forces and published doctrine. Of course, the Japanese used their interventions in China as a testing ground for amphibious warfare, and they made major adjustments to their doctrine and forces as a result of those experiences, in particular after the Shanghai Incident of 1932 and the initial landing operations of 1937 in support of their invasion of northern China.

As the threat of war with Russia and the press of operations in China consumed more of the attention of the Japanese army, the leadership in refining amphibious warfare doctrine and planning shifted to the navy, which still focused upon American possessions in the Pacific. The navy's development of amphibious landing forces, therefore, concentrated on small, highly mobile, and lightly armed naval infantry regiments adequate for the capture of the weakly defended islands of Guam and Wake. Any major operations, i.e., those involving a division, would require the participation of the Imperial Japanese Army, which controlled not only the landing forces, but most of the landing craft and transports of the amphibious force itself. Japan, therefore, had two amphibious forces—one each for the army and the navy—capable in many ways of operating together, but also independent of one another. As these forces proved in 1941 and 1942, Japan could conduct amphibious landings with dazzling success, but these forces also had inherent weaknesses eventually exploited by the Allies after 1942.[19]

Like their Western counterparts, the general staffs of the Imperial Japanese Army and Imperial Japanese Navy—both of which operated independently from the Ministries of War and Navy—conducted a major reassessment of their contingency plans, embodied in the *Yohei koryo* ("Outline of the Employment of Forces") and its companion *Teikoku Kokubo Hoshin* ("Imperial Defense Policy"). The 1918 version of *Yohei koryo* stressed preparations for an opposed landing on Luzon as part of a general war plan for a conflict with the United States. Another review completed in 1923 retained the Luzon operation and added another landing to capture the American base at Guam. Having also studied the Gallipoli operation, general staff planners of the Japanese army concluded that the army could no longer rely principally on the navy to organize amphibious operations. Even before they worried about the menace of airplanes and submarines to amphibious forces, army officers believed they would require more rapid landings than in the past and that they should have to come ashore prepared to fight. The general scenario of past operations would not meet the test of modern battle: light naval infantry securing an undefended landing site, soldiers towed to shore in unpowered barges by navy boats, the slow disembarkation of artillery and horse-drawn transport, the measured organization of army field forces ashore before an overland campaign against the enemy. Army planners reached the peak of their concern in 1926 when they estimated that the conquest of the Philippines would require three divisions.[20]

The army, the dominant military service in Japan, already had an institutional base upon which to build an amphibious warfare force. Powerless to halt the army's amphibious warfare crusade, the navy provided substantial support when the army began to develop its force in the 1920s. The Army Transportation Department, located at Ujina in Hiroshima prefecture, provided all land and sea transportation services much as the Quartermaster Department did for the U.S. Army during the same period. This department (*Rikugun un'yubu*) contained a special bureau for sea transportation with its own fleet of transports and had well-developed ties to the Japanese shipping industry. Policy guidance and supervision for the Transportation Department came from the Eighth Section (Land and Sea Transport) of the Third Bureau (Communications and Transport) of the army

general staff, and budgetary and logistical support came from the Economic Mobilization Bureau of the Ministry of War. The department received, in addition,
the helpful attention of the army's liaison office with the navy general staff and
the dominant First Bureau (Operations) of the army general staff. At the operational level the Army Transportation Department developed an elaborate system
to provide waterborne transport services. Shipping engineer regiments were the
most useful, with 1,200 officers and men in a headquarters and three companies
(*sempaku kohei rentai*) to operate landing craft and barges. The army also maintained debarkation units, trained and equipped to load or unload transports in
port or over a beach, while shipping transport commands managed port facilities
used by army vessels. The Army Transportation Department had both the money
and supervisory staff to conduct research and development for amphibious operations without navy interference, especially in the design of landing craft, although in practice the two services worked closely together.

Although the army did not begin formal training and instruction in amphibious operations until 1921, one division (the 5th, also stationed at Hiroshima)
conducted a minor landing exercise in 1918. Under general staff supervision, the
army conducted a major exercise in 1920, observed by the chief of the general
staff, General Uehara Masasaku, who then attached a high priority to development of a steel, armored, and self-propelled landing craft. Although it did not put
the new landing craft into operation until 1925, the army conducted landing exercises with one division in 1921 and in 1922 with three divisions, out of a total
force of seventeen divisions. In addition, it undertook command post exercises
with amphibious operations scenarios, not only in its staff college but also in tactical armies and divisions. In 1926 the army general staff identified the 5th, 11th,
and 12th Divisions as "especially designated" divisions for the Luzon landing and
assigned them to the Army Transportation Department to develop ready forces
and appropriate doctrine for amphibious assaults. The home stations of all three
divisions placed them close to army waterborne transportation units and shipping engineer regiments.[21]

The army also soon supplanted the navy as the authoritative source of amphibious warfare doctrine, although navy participation in writing and testing doctrine, especially for naval gunfire support, did not end. In 1924 the army published
its authoritative "Summary of Amphibious Operations and Operations Defending
Against Amphibious Attacks" (*Joriku oyobi joriku bogyo sakusen koyo*), which

represented a major improvement of the navy's *Kaisen yomurei*, the equivalent of the *U.S. Landing Party Manual* or guide for naval infantry units drawn from the fleet. Additional doctrinal pronouncements flowed from the army and navy and culminated in publication of "Outline of Amphibious Operations" (*Joriku sakusen koyo*) in 1932 after five years of collective drafting by both service staffs. This doctrinal manual carried the force of general staff endorsement and informal imperial sanction. The 1932 publication reflected major advances in ship-to-shore movement, as tested in major joint landing exercises in 1929 and 1931.[22]

Developed in the hard school of the real ocean, which extracted payment in the form of wrecked landing craft and dead soldiers, Japanese amphibious warfare doctrine by 1932 showed a sound appreciation in both army and navy of the fundamental requirements for a successful opposed landing. First, the navy's covering forces would have to ensure that an enemy air attack, submarine attack, or warship raid did not disrupt the amphibious force convoy or ship-to-shore movement. Moreover, the navy would provide sufficient naval gunfire to destroy beach defenses and suppress enemy mobile defense forces. Toward that end the navy conducted naval gunfire experiments with army participation beginning in 1926. The results were not heartening but suggested that cruisers and destroyers could—with better range-finding and communications equipment—fire on targets identified by ground forces. The nature of naval ordnance was another limitation; shells designed to penetrate steel ships did not work well against field fortifications unless the structure was made of concrete and offered a vertical face toward the firing ship. In mounting an efficient ship-to-shore movement, the army, equipped with progressively larger numbers of motorized landing craft, showed increased ability to move troops and equipment rapidly from transports to landing craft—even at night—and then land them at the intended objective. After the Shanghai punitive landing (1932), when the navy landed soldiers without adequate ammunition and heavy weapons, the army planned to use only its own transports and landing craft for future operations.[23]

The army's domination of amphibious operations development, however, had two effects that reduced Japanese primacy in this operational specialty. The war in China after 1937 forced the army to release its three "amphibious divisions" for service on mainland Asia, and one of them, the 11th Division, suffered grievous casualties near Shanghai, which suggested that "especially designated" divisions had spent too much time in learning to land and too little in learning to fight. Therefore, the army made no special effort to make other divisions ready for amphibious landings on a routine basis. The Chinese campaign also demonstrated to the army that the old method of landing at undefended sites still worked; with no significant Chinese naval or air forces to disrupt operations, the Japanese could revert to landing with all deliberate speed and without any essential dependence on naval gunfire. No close air support was required. Although the army and navy could and did disagree about optimal landing sites, the Chinese campaign buttressed the traditional view in both services of amphibious operations, i.e., that landings at night by battalion-size units, widely dispersed yet concentrated at the point of attack (Japanese landing craft tended to deploy in columns, not waves) brought startling success in terms of enemy confusion and demoralization. The emphasis upon preparing for an Asian Gallipoli disappeared in operations at Shanghai, Tsingtao, Canton, and Hainan Island. This form of "victory disease" was then reinforced by the army's preoccupation with a mechanized, artillery-dominated war on its northern border with a revived Red Army.[24]

Although the use of naval landing parties to secure initial landing sites waned in theory and practice in the 1920s, the Imperial Japanese Navy maintained its own landing forces and its own programs, funded from its own budget. This naval program met no serious resistance from an army general staff absorbed with operations in China and training for a war with the Soviet Union. The navy's senior officers wanted to reduce "military" training for sailors (which closely followed rigorous army standards of physical fitness, weapons training, drill, and discipline) and to stress technical training, a fact that spurred development of specialized amphibious forces. More importantly, the use of sailors for landing parties reduced the readiness of a ship for fleet action to dangerous levels. However, if the navy still aimed to undertake initial landings, which it did, then it wanted better tactical performance than naval infantry showed in the Shanghai operation of 1932. Japanese naval infantry looked very smart in Shanghai but fought ineptly, and their performance accelerated the movement to establish permanent Special Naval Landing Forces (*rikusentai*) of reinforced battalion strength (1,069 officers and men) that included two infantry companies of six platoons and a heavy-weapons unit armed with light artillery. The *rikusentai* formed at the four major naval bases at Sasebo, Kure, Yokosuka, and Maizuru and bore numbers related to their activation at each base, e.g., Sasebo 2nd Special Naval Landing Force. By 1941 the Special Naval Landing Force battalions looked and fought like the army battalions, distinguishable mainly by their use of navy rank and the anchor insignia rather than the army's star.[25]

The navy, however, saw no reason to create a second amphibious army. It did not buy or develop specialized transports, and its only specialized landing craft (as compared with the lifeboats and cutters found on any warship) was essentially the army's Type A landing craft. Instead, it planned to embark the Special Naval Landing Force battalions on warships, most often older destroyers, which carried naval guns and rapid-fire cannon that could support the landing force. This force, however, had a limited and highly specialized mission appropriate only to the central Pacific. Emphasizing surprise with night landings or the use of smokescreens, it could enter the narrow passages in coral barrier reefs, navigate in shallow lagoons and put forces ashore on small atolls like Wake or poorly defended larger islands like Guam. It was not a substitute for the army divisions that would have to seize the Philippines, Hong Kong, and Malaya. Realizing the degree of risk they were accepting by assuming a rapid night landing, the Special Naval Landing Forces attempted to complicate the defender's problem by creating parachute-trained battalions, which land-based naval aviation might drop. In a sense the new Special Naval Landing Forces simply provided the navy with a capability to perform traditional limited missions with greater efficiency.

By December 1941, the Japanese army and navy had acquired a substantial body of theoretical expertise and operational wisdom about the conduct of amphibious operations. Although interservice rivalry in strategic planning complicated collaboration of the army and navy, it did not prevent the Japanese armed forces from conducting successful landings when and where they wanted after December 7, 1941. Their limitations came only after the American armed forces altered the conditions under which the Japanese had to operate, and that change did not occur until the counteroffensives on New Guinea and the southern Solomons in late 1942.

THE UNITED STATES

The American forces that eventually defeated the Japanese were already well developed by the time the Japanese army began its China campaign in 1937, and in the four remaining years of peace the navy and the marine corps brought their understanding (if not their capability) of amphibious operations to a level unequaled in Britain or Japan. By 1940 the U.S. Army had joined the annual fleet

At New Georgia, U.S. troops come ashore from LCI, July 1943.

landing exercises and accepted the naval services' published doctrinal guidance. The relative success of the American armed forces in creating both doctrine and forces for amphibious operations stemmed from a complex interaction of strategic guidance, service roles and missions, interservice and civil-military politics, and military-industrial collaboration.[26]

The challenge of solving the problems of an opposed amphibious landing fell to the United States Marine Corps. The marine corps was an organization so small and obscure before World War I that most political leaders and the public did not know that it was an independent service administered by the Navy Department, but not part of the navy itself. Association with the navy, however, had been close since the corps's founding in 1798, for marines served principally as ships' guards and naval base security forces throughout the nineteenth century and then became the cutting edge of the naval landing parties that went ashore with increasing frequency and growing size throughout the Western Hemisphere and in Pacific and Asian waters.

Marines also participated in land campaigns. During World War I, a single marine brigade serving with the American Expeditionary Forces in France made the corps the darling of the media and public and cemented its already close ties with Congress. With the review of War Plan ORANGE in 1919–1920, the marine corps recognized that it could fill an important void in its organizational existence. An amphibious assault specialty represented a focused wartime mission that would once and for all make it distinct from the army. The army, however, doubted the need for an independent marine corps, especially a large one with an important wartime function, but Congress routinely rebuffed army initiatives to absorb the corps. In truth, the interwar army had plenty of problems with modernization and aviation development, and its interest in amphibious operations of any kind remained limited, beyond negotiations over the language of Joint Action of the Army and Navy.

Four centers of study for amphibious operations existed in the Navy Department—the Office of the Chief of Naval Operations, the General Board of the Navy, the Naval War College, and the Commandant of the Marine Corps—but only the marine corps viewed the mission as critical. In 1920 the commandant, Major General John A. Lejeune, grasped the importance of the amphibious assault mission and its contribution to War Plan ORANGE planning. He ordered a trusted staff officer, Major Earl H. Ellis, who had studied the defense and capture of advanced naval bases before World War I, to analyze the requirements of amphibious operations across the central Pacific. In seven months of manic work, Ellis wrote Operations Plan 712, "Advanced Base Force Operations in Micronesia," a detailed and prescient study that Lejeune endorsed on July 23, 1921, as the basis for future training and wartime mobilization planning in the marine corps.

Lejeune's commitment did not come from cosmic inspiration, but from a further refinement of his own and his predecessor's views that the marine corps ought to be the organization to take and hold advanced naval bases in any future naval campaign. Ellis's work, however, gave the commitment a valid geographic theater and a specific enemy as well as some refined, if speculative, thoughts about the exact equipment and training required for an opposed assault. Moreover, it represented a natural extension of the marines' "advanced base force" mission (established in 1900) and the existence of forces to fulfill that role, a mix of mobile and fixed defense battalions created before World War I and reconstituted after the war with a paper strength of sixty-five officers and 1,502 men. (Its actual numbers were less than fifty in 1920.)[27]

In the extended and continuing debates over service functions in the interwar period, army and navy planners wrestled with the general problems of mutual cooperation in coast defense and overseas expeditions. The development of aviation in both services complicated the exchanges. Influenced by War Plan ORANGE, the members of the Joint Army-Navy Board did not exclude the army from amphibious operations. Historical precedents and good sense argued that the army might have to make an amphibious assault to retake Luzon, and it certainly needed a plan to embark overseas expeditions to the Philippines, Alaska, Hawaii, Puerto Rico, and the Panama Canal Zone, all of which it defended.

Only Guam and Samoa were sole navy responsibilities. In all the textual interpretation that accompanied revising Joint Action (1927), the marine corps still received the mission "to provide and maintain forces in support of the fleet for the initial seizure and defense of advanced bases and such limited auxiliary land operations as are essential to the prosecution of the naval campaign."[28] Nevertheless, the army and navy continued to discuss mutual cooperation in amphibious operations and set forth their general operational approach in Joint Overseas Expeditions (1933), which included concepts for making an opposed beach assault. The document, however, left open the question of which service's senior officer would command a joint expeditionary force, a nonexistent problem for the navy and marine corps because the senior navy officer afloat always held overall command. In any event, all the services recognized the requirement to test amphibious warfare theory in their classrooms and in field exercises.[29]

Resembling their counterparts in Britain and Japan, the staff and war colleges of the American armed forces turned greater attention to instruction on amphibious operations in the interwar period. By 1930 the navy and army war colleges conducted annual exercises, largely drawn from Plan ORANGE. The Naval War College made its exercise a grand production that included navy and marine faculty and student officers from Newport and Quantico, site of the Marine Corps Schools. Although army instruction on joint operations came largely through lectures by naval officers, the Command and General Staff College at Ft. Leavenworth devoted approximately one week's work to an amphibious exercise in the mid-1930s. Nevertheless, the Marine Corps Schools (principally the senior course for majors and captains) made the greatest institutional commitment. From the mid-1920s through the mid-1930s the curriculum on amphibious operations increased from about 25 percent to 60 percent of the total hours of instruction. The professional interaction on amphibious operations drew additional strength from the fact that marine officers attended the Naval War College, the Army War College, and the Army Command and General Staff College, and these officers carried on missionary work for the specialization.[30]

Beyond school problems, officer instruction profited from the conduct of real amphibious exercises, however limited in scope. The navy and marines attempted their first offensive landing problem in 1924. Called the "Expeditionary Force" by Commandant Lejeune, 3,300 officers and men split into two reinforced regiments (one on the defense, the other in a beach assault) that battled each other at Culebra, Puerto Rico, and the Canal Zone. With the army taking the defensive role the following year, 2,500 marines assaulted Oahu, but the deployment of the two available marine brigades to Nicaragua and China in 1927 brought large-scale interservice exercises to a halt. In 1932, however, the navy, army, and marine corps resumed amphibious exercises in Hawaii. All of these operations revealed a basic problem: moving combat troops to landing boats from transports was too slow and disorganized. The limitations of landing craft also meant that supporting artillery and light tanks could not get ashore to help breach beach defenses, and assault engineering, close air support, and naval gunfire were clearly inadequate. Compared to Japanese fleet exercises at the same time, American efforts did not represent a meaningful current capability, but grist for doctrinal assessment.[31]

The void in amphibious exercises after 1925 helped clarify the general interest in amphibious operations within the three services. Interest in the navy was directly proportional to the availability of a landing force; unless scheduled as part of a general set of amphibious exercises, the fleet showed little concern for developing naval gunfire techniques against land targets. With transports of its own, operated by the Quartermaster Department, the army might have volunteered to fill the gap in landing forces between 1925 and 1932, but, despite the navy's desire for greater participation, the War Department General Staff showed more interest in attacking the marine corps than in attacking defended beaches.

Alarmed by rumors of its own demise—a realistic fear, given the budget crisis of the Great Depression—the marine corps turned to writing doctrine. What started as an effort to produce a textbook on amphibious operations at the Marine Corps Schools in 1931 turned into an organizational "barn-raising" of doctrine-writing and resulted in the *Tentative Manual for Landing Operations* in 1934, the most comprehensive and detailed effort to think through the numerous problems involved in amphibious operation. Spurred by navy and marine corps interest in the *Tentative Manual*, the Marine Corps Schools continued the review and revision process into 1935, when the *Manual* became the authoritative guide within the Navy Department for future amphibious exercises and for research and development. The writing of the *Tentative Manual* coincided with the withdrawal of the marine brigades from overseas, the redesignation of the "Expeditionary Force" as the "Fleet Marine Force" in 1933, and the preparation for the resumption of fleet landing exercises.[32]

With a landing force in hand—the 1st Marine Brigade on the East Coast and 2nd Marine Brigade on the West Coast—the navy renewed annual Fleet Landing Exercises (FLEXs) in 1935. The first two FLEXs in 1935 and 1936 proved to be limited undertakings with most of the landing force (two infantry battalions and a mixed artillery battalion) embarked on two old battleships or on the light cruisers and gunboats of the Special Service Squadron, the navy's nautical constabulary for Latin America. At the navy's initiation, FLEX-3, held off the Southern California coast in 1937, embraced all three services. The army provided an expeditionary brigade (built around one understrength infantry regiment and its own transports), and the navy embarked 2,500 marines, the full strength of the Fleet Marine Force. FLEX-4 ran from January to March 1938 in Puerto Rico and involved a similar number of marines, with an army expeditionary brigade participating again at the navy's request. This time the army's contribution climbed to three infantry regiments with supporting arms, and the brigade alternated in the part of landing force and shoreline defender. FLEX-5 in 1939 and FLEX-6 in 1940 reflected both increasing tension in world affairs and the first stirrings of rearmament and expansion in the armed forces. Not until February 1940, however, did the Fleet Marine Force's 1st Brigade approach anything close to a wartime expeditionary force in numbers and armament, and the navy had yet to provide a transport squadron for the marine brigade. Instead it used a group of old battleships and converted destroyers (APDs or fast attack transports). Neither FLEX-5 nor FLEX-6 involved army landing units.[33]

The fleet exercises provided additional experience to refine the concepts in the *Tentative Manual*, now raised to the august status of Fleet Training Publication 167 (Landing Operations Doctrine, USN, 1938) and issued under the imprimatur of the chief of naval operations. The exercises also refined the mobilization concepts and training embodied in Marine Corps Contributory Plan, C-2, ORANGE, reviewed and revised each year in the 1930s. The marine corps in 1932 anticipated a minimal wartime requirement of one division and one brigade for amphibious assaults and one base defense force, in all 20,400 officers and men drawn from a wartime corps of 119,000. Clearly traditional naval duties still influenced planning. The total strength of the marine corps was only 17,000 in 1936 and 28,000 in 1940.

In the course of the FLEXs the navy and marine corps experimented with about every imaginable amphibious technique and tactical approach allowed by their equipment. They tried day and night landings, smokescreens, varieties of air and naval gunfire support, concentrated assaults and dispersed infiltrations, the firing of all sorts of weapons from landing craft, and an array of demonstrations, feints, subsidiary landings, and broad-front attacks. Although rich debate went on within the navy and marine corps over the "lessons learned" from the FLEXs, FTP-167 reflected the growing consensus that amphibious assaults might be possible, but not easy. Army opinion remained even more skeptical.

The amphibious force would have to isolate the objective area, then pound the defenders into a stupor with naval gunfire and close air support. The landing itself would require a violent assault by a combined arms team, probably over a broad front, perhaps a beach of a thousand yards' width or more. To secure the beachhead, the landing force would need rapid reinforcement, complete with artillery and tanks. The greatest threat to a landing was a disruptive air and naval attack, which might pull critical fleet units from the objective area, but a combined air and ground counterattack was the most immediate concern. A counterlanding might give the enemy a striking advantage because it would be difficult for a landing force to protect its supply line and logistics support areas as well as defend the perimeter of its enclave. An amphibious expeditionary force could not rely on guile for success, but would require local superiority in every element of air, naval, and ground combat power.[34]

The debate over amphibious warfare in the United States did not have the same closed character that it had in Britain and Japan. Articles on the subject appeared with regularity in service journals and even occasionally in civilian magazines. More importantly, Congress followed the discussions in the annual reports of the service secretaries and in Congressional hearings. Whenever someone in or out of office, military or civilian, criticized the marine corps, it opened the issue of readiness for amphibious warfare. It required no championing from the chief of naval operations or secretary of the navy for the marine corps to keep alive its operational specialty, even if underfunded. Amphibious operations and the marine corps received a special boost when Franklin D. Roosevelt became president because he had spent seven years as assistant secretary of the navy in the Wilson administration and viewed himself as an honorary marine. His personal friendship with Commandants John H. Russell and Thomas Holcomb could not have been more congenial. Yet Navy Department funding decisions, approved by Congress, still presented problems of materiel procurement that limited amphibious development.[35]

The Procurement Issue: Amphibious Shipping and Landing Craft

The materiel requirements of the amphibious assault—or less deadly types of landings—did not demand exciting or high-risk investments in new military technology but, rather, special adaptations of shipping, aircraft, vehicles, and weapons to maritime service and the conditions of amphibious combat. Such adaptations often involved tradeoffs in capability. In a landing craft, for example, the designers needed to consider size, speed, degree of armored protection and armament (all weight builders), maneuverability, the ease of disembarking troops and equipment, engine noise, buoyancy and freeboard (a high profile invited accurate shore battery fire, but a low profile swamped in heavy seas), ease of retraction from the beach, transportability by carrier ships, and seaworthiness under a variety of conditions. As practical experience mounted in the 1920s, the materiel requirements of amphibious operations became increasingly clear to British, Japanese, and American planners, who displayed some degree of similarity in their responses. All three groups, for example, saw the need to develop a special landing craft suitable for disembarking infantry, light artillery, vehicles, and even light tanks over a sand beach.

In some cases, the principal challenge was to identify army weapons, equipment, and vehicles most suitable for amphibious operations and then make modifications to protect the equipment from saltwater corrosion, flooding, foreign matter (mostly sand), and the amphibious troops themselves. For example, the Japanese Special Naval Landing Force assault battalions and U.S. Marine Corps's artillery battalions armed themselves with light pack howitzers designed for mountain warfare. In terms of weapons, the small specialized landing forces of Japan and the United States could hardly afford armaments not already designed and manufactured for army use. Yet in two areas—specialized ship and craft development and aviation development—the amphibious forces of the interwar period attacked their common material problems with results that had more to do with military politics and the procurement culture of each nation than with technology.

In Britain, the requirement for a motorized landing craft emerged out of the operational analysis and limited testing of the 1920s. The Royal Marines Mobile Naval Base Defense Organization project gave the development of such craft a modest sponsor, but landing craft development remained essentially an Admiralty responsibility and not one with a high priority. In theory, a joint Landing Craft Committee, composed of military and industrial representatives and eventually answerable to the Chiefs of Staff, should have developed landing craft programs, but landing craft development received little attention until the formation of the Inter-Service Training and Development Centre in 1938 under the patronage of the Deputy Chiefs of Staff. The Landing Craft Committee, however, let contracts that produced two prototype Motor Landing Craft (MLC) in 1926 and 1929. These MLCs were principally for transporting tanks and cargo as well as troops, carried armor and weapons, moved slowly despite jet-assisted engines, and weighed 36,000 and 45,200 pounds. Civilian passenger liners, which doubled as transports in wartime, could carry neither craft because their davits and cranes could not handle boats over 20,000 pounds. By 1939 eight MLCs had been built; the estimates at the time suggested that several hundred would be necessary to land just one division.[36]

Under Captain Maund's direction, the Inter-Service Training and Development Centre (ISTDC) gave landing craft development its highest priority. Working with shipbuilders White of Cowes and Thornycroft, ISTDC sponsored the construction of a prototype Landing Craft Assault (LCA), which went through several modifications but remained under forty feet in length and ten tons in preloaded weight. The ISTDC also had Thornycroft produce a twenty-ton Landing Craft Mechanized that could be carried by the sturdier cranes and davits of converted cargo ships. After World War II began, the British urgently accelerated work on LCA and LCM variants, but also began increasingly to concentrate on beaching ships and craft that did not require carrier ships and could sail from a friendly harbor to an unfriendly beach with embarked vehicles and supplies.

Spurred by the disaster of 1940 and a new prime minister, Winston Churchill, the Admiralty pushed a crash program to develop a fleet of beaching ships and craft that could carry heavy tanks, artillery, and preloaded supplies. From this program came the Landing Ship Tank (LST), a World War II workhorse in every maritime theater, and an associated family of smaller amphibious assault ships. These beaching ships featured bow ramps and a shallow draft forward, had at least one deck for loading (LSTs had two), and carried the weight of their engineering plants, living areas, and work spaces aft, allowing them to beach and retract under their own power.

British ship designers found a working prototype in commercial vessels designed to serve Latin American oil fields, but the "Maracaibo oilers" needed a great deal of reworking to strengthen and balance them for military loads, and their bow doors and ramps required fundamental reengineering. Nevertheless, the British development of beaching ships and craft after 1940 produced a variety of vessels noteworthy for their durability and ease of construction. The British problem was timing, not ingenuity. Not until mid-1942 had their amphibious ship and craft inventory reached an operational level to support major raids, let alone a full amphibious assault. The British amphibious fleet also had short legs: It could not stay at sea for long periods and long distances, and it was vulnerable to air and naval attack. In operational terms it was a fleet designed for one purpose—an Allied amphibious return to northern Europe under the cover of land-based air power.

The Japanese attacked the problem of specialized shipping principally

FULLY LOADED LST STEAMS TOWARD NEW GUINEA INVASION, 1944.

through the shipping bureau of the Army Transportation Department, and their successes and failures stemmed directly from the army's definition of amphibious requirements. Neither the Japanese shipbuilding industry nor the navy had much impact until after Pearl Harbor. Less than a year after their devastating amphibious successes in the Philippines and Malaya, the Japanese armed forces had surrendered the strategic initiative and switched their interest from amphibious assaults to general logistical problems in the Pacific war in determining their shipping requirements.

The army's initial interest—as in Britain and the United States—centered on the development of self-powered landing craft that army transports or converted merchantmen could carry, which meant ships in the 10,000-ton range with limited convertability to accept davits and cranes strong enough to handle landing craft. By 1930 the Japanese Army had developed and adopted two basic landing craft, the Type A (*daihatsu*) craft, a forty-nine footer with a bow ramp designed to carry one hundred assault infantry, and the Type B (*shohatsu*), a thirty-footer with no bow ramp for thirty troops. Both craft carried armor and machine guns.

In 1937, British and American observers watched *Shinsu-maru* at work and immediately recognized a significant development in amphibious warfare.

Although satisfied with the craft, the army's shipping engineers did not like the limitations of army transports or navy warships, also still used for landings. At army request the navy's design bureau began work on a specialized carrier ship, construction of which began in March 1933. In spring 1935, the army accepted the *Shinsu-maru* and initiated a new era in amphibious shipping.[37]

With the deployment of the 8,000-ton *Shinsu-maru* and a further refinement, the 9,000-ton *Akitsu-maru* (1941), the Japanese amphibious forces had in hand prototypes for all-purpose amphibious ships. Today the U.S. Navy and Marines use this fundamental concept to the exclusion of all others in their LHA and LHD class amphibious assault carriers. In 1937, British and American observers watched the *Shinsu-maru* at work off Shanghai and immediately recognized a significant development in amphibious warfare. The *Shinsu-maru* carried landing craft in a well deck that could be flooded, which allowed the landing craft to float free from an open stern gate. The ship could also hold additional craft on davits, but its next most impressive function was an ability to discharge vehicles from a deck-level parking garage directly onto a pier. It also carried two catapults for aircraft but did not embark operational seaplanes. It could, however, transport and unload aircraft if necessary, a capability further developed in the *Akitsu-maru*, which even had a short takeoff flight deck.

The army, however, halted construction of this class of vessel and turned to another type, a beaching craft for tanks and other vehicles, which it designated a *koryu* or general-purpose vehicle-carrying ship. Seeing the potential requirement for such a ship in Pacific operations, the navy persuaded the army to accept smaller ships (800 tons) that could be built more quickly, a proposal facilitated by naval control of the shipbuilding program and construction yards. The first Type 101 2nd Class Transport (*Nito yuso-kan*) did not enter service until 1942; of the sixty-nine Type 101 transports built, twenty went directly under army control, and the navy obtained all the others.

The navy's modest interest in the Type 101 (the Japanese version of the LST) reflected its greater commitment to its own 1st Class Transport (*Itto yuso-kan*), which had most of the operating characteristics and some of the armament of a destroyer. Its amphibious capability came from its troop and cargo compartments within the hull and from its aft deck configuration, stripped bare to carry four *daihatsu* or larger numbers of amphibian tanks or small craft. Almost 2,000 tons and capable of twenty-two knots, the 1st Class Transport, which entered service in 1943, had almost a 4,000-mile steaming radius. Its design reflected the navy's assumption that it would conduct emergency resupply missions to areas not necessarily under Japanese sea and air control and that it would depend upon its speed and antiaircraft armament to survive. In the U.S. Navy inventory the 1st Class Transport would have received designation as a fast attack transport (APD), which in reality was a converted destroyer that could conduct amphibious raids and reconnaissance missions. Such a vessel did not have the capability to provide adequate lift for a major landing.

In many ways the American development of amphibious warfare ships and craft appears far less innovative than the prewar and early war efforts of Britain and Japan. One looks in vain for American equivalents of the LST or the *Shinsu-maru*. Yet the American effort had a certain simplicity that allowed it to create an adequate amphibious force in 1943 and a truly global force by 1944. Part of the wartime capability rested in the industrial means to mass produce British designs like the LST, and part of it came from the ingenuity to exploit the *Shinsu-maru* design and build a fleet of twenty-six Landing Ships Dock (LSD). (The British, for comparison, converted two Channel ferries to well-deck, stern-gate configurations and accepted one LSD from the United States.) Amphibian tractors were the one truly novel American development, a vehicle the marine corps sought with the determination of the quest for the Holy Grail. Nevertheless, the marine corps collaborated with the navy to create the foundations of an amphibious force before Pearl Harbor.[38]

The major barrier to building an amphibious force before 1941 lay in the tacit agreement between the navy and Congress to focus the most important shipbuilding programs on warships, not auxiliaries. Real capability for fleet support lagged badly behind the need; in 1934 the navy had seven support ships to service a combatant force of 149 warships, excluding submarines. Two transports served navy and marine corps bases worldwide and were unavailable for amphibious operations. Even when Congress approved a fleet expansion program in 1938 that would allow the navy to modernize and enlarge its numbers past treaty limitations, transports did not appear in the program. The same thing happened in 1939, in part because even pro-navy Congressmen like Carl Vinson thought that transports were not "defensive" enough for the times. Not until the "Two Ocean Navy" Act of 1940—after the fall of France—did Congress approve the addition of twelve transports, all converted merchant vessels already in service, some for twenty years. By way of comparison, the transport fleet of the army numbered nine vessels in commission in 1940.

At the end of 1941, however, the navy had thirty-eight transports and a like number of cargo ships, but no specialized amphibious shipping. It commissioned its first LST in 1942, its first LSD in 1943. The navy followed the prewar assumption that it could create an amphibious transport force by converting merchantmen and liners to military service, including the installation of davits and cranes capable of handling landing craft. Such a policy, however, demanded development of an optimum landing craft, not only for assault infantry, but for tanks and vehicles essential for establishing a beachhead. Marine officers assigned to amphibious development in the 1930s despaired that the Navy Bureau of Construction and Repair (later called the Bureau of Ships) would ever produce or accept an adequate landing craft. After a decade of experimentation with bureau boats, the Navy Department accepted a marine corps proposal to invite commercial competition, but this ploy produced nothing much better by 1938.

In the meantime, the small navy office for landing craft reopened stalled negotiations with Andrew Higgins, a Louisiana boat-builder who had offered one of his thirty-foot "Eureka" boats for testing in 1936. Although it did not have a bow ramp, the "Eureka" had excellent power for its size, even after being enlarged to thirty-six feet, and it had shallow-water handling capabilities. Its light draft forward allowed it to ground and retract without difficulty, an essential feature in the Gulf bayous. Tested by the navy and marines in 1939 the Higgins boat won universal praise, and in 1941 Higgins, prompted by pictures of the *daihatsu*, redesigned the boat for a bow ramp. At the same time, the navy worked with its own and Higgins's designs to produce a tank lighter, the Landing Craft Mechanized (LCM), which was also transport-borne.[39]

Confronted by coral reefs in the central Pacific, the marine corps pursued a special piece of equipment to supplement the navy's landing craft, an amphibian tractor. In 1923 the marine corps investigated an amphibian tank complete with propellers, designed by J. Walter Christie, the brilliant but stubborn inventor of armored vehicles. Tested by the marines in the landing exercises of 1924, the Christie amphibian, although rejected for mechanical fragility, whetted marine appetites for an amphibian tractor. They found another possibility in a tracked vehicle designed by Donald Roebling, a third-generation inventor and engineer then living in Florida. Roebling had built his prototype "alligator" for swamp rescues; the vehicle featured a lightweight buoyant aluminum hull and tracks that could propel it through water or across soft earth. It had to be strengthened and reengineered for military use, but from 1937 until its adoption in 1940, the Roebling tractor and its inventor developed a sound working relationship with the marine corps. Both the Japanese and British armies investigated amphibian tanks, but neither pursued the equipment with any sense of urgency, and neither had an operational amphibian in service in 1941 when the marine corps took delivery of the first one hundred of the eighteen thousand American-built amphibian tractors that served in World War II.[40]

The use of aviation for amphibious operations fell into two functional categories: 1) to gain air superiority over the land and sea space of the amphibious objective area, and 2) to conduct offensive air operations against enemy targets. While the development of naval aviation moved slowly in interwar Britain, Japan and the United States created carrier-based and land-based naval aviation units to protect or to attack fleets and naval bases. No responsible amphibious planner thought air superiority an insignificant requirement; the persistent challenge came in finding a way to move air units within range of an amphibious objective area, whether by establishing air bases nearby or deploying carriers. Understandably nervous about air attacks on transports laden with troops and equipment, naval aviation planners tended to think of amphibious operations in the same way they thought about fleet action. The naval aviators of Japan and the United States, for example, approached this problem with similar doctrinal conclusions. In contrast, Army Air Corps and Japanese army aviators or those of the independent Royal Air Force gave hardly any thought to amphibious operations.

The aviation component of the U.S. Marine Corps—two small air groups in the 1930s—thought a great deal about amphibious operations because the commandant told them to do so. Marine aviation had a unique status. Its personnel were marines, subject to corps policies on assignments, promotions, training, and operations, but its officers were designated "naval aviators" and certified as such by the navy's aviation training establishment. Aircraft were the same as in the navy, and money for them came from the navy budget, known in Washington as "blue dollars." The marine corps, for example, had a senior aviator stationed at Headquarters Marine Corps, to manage aviation policy, but he paid as much or more attention to the chief of the navy's Bureau of Aeronautics as to the commandant. The navy's aviation leaders maintained fairly clear expectations for marine aviation squadrons: they were to carry the burden of aerial defense for established and advanced naval bases. This mission thus freed the navy's aviation squadrons for the role they most sought, service aboard fleet carriers and participation in great naval battles. Marine aviators received carrier training, too, but they did not expect to participate in fleet engagements, only to perform their "advanced base mission" of land-based air defense.

Even before 1914 some marine aviators began to recognize the contribution that air attacks could make to land campaigns, and they viewed World War I as an important example of the potential of tactical aviation in ground combat. These lessons, they thought, received validation in marine aviation operations in Haiti, the Dominican Republic, China, and Nicaragua. But marine aviators focused on their place in a naval campaign, and some of them concluded that they should create a unique role, close air support of amphibious assaults. This fire support role did not necessarily release marine squadrons from the performance of the general tasks of all naval aviation; it simply gave them a task in which only they specialized. Their aircraft were no different from the navy's, but their pilots had to serve in the ground marine corps before attending flight school, and their special interest was the attack upon beach defenses.

The problem of fire support for an opposed amphibious landing could not be wished away. Unless an enemy proved especially inept, poorly led, and deployed with carelessness, an established defense had inherent advantages over a landing force that only overwhelming supporting fire could negate. The artillery and small arms of the landing force, fired from boats or neighboring islands, was not likely to provide the edge. Naval gunfire—high velocity and flat trajectory fire whatever the weight of shell—could not destroy reverse slope defenses and might easily miss well-constructed fortifications on flat terrain like Pacific atolls. Dive-bombing by marine aviation provided the answer to targets that eluded naval gunfire.

The problem then became the proper identification of targets, which might require radio communications from ground observers to attack aircraft, and the co-ordination of ground tactical action with supporting air attacks. The navy solution to the problems of close air support reflected practices designed to control naval gunfire in fleet engagements: airborne spotters in seaplanes or carrier aircraft con-trolled the strikes. The marine corps did not believe that this method was adequate and thought instead of control from the ground, exercised by aviation experts at-tached to the ground forces. This system would only work, however, with marine-trained squadrons that could reach the amphibious objective area, and such squadrons were likely to be navy units based on carriers, not marines.[41]

The Test of War: Amphibious Operations, 1940–1942

The audit of war found all three of the great naval powers wanting in varying de-grees, and their experience with amphibious warfare brought disappointments proportional to their prewar interest. With less time to adjust and a greater gap between requirements and forces, the British had the most unpleasant tutorial, but under the whip of Churchill they pioneered the creation of specialized tech-niques and equipment for landings, the creation of joint staffs, and the orches-tration of naval forces and landing forces. Yet the British experience demonstrated that one cannot conjure specialized military capabilities into existence overnight, not just in scale but in the expertise and quality of operational management that comes only from time-consuming hard training and trial and error.

For the Japanese, their prewar investment paid off in the dazzling success of the Western Pacific campaign of 1941. Yet within that campaign the Japanese demonstrated liabilities that cost them the initiative in the South Pacific in 1942. For the United States the outbreak of war in 1939, especially the Allied collapse of 1940, galvanized rearmament and preparedness, which included the amphibious forces. The lack of Allied air and naval control in every theater until late 1942 meant that the first prerequisite for amphibious operations could not be met, thus sparing the premature commitment of American amphibious forces. Even their first use in the South Pacific and North Africa dramatized that much remained to be done to bring those forces to full efficiency.

Among its many defeats in 1940, Britain could attribute only one to its ineptness in joint operations, the Norwegian campaign of April–June 1940. Ironically the campaign paved the way for its author, Winston Churchill, to move from his position as First Lord of the Admiralty to prime minister. The Allied difficulties in taking and holding Narvik, Trondheim, Stavanger, and Bergen were more a result of the air and the naval aspects of the campaign, not its amphibious elements. Yet German aerial superiority represented the key to the campaign, and it rested on the daring and rapidity with which German seaborne and airborne ground forces seized and held Norwegian airfields. The Royal Navy had ravaged the surface forces of the Kriegsmarine before the Luftwaffe established itself ashore, and amphibious counterlandings might have slowed or even halted the German conquest of southern and central Norway. Moreover the Allies squandered their forces in the extended battle to seize and then defend Narvik, which they ultimately abandoned. Of all the campaigns in Western Europe in 1940, only the Norwegian campaign was inherently naval, and the British amphibious participation proved too little and too late. Although German naval losses were severe and the flight of the Norwegian merchant fleet was a plus for the Allies, the long-term impact of Norway's loss meant that German air and naval forces could operate with much greater freedom and effectiveness against Allied convoys in the North Atlantic and along the route to the Soviet Union. The seizure of Norway also ensured the closing of the Baltic and the free transportation of Scandinavian raw materials to German factories.[42]

Stung by his Nordic Dardanelles, Churchill, ever given to unconventional thought and action, dramatically elevated amphibious operations within the British command hierarchy. He changed a Chiefs of Staff initiative to appoint an adviser for raiding and amphibious operations into a full-blown independent command, Combined Operations Headquarters, first under Admiral of the Fleet Sir Roger Keyes and then Admiral Lord Louis Mountbatten. With Churchill's firm support, the Combined Operations Headquarters developed an empire of construction programs, landing force development schemes for equipment and techniques, and training centers for commando raiding units and more conventional ground forces, including specialized armored units.

Although Churchill stressed the raiding mission early in the war, he expected Combined Operations Headquarters to prepare for a major invasion of France. In the meantime, Churchill and the bolder souls among his military family considered all sorts of amphibious raiding operations along the entire rim of the Axis—from Norway to Western Africa, from Vichy France to the Middle East, and from the Mediterranean to Madagascar. The most memorable of these enterprises, so worthy of Pitt in conception and of Falstaff in execution, was the amphibious raid on Dieppe in August 1942, a learning experience to be sure, but an extremely costly one to the Allied landing force, which lost more than half its 6,000 officers and men. The vigor of the German defense froze amphibious tactics; direct assaults upon prepared defenses still looked suicidal, and the "commando concepts" of stealth, surprise, and small forces still dominated planning.[43]

The Japanese armed forces proved in their southern operations that they could 1) establish air and naval supremacy and 2) put landing forces ashore in speed and numbers that confounded Allied defenders. Whether the landing forces came from the Japanese army, which put about one-third (eleven divisions) of its field force in the campaign, or from the Naval Special Landing Forces, the results were largely the same. Japanese aircraft, warships, and sometimes land-based artillery pummeled the objective area but not for long if surprise was a major factor. Then landing craft and barges brought the troops ashore at multiple, narrow landing sites distributed along long stretches of coastline. When possible, small craft brought troops inland by rivers and estuaries, and the Japanese launched parachute assaults in conjunction with the landings when the transports and troops were available. The grand tactics of infiltration and exploitation worked well, especially in confounding a larger Commonwealth army in Malaya. The Japanese continued to regard night operations as preferable unless they were confident of overwhelming fire superiority and adequate troop numbers. Japanese landing forces swept into Luzon, Hong Kong, Malaya, Guam, the Solomons, New Britain Island, and the Dutch East Indies and routed Allied defense forces without serious losses to their amphibious expeditionary forces and their supporting air and naval units.[44]

The Japanese, however, also proved that they could make errors, usually resulting from their own hubris as warriors and their contempt for the soldierly qualities of their Western opponents. They tended to attempt too much, with too little, and too quickly, and their operational plans accepted a degree of risk and planned complexity that carried within them the seeds of interservice conflict and disaster. Amphibious operations were no exception, and in fact, official doctrine reinforced this tendency. The campaign against the Americans had its upsets, however temporary. The first landing attempt against Wake Island on December 11, 1941, resulted in disaster; U.S. Marine aircraft and coast defenses ruined the landing before the troops could even disembark from their transports. Even when the atoll finally fell, the marines on one island wiped out the Japanese landing force. During the Philippines campaign, the Japanese attempted to envelop the American defense line on Bataan in January–February 1942 with four different amphibious landings, all of them unsuccessful. In part because of the limitations of their amphibious ships and craft, the Japanese landed too few troops and failed to give them close artillery and armored support, a problem the British, for example, fully appreciated.

In summer 1940, the United States had to face the twin possibilities of war with Japan and the spillover of the war in Europe into the Western Hemisphere. American planners needed to consider the prospect that the victorious Nazis might jump the Atlantic to Vichy French bases or to ones deeded them by their Latin American admirers. The possibility of preemptive amphibious landings in the Caribbean or the Western Atlantic brought requirements for troops that the marine corps alone could not fill. By 1941 the Army-Navy Joint Board, which became the Joint Chiefs of Staff, had arranged for army divisions to undertake amphibious training as part of two joint training forces, one on each coast.

Eventually, four army infantry divisions participated in the training, and the Joint Training Force Headquarters, commanded by two marine major generals, evolved into amphibious corps headquarters for the Atlantic and Pacific Fleets. The training exercises revealed serious shortcomings in transports and landing craft, and the army, in fact, established its own amphibious training command in June 1942 and began to organize specialized engineer brigades for amphibious operations. The navy, in all truth, had its mind on its own inadequacies in gunnery, antisubmarine warfare, and aerial combat rather than amphibious landings. Senior marine officers like Major General Holland M. Smith had ample knowledge but little influence over fleet exercise priorities.[45]

After Pearl Harbor and six months of frustration, the Joint Chiefs of Staff accepted the idea of an African "Second Front" after prodding from Roosevelt and from the British and started planning for landings in Morocco and Algeria. At the same time, Admiral Ernest J. King decided that the time was ripe after the Japanese defeat at the Battle of Midway in June 1942 to open a Pacific war "Second Front" along the eastern edge of the Malay Barrier. He planned to use navy units and a marine division to defend the logistical lifeline to Australia and New Zealand. This idea was also dear to the heart of General Douglas MacArthur, who longed to start his return to the Philippines from Australia. From August 1942 through the rest of the year, American expeditions explored the state of their combat readiness under fire and experienced a severe test from the Japanese but less so from the Vichy French forces that greeted the November landings in North Africa. In the South Pacific an Australian-U.S. Army task force of two divisions invested the Buna-Gona area on New Guinea's north shore and engaged the Japanese defense forces in a brawl reminiscent of World War I. Both forces suffered serious losses from disease and logistical hardships over the course of the offensive. The same conditions developed on Guadalcanal in the southern Solomons, invaded by a marine division on August 7, 1942, except this time Americans held the base area, the Henderson Field enclave. The Americans in both places proved they could move forces by sea and support them and, in the case of Guadalcanal and companion landings on Tulagi and Gavutu, stage an amphibious assault against opposition.[46]

The two Allied offensives in the South Pacific are important for revealing what the Japanese could not do. To relieve forces meagerly supplied and reduced by casualties and illness, they could not stage a major counterlanding, a concern for the local American commanders. The Japanese capability to introduce new troops to both battles was strictly a logistical effort and did not represent an operational concept that altered the fundamental battle of attrition. At Buna-Gona the Japanese reinforced their established defensive positions, and on Guadalcanal they deployed the better part of two army divisions to launch three separate attacks over three months at the 1st Marine Division's perimeter. In both campaigns the Japanese had or could have employed ample naval and air forces to cover an amphibious landing; in the Solomons, in fact, they badly mauled the U.S. Navy, especially in night actions. But the Japanese did not have an amphibious force that could make an opposed landing against determined troops. Had they possessed such a force they might have struck a decisive blow against the American landing forces. In the dark days of 1942 an Allied defeat in the South Pacific might have influenced the course of the Pacific war.

U.S. Sherman tank drives ashore from LST, Normandy, June 8, 1944.

The landings in North Africa provided somewhat different lessons because the French mounted no serious air and naval challenge and waged only sporadic battle against the landing forces. British thinking about amphibious warfare influenced the landings, which focused on the ports of Casablanca, Oran, and Algiers. All three succeeded but not without significant problems. Night landings and airborne drops dispersed forces and injected unnecessary confusion. Artillery and tanks arrived slowly, and, two Dieppe-style *coups de main* ended in disaster in the inner harbors of Oran and Algiers. In Oran almost 500 American soldiers died or fell prisoner, and a similar size group of proto-rangers surrendered in Algiers. In actions elsewhere an errant shell from an American cruiser inflicted the only Allied casualties, while disarray among the troops and beach control groups allowed landed supplies to go astray.

The commanding general of the Moroccan force, Major General George S. Patton Jr., landed on the beach at Fedala and immediately wished he were somewhere else, but he did report that "the performance of the navy...has been of the highest order. I am amazed at their efficiency, and I am delighted at the wholehearted spirit of cooperation they have evinced."[47] All told, the Allies lost 1,181 killed and missing in two days of fighting, about the same number of combat deaths suffered by the 1st Marine Division in four months on Guadalcanal. Clearly the Allies had some more training to do to improve their amphibious skills, which in any event were not needed for almost a year until the landings on Sicily. The marines and navy also had an additional learning experience ahead, the first atoll assault at Tarawa in November 1943. This bloodletting led to a revision of naval gunfire and air support doctrine as well as to the accelerated production of amphibian tractors.

The amphibious forces of Britain, Japan, and the United States all required changes in doctrine, organization, communications, weapons, equipment, training, and logistics as World War II dragged on, but the Allies, shifting to the offensive in 1942, had greater incentive and greater resources to make the necessary changes. Neither Germany nor Japan could be defeated without major landings in 1944, operations of a scale and scope unimaginable in 1942. Japan, by contrast, lost its amphibious capability in 1942 when it shifted to the strategic defensive. Japanese planners in both services apparently did not believe that amphibious forces had an offensive role to play within the general concept of a defensive naval campaign. Made by default, such a decision probably was ill advised before 1944 because an amphibious counterstroke in the South Pacific in summer 1943 might have seriously disrupted Allied planning in that theater, a theater already plagued by inter-Allied and interservice disputes. In the end, British and American amphibious knowhow provided an operational capability that was unmatched by the Axis and essential to the Allied victory.

Conclusion

The interwar development of amphibious warfare by Britain, Japan, and the United States reveals several common threads that explain each nation's limited success in developing the operational specialty. Each nation shared the same geopolitical characteristics: all avoided alliances until the eve of World War II, so they had to provide all of the armed forces they thought they required. As maritime nations all three exercised military power to a large degree through their navies; all three nations, however, had to maintain ground forces to occupy and defend territories away from their home territory. All three nations recognized the potential of military aviation; Britain's need for both offensive and defensive air forces, of course, became critical in the face of the Luftwaffe's challenge in the 1930s. Neither Japan nor the United States faced a comparable threat. Judged against the course of

Japan, on the other hand, entered World War II with too little navy for the protracted war that lay ahead.

World War II, all three nations might have profited from a greater investment in amphibious forces, but such investment depended upon the relative balance of investment in air, naval, and ground forces, and it is difficult to fault the broad investment strategies adopted by the United States and Britain, both of which viewed air and naval forces as more important than a large ground force. Japan, on the other hand, entered World War II with too little navy for the protracted war that lay ahead. The quick war assumptions and "continentalism" of Japanese planners doomed the imperial armed forces to a high-risk strategy with limited forces, which produced dramatic operational swings from caution to impetuousness.

Yet given the nature of the resistance they faced in 1941–1942 and the success of their landing operations, the Japanese demonstrated they had more than adequate amphibious forces against the opposition they then faced. But their successes depended upon inept Allied air and naval defensive operations, which could be and were corrected with larger forces, better planning, and more skilled commanders. As Allied operations improved in late 1942, the Japanese found they were not fully prepared to conduct an aggressive strategic defense of their newly won islands. Time and again they suffered prohibitive losses because they could not seize or recapture bases for their air and naval forces. The ground forces they committed had little mobility and endured crippling logistical shortages. The Japanese had the novel experience of losing their amphibious assault capability after using it so well in the early stages of the Pacific war. This capability disappeared by institutional decisions, not enemy action. The landing forces became land-bound defensive garrisons, and the amphibious shipping shifted to general resupply missions. In a sense the Japanese amphibious forces committed collective *sepuku* after their greatest victories, largely for want of prudent strategic vision.

The Japanese experience reflected a factor that applied to the British and Americans, i.e., having staff proponency, doctrinal publications, and a place in the officer educational system does not ensure that a developing operational specialization will reach fruition. There must be a foundation in institutional commitment, and a major organizational embrace of a new mission. The British military establishment had many splendid ideas about amphibious operations, but the thinkers had no firm organizational foundation and funding. Only the Inter-Services Training and Development Centre met this requirement, and its modest mandate produced equally modest accomplishments in landing craft design. The Japanese problem was different and the same; its amphibious development followed service lines and service conceptions of strategic requirements. The Japanese army had its own amphibious force, the navy another. Only in the opening campaign of the Pacific war did they work in concert, and even then these forces worked simultaneously but most often separately in accordance with their missions and operational capabilities. Neither the army nor the navy viewed amphibious forces as anything other than forces of limited utility and peripheral to their services' real functions, which were war on land and sea, not the projection of military power from the sea to the land. The United States, on the other hand, had a politically potent independent service—the U. S. Marine Corps—that had a clear wartime mission after 1921, which was to seize advanced naval and air bases in a war with Japan across the Pacific Ocean. Even with the foreign intervention disruptions of 1927–1933 the marine corps did its best to train and equip itself for amphibious assaults, and the senior leadership of the U.S. Navy (at least some of it) cooperated because the admirals understood that base-seizure was a real mission. Few admirals or generals in Britain and Japan showed equal clarity of vision. Only American planners fully appreciated that if amphibious forces prepared to capture and defend objectives of critical military utility (air and naval bases), they could also conduct amphibious operations of a less demanding nature.

Notes

1.   Harold and Margaret Sprout, *Toward a New Order of Sea Power: American Naval Policy and the World Scene, 1918–1922* (Princeton, NJ: Princeton University Press, 1943), 149–160.

2. Article XIX, Treaty for the Limitation of Armament, February 6, 1922, reprinted in ibid., 302–311. For an introduction to the naval history of World War I and the interwar period, see Clark G. Reynolds, *Command of the Sea: The History and Strategy of Maritime Empires* (New York: Morrow, 1974), 445–500.

3. For the history of the Royal Navy at the height of the Victorian era, see Arthur J. Marder, *The Anatomy of British Sea Power: A History of the British Naval Policy in the Pre-Dreadnought Era, 1880–1905* (New York: Knopf, 1940).

4. For the interrelationship of British, American, and Japanese naval policy in the interwar years, see Harlow A. Hyde, *Scraps of Paper: The Disarmament Treaties between the World Wars* (Lincoln, NE: Media Publishing, 1988); Arthur J. Marder, *Old Friends, New Enemies: The Royal Navy and the Imperial Japanese Navy* (Oxford: Clarendon, 1981); Gerald E. Wheeler, *Prelude to Pearl Harbor: The United States Navy and the Far East* (Columbia: University of Missouri, 1963); Stephen E. Pelz, *Race to Pearl Harbor: The Failure of the Second London Conference and the Onset of World War II* (Cambridge, MA: Harvard University, 1974); Mark R. Peattie, "The Nan'yo: Japan in the South Pacific, 1885–1945," in Ramon H. Myers and Mark R. Peattie, eds., *The Japanese Colonial Empire, 1895–1945* (Princeton, NJ: Princeton University Press, 1984), 172–210; David C. Evans and Mark R. Peattie, *Kaigun: Doctrine and Technology in the Japanese Navy, 1887–1941* (Annapolis: U.S. Naval Institute, 1997), chapter 14; Michael A. Barnhart, *Japan Prepares for Total War: The Search for Economic Security, 1919–1941* (Ithaca, NY: Cornell University Press, 1987); Jonathan G. Utley, *Going to War with Japan, 1937–1941* (Knoxville: University of Tennessee Press, 1985); Thaddeus Tuleja, *Statesmen and Admirals: Quest for a Far Eastern Naval Policy* (New York: Norton, 1963).

5. Much of the best writing on Gallipoli predates World War II: John Masefield, *Gallipoli* (London: Macmillan, 1916); C. E. W. Bean, *The Story of Anzac*, 2 vols. (Sydney: Angus and Robertson, 1921, 1924); Capt. William D. Puleston USN, *The Dardanelles Expedition* (Annapolis: U.S. Naval Institute, 1927). The author visited Gallipoli in July 1993.

6. Admiral Sir Herbert Richmond, *Amphibious Warfare in British History* (Exeter: Wheaton, 1941); Admiral the Lord Keyes, *Amphibious Warfare and Combined Operations* (Lee Knowles Lectures, 1943) (Cambridge: Cambridge University Press, 1943); James L. Stokesbury, "British Concepts of Amphibious Warfare, 1867–1916," PhD dissertation, Duke University, 1968.

7. For general histories of the Japanese armed forces, see Stephen Howarth, *Fighting Ships of the Rising Sun: The Drama of the Imperial Japanese Navy, 1895–1945* (New York: Atheneum, 1983) and Meiron and Susie Harries, *Soldiers of the Sun: The Rise and Fall of the Imperial Japanese Army* (New York: Random House, 1991).

8. In addition to the sources already cited, see Paul Kennedy, "British 'Net Assessment' and the Coming of the Second World War," in Williamson Murray and Allan R. Millett, eds., *Calculations: Net Assessment and the Coming of World War II* (New York: Free Press, 1992), 19–59; Brian Bond and Williamson Murray, "British Armed Forces, 1918–1939," in Allan R. Millett and Williamson Murray, eds., *Military Effectiveness* (London and Boston: Cambridge University Press, 1988), vol. 2, 98–130; H. P. Willmott, *Empires in the Balance: Japanese and Allied Pacific Strategies to April 1942* (Annapolis: U.S. Naval Institute, 1982), 39–66, 95–129; Peter Lowe, "Great Britain's Assessment of Japan Before the Outbreak of the Pacific War," in Ernest R. May, ed., *Knowing One's Enemies: Intelligence Assessment before the Two World Wars* (Princeton, NJ: Princeton University Press, 1984), 456–475.

9. Fujiwara Akira, "The Role of the Japanese Army," and Asada Sadao, "The Japanese Navy and the United States," in Dorothy Borg and Shumpei Okamota, eds., *Pearl Harbor as History: Japan-*

ese American Relations, 1931–1941 (New York: Columbia University Press, 1973), 189–195 and 225–259; Louis Morton, "The Japanese Decision for War," *U.S. Naval Institute Proceedings*, 80, December 1954, 1324–1334; Carl Boyd, "Japanese Military Effectiveness: The Interwar Period," in Millett and Murray, eds., *Military Effectiveness*, vol. 2, 131–168; Alvin D. Coox, "Japanese Net Assessment in the Era before Pearl Harbor," in Murray and Millett, eds., *Calculations*, 258–298; Michael A. Barnhart, "Japanese Intelligence before the Second World War," in May, ed., *Knowing One's Enemies*, 424–455; and Rear Adm. Yoichi Hirama, "Japanese Naval Preparations for World War II," *Naval War College Review*, 44, Spring 1991, 63–81.

10. Edward S. Miller, *War Plan ORANGE; The U.S. Strategy to Defeat Japan, 1897–1945* (Annapolis: U.S. Naval Institute, 1991); David Kahn, "United States Views of Germany and Japan by 1941," in May, ed., *Knowing One's Enemies*, 476–501; Fred Greene, "The Military View of American National Policy, 1904–1940," *American Historical Review*, 65, January 1961, 354–377; Calvin Christman, "Franklin D. Roosevelt and the Craft of Strategic Assessment," in Murray and Millett, eds., *Calculations*, 216–257; Russell F. Weigley, "The Role of the United States Navy," in Borg and Okamoto, eds., *Pearl Harbor as History*, 165–188, 197–223; Philip T. Rosen, "The Treaty Navy, 1919–1937," and John Major, "The Navy Plans for War, 1937–1941," in Kenneth J. Hagan, ed., *In Peace and War: Interpretations of American Naval History, 1775–1984* (Westport, CT: Praeger, 1984), 221–262; and Ronald Spector, "The Military Effectiveness of the U.S. Armed Forces, 1919–1939," in Millett and Murray, eds., *Military Effectiveness*, vol. 2, 70–97.

11. Miller, *War Plan ORANGE*, 100–249; Chief of Naval Operations (Adm. R. E. Coontz) to Major Gen. Cmdt. Marine Corps (Major Gen. G. Barnett), "Function of Marine Corps in Wartime," memo, 28 January 1920, File 221 (1920), Secretary of the Navy/CNO Correspondence, 1920 General Records of the Navy Department, RG 80, NA.

12. This analysis depends primarily upon the following sources: Kenneth J. Clifford, *Amphibious Warfare Development in Britain and America from 1920–1940* (Laurens, NY: Edgewood, 1983); Stephen Roskill, *Naval Policy Between the Wars*, vol. 1, *The Period of Anglo American Antagonism, 1919–1929* (New York: Walker, 1968) and vol. 2, *The Period of Reluctant Rearmament, 1930–1939* (London: Collins, 1976); Bernard Fergusson, *The Watery Maze: The Story of Combined Operations* (New York: Holt, Rinehart, and Winston, 1961).

13. Clifford, *Amphibious Warfare Development*, 30–41.

14. Roskill, *Naval Policy*, vol. 1, 539–540; Fergusson, *Watery Maze*, 36; Clifford, *Amphibious Warfare Development*, 48–49.

15. Clifford, *Amphibious Warfare Development*, 5–19.

16. Donald F. Bittner, "Britannia's Sheathed Sword: The Royal Marines and Amphibious Warfare in the Interwar Years—A Passive Response," *Journal of Military History*, 55, July 1991, 345–364; Maj. Gen. J. L. Moulton, *The Royal Marines*, vol. 1 (London: Cooper, 1982), 86–90.

17. Clifford, *Amphibious Warfare Development*, 46–84.

18. Ibid., 72–84; Rear Adm. L. E. H. Maund, *Assault From the Sea* (London: Methuen, 1949), 1–23.

19. Clifford, *Amphibious Warfare Development*, 76.

20. Unless otherwise noted, this analysis of Japanese amphibious warfare development is drawn primarily from two studies written from official Japanese archival sources by Capt. Masao Suekuni, JMSDF (Ret.), chapter 1, "Landing Operations of the Japanese Military," of the larger work, "Amphibious Operations in Military History and Their Background," (*Senshi ni miru joriku sakusen to sono urakata*) published as a volume in the series Reference Materials in Naval History (*Kaigun senshi sanko shiryo*) (Toyko, 1982) and "The Evolution of Joint Landing Operations of the Japanese Military," *Journal of Military History* (*Gunji Shigaku*), 27, July 1991, 55–56. I want to

thank Prof. Hisao Iwashima of Iwate University for serving as my liaison with Captain Suekuni, and I am indebted to Professor David Evans of the University of Richmond for his translation of Captain Suekuni's pamphlet and to Mr. Walter Grunden for additional translations from the same text. Mr. Grunden also translated "The Evolution of Joint Landing Operations." I also acknowledge the contribution of Dr. Edward J. Drea, U.S. Army Center of Military History, also a Japanese linguist, for the use of his "The Development of Imperial Japanese Army Amphibious Warfare Doctrine," 18 February 1992, also based on Japanese archival documents and official histories. See also Saburo Hayashi and Alvin D. Coox, *Kogun: The Japanese Army in the Pacific War* (Quantico, VA: Marine Corps Association, 1959), 1–28.

21. Suekuni, Sections 4 and 6, "Landing Operations"; Drea, "The Development of Imperial Japanese Army Amphibious Warfare Doctrine"; Suekuni, "The Evolution of Joint Landing Operations of the Japanese Military," *Gunji Shigaku*, previously cited; G-2, War Department General Staff, *Handbook on Japanese Military Forces*, 1 October 1944, TM E30–480 (reprint, Novato, CA: Presidio, 1991), 10–18, 49–50.

22. Suekuni, Section 3, "Landing Operations"; War Department General Staff, Japanese Military Forces, 166–169; "Japanese Landing Operations and Equipment," May 1943, Office of Naval Intelligence, reprinted in A.D. Baker III, ed., *Japanese Naval Vessels of World War Two* (Annapolis: U.S. Naval Institute, 1987); Hans G. Von Lehman, "Japanese Landing Operations in World War II," in Lt. Col. Merrill L. Bartlett, ed., *Assault From the Sea: Essays on the History of Amphibious Warfare* (Annapolis: U.S. Naval Institute, 1983), 195–201.

23. Suekuni, Sections 8 and 9, "Landing Operations."

24. Ibid.; Hayashi and Coox, *Kogun*, 6–27.

25. War Department General Staff, Japanese Military Forces, 76–79; David C. Evans, "The Japanese Navy in the Invasion of the Philippines," *American Historical Association Meeting*, 1991.

26. This section is drawn from the following sources, all of which represent extensive research in official archives and private papers collections: Allan R. Millett, *Semper Fidelis: The History of the United States Marine Corps* (New York: Macmillan, 1980, rev. 1991), 319–343; Jeter A. Isley and Philip A. Crowl, *The U.S. Marines and Amphibious War* (Princeton, NJ: Princeton University Press, 1951), 14–71; and Clifford, *Amphibious Warfare Development*, previously cited. For a partisan, but "inside" account, see Gen. Holland M. Smith, USMC (Ret.), "The Development of Amphibious Tactics in the U.S. Navy," Occasional Paper, History and Museums Division, HQUSMC, 1992, which is a reprinting of General Smith's 1946–47 articles for the *Marine Corps Gazette*. Given the relationship of the U.S. Marine Corps and the function of amphibious warfare, it is unsurprising that the analysis is done from the marine corps' perspective, but these works, as well as other sources in this section, represent navy perspectives as well. For an effort to diminish the institutional role of the marine corps, see William F. Atwater, "United States Army and Navy Development of Joint Landing Operations, 1898–1942," PhD Dissertation, Duke University, 1986. The most articulate and forceful presentations (other than Smith's) of the Marine Corps perspective may be found in Lt. Gen. Victor H. Krulak, *First to Fight: An Inside View of the U.S. Marine Corps* (Annapolis: U.S. Naval Institute, 1984) and Col. Robert D. Heinl Jr., *Soldiers of the Sea: The U.S. Marine Corps, 1775–1962* (Annapolis: U.S. Naval Institute, 1962).

27. Lt. Col. Merrill L. Bartlett, *Lejeune: A Marine's Life, 1867–1942* (Columbia: University of South Carolina Press, 1991), 190–202; Dirk A. Ballendorf, "Earl Hancock Ellis: The Man and His Mission," *U.S. Naval Institute Proceedings*, 109, November 1983, 53–60; Major Gen. Cmdt. to CNO, September 30, 1919, File 223, and Major Gen. Cmdt. to Sec. Nav. (Operations), Quarterly Readiness Reports, October–March 1920, File 221 (1919), Confidential Correspondence, 1919–1926,

RG 80, NA; both Col. B. H. Fuller to Major Gen. Cmdt., "Advanced Base Plans," August I, 1921, File 2515, CMC General Correspondence, 1913–1932, RG 127, NA; Major Gen. Cmdt., memorandum for the General Board of the Navy, "Future Policy for the Marine Corps and as Influenced by the Conference on Limitation of Armament," February 11, 1922, File 432, General Board Records, Naval Historical Center, Washington, DC; Office to the Commandant, "History of Advanced Base Training in the Marine Corps," August 28, 1931, File 432, General Board of Records, Naval Historical Center, Washington, DC.

28. *The Joint Board, Joint Action of the Army and the Navy* (Washington, DC: USGPO, 1927), 3. This publication superseded *War Department and Navy Department, Joint Army and Navy Action in Coast Defense* (Washington, DC: USGPO, 1920).

29. *The Joint Board, Joint Overseas Expeditions* (Washington, DC: USGPO, 1933); Col. A. T. Mason, "Special Monograph on Amphibious Warfare," 1949–1950, mss. history, Command File World War II, Operational Archives, Naval Historical Center, Washington, DC. The 1933 manual followed a complete review of the marine corps missions within the Department of the Navy and the validation of the requirement for an amphibious assault force under unambiguous naval control. This correspondence and set of records is in File 432, 1931–1933), General Board Records. The basic navy justification for the marine corps is expressed in Chairman General Board to Secretary of the Navy, memo, "Examination of the Organization and Establishment of the U.S. Marine Corps," August 10, 1932, CMC Correspondence, 1913–1932, RG 127, NA.

30. Atwater, "United States Army and Navy Development of Joint Landing Operations," 45–47; Michael Vlahos, *The Blue Sword: The Naval War College and the American Mission 1919–1941* (Newport: Naval War College, 1980); Anthony A. Frances, "History of the Marine Corps Schools," 1945, Breckinridge Library, Marine Corps University, Quantico, Va.; Research Section, Marine Corps Schools, "A Brief Historical Sketch of the Development of Amphibious Instruction and Doctrine at the Marine Corps Schools during the Years Prior to World War II," 1949, "Schools" File, Reference Section, Marine Corps Historical Center; Lt. Col. Donald F. Bittner, "Curriculum Evolution Marine Command and Staff College 1920–1988," Occasional Paper (Washington, DC: USGPO, 1988), 1–30; Boyd Dastrup, *The U.S. Army Command and General Staff College: A Centennial History* (Ft. Leavenworth, KS: Sunflower University Press, 1981).

31. General Board, "History of Advanced Base Training in the Marine Corps," 1931, previously cited; Mason, "Special Monograph on Amphibious Warfare," 1949–1950, previously cited; Col. Dion Williams, "The Winter Maneuvers of 1924," Marine Corps Gazette, 9, March 1924, 1–25; Brig. Gen. Dion Williams, "Blue Marine Expeditionary Force," Marine Corps Gazette, 10, September 1925, 76–88.

32. Clifford, *Amphibious Warfare Development*, 92–108; Maj. Gen. John H. Russell, "The Birth of the Fleet Marine Force," *U. S. Naval Institute Proceedings*, 72, January 1946, 49–51; Maj. Gen. Cmdt., to CNO, August 17, 1933, and CNO to Maj. Gen. Cmdt., September 12, 1933, File 1975–10, CMC General Correspondence, 1933–1938, RG 127, NA; Joint Board to Secretary of the Navy and Secretary of War, "Further Consideration of Joint Operations," April 5, 1927, File 350, Joint Board Records, RG 225, NA; Maj. Gen. J. T. Myers to Col. L. McC. Little, June 30, 1931, Maj. Gen. Louis McCarty Little Papers, Library of Congress.

33. Lt. Col. Benjamin W. Gaily, "A History of U.S. Fleet Landing Exercises," 1939, Discontinued Command File, Operational Archives, Naval Historical Center, Washington, DC; Mason, "Special Monograph of Amphibious Warfare," 4–16; Headquarters, 1st Mar Brig, FMF, "Notes on the Organization and Activities of the Fleet Marine Force in Connection With Landing Operations," 1938, File 1975–10, CMC General Correspondence, 1933–1938, RG 127, NA; The Adjutant's Gen-

eral's Office, USA, "Notes on Fleet Landing Exercise No.2," March 30, 1936, copy in Archives, U.S. Army Military History Institute; Major Gen. L. McC. Little to Maj. Gen. Cmdt. T. Holcomb, April 7, 1938 (review of FMF exercises), Little Papers; Commander in Chief U.S. Fleet, "Report on Fleet Landing Exercise No.6," June 13, 1940, Atlantic Fleet Command Files, Operational Archives, NHC; HQ 1St Mar Brig, reports of FLEX 6 problems and critiques, File 1975–10, RG 127, NA.

34. *Office of Naval Operations, Landing Operations Doctrine United States Navy* (FTP167) (Washington, DC: USGPO, 1938); Marine Corps Contributory plan, C-2, ORANGE, MCWP–2, October 1932, MCS Files, RG 127, NA; Isley and Crowl, U.S. Marines and Amphibious War, 37–58. In naval gunfire, for example, opinion divided on two concepts, one borrowed from World War I artillery practice of saturation bombardment measured by the tonnage per area, the other conceived during testing of firing slowly with pinpoint fire at exact targets like pillboxes. At issue was the question of effective range and warship vulnerability to shore batteries. The arguments are summarized in Cmdr. C. G. Richardson, "Naval Gunfire Support of Landing Operations," MCS lecture, 1938–1939, and Lt. Cmdr. David L. Nutter, "Gunfire Support in Fleet Landing Exercises," September 1939, both Breckinridge Library, MCV.

35. The observations on Chiefs of Naval Operations and Secretaries of the Navy are based on the essays on these men in Robert W. Love Jr., ed., *The Chiefs of Naval Operations* (Annapolis: U.S. Naval Institute, 1980) and Paolo Coletta, ed., *American Secretaries of the Navy* 2 vols. (Annapolis: U.S. Naval Institute, 1980).

36. The development of British amphibious ships and craft are described in J. D. Ladd, *Assault from the Sea, 1939–1945* (New York: Hippocrene, 1976); Fergusson, *The Watery Maze*, 35–85; Clifford, *Amphibious Warfare Development*, 72–82; *Division of Naval Intelligence, Allied Landing Craft of World War II* (1944 edition, reprinted Annapolis: U.S. Naval Institute, 1985).

37. The account of Japanese amphibious and ship and craft development comes from the following sources: Seukuni, Sections II and 12, "Landing Operations," previously cited; Administrative Division, Second Demobilization Bureau as compiled by Shizuo Fukui, ex-Lt. Cmdr. IJN, *Japanese Vessels at the End of the War*, issued 25 April 1947 for the American occupation authorities, copy in the author's possession; Hans Lengerer, Sumie Kobler-Edamatsu, and Tomoko Rehm-Takahara, "Special Fast Landing Ships of the Japanese Navy," in three parts in vol. 10 of Andrew Lambert, ed., *Warship* (Annapolis: U.S. Naval Institute, 1986); Ladd, *Assault from the Sea*, 65–76; War Department General Staff, Japanese Military Forces.

38. This analysis is based on data from the four editions (1939, 1941, 1942, and 1945) of James C. Fahey, comp., *The Ships and Aircraft of the U.S. Fleet* (reprinted, Annapolis: U.S. Naval Institute, 1976); Calvin W. Enders, "The Vinson Navy," PhD dissertation, Michigan State University, 1970; DNI, *Allied Landing Craft of World War II*; Lt. W. E. Royall USN, "Landing Operations and Equipment," 1939 study with data and photographs, Breckinridge Library, MCU; Proceedings, Conference Concerning Various Types of Landing Craft, Their Capabilities and Limitations, 1943, Breckinridge Library, MCU; Michael Vlahos and Dale K. Pace, "War Experience and Force Requirements," *Naval War College Review*, 41, Autumn 1988, 26–46; Thomas C. Hone, "The Navy, Industrial Recovery, and Mobilization Preparedness, 1933–1940," 1990, author's possession.

39. Clifford, *Amphibious Warfare Development*, 108–117; Maj. John W. Mountcastle, "From Bayou to Beachhead: The Marines and Mr. Higgins," *Military Review*, 70, March 1980, 20–29; Jerry E. Strahan, *Andrew Jackson Higgins and the Boats That Won World War II* (Baton Rouge: Louisiana State University Press, 1994).

40. Col. Victor J. Croizat, *Across the Reef: The Amphibious Tracked Vehicle at War* (London: Arms and Armour Press, 1989); Maj. Alfred D. Bailey, "Alligators, Buffaloes, and Bushmasters: The

History of the LVT through World War II," Occasional Paper, (Washington, DC: History and Museums Division, HQUSMC, 1986); George F. Hofmann, "The Marine Corps and J. Walter Christie: The Development of the Amphibian Tractor in the 1920s," paper presented to the Society for Military History, April 1992.

41. Archibald D. Turnbull and Clifford L. Lord, *History of United States Naval Aviation* (New Haven: Yale University Press, 1949); Lt. Col. Edward C. Johnson with Graham A. Cosmas, *Marine Corps Aviation: The Early Years 1912–1940* (Washington, DC: History and Museums Division, HQUSMC, 1977); Marine Corps Schools, *A Text on the Employment of Marine Corps Aviation* (Quantico, VA: Marine Corps Schools, 1935); Headquarters Marine Corps, *Marine Corps Aviation General* (Washington, DC: USGPO, 1940).

42. Arthur J. Marder, "'Winston is Back': Churchill at the Admiralty, 1939–1940," in Arthur J. Marder, *From the Dardanelles to Oran: Studies of the Royal Navy in War and Peace, 1915–1940* (London: Oxford University Press, 1974), 105–178. For a recent assessment of the Norwegian campaign, see H. P. Willmott, *The Great Crusade: A New Complete History of the Second World War* (New York: Free Press, 1989), 70–80.

43. Fergusson, *The Watery Maze*, 70–185.

44. Hayashi and Coox, *Kogun*, 29–46; Suekuni, Section 13, "Landing Operations"; Willmott, *Empires in Balance*, 130–397; Paul S. Dull, *The Imperial Japanese Navy (1941–1945)* (Annapolis: U.S. Naval Institute, 1978), 21–29.

45. Isley and Crowl, *The U.S. Marines and Amphibious War*, 58–67; Atwater, "United States Army and Navy Development of Joint Landing Operations," 118–165; Clifford, *Amphibious Warfare Development*, 144–159; Norman V. Cooper, *A Fighting General: Gen. Holland M. "Howlin' Mad" Smith* (Quantico, VA: Marine Corps Association, 1987), 57–86; Joint Landing Force Board, "Study of the Conduct of Training Landing Forces for Joint Amphibious Operations during World War II," May 1953, Command File World War II, Operational Archives, Naval Historical Center; Operations Division Correspondence, 1942–1945, Records of the War Department General Staff, RG 165, NA.

46. Richard B. Frank, *Guadalcanal* (New York: Random House, 1990); Jay Luvaas, "Buna, 19 November 1942—2 January 1943," in Lt. Col. Charles E. Heller and Brig. Gen. William A. Stofft, eds., *America's First Battles, 1776–1965* (Lawrence, KS: University of Kansas Press, 1986), 186–225.

47. Maj. Gen. George S. Patton Jr., to Maj. Gen. A. D. Surles, 6 November 1942 and diary entries, 8 November 1942 and 9 November 1942 in Martin Blumenson, ed., *The Patton Papers* (Boston: Houghton Mifflin, 1974), vol. 2, 101–102, 103–106, 108. The landings are described in detail in George F. Howe, *Northwest Africa: Seizing the Initiative in the West* in the official history series *The U.S. Army in World War II* (Washington, DC: Office of the Chief of Military History, Department of the Army, 1957).

NAZI PARTY RALLY IN NUREMBERG, 1936: AN EARLY EXPRESSION OF HITLER'S VISION FOR A GREATER GERMANY

CHAPTER THREE

GERHARD L. WEINBERG

"Adolf Hitler" and "Franklin D. Roosevelt"
from *Visions of Victory*

IN HIS GROUNDBREAKING *VISIONS OF VICTORY*, Gerhard L. Weinberg, professor emeritus of history at the University of North Carolina in Chapel Hill, explores the postwar futures that eight World War II leaders imagined for their presumably victorious nations. Though each man focused on the strategies and policies that would achieve that anticipated victory, the ways in which they conducted their wars were shaped by their beliefs, aspirations, and personalities. As Weinberg explains, the realities of warfare inevitably led each man's vision to evolve throughout the conflict.

While each person Weinberg covers is worthy of scrutiny, the two chosen for this excerpt offer perhaps the best illustrations of how a wartime leader's vision of the future can shape, if not determine, the fate of both his nation and the world. Hitler, who envisioned the extermination of the Jews and a thousand-year global Reich, led Germany to disaster. Roosevelt, conversely, helped lay the foundation for an Allied victory and a dramatically reshaped postwar world.

Adolf Hitler

Born into the family of an Austrian customs official in 1889, Adolf Hitler had moved to Munich as a young man and served in a Bavarian unit of the German army on the western front in World War I. After the war he was briefly assigned to give indoctrination lectures to soldiers and then to observe a small new political party in Munich, which he soon joined and came to dominate. A failed coup attempt in 1923 brought him a short time in jail but also his first national publicity. After his release, he devoted himself to building up the National Socialist German Workers Party. A temporary alliance with the extremist German Nationalist Party gave him the opportunity to bring his message to large numbers of Germans, and by the beginning of the 1930s his party was the largest in the country. A small group of individuals around the president of Germany, Paul von Hindenburg, persuaded the latter to appoint Hitler chancellor of Germany on January 30, 1933. Over the subsequent year and a half, Hitler succeeded in consolidating his hold on the country by ending all civil liberties, dissolving all other political parties, and establishing an effective police and terror apparatus.

In any review of Hitler's hopes and aims in World War II, two interrelated aspects of his thinking before he became chancellor of Germany must be noted. Already in the 1920s he was clear in his own mind that Germany deserved to conquer the globe and would be able to do so if only he were given the opportunity to lead it in the manner he thought appropriate—and his closest followers fully understood this concept.[1] In November 1930 he explained to the faculty and students of Erlangen University in a subsequently published speech that no people had more of a right to fight for and attain control of the globe (*Weltherrschaft*) than the Germans.[2]

Hitler was under no illusion that a goal that others might consider preposterous could be attained without a great deal of fighting. He recognized that this privately held and publicly proclaimed ambition would necessarily require a series of wars. This makes it more understandable why he explicitly asserted in the book he dictated in the summer of 1928 that each war Germany fought would merely provide the starting point for the next one in the series.[3] Dictated after his party's very poor showing in the May 1928 election, this text, which was designed

HITLER SAW THAT HIS VISION COULD NOT BE REALIZED WITHOUT GOING TO WAR.

to reinforce the very views on foreign policy that he thought had cost his party votes, offers support for the increasingly accepted view that Hitler was a man driven by ideology rather than simply an opportunist seeking power.

It had been within the framework of a series of wars that he had decided in early May 1938 to attack Czechoslovakia that fall. The purpose of conquering Czechoslovakia was to strengthen Germany's strategic position in central Europe and also provide, through the utilization of the portion of the country's population that was of German cultural background, the opportunity to raise additional divisions for the German army to employ in the next war. Such potential reinforcement was especially important to Hitler because the second war in the series was the one he believed likely to be the most difficult, namely, the war against the Western powers, Great Britain and France.[4] At the last moment Hitler changed his mind, called off that war, and settled at the Munich Conference for what others thought of as a German triumph but Hitler came to consider the worst mistake of his career. The "lesson" he derived from Munich was that no one was ever going to cheat him of war again, and in 1939 he so conducted German diplomacy and military preparations that there would be no possibility of a negotiated settlement.[5] Since the Polish government, unlike those of Hungary and Lithuania, had not been willing to subordinate itself to Germany as the latter moved to initiate its war against France and Great Britain, Poland would be crushed first, with the two Western powers either attacked thereafter or in the same conflict if they supported Poland.

The war started by Germany on September 1, 1939, was therefore seen as the next in a series; it was to pave the way for a war against the Soviet Union, thereafter a war against the United States, and another to follow, as explained below. For a number of reasons growing out of the actual course of hostilities, the wars against the Soviet Union and the United States were initiated by Germany before the preceding one in the expected series had been finished, and therefore Hitler's aims in this wider conflict are what must be examined.[6] But before that can be done, something has to be said about the war that was expected to be next in the series.

Because Britain had not surrendered after the fall of France to the German invasion of that country in May and June of 1940, Hitler was most anxious for Japan to enter the war. That country could provide a great surface navy, and although Hitler had planned such a navy and the Germans had begun construction on it, they had not yet had the time to complete it. When the Japanese explained to their German ally that they could not attack the British base at Singapore without going to war with the United States to protect the flank of their advance south, Hitler personally promised Japanese Foreign Minister Matsuoka Yosuke that in that case Germany would immediately join in war against the United States. He worried all during the negotiations between Japan and the United States in 1941 that an agreement might be reached between them, and he did what he could to encourage the Japanese to strike. Furthermore, to judge by recently published evidence, he had driven the German army forward toward Moscow in December 1941 in part for fear that the authorities in Tokyo might get cold feet at the last moment and not plunge into the wider war about which they had just asked Berlin and Rome for reassurance.[7]

The enthusiasm for Japan to enter the war, strike at the British base at Singapore, and assure Germany of the participation of a major navy on the side of the Axis did not, however, imply any great German fondness for that power. Relations between Berlin and Tokyo during their joint war against the Western powers were distant, and not only in miles. All the evidence points to the assumption that Hitler, just as he had been willing to make extensive concessions to the Soviet Union in 1939 to obtain its assistance until Germany was ready to move East, was willing to sign over to the Japanese whatever they wanted in order to get them into the war on his side until they in turn could be conquered in a subsequent war.[8] Since Germany was defeated by the Allies rather than victorious, that subsequent war did not take place; and it is to the aims of the one that Germany began on September 1, 1939, that attention must be devoted.

It seems appropriate to examine first the territorial dimensions of the empire Hitler intended to take over for Germany as a result of the conflict he had been so insistent on starting before dealing with the developments he expected to bring about within that empire. Inside Europe, the countries of Scandinavia—Denmark, Norway, Sweden, and Finland—were to be annexed to Germany. In the West, Luxembourg had already been incorporated into Germany in 1940. The small pieces of land Germany had lost to Belgium and the tiny jointly ruled area of Moresnet had also been annexed that year; the rest of Belgium was expected to follow. Alsace and Lorraine were placed under the direct administration of the adjacent portions of prewar Germany, and much of northern France and all of Burgundy were to be annexed later. The evidence suggests that some sort of dependent French puppet state, possibly with Brittany detached as a special region aligned with Germany, would make up the remainder, although a portion adjacent to Italy might be allocated to the latter.[9] Most of Switzerland would be annexed to Germany; the southeastern quarter was to be Italy's share.

The whole United Kingdom, including all of Ireland, was to be under German control. No decision had been reached as to whether this would be a puppet state of some sort, possibly under the Duke of Windsor, or incorporated directly as a Gau, a German province. The preparation and printing of a lengthy arrest list, a detailed administrative plan that included provisions for the deportation of all males between seventeen and forty-five, the designation of the police chief who would subsequently be nominated for Moscow, and the assumption that all Jews living in Great Britain would be killed point in the direction of a form of control that very few on the islands would have found acceptable.[10]

Spain and Portugal were expected to become subordinate satellite states, with Spain turning over bases on and off the coast of northwest Africa to German sovereignty. Portugal was expected to surrender to the Third Reich much of its African colonial empire as well as bases on the Azores in the Atlantic. The precise details of the future division of the French colonial empire in northwest Africa between Spain and Italy constituted a subject on which Hitler preferred not to commit himself during the war. In any case, Spain was to receive Gibraltar, which in Hitler's eyes was always seen as a steppingstone on the way to northwest Africa rather than as a base commanding the western entrance to the Mediterranean.[11]

In southeast Europe, there would be governments subordinate to Germany in Slovakia, Hungary, Romania, and Bulgaria. A piece of northern Yugoslavia was annexed to Germany during the war; other portions would be under German control, as would parts of Greece, with the bulk of the latter country, including the strategically important island of Crete, allotted to Italy. The future of Turkey had not been decided, but Hitler would probably have been willing for it to go to Italy, along with the Middle East. Austria had been annexed in 1938, and the main portions of the former Czechoslovakia were destined to be incorporated into Germany and populated exclusively by Germans as well.

Once Italy left the war in 1943, the aggrandizement that Hitler had earlier been willing to allow the Italy of Benito Mussolini was canceled. The Italian zones of occupation in France, Yugoslavia, and Greece were taken over by Germany, and the puppet state of Croatia, formerly under partial Italian control, came completely under German control. Furthermore, Italy was to lose not only Albania but also very large portions of prewar Italian territory. Hitler now intended to annex to Germany the southern Tyrol area that he had once been willing to see Italianized, along with very extensive additional parts of northeast Italy, including the port city of Trieste on the Adriatic.[12] In view of this enthusiasm for taking into the Third Reich purely Italian parts of Italy proper, it can be assumed that the Italian empire once scheduled for North Africa and the Middle East would now be added to Germany's spoils of war instead.

In eastern Europe, a victorious Germany would extend to the Ural Mountains, thus including all of prewar Poland, the Baltic States, and European Russia, including the Caucasus, whose oil was expected to fuel Germany's navy and air force in war with the United States. Looking east beyond the Urals into Asia, Hitler was, for the time being, willing to be content with a border that extended some two to three hundred kilometers into Central Asia, as he explained in late July 1941.[13] This relative modesty compared with the suggestion from the High Command of the German Armed Forces (OKW) that Germany take all of central Siberia to the Yenisei River may help explain Hitler's willingness to accept the Japanese proposal for a division at the seventieth degree longitude. This meant that for the time being Japan would seize more of central and eastern Siberia while Germany would control less of Siberia but take Afghanistan and the part of British India that is now Pakistan instead.[14] As for Germany's former colonial possessions in the central and southwest Pacific, which had been turned over to Japan, Australia, and New Zealand as mandates after World War I, these were—at least temporarily—to be sold to Japan on the basis of highly complex negotiations and exchanges of notes that do not, from the available evidence, appear to have been reviewed with Hitler.[15] What Hitler actually thought about these former German colonies is not discernible from the record; his thinking about colonies—as contrasted with land for German agricultural settlement—was concentrated entirely on Africa.

Hitler's view of the future of Africa saw that continent divided essentially into three parts; since he had not studied Latin, this was not an echo of Caesar's Gaul. The northern segment of the continent was to be Italy's. This would include Egypt, the Anglo-Egyptian Sudan, Kenya, French and British Somaliland, Tunisia, Algeria, and at least some portions of French Morocco, to use the names then current for the territories to be added to the prewar Italian colonies of Libya, Eritrea, and Italian Somaliland and the recently conquered Ethiopia. Some of French Morocco and Mauretania might be added to Spain's colonial empire. In any case, the German troops sent to fight in North Africa were to help maintain Mussolini's empire and status and to implement Germany's policy of killing all Jews, not establish a German colonial presence in this portion of the continent. Only naval and air bases on and off the coast of northwest Africa were to be under Germany's complete control, as already mentioned.

Germany's own colonial empire was expected to be a broad swath of land in central Africa from the South Atlantic to the Indian Ocean. It was expected to include the former German colonies in that part of Africa, that is, Togo, Cameroon, and German East Africa, as well as the French and British colonial possessions

The German troops sent to North Africa were to help maintain Mussolini's empire and status.

from Senegal to Uganda and Northern Rhodesia. This would entail the inclusion in Germany's empire of the British colonies of Gambia, the Gold Coast, Sierra Leone, and Nigeria; the French colonies of Dahomey, French Guinea, the Ivory Coast, and French Equatorial Africa; the Belgian Congo; and at least the northern parts of the Portuguese colonies Angola and Mozambique. It was expected that the land south of this huge German colonial empire would be controlled by a pro-Nazi Afrikaner government that would take over in the Union of South Africa as one result of Germany's victory over Britain and the British Dominions. That regime might be asked to return former German Southwest Africa, today's Namibia, in exchange for the British protectorates of Bechuanaland, Basutoland, and Swaziland, plus possibly southern portions of the Portuguese colonies.[16]

The thinking of Hitler about future German control of the Western Hemisphere was, from all evidence, far vaguer than his plans for Europe and parts of Asia and Africa. Always assuming that the United States would not give up its independence without a fight, he had asserted as early as the summer of 1928 that preparing for war with that country would be one of the main responsibilities of a National Socialist government of Germany.[17] While there would presumably have to be some agreement on a division of the area with Japan, as there had been on the division of Asia, Hitler remained vague on details. It is by no means clear how he saw the role of the large German settlements in Brazil, Argentina, and elsewhere in Latin America. There is no evidence to show whether he was aware of the interest of segments of the Nazi Party in Patagonia, the southernmost portion of Argentina.[18] At times he referred to the possibility of encouraging the descendants of German settlers in the Western Hemisphere to move back to European lands under German control,[19] but the whole issue of the disposition of the Americas once Germany obtained world domination remained cloudy in Hitler's thinking and comments.

Hitler assumed that during the course of then current hostilities Australia and New Zealand would come under Japanese control. Whether that was to be their final disposition will have to remain an open question. His only lengthy recorded discussion of New Zealand reveals a degree of ignorance that exceeds anything foolish he said and believed about the Soviet Union and the United States. He seriously argued that the people there lived in trees and had not learned to walk upright.[20] Had they ever come under German control, their fate would presumably not have been very nice.

If these were the dimensions of the vast empire Hitler expected to conquer in the war then underway as a first major installment of Germany's march to world conquest, how did he envision what was to happen inside that domain? How were its peoples to be structured and controlled; what sort of life—or death—could they expect? Perhaps the best way to try to picture the German-controlled areas of the post–World War II era as the German leader imagined it during the war would be to divide Hitler's plans into three separate but closely interrelated aspects. First, there are the intentions that were to affect all of Germany's empire regardless of the individual status of its parts. Second, it will be important to look at his plans for the lands that were already or were in the near future expected to be inhabited by Germans and what he considered Germanic peoples (whether the latter recognized their own Germanic character or not). Third, it will be necessary to examine plans for those conquered lands that were not settled or expected to be settled by Germans.

There was no doubt in Hitler's mind that the German empire would be ruled from its capital, "Germania," as Berlin was to be renamed. The changes intended, and in part already being implemented, for that city will be reviewed below, but that all people would look to Germania as the seat of absolute power—as they had once looked to Rome—was beyond question. Certainly two categories and per-

haps a third category of persons were to be exempted from this perspective because they would be killed wherever they were located. The German government
had in 1939 initiated a program for the systematic killing of all handicapped people as the German government from time to time defined that category. Although
begun inside the country, this process had been applied in German-occupied
Poland right away and, while the subject has not received the scholarly attention
it deserves, was clearly designed to be of universal application. Hitler's personal
authorization for this, the first bureaucratically organized mass murder program,
was signed by him in late October 1939; he had backdated it to September 1,
1939—a clear sign of his sense of the connection between the killing and the war.[21]
It was in this terrible step into uncharted territory that the Germans experimented with the social and mechanical procedures for defining, identifying, and
murdering large numbers of people, disposing of enormous numbers of corpses,
and recruiting individuals who would kill others from morning to lunchtime and
then all afternoon, six days a week, as their steady occupation.

Although inside Germany the killing of the handicapped was decentralized
after August 1941, the process continued until the German surrender of 1945, and
those involved in it attempted to maintain it afterward until physically halted by
the occupation authorities. By that time, among the tens of thousands murdered
were numerous German World War I veterans, and a start had been made on German World War II veterans as well. Germany's defeat saved not only innumerable
old people, handicapped individuals, and persons in mental institutions inside and
outside the prewar Third Reich but also tens of thousands of Germany's own seriously wounded veterans from death at the hands of their own government.

A second group targeted for total extermination was the Jewish population
not only of Germany and all of Europe but also of all other parts of the globe, what
Hitler, with his Eurocentric perspective, referred to as Jews living "among non-
European peoples."[22] Whether this had always been his intention is beyond our
knowledge, though reference to total extermination as a desired fate for the Jews
appears in his public speeches as early as April 1920.[23] When urged by German
medical professionals and so-called racial scientists to initiate the killing of the
handicapped in the 1930s, Hitler had explained that this could be done only when
the country was at war; as has just been shown, it was in this context that he authorized it. After deciding to go to war for certain in 1939, he mentioned the forthcoming destruction of Germany's Jews to the Czechoslovak minister for foreign

NAZI INDOCTRINATION OF GERMAN YOUTH WAS A CORNERSTONE OF HITLER'S VISION.

affairs on January 21, 1939, and soon after, on January 30, he predicted in a speech to the German parliament that in another war—on which he had already decided—all Jews in Europe would be killed.[24] In numerous public speeches in subsequent years, Hitler would refer to the prediction he had made in this speech, always misdating it to September 1, 1939, just as he had misdated his directive for the killing of the handicapped.[25] These shifts forward and backward to the same date—the one on which he started the war—surely provide some insight into the way he saw those killing programs, that is, as part and parcel of a conflict fought not so that he could see the Eiffel Tower but to bring about a total demographic and racial reordering of the globe.

In this field, as in that of killing the handicapped, the Germans were entering new territory, and it should therefore occasion no surprise that the actual implementation of the policy came in steps, some of them tentative, to see what the reaction would be. Systematic killing of Jews in newly occupied Yugoslav and Soviet territory began in the spring and summer of 1941. As there appeared to be no substantial opposition from the German military, but considerable assistance and support instead, by the end of July Hitler clearly believed it possible to expand the killing to all areas under German control or influence, as he explained in late July 1941 to the visiting war minister of Croatia.[26] The details of the way in which Hitler's aspiration was implemented do not belong in this context and are adequately described and analyzed elsewhere.[27] The critical point is that application of this policy, like the killing of the handicapped, was initiated on a large scale during hostilities.

Hitler's earlier support of the emigration of Jews from Germany to Palestine and elsewhere must be understood as a product of his genuine belief in the legend that Germany had lost World War I, not because of defeat at the front, but as the result of a stab in the back by Jews and others. Driving as many Jews as possible out of the country—after stealing most of their assets—before he started the first of his wars, therefore, was motivated by his concern to ensure a solid home front in that war, a solidity that would guarantee victory this time. As for the Jews in Palestine, they were to be killed along with all other Jews then living in the Middle East, as he assured the grand mufti of Jerusalem in November 1941.[28] Hitler did not feel confident that he could rely on the Italians to carry out such a program when they received the area, an assessment that was as correct as the expectation he had voiced earlier in his conversation with the Croatian minister of war, that the Hungarian government would be the last in German-controlled Europe to surrender its Jews for killing. The murder of the very large number of Jews then living in the areas of North Africa and the Middle East was a task of such importance to Hitler that he intended his own forces to carry it out. Naturally he did not explain to the grand mufti, who wanted a German declaration favoring Arab independence, that the area would instead be incorporated into Italy's colonial empire when the Axis powers had won the war.

The question of whether the total elimination by systematic killing of the Sinti and Roma, the Gypsies, was also intended by Hitler is still not entirely clear. The available evidence and the scholarship devoted to the subject show very extensive persecution, large-scale killing, and some indications of additional killing intended for later. At the same time, there are signs that some Roma who had not intermarried with others and also the descendants of "pure" Roma were expected to survive, if under special restrictions.[29] This author inclines to the interpretation that there was to be no room for Roma in the thousand-year Reich, but the evidence, especially on Hitler's role, views, and expectations, is not nearly as definite as it is for the other two categories, the handicapped and the Jews.

I f these were the groups that Hitler expected to remove from the face of the earth, what fate was in store for those whom he considered Germans or sufficiently Germanic to qualify for membership in the superior race that would inhabit the territories Germany already held or planned to annex? First, an issue has to be dealt with that concerns the population itself. The killing of all considered handicapped, whatever the changing definition of this category over time, has already been mentioned. In addition, male and female Germans who Germany's doctors and so-called racial experts believed might beget defective babies would be subjected to surgical sterilization. This had been provided for in the first piece of legislation in the field of family law that the new Nazi government had included in the mass of laws enacted by the cabinet on July 14, 1933, a date chosen for its symbolic significance.[30] By the end of the Nazi regime, some 400,000 Germans had been forcibly subjected to this procedure; it was assumed that a victorious Germany would continue this practice. An additional precaution, alongside extensive education in what was called "racial hygiene," was a series of legal provisions requiring permission to marry, permission that could be withheld for a number of reasons that the Nazi leadership imagined as having racial significance.[31] Although there appears to be no evidence on the subject, such procedures, both sterilization of certain categories and marriage permission for others, would no doubt have been applied to all those whom the Nazis were prepared to include in their definition of "Aryans" and hence to include among those favored by the regime.

If the measures that have been mentioned and that were already being applied to the German population before the war were designed to eliminate the imagined danger of racial degeneration—and hence might be called the negative side of racial policy—what about the positive side of racial policy? On this side of the ledger there was a whole series of special measures. Beginning in 1933, a variety of laws and other procedures were instituted to encourage the "right" people to marry and have lots of children. Simultaneously, the legal and social status of illegitimate children was to be improved. Women were pushed toward staying at home and away from higher education, from the professions, and, to the extent possible, from any gainful employment outside the home (other than on farms). During World War II, the regime was obliged to make temporary adjustments to several of these policies, but there was the clear intent to reverse these concessions to the needs of the moment. Furthermore, women, as Hitler personally emphasized, were under no circumstances to be paid equally for equal work.[32]

Additional steps were taken during the war to increase the number of Germans for the planned expansion of German settlement. The families of soldiers were provided with substantial support payments. To increase the number of Germans as well as decrease and weaken the peoples in occupied eastern Europe, the regime instituted a procedure called "hay action" (*Heuaktion*). This was a massive kidnapping operation in which thousands of young children who looked "Germanic" to the Nazis were stolen from their parents and entrusted as so-called orphans to German families for raising.[33] Special procedures were introduced to offset the losses in men that warfare entailed. For example, Germany established a form of marriage to a dead man (*Totenehe*). If a woman could demonstrate by letters, testimony, or other evidence—such as a pregnancy—that a soldier who had in the meantime lost his life at the front had intended to marry her, she would be retroactively legally married to him. She would then be eligible for a widow's pension, any baby would be legitimate, and her prospects for a later marriage and the bearing of additional children would be greatly enhanced.[34] More dramatic procedures were planned for the postwar era but were deferred because of the possible morale implications on the home front if enacted in wartime. Just as systematic mass killing had had to be postponed until it could be covered by hostilities, so a drastic revision of the whole concept of marriage had to await the end of the war because victory would further empower the regime and increase popular support for its policies.

By January 1944 Hitler realized that the losses of the German armed forces during the war would be so large that millions of women would become widows and, like the young unmarried, would be unable to find a husband after Germany's expected victory. At that time Hitler estimated that the number would be between three and four million; in fact the number substantially exceeded five million.[35] Having previously strongly endorsed the 1939 order of Heinrich Himmler calling on the men of the SS to have children inside and outside marriage, Hitler now laid out the reasons for the surviving men to have multiple marriages. The provision of various forms of state support for the women in these marriages would assure them that they could have children with firm legal status and without fear for their own and their children's financial future. Those children would be needed, if for no other reason, for the future wars in the series Germany would have to fight for world dominion. It must be noted that, just as the reason for his interest in the Germans inside Czechoslovakia was the additional number of army divisions their incorporation into Germany would make possible, the main reason for his con-

The discharged German soldier would be obliged to marry a woman willing to be a farm girl.

cern with the multiple-marriage concept was the number of army divisions that Germany would lose in the future without such a measure.[36] Since, as will be reviewed subsequently, there were to be no churches in the Germany of the future, religious objections to such a dramatic alteration in the concept of marriage would have no institutional support at a time when, as Hitler insisted, the Nazi Party, and especially its women's organization, preached the new approach to childbearing.

The style of life he planned for Germans in the future was also expected to lead to the raising of large families. Since very many, if not all, German males were expected to serve in the postwar army, their situation after service would, in effect, apply to a substantial proportion of couples. As Hitler explained in detail on July 27, 1941, the term of service for an army of one and a half to two million would be twelve years, with much attention during the last two years spent on preparing the man to be a farmer. Each man upon discharge would be given a farm for free since the land would have been stolen from those "inferior" Slavic people living on it previously. The fate of the displaced Slavic residents is reviewed below. The discharged German soldier would be obliged to marry a woman willing to be a farm girl and to raise a large family, whose members in turn would be provided with additional stolen land. Hitler was always strongly opposed to the system of dividing farms among children. That system, which had long obtained in much of Germany, would be abolished inside the country as well as for these settler families, with all given land in the East whose inhabitants had been killed or expelled. Hitler suggested further that in the territory of the Baltic States, Dutch, Norwegian, and possibly Swedish settlers would be included in this expansion of Germanic settlement.[37]

As for those who were to be displaced, three options were under review by Hitler and his associates. There were the two obvious options of either killing them or letting them starve to death, both much discussed and in part already practiced during the war. The other possibility, in this author's opinion, was to implement a policy related to the extensive horrendous experimentation initiated in the concentration camps in 1942 in an effort to discover some means of mass sterilization. Unlike the individual surgical sterilizations that had been carried out on a large scale in Germany since 1933, these were to be applied wholesale to masses of people without their immediately knowing what had been done to them.[38] In a world that would be without the handicapped, without Jews, and entirely or almost entirely without Roma, who was to be mass sterilized? Is it not most likely that this process, if ever developed, was designed for Slavic people whose work on farms and in factories, mines, and construction was still needed—but who would have no progeny?

While a larger percentage of Germans than in the past were to live on farms and in new villages in eastern Europe, what about those who continued to live in cities? They, too, would see major changes. One of Hitler's main interests in life had always been architecture. His focus, however, was not on a particular style or only on projects for specific buildings but rather on the role of architecture in the fashioning of a total urban environment that would be conducive to the political aims of his regime. Long before he became chancellor of Germany, he had begun to sketch designs for buildings of colossal size to be built in the world capital that Berlin was to become. Designed to overawe all those living in the city and those who would visit it, these buildings would give Berlin an entirely new form. As the most comprehensive study of the subject shows, once in power Hitler not only moved forward with these projects, utilizing a totally compliant Albert Speer to implement his plans, but decided to have a large number of German cities altered to fit his concept of broad avenues for mass demonstration marches, large central Nazi Party headquarters, and other common features.[39]

Two special features characterized all the plans for Germany's cities of the postwar era. In the first place, because Christianity was to disappear, no space was provided for churches in the plans for the German cities, towns, and rural settlements of the future. Second, the public structures that were to dominate the cities of the future Germany were to be constructed of stone so that even after the end of the thousand-year Reich their ruins would impress subsequent generations the way the ruins left behind by the Roman Empire still impressed observers more than a thousand years after its fall. Contracts for the enormous quantities of stone needed for the huge building program were still being worked on in Scandinavia in 1943. Both before and after that date, the concentration camp system was directly structured and located so that the inmates could help provide the stone and brick required for the buildings.[40] The people living in these cities would know from their surroundings that they were small, indistinguishable elements in a huge empire that from time to time would expect them to assemble either to participate directly in the ritual parades or to observe them cheering from the sidewalks.

A few cities were each expected to have a special feature. Berlin would have the world's largest structures and widest streets. Hamburg was to have the world's largest bridge, one deliberately constructed in such a fashion that its total surface would exceed that of the Golden Gate Bridge in San Francisco, then the world's longest suspension bridge.[41] Linz, the city in what had been Austria with which Hitler most identified personally, was to receive, in addition to other major structures, a museum that was to put all others on earth in the shade.[42] Partly as a product of systematic looting all over Europe, Linz would obtain a huge art collection—including Hitler's own—a great art library, a coin collection, and a collection of weapons. This project was actually initiated in 1938, and in February 1945 Hitler had the models for the future of the whole city of Linz brought to Berlin for him to study as the Red Army closed in on the German capital. The city of Königsberg (now Kaliningrad) in East Prussia was to have an art museum for the eastern portion of the German empire, but Hitler was considerably more explicit about his plans for the Norwegian city of Trondheim.

The North Sea port of Trondheim was to become one of Germany's most important cities and naval bases. This city was to have a German population of 300,000 and was to have the art museum for the northern part of Germany's empire, a museum that would hold only works by German masters.[43] As Germany's major naval base, designed to be the home port of the superbattleships and aircraft carriers of the postwar German navy, the city would be completely reconstructed by Albert Speer with the enthusiastic cooperation of the leadership of the German navy.[44] At the time, Trondheim was, of course, separated from Germany by the exits from the Baltic to the North Sea. This minor detail would be overcome by one portion of Hitler's plans for the transportation system of the future: a four-lane highway running from Klagenfurt in the south by bridges across the Little and Great Belts all the way to Trondheim. All other areas of German-controlled Europe would be similarly joined by highways, while the Danube and the Main and the Danube and the Oder would be joined by canals.[45]

The one German city that was to be downgraded in the future was Vienna. Unlike such cities as Moscow and Leningrad, it was not to be razed, but its importance, especially as a cultural center, was to be drastically reduced. Hitler did not want the old capital of the Habsburg Empire that he so hated to be in any position to challenge Germania for predominance in any segment of life. The very structures in Vienna that had so impressed him in his youth when he lived there now contributed to his concern about the possibility of a challenge to the primacy of Germania and inspired a strongly negative attitude toward the city.[46]

The cities of the enlarged postwar Germany would be connected not only by four-lane highways and the existing railway and airplane routes. In addition, Hitler was planning a network of super-railways running on tracks about double the width of the west and central European standard track (and hence substantially wider than the Russian railway track). Colossal engines would haul large and luxurious coaches for the members of the master race, while something like barracks on wheels would haul slave laborers from eastern Europe to whatever destinations seemed desirable. Hitler worked on the details of this project with Fritz Todt, his favorite construction engineer, and kept it secret from the German railway administration as well as Albert Speer.[47] In his enthusiasm for the wide-track railway, Hitler even anticipated extending this system of tracks to India and Vladivostok, the Soviet port on the Pacific![48] This final destination was not explained to the Japanese, to whom the Far East provinces of the Soviet Union were assigned in the territorial division previously mentioned.

What kind of education and what sort of life could the favored members of the master race expect in the cities and on the farms to which they had been sent? On the subject of education, Hitler's views were clear in outline if lacking in detail. There was to be a very strong emphasis on physical education, on the one hand, and an end to anything resembling a liberal arts education, on the other. How the latter was to be implemented in view of Hitler's very great interest in and emphasis on wide popular participation in and attendance at cultural activities like concerts and theater performances was left open.[49] The major emphasis in the school system was to be on the physical rather than the mental aspect of development of the young, thus preparing the young males for their role as warriors and the young females for their role as bearers of numerous children. It was presumably within the education system that Hitler expected to further his goal of convincing everyone of the advantages of a vegetarian diet and nonuse of tobacco.[50] Until after the war, he reserved his lectures on these subjects, endless and repetitive though they were, to his evening companions at headquarters.

An aspect of the life of Germans in the future on which he was consistently emphatic was the disappearance of religion, a subject already touched on in connection with his concepts of city planning. In order to avoid any danger of lowering morale during the war, he repeatedly deprecated overt steps in this direction while hostilities were in progress, and at times he felt obliged to restrain his more exuberant followers in this regard. But once the war had been won, all would change. Since he was certain that National Socialism and Christianity were incompatible, he was never seriously interested in the experiments of those who

THE SUBJUGATION OF FRANCE, THE HATED VICTOR OF WORLD WAR I, OBSESSED HITLER.

called themselves "Germanic Christians." Constituting a major faction within the German Protestant Church, these believers tried to remove the Jewish core from Christianity, make Jesus into a non-Jew, and create a form of Christianity that was compatible with the racial concepts of the new Germany.[51] Whatever temporary support this widely popular tendency within the German Protestant Church received from the Nazi government in 1933–34, there was no doubt in Hitler's mind as to the future. In July 1941 he explained to his immediate associates that there was no possible coexistence of National Socialism and the Christian churches; in May of the following year he explained this in some detail to the leaders of the Nazi Party and to its regional chiefs.[52] As he confided to Joseph Goebbels, when Christianity was terminated in postwar Germany, many of the bishops would be executed.[53]

As for the physically rather than mentally educated German citizens, they not only were to enjoy their superior status as compared with any surviving members of "inferior races" but also were to have great opportunities for personal and social advances (though these would always be restricted to males). Hitler was a strong opponent of hereditary status and class divisions and privileges. A vehement critic of both the old royal and princely houses as well as of nobility in general, he wanted a society in which any man of any background could aspire to any position or rank. His disdain for the old class divisions in Germany made him something of a social revolutionary in his own estimation and practice, though few historians have paid much attention to this aspect of National Socialism.[54] And, of course, all German families could expect to have cars—no longer a sign of class status—and could use them to roar across the thousands of kilometers of the thousand-year Reich's superhighways.

Hitler's views about the future of the German economy were dominated by his invariable insistence that politics dictated economics, not the other way around. Where and when private initiative and ownership operated in accordance with the political directives from the top, they could continue and, within limits set by the government, be rewarded. But whenever the political demands established by the leader were not met by the private sector, then there would be government-owned factories and other such controlled operations. The establishment in 1936 of the Four Year Plan, with its massive mining and industrial complexes, can be seen as something of a model for any field where private industry might be recalcitrant in the future. On the one hand, when in 1933 IG Farben had been willing, even enthusiastically willing, to pursue a vast expansion of synthetic fuel production as long as it was guaranteed a cost-plus price, the government had been happy to oblige. On the other hand, when the leaders of Germany's steel industry had declined to become involved in the processing of low-grade iron ore, which they disdainfully called "potting soil" (*Blumenerde*), they next read in the news-

Anyone could now be punished in a way the leader thought appropriate, without the affected person having any rights.

papers about the establishment of a government-owned organization that would do that work and was to begin with capital the steel industry itself was to provide. There was, in other words, not a doctrine of either a free market or government ownership; the regime would give direction and set goals that could be reached by any means, either private or public.[55]

It was Hitler's intent that the administrative apparatus of the Germany of the future would, under a genius like himself, undergo an important change—a change that would, ironically, be implemented by the actual victors of World War II. Hitler wanted the state of Prussia to be broken up into smaller units. As for taxes, these would be collected centrally and then allocated to the provincial governments. Since these would be controlled by the Nazi Party, there could be regional variations within the system, for the party would maintain a common ideological orientation.[56] Hitler was always skeptical about the ideological purity of judges and administrators and insisted that they be appointed or dismissed as he thought appropriate. He had arranged for the last session of the Reichstag, the German parliament, to enact in April 1942 a law that stripped all Germans of procedural protections of any type. Anyone could now be punished in any way the leader thought appropriate and without the affected person having any rights on the basis of prior service, position, or established procedure. The unanimous enthusiasm with which this stripping of all Germans of any and all rights had been greeted by the Reichstag reflects the careful way in which its members had been selected before Nazi Germany's last "election" in April 1938. A joke current in Germany at that time—that someone had stolen the results of the next election—provided a grim but accurate foretaste of what the future would bring to German voters.

Before we turn to the two other categories of territory controlled by Germany—those believed suitable for German settlement and those destined for permanent colonial status—it is important to recall that the "Germans" whose future land and lives have been reviewed would include numerous persons from states outside the borders of the Germany of 1939. As they grasped the concept that they were indeed Germans, or at least members of a branch of the so-called Aryan family of tribes, Norwegians and other peoples of Scandinavia, as well as the Dutch, at least the Flemish segment of Belgium's population, and the inhabitants of the already annexed Luxembourg, were expected to become full-fledged citizens of the Reich. Furthermore, in addition to the large program of kidnapping children believed to be Germanic-looking mentioned earlier, there were complicated procedures, some already implemented, some only in the planning stage, to separate out individuals and families thought to be suitable for Germanization who were living in the occupied territories of central and eastern Europe. The area where such a procedure was first developed and experimented with was Poland, the first country overrun by the German army, partially in 1939, with the rest occupied in 1941. It was in Poland that the Germans not only first exported their policy of killing the handicapped but also experimented with a variety of schemes for transforming conquered territory into settlement land for Germans.

Already in October 1939, Hitler had appointed Heinrich Himmler to the newly created position of Reich Commissioner for the Strengthening of Germandom, an office that was to play a central role in the settlement and resettlement of Germans, the displacement and murder of non-Germans, and the development of long-range plans for the transformation of conquered lands into areas of German living space.[57] Although Hitler and Himmler frequently conferred in the subsequent years of the war, the extent to which Hitler personally directed or approved all the measures that Himmler implemented in his new capacity cannot be stated with certainty; what is clear is that Poland provided the place for all manner of horrendous experiments in the establishment of the New Order, as the Nazis chose to call it. Germanization would mean massive expulsions, killings, resettlements, deportations, and other procedures. Hitler had every reason to feel confident that Himmler understood what was intended in general and would move in the desired direction as quickly and as ruthlessly as possible. If these measures involved temporary reversals, for example, the repeated shuffling of the same people back and forth, that was simply a part of great new experiments with few if any precedents. What worked could be applied subsequently farther east; what did not work could be changed in the next area to be affected. What mattered in Hitler's eyes were the direction and the sense of pressure; the details could safely be left to Himmler, who would, as his official record of daily activities shows, check with Hitler whenever a major appointment or new project needed approval.[58] What Hitler had specified at the beginning of the occupation of Poland was that the clergy and the intelligentsia were to be killed, that the standard of living of the remaining Poles was to be kept as low as possible while they were still alive, and that many would be made into slave laborers for the Germans.[59]

I f German policies in Poland and the first parts of pre-1941 Soviet territory seized by the Germans provide a useful glimpse into how Hitler envisioned the transformation of eastern Europe into an area of German settlement, what were his plans for those lands that he did not believe suitable for such settlement but intended to hold as colonial possessions instead? From the beginning of his interest in colonies, Hitler thought about the central segment of the African continent. He had entrusted German planning for that colonial empire to an old associate, Franz Ritter von Epp, a retired army general who had been an ardent supporter of the Nazi Party already in the 1920s, had been rewarded with an appointment as Hitler's representative in Bavaria, and was to be the new minister for the colonies.[60] In these colonies there would be German administrators and police, but Hitler did not envision massive German settlement, as he intended for eastern Europe. There was a serious argument between him and his military advisers on the subject of recruiting Africans to serve in a military role, as they had done in World War I, especially in the German force that had held out in German East Africa until the end of that war. Against the preference of his highest military advisers, Hitler opposed such recruitment and instead intended each German division to spend one year in colonial service on a rotating basis.[61] The colonies were expected to provide Germany with raw materials that were otherwise difficult, expensive, or impossible to obtain; their populations would be required to work for the benefit of their German masters. Under von Epp's direction, and with the enthusiastic support of many German ministries and agencies, especially the German navy, there was very extensive planning for the great colonial empire Germany

HURRYING TOWARD THE FUTURE, HITLER GREETED FOLLOWERS IN NUREMBERG IN 1936.

assumed it would control. Other than vast quantities of paper, only the proof coins for use in the colonies remain as silent testimony to great schemes that the peoples of a wide swath of Africa, unlike those of eastern Europe, were spared.[62]

A word needs to be said about those territories in western Europe that Germany either had conquered during the war or expected to occupy in the course of winning but that were neither lands destined for German settlement in the near future nor included in the anticipated African colonial empire. As mentioned, the Norwegians, Danes, and Dutch were assumed to be of kindred "Aryan" background and would be treated like Germans once they had come to understand their true racial nature. A similar future was accorded to the German-speaking Luxembourgers and Swiss as well as the Flemish population of Belgium. The French-speaking populations of Belgium and Switzerland as well as the bulk of the population of France itself could expect a harsh rule. The long-term fate of these people was not, however, clear.[63] The dreams possessed by such leaders of Vichy

France as Henri Pétain and Pierre Laval of a secure if subordinate place in a German New Order were most likely to remain mere dreams in view of Hitler's hatred for and distrust of the French.[64] The policy of collaboration was always a one-way street: the French could collaborate but there was no way that in German eyes they could remove the stain of being French.[65] As already mentioned, the Germans had made extensive detailed preparations for the administration of the occupied United Kingdom, but other than extreme harshness, these provide few clues as to the longer term future. Only two aspects are clear: all Jews on the two islands would be killed, and the enthusiastic admirers of Hitler's Germany in the IRA would see Ireland united into one territory under Nazi rule.[66]

It had been Hitler's initial expectation that the process of attaining world control as a result of a series of victorious wars would be completed by 1950.[67] Although he would not accept the reality that Germany was about to be totally defeated until late April 1945, he did begin to recognize by the fall of 1943 that total German victory was unlikely in the near future. In a lengthy discussion with Goebbels on October 26, 1943—after the defeat of Germany's summer offensive on the Eastern Front and the landing of the Western Allies on the Italian mainland—Hitler took the position that Germany might best make a deal with Stalin on the 1939 model, returning to its 1941 border in the East and then concentrating on crushing Great Britain. A renewed war against the Soviet Union to seize land in the East would have to come later, and it might have to be led by a successor, as he himself might by then be too old.[68] Two aspects of the discussion should be noted: Hitler held to his preference for a temporary peace in the East to facilitate the defeat of Britain in the face of the lengthy arguments of Goebbels for the opposite course, and any reference to the United States was completely absent from the record Goebbels made of this conversation. The first aspect confirms the reality of Hitler's obsessive hate for England—contrary to the often repeated fables on this topic.[69] The absence of any reference in this record to the fact that Germany was at war with the United States demonstrates once again the extent to which preposterous assumptions about that country remained a fixture of the thinking of the highest authorities in Germany during the war.[70]

It was in part this underestimation of the United States that led Hitler to throw his last major reserves into an offensive against American forces in December 1944—what the Germans called the Ardennes Offensive and others have come to call the Battle of the Bulge. Partly in anticipation of that operation, partly as a result of the attempted coup against him and his government by domestic opponents of the Nazi regime on July 20, 1944, Hitler focused his attention on domestic German affairs following Germany's assumed winning of the war. As he explained to Goebbels at the beginning of December, both the aristocracy and the traditional officer corps would vanish. New elections were to be held for the Reichstag, it being assumed that the law allowing only the Nazi Party to nominate candidates would continue to be in effect. This Reichstag would establish a new constitution that would provide for a dictatorially ruled republic but contain only a minimum of other provisions. The Nazi Party would be the main pillar of the regime. Stringent measures would be taken against the churches. On the other hand, Hitler reiterated his interest in the furthering of what he considered the appropriate cultural life of the German people. The enormous bribes already distributed to the highest military and political leadership of Germany were to be further increased, especially with the allocation of landed estates (though he failed to specify where at this stage of the war that land was to be located).[71]

By March 11, 1945, the development of the military situation led Hitler at least to mention the United States; perhaps the crossing of the Rhine by American troops at Remagen a few days earlier had reminded him of that allegedly inferior country's role in the war. He now hoped that a local defensive victory on the Eastern Front, preferably accompanied by massive Red Army casualties, would facilitate a separate peace agreement with Stalin. Hitler hoped that such an agreement would yet allow for a partition of Poland with the Soviet Union and leave Hungary and Croatia under German control. (At this time, Croatia was still under German occupation.) Germany would then defeat the Western powers of Britain and the United States. The administration of a deathblow (*Todesstoss*) to England would give the war its real meaning.[72] It was Hitler's thinking along these lines that contributed to his choice of Admiral Karl Dönitz as his successor—here was an enthusiastic Nazi among the highest military leaders, one whose whole career had been devoted to the crushing of Great Britain.[73]

The trajectory of Hitler's view of the future was as terrible as it was consistent. He began with an aspiration for world domination by the racially superior Germans and ended, in his last testament, with an admonition to the German people to adhere to his racial concepts. And he wanted them to do so under the leadership of a successor whose vision reached across the oceans to the wider world that might yet, some day, fall under German domination.

When in his last days Hitler pondered what had gone wrong, he thought that his major mistake was not the terrible war that he had been so insistent on unleashing but rather his drawing back from war in 1938; for it was then, he was certain in retrospect, that Germany's prospects had been at their best.[74] He thought of himself thwarted in his hopes, not by Churchill or Stalin or Roosevelt, but by Neville Chamberlain. In his speech to Germany's military leaders on August 22, 1939, he had explained that his only concern was that at the last moment some "pig-dog" (*Schweinehund*) might propose a compromise that would make it difficult for him to start a war this time, leading him to duplicate the failure of the year before. No one will ever know how a war started by Germany in 1938 would have developed; Hitler did get war in 1939, and it ended in Germany's utter defeat. Although Admiral Karl Dönitz believed after his arrest, trial, and imprisonment that he was in fact still the country's legal head of state, no other German is known to have shared this view.[75]

Notes

1. Rudolf Hess to Walter Hewel, 30 March 1927, National Archives, Nuremberg, document 3753-PS; English translation in Gerhard L. Weinberg, *Germany, Hitler, and World War II* (New York: Cambridge University Press, 1995), chap. 2. Very important on this issue, Jochen Thies, *Architekt der Weltherrschaft: Die "Endziele" Hitlers* (Düsseldorf: Droste, 1976), 32–33, 152. The English-language version is *Hitler's Plans for Global Domination: Nazi Architecture and Ultimate War Aims* (New York: Berghahn Books, 2013).

2. Constantin Goschler, ed., *Hitler: Reden Schriften, Anordnungen Februar 1925 bis Januar 1933*, vol. 4, pt. 1 (Munich: Saur, 1994), no. 28, p. 95.

3. Gerhard L. Weinberg, ed., *Hitlers Zweites Buch: Ein Dokument aus dem Jahr 1928* (Stuttgart: Deutsche Verlags-Anstalt, 1961), 77. In the English language edition, Gerhard L. Weinberg, ed., *Hitler's Second Book: The Unpublished Sequel to Mein Kampf* (New York: Enigma Books, 2003), translated by Krista Smith, the passage is on p. 47.

4. Gerhard L. Weinberg, *The Foreign Policy of Hitler's Germany: Starting World War II, 1937–1939* (Atlantic Highlands, NJ: Humanities Press, 1994), chaps. 10–11.

5. Ibid. chap. 14.

6. Weinberg, *Germany, Hitler, and World War II*, chaps. 12, 14–15.

7. Marianne Feuersenger, *Im Vorzimmer der Macht: Aufzeichnungen aus dem Wheremacht-führungsstab und Führerhauptquartier 1940–1945* (Munich: Herbig, 1999), 110.

8. Thies, *Architekt der Weltherrschaft*, 162–63, 174ff.

9. An excellent introduction in Norman Rich, *Hitler's War Aims: The Establishment of the New Order* (New York: Norton, 1974), chaps. 7–8.

10. A satisfactory and thorough study of the German plans for the invasion and occupation of the United Kingdom remains to be written. There is a useful summary in Basil Collier, *History of the*

Second World War: The Defense of the United Kingdom (London: Her Majesty's Stationery Office, 1957), chap. 11. A good early account is in Peter Fleming, *Operation Sealion* (New York: Simon & Schuster, 1957). A German survey of the purely military side is in Karl Klee, *Das Unternehmen "Seelöwe": Die geplante deutsche Landung in England 1940* (Göttingen: Musterschmidt, 1956). The Imperial War Museum in London has reprinted the arrest list and the administrative handbook for the planned German occupation.

11. See Norman J. W. Goda, *Tomorrow the World: Hitler, Northwest Africa, and the Path toward America* (College Station: Texas A&M Press, 1998).

12. A preliminary account is provided in Rich, *Hitler's War Aims*, 320–25. For a detailed, more recent study, see Lutz Klinkhammer, Zwischen Bündnis und Besatzung: Das nationalsozialistische Deutschland und die Republik von Salo, 1943–1945 (Tübingen: Mohr, 1993).

13. Werner Jochmann, ed., *Adolf Hitler Monologe im Führerhauptquartier 1941–1944* (Hamburg: Knaus, 1980), no. 10.

14. Johanna M. Meskill, *Hitler and Japan: The Hollow Alliance* (New York: Atherton, 1966), chap. 3.

15. Gerhard L. Weinberg, "Deutsch-japanische Verhandlungen über das Südseemandat 1937–1938," *Vierteljahrshefte für Zeitgeschichte* 4 (1956): 390–98; Johanna M. Menzel, ed., "Der geheime deutsch-japanische Notenaustausch zum Dreimächtepakt," *Vierteljahrshefte für Zeirgeschichte* 5 (1957): 184–93.

16. Klaus Hildebrand, *Vom Reich zum Weltreich: Hitler: NSDAP und koloniale Frage 1919–1945* (Munich: Fink, 1969); Gerhard L. Weinberg, "German Colonial Plans and Policies, 1938–1942", in *World in the Balance: Behind the Scenes of World War II* (Hanover, NH: University Press of New England, 1981), 96–136; Dietrich Eichholtz, ed., "Die Kriegszieldenkschrift des Kolonialpolitischen Amtes der NSDAP von 1940," *Zeitschrif für Geschichtswissenschaft* 22 (1974): 308–23.

17. Hitler's views of the United States before he became chancellor may be found in Weinberg, *Hitler's Second Book*, chap. 9. For a general survey, see Gerhard L. Weinberg, "Hitler's Image of the United States," in *World in the Balance*, 53–74.

18. Alton Frye, *Nazi Germany and the American Hemisphere, 1933–41* (New Haven, CT: Yale University Press, 1967), 93–94, 122–23. The authenticity of the incident is confirmed by the document published as no. 137 in *Documents on German Foreign Policy 1918–1945*, ser. C, vol. 4 (Washington, DC: USGPO, 1983).

19. Weinberg, *Hitler's Second Book*, chap. 9; Thies, *Architekt der Weltherrschaft*, 171.

20. Clemens Vollhans, ed., *Hitler Reden, Schriften, Anordnungen Februar 1925 bis Januar 1933*, vol. 1 (Munich: Saur, 1992), 127.

21. Henry Friedlander, *The Origins of Nazi Genocide: From Euthanasia to the Final Solution* (Chapel Hill: University of North Carolina Press, 1995).

22. *Documents on German Foreign Policy 1918–45*, ser. D, vol. 8 (Washington, DC: USGPO, 1951), no. 515. See also Zvi Elpeleg, *The Grand Mufti: Haj Amin Al-Husseini, Founder of the Palestinian National Movement* (London: Frank Cass, 1993).

23. Hitler's speech of 6 April 1920 in Eberhard Jäckel, ed., *Hitler: Sämtliche Aufzeichnungen 1905–1924* (Stuttgart: Deutsche Verlags-Anstalt, 1980), 120; cf. ibid., 28 June 1920, 152.

24. *Documents on German Foreign Policy 1918–45*, ser. D, vol. 4 (Washington, DC: USGPO, 1951), 193. The original German reads: "*Die Juden würden bei uns vernichtet.*"

25. On this deliberate misdating, see Gerhard L. Weinberg, "The Holocaust and World War II: A Dilemma in Teaching," in *Lessons and Legacies*, vol. 2, ed. Donald G. Schilling (Evanston, IL: Northwestern University Press, 1998), p. 27 and the detailed references for n. 3.

26. No complete text of the record of this meeting appears to have survived. The best text is in *Akten zur deutschen auswärtigen Politik 1918–1945*, ser. D, vol. 8, pt. 2, app. 3.

27. Excellent surveys include Raul Hilberg, *The Destruction of the European Jews*, 3rd ed. (New Haven, CT: Yale University Press, 2003), and Christopher Browning, *The Origins of the Final Solution: The Evolution of Nazi Jewish Policy, September 1939–March 1942* (Lincoln: University of Nebraska Press, 2004).

28. See. n. 22.

29. Guenter Lewy, *The Nazi Persecution of the Gypsies* (New York: Oxford University Press, 2000); Michael Zimmermann, *Rassenutopie und Genozid: Die nationalsozialistische "Lösung der Zigeunerfrage"* (Hamburg: Christians, 1996).

30. Karl Dietrich Bracher et al., *Die nationalsozialistische Machtergreifung*, 2nd. ed. (Cologne: Westdeutscher Verlag, 1962), 214; Dorothee Klinksiek, *Die Frau im NS-Staat* (Stuttgart: Deutsche Verlags-Anstalt, 1982), 72–74.

31. See Michael Burleigh and Wolfgang Wippermann, *The Racial State: Germany 1933–1945* (Cambridge: Cambridge University Press, 1991).

32. Elke Fröhlich, ed., *Die Tagebücher von Joseph Goebbels* (hereafter cited as *Goebbels Tagebücher*), pt. 2, vol. 12 (Munich: Saur, 1995), 14 May 1944, 289–90.

33. For a summary of this operation in Poland, see Richard C. Lukas, *Forgotten Holocaust: The Poles under German Occupation 1939–1944* (Lexington: University Press of Kentucky, 1986), 25–27. See also the same author's *Did the Children Cry? Hitler's War against Jewish and Polish Children, 1939–1945* (New York: Hypocrene, 1994). There appears to be no study of the kidnapping operation as a whole. The "Lebensborn" organization of the SS was involved in it.

34. There are scattered references to the subject of *"Totenehe,"* but there appears to be no systematic study. A summary is presented in Elizabeth D. Heineman, *What Difference Does a Husband Make? Women and Marital Status in Nazi and Postwar Germany* (Berkeley: University of California Press, 1999), 47–48.

35. For an excellent scholarly analysis, see Rüdiger Overmans, *Deutsche militärischen Verluste im Zweiten Weltkrieg* (Munich: Oldenbourg, 1999).

36. Oron J. Hale, ed., "Adolf Hitler and the Postwar German Birthrate," *Journal of Central European Affairs* 17 (1957): 166–73. This is a record by Martin Bormann of Hitler's comments on the subject.

37. Jochmann, *Adolf Hitler Monologe*, 49: see also ibid., 8–11 August 1941, 55; Ian Kershaw, *Hitler 1936–1945: Nemesis* (New York: Norton, 2000), 57; Bruno Wasser, *Himmlers Raumplanung im Osten: Der Generalplan Ost in Polen, 1940–1944* (Basel: Birkhäuser, 1933), 33–34.

38. Still useful is the pioneering work of Alexander Mitscherlich and Fred Mielke, *Das Diktat der Menschenverachtung* (Heidelberg: Lambert Schneider, 1947), 149ff.

39. The best study is Jost Dülffer et al., *Hitlers Städte: Baupolitik im Dritten Reich* (Cologne: Böhlau, 1978).

40. An excellent survey of the subject is presented in Paul Jaskot, *The Architecture of Oppression: The SS, Forced Labor and the Nazi Monumental Building Economy* (New York: Routledge, 2000).

41. This ridiculous aspect of the project is summarized in Thies, *Architekt der Weltherrschaft*, 80. The plans for the world's biggest stadium, hall, street, airport, etc., are discussed in ibid., 80–81, 90–91.

42. On the planned museum for Linz, see the popular but useful account in Charles de Jaeger, *The Linz File, Hitler's Plunder of Europe's Art* (Exeter, England: Webb & Bower, 1981); for more details, see Günther Haase, *Kunstraub und Kunstschutz: Eine Dokumentation* (Hildesheim: Georg Olms, 1991), 30–31, 62–66, and relevant documents. See also Anja Heuss, *Kunst- und Kulturraub: Eine vergleichende Studie zur Besatzungspolitik der Nationalsozialisten in Frankreich und*

der Sowjetunion (Heidelberg: C. Winter, 2000), 29–72.

43. Jochmann, *Adolf Hitler Monologe*, 19/20 February 1942, 285.

44. Thies, *Architekt der Weltherrschaft*, 131; Hans-Dietrich Loock, *Quisling, Rosenberg und Terboven: Zur Vorgeschichte und Geschichte der nationalsozialistischen Revolution in Norwegen* (Stuttgart: Deutsche Verlags-Anstalt, 1970), 457.

45. Jochmann, *Adolf Hitler Monologe*, 13 October 1941, 78; Thies, *Architekt der Weltherrschaft*, 77.

46. *Goebbels Tagebücher*, pt. 2, vol. 4, 31 May 1942, 406–7; ibid., 23 June 1942, 583–84. Hitler ordered that Vienna, including all its historic buildings, was to be totally demolished in 1945 to "spare" it occupation; see Susan Schwarz, "Operation Radetzky," *Austria Kultur* 3, no. 4 (1993): 1–2.

47. Anton Joachimsthaler, *Die Breitspurbahn Hitlers: Eine Dokumentation über die geplante 3-Meter Breitspureisenbahn der Jahre 1942–45* (Freiburg: Eisenbahn-Kurier Verlag, 1981), 42, 58–60, 80ff., 250–51, and maps on 295, 306, 307.

48. Ibid., 380 n. 441.

49. Jochmann, *Adolf Hitler Monologe*, 29 August 1942, 375; Thies, *Architekt der Weltherrschaft*, 179–80; *Goebbels Tagebücher*, pt. 2, vol. 5, 24 August 1942, 308. It was for this very reason that he wanted Bavaria to be continued as a unit, because Munich, unlike Vienna, would not become a rival for Berlin; ibid., vol. 4, 22 June 1942, 583–84.

50. *Goebbels Tagebücher*, pt. 2, vol. 4, 26 April 1942, 177.

51. See Doris L. Bergen, *Twisted Cross: The German Christian Movement in the Third Reich* (Chapel Hill: University of North Carolina Press, 1996).

52. Jochmann, *Adolf Hitler Monologe*, 11/12 July 1941, 40–41; 14 October 1941, 82–85; 13 December 1941, 150–51; *Goebbels Tagebücher*, pt. 2, vol. 4, 24 May 1942, 360, and 30 May 1942, 410. The effort to refute the assessment of Hitler as intending to find a way to do away with Christianity in Richard Steigmann-Gall, *The Holy Reich: Nazi Conceptions of Christianity, 1919–1945* (Cambridge: Cambridge University Press, 2003), is entirely unconvincing.

53. *Goebbels Tagebücher*, pt. 2, vol. 5, 20 August 1942, 359–60. In March 1944 Hitler explained that after the war he would get rid of the generals and the priests as Stalin had done and as he himself had done with the Jews. Ibid., vol. 11, 4 March 1944, 403–4.

54. Among the few, Rainer Zitelmann, *Hitler: Selbstverständnis eines Revolutionärs* (Hamburg: Berg, 1987); Ronald M. Smelser, *Robert Ley: Hitler's Labor Front Leader* (Oxford: Berg, 1988). Also helpful, David Schoenbaum, *Hitler's Social Revolution: Class and Status in Nazi Germany* (New York: Doubleday, 1966), and Shelley Baranowski, *Strength through Joy: Consumerism and Mass Tourism in the Third Reich* (Cambridge: Cambridge University Press, 2004).

55. Dieter Petzina, *Autarkiepolitik im Dritten Reich: Der nationalsozialistische Viernjahresplan* (Stuttgart: Deutsche Verlags-Anstalt, 1968).

56. *Goebbels Tagebücher*, pt. 1, vol. 4, 5 February 1941, 491–92.

57. Still very useful is Robert L. Koehl, *RKFDV: German Resettlement and Population Policy 1939–1945: A history of the Reich Commissioner for the Strengthening of Germandom* (Cambridge, MA: Harvard University Press, 1957). For a fine new survey, see Werner Röhr, ed., *Europa unterm Hakenkreuz: Analysen, Quellen, Register* (Berlin: Hüthig, 1996), 8:129–30; see also Norman Rich, *Hitler's War Aims: The Establishment of the New Order* (New York: Norton, 1974), 27–55.

58. See Peter Witte et al., eds., *Der Dienstkalender Heinrich Himmlers 1941/42* (Hamburg: Christians, 1999); Wasser, *Himmlers Raumplanung im Osten*, 47–52.

59. Lukas, *Forgotten Holocaust*; Gerhard Eisenblätter, "Grundlinien der Politik des Reiches

gegenüber dem Generalgouvernement, 1939–1945" (Ph.D. diss., University of Frankfurt, 1969); Czeslaw Madajczyk, *Die Okkupationspolitik Nazideutschlands in Polen 1939–1945* (Berlin: Akademie Verlag, 1987).

60. See n. 16.

61. Thies, *Architekt der Weltherrschaft*, 126.

62. The proof coins are in Koblenz, Bundesarchiv, R 2/30737.

63. Rich, *Establishment of the New Order*, chaps. 6–8.

64. Eberhard Jäckel, *Frankreich in Hitlers Europa* (Stuttgart: Deutsche Verlags-Anstalt, 1966).

65. Robert O. Paxton, *Vichy France: Old Guard and New Order 1940–1944* (New York: Columbia University Press, 1972).

66. See n. 10.

67. Thies, *Architekt der Weltherrschaft*, 152.

68. *Goebbels Tagebücher*, pt. 2, vol. 10, 27 October 1943, 184–85.

69. Gerhard L. Weinberg, "Hitler and England: 1933–1945: Pretense and Reality," in *Germany, Hitler and World War II*, chap. 6.

70. See n. 17.

71. *Goebbels Tagebücher*, pt. 2, vol. 14, 2 December 1944, 324–27. On the systematic bribery of German military leaders by Hitler, see Gerd R. Ueberschär and Winfried Vogel, *Dienen und Verdienen: Hitlers Geschenke an seine Eliten* (Frankfurt am Main: S. Fischer, 1999); Norman J. W. Goda, "Black Marks: Hitler's Bribery of His Senior Officers in World War II," *Journal of Modern History* 72 (2000): 413–52.

72. *Goebbels Tagebücher*, pt. 2, vol. 15, 485–86.

73. Gerhard L. Weinberg, "Der Überfall auf die Sowjetunion im Zusammmenhang mit Hitlers diplomatischen und militärischen Gesamtplanungen," in *"Unternehmen Barbarossa": Zum historischen Ort der deutsch-sowjetischen Beziehungen von 1933 bis Herbst 1941*, ed. Roland G. Foerster (Munich: Oldenbourg, 1993), 184; David Grier, "The Appointment of Admiral Karl Dönitz as Hider's Successor," in *The Impact of Nazism: New Perspectives on the Third Reich and Its Legacy*, ed. Alan E. Steinweis and Daniel E. Rogers (Lincoln: University of Nebraska Press, 2003), 182–98.

74. Weinberg, *Foreign Policy of Hitler's Germany*, 462–63.

75. Albert Speer, *Spandauer Tagebücher* (Frankfurt am Main: Ullstein, 1975), 20 January 1953, 335.

Franklin D. Roosevelt

Franklin Delano Roosevelt was born into a patrician New York family in 1882. He entered politics as a Democratic state senator in New York and became the second man in the Department of the Navy in the administration of President Woodrow Wilson. After his defeat as the vice-presidential candidate of the Democratic Party in the 1920 election, he suffered a serious polio attack. While still recovering from that, he was elected governor of New York, was reelected to a second term, and then won election to the presidency in 1932. In 1936 he was reelected. In 1940, in the world crisis resulting from Germany's great victories in western Europe, Roosevelt created something like a coalition government before being reelected to an unprecedented third term. It was during this third term that the country was thrust into the war by the decisions of Japan, Germany, and Italy.

Any review of President Roosevelt's vision of the world after World War II must begin with the way in which his experiences in and after World War I affected his perspective. There will be repeated references in this chapter to aspects of the "lessons" he believed he had learned, but certain points have to be made clear at the beginning. Unlike the majority of Americans, who had come to believe by the late 1920s and early 1930s that it had been a mistake for the United States to enter what was then called the Great War, Roosevelt had been in favor of an earlier entrance and did not share the opinion that the United States should have stayed out altogether. He had also thought it a mistake to grant Germany an armistice in 1918 rather than fight on until Germany surrendered unconditionally, but on this issue, like the preceding one, he had kept quiet at the time as a loyal member of the administration of President Wilson.

He had been a supporter of the 1919 peace settlement, including the League of Nations, and had suffered his only electoral defeat when running for vice president of the United States on the Democratic Party ticket in 1920. Not surprisingly, he believed that it was the American repudiation of the peace settlement and his political party that had contributed to the dangerous international situation that the country faced during his time as president.

ASSISTANT SECRETARY OF THE NAVY FRANKLIN D. ROOSEVELT IN WASHINGTON, D.C.

It had been Roosevelt's hope that the United States could keep out of another war with Germany by assisting the Allies to defeat the Axis powers. That meant initially helping France and Britain and, from the summer of 1941, assisting Britain and the Soviet Union. He hoped that by lengthy negotiations, to which he devoted an enormous amount of his own time, the Japanese could be delayed from entering the wider war on Germany's side long enough to allow the authorities in Tokyo to see that Germany was likely to lose rather than win and was therefore not worth joining. This tactic came within two weeks of working, but the Japanese took the plunge before they could assimilate the lesson of German defeats on the Eastern Front and in North Africa in the last days of November and the first days of December 1941. With Germany and Italy hastening to go to war with the United States, the only remaining possibility was to restrain Germany's European satellites from joining in. For half a year the president had the State Department try to persuade the governments of Hungary, Romania, and Bulgaria to withdraw their declarations of war. Only in June 1942 did Roosevelt give up and ask the Congress to declare war also on these countries, which absolutely insisted on fighting the United States.

In the face of the determination of Japan, Germany, Italy, and their associates to go to war with the United States, Roosevelt was equally determined that this would be the last time that they did so. If this was not the most important objective of the president for the postwar world, it was certainly very high on his agenda and would stay there as long as he lived. The demand for the unconditional surrender of the Axis powers was designed to make certain that Americans would not have to fight Germany a third time and the other countries a second time. The evidence is clear that this was Roosevelt's view from December 1941 on.[1] The United Nations Declaration of January 1, 1942, issued from Washington, called for a "complete victory."[2] Though the words "unconditional surrender" were not used in this document, that was certainly implied.[3] The public proclamation of the demand for unconditional surrender did not come until the Casablanca Conference of January 1943. It was openly announced with great emphasis at that time for both domestic political and international diplomatic reasons, but the policy itself was the president's from December 1941.[4]

In Roosevelt's view, the failure to make such a demand in 1918 had contributed to the German arguments that they had not really been defeated and/or that they had stopped fighting on the basis of promises that had not subsequently been kept by the Allies. No one was going to make that kind of argument in the future as a way to persuade the Germans or Japanese that they might do better the next time. Roosevelt resisted all arguments supporting a modification of the unconditional surrender policy in subsequent years.[5] In a speech on August 25, 1943, at a time when the Allies had recently won great victories, Roosevelt reassured the peoples of the Axis about their fate after unconditional surrender but insisted that surrender would have to come first.[6]

Like Churchill and Stalin, the experience of Germany starting another world war so soon after its prior defeat left Roosevelt impressed with the danger that in another twenty years the Germans might try again. Such a contingency had to be prevented if at all possible. In Roosevelt's view, that meant surrender first and substantial precautions thereafter. Because of the descent into barbarism that accompanied Germany's insistence on another world war, only when major changes had been made in the thinking of Germany's population under strict supervision could any German state or states rejoin the civilized world.

What kinds of precautions did the president believe should be taken after Germany had surrendered and was occupied by the Allies? Roosevelt was not in favor of the preparation of extensive and detailed plans for the future of Germany. As he put it in a memorandum of October 20, 1944, "I dislike making detailed plans for a country which we do not yet occupy."[7] If this was his view the day before American troops seized Aachen, the first large German city occupied by the Allies, it obviously characterized his approach earlier in the war. That does not mean, however, that there were not a number of issues on which the president had quite definite ideas.

From an early date Roosevelt believed that substantial territory should be taken from Germany, especially in the East. The unbelievably stupid German propaganda of the interwar years had convinced Roosevelt—as it had Churchill—that East Prussia was to be taken from Germany and turned over to Poland, most likely together with Danzig. He would subsequently agree to the partition of East Prussia between the Soviet Union and Poland, but there was under no circumstances to be a return to the situation of prior centuries, when Germans had lived on both sides of Poles. The Germans had declaimed endlessly against the Versailles concept of adjusting boundaries to people and had insisted instead on adjusting people to new boundaries. The Germans living in East Prussia would have this German principle applied to them. Furthermore, the president was agreeable to substantial but not precisely defined portions of German Silesia also to be assigned to Poland. While these territorial changes would come to play a role in Roosevelt's view of the future eastern as well as western borders of Poland, they must be seen in the first place as part of what he believed should happen to Germany.[8]

In addition to these external territorial changes, the president repeatedly considered a major internal territorial change: the "dismemberment" (as it was called) of Germany into several separate states. This project was discussed in Washington as well as London and Moscow during the war, only to be abandoned in theory but adopted in practice. No theoretical division of Germany into a set number of separate states was ever worked out and agreed upon either within any of the three major Allies or between them. Nevertheless, in practice the allocation of supreme power to the commanders-in-chief in each of the zones of occupation by the leaders of the Allies would actually make for a division of Germany, at first into four and then for decades into two states. Roosevelt had quite concrete ideas on this subject, and those ideas greatly influenced American strategy in the European war.

Starting in 1942, Roosevelt had stated his preference for American troops to enter Berlin first, and in 1943 he had sketched out a division of Germany into occupation zones, with the American and Soviet zones meeting at Berlin and with the British taking up a zone in the south.[9] Since the British government obtained Soviet agreement to a division that placed Berlin deep inside the Soviet zone, that foreclosed this particular issue, though Roosevelt argued from the fall of 1943 to the fall of 1944 for an American zone in the northwest rather than the south of Germany. This aspect of occupation zone allocation was eventually worked out in a new agreement,[10] and in early 1945 the president was persuaded to agree to an occupation zone for France carved out of the territory allotted to Britain and the United States.[11] This issue is reviewed below in connection with Roosevelt's views on the future of France. The critical point is that his belief in the need for an American presence in the center of Europe dovetailed with his and his advisers' insistence on an invasion of northwest Europe across the English Channel. This was both the best way to defeat Germany and simultaneously the way to ensure the United States a role in the postwar disposition of Germany and of European questions in general.

Whatever the external and internal boundaries of Germany after that country had surrendered unconditionally, there were also some internal changes that Roosevelt believed essential to keep the Germans from starting still another big conflict. Based in part on his views of Germany from the time before he became president, Roosevelt was certain that any future Germany should be allowed no aircraft, no uniforms, and no marching.[12] It was the president's concern about a future possible threat from Germany that led him, like Churchill, to agree with the concept of transforming Germany into a country rather like Holland and Denmark, with a high standard of living but no heavy industry, as suggested by his secretary of the treasury and friend, Henry Morgenthau. Most of the literature on

> ## Roosevelt was certain that any future Germany should be allowed no aircraft, no uniforms, and no marching.

this subject has ignored the relevant maps and echoes the propaganda of Joseph Goebbels instead. A change of Germany in the direction of what Churchill called "fat but impotent" required the retention by Germany—whether in one state or several—of much of its eastern agricultural area. If, however, the whole territory east of the line of the Oder and Western Neisse Rivers was taken away and the Germans living there were pushed into the remainder of the country, such a project would become impossible. One could either try to shift some five million Germans from industry into agriculture or drive some additional six or seven million Germans out of their homes in the agriculturally important parts of the country into the remainder, but one could not do both.

In view of Stalin's insistence on the Oder–Western Neisse border, the Morgenthau plan had to be dropped as too soft on the Germans, not too hard as some still imagine.[13] This abandonment of Roosevelt's preference in the face of Soviet insistence did not mean that his view of Germany had fundamentally changed. His approval of harsh directives for the occupation in March 1945 shows that his view that the Germans should not starve but must accept the fact that they had been defeated and were responsible for their own sad situation had in no way changed.[14] The whole country would be occupied by the Allies. The unconditional surrender at the end of the fighting, this time signed by leaders of the German armed forces instead of civilians who could be blamed for defeatism by military figures afterwards, would come after Roosevelt's death. By that date the German surrender was obviously very near—and American troops were in the center of Germany and of Europe, as Roosevelt had wanted.

As for planning for the peace in general terms, Roosevelt wanted that done in the Department of State. This was quite different from President Wilson's establishment of an agency outside the State Department for the planning of the World War I peace treaties. Other agencies would be called on for membership and expertise, but the lead was to be taken by State.[15] The president had his own view on many subjects, but he wanted any detailed planning to be coordinated by a staff housed in the State Department, an approach that his secretary of state certainly appreciated. And this insistence very much affected his handling of the second subject he believed had not been handled properly at the end of World War I. This time the State Department would be very much involved in the planning of a new international organization.

If one mistake that Roosevelt believed had been made at the end of World War I was the failure to insist on Germany's unconditional surrender, another very big one, very much in his mind because of its link to his own 1920 candidacy, had been the American repudiation of the peace settlement and of the Democratic Party—the party identified with the settlement. Central to this was the issue of an international organization to preserve the peace. Whatever the merits and defects of the League of Nations, Roosevelt had no doubt that the refusal of the United States to participate in it had greatly reduced the chances for its success. A new international structure would be needed after World War II. The president was determined that such a structure should grow out of the alliance against the Axis along lines developed at State, that it should include the major powers, that the American public should become accustomed to the idea of playing an active role in it, and that such an organization should be supplemented by a series of specialized organizations that would assist both in their fields of activity and also contribute to more harmonious international relations. Although all four of these aims were interrelated in his mind and in their implementation during the war, it may be simplest to review his ideas about them and the steps he took toward their implementation one at a time.

At Roosevelt's meeting with Churchill at the Atlantic Conference of August 1941, the president still thought it best to postpone public discussion of a future world organization, presumably for fear of arousing too much opposition at home. As urged by Undersecretary of State Sumner Welles, he already looked toward a new structure but one that would depend on power rather than formal structure.[16]

Very early, however, at least by the fall of 1941, Roosevelt was thinking of an association of the four major powers to keep the peace, a concept that he referred to as the "Four Policemen."[17] These four, in his thinking, always meant the United States, Great Britain, the Soviet Union, and China. The future role of France as seen by Roosevelt is discussed subsequently. A critical issue for examination in this context is the president's view of China as one of the great powers, a concept that others, especially Churchill, found extremely difficult to accept.

There is substantial contemporary evidence that the future role of China as a major world power was a significant element in Roosevelt's vision of the postwar world. Whether he really had long favored the end of extraterritorial rights by other powers in China, as he subsequently claimed,[18] he certainly saw to it that they ended during the war. Over and over Roosevelt insisted during the war that China, after a

RIGHT IN DER FUEHRER'S FACE

MESSAGE TO HITLER FROM CASABLANCA: "UNCONDITIONAL SURRENDER" period of reconstruction, would play a major role in world affairs. He saw it as a future counterweight against any effort of the Soviet Union to dominate Asia once Japan's military power had been crushed. With the agreement of his secretary of state, Cordell Hull, he wanted China associated with the October 1943 Moscow Declaration. He held to this general perspective in the face of strong objections from Churchill and the less vehement reluctance of Stalin.[19] As the new Australian minister to Washington reported on the president's views after their first meeting in his diary, "He said that he had numerous discussions with Winston about China and that he felt that Winston was 40 years behind the times on China and he continually referred to the Chinese as 'Chinks' and 'Chinamen' and he felt that this was very dangerous. He (Roosevelt) wanted to keep China as a friend because in 40 or 50 years' time China might easily become a very powerful military nation."[20] Eventually both Churchill and Stalin would yield to Roosevelt's insistence on China's formal status, but more to please him than out of a change in conviction.[21]

Although the president was by no means prepared to specify details about a new international organization, he was happy to have originated its name and was willing to share his concept of the four policemen with both the British and the Soviets early in the war.[22] At some time in the early years of the war, he thought that no others should have armaments at all, but he dropped that idea. He was at all times opposed to Churchill's idea of giving primacy to a European council of some sort, with American participation, since he was certain that any such arrangement would again arouse the strongest opposition in the United States. There could be regional associations within the United Nations—and he always thought of the countries of the Western Hemisphere in this connection—but the main structure was to be an international one. He was influenced in this not only by the experience of 1919–20 but also by the advice of Sumner Welles, whose judgment the president respected. Roosevelt was dubious about the concept of an international police force but believed that the new organization should be able to call on the major powers for armed forces.[23]

Before the Quebec Conference of August 1943, he agreed to Hull's draft of a four-power agreement for a new postwar organization, obtained Churchill's agreement there, and then sent Hull to Moscow in October to secure Stalin's assent.[24] It should, therefore, not be surprising that he returned to the subject when for the first time he met with Stalin at Teheran later that year.[25] The United States and the Soviet Union had both been absent from the League of Nations in its formative years; Roosevelt wanted to make sure that this situation would not be repeated. In addition to urging his allies to agree to a new organization for the postwar world, Roosevelt simultaneously worked to build up public support for such an organization in the United States, especially in the Senate, where the prior effort had failed. He was obviously extremely sensitive on this point and devoted considerable time and effort to make sure that there would be substantial support in the country long before the issue was likely to come to a vote.[26] In the winter of 1943–44, Roosevelt and his advisers discussed the details of the proposed new organization, and the president obviously was very much interested in these.[27] It may be seen as a sign of his concern that he paid such attention to the details in this case, for his usual practice was to leave the finer points to others. Micromanaging was simply not his style.

When the issue of the extent to which each of the major powers with a permanent seat on the Security Council of the planned new international organization could exercise a veto over procedural as well as substantive matters came to the fore at the Dumbarton Oaks Conference and in subsequent months, Roosevelt became very much involved in this question. Although he wanted the veto reserved for substantive matters so that no power could keep the Council from at least discussing an issue, both he and Churchill, independently of each other, came to a significant conclusion. Both were of the opinion that the participation of the Soviet Union, which insisted on a right to veto procedural matters, was so important to the successful establishment of the United Nations Organization and the future peace of the world that each went to the Yalta Conference prepared to yield on this issue to Stalin if necessary. At that meeting, Stalin came around to their point of view, and the fact that neither Churchill nor Roosevelt had been obliged to make the concession that they had been reluctantly willing to make has to be seen as one factor in their feeling of satisfaction with the conference in its immediate aftermath.[28]

One further issue that greatly interested Roosevelt was the need for speed in setting up the new organization. He was of the opinion that the delay in actually forming the League of Nations at the end of World War I had been a mistake.[29] The pressure from the United States for both the preliminary meeting at Dumbarton Oaks and the formal organizing meeting in San Francisco to take place already during hostilities has to be seen as a result of Roosevelt's concern over this question, even though he himself had died before the second of these meetings could be held. And again in order to avoid what he believed was a mistake made by President Wilson earlier, Roosevelt made sure that the American delegation to the San Francisco

Conference included prominent Republican as well as Democratic Party members.

The United Nations Organization was by no means the only postwar structure in which Roosevelt was personally interested. Knowing in general terms of the destruction and suffering caused by the war, Roosevelt played a major role in the establishment of an international relief organization, what came to be called UNRRA, the United Nations Relief and Rehabilitation Administration. He was concerned that the task was so large that private charities, however helpful, simply could not be expected to cope, as had largely been true in World War I. On the other hand, he did not want the United States to carry the whole burden alone, however large its share might have to be. With these views, he played a major part in setting up UNRRA, had the basic instrument for it signed in the White House, and insisted on appointing his former lieutenant governor and successor as governor of New York, Herbert Lehman, as its first head, even before the end of the latter's current term of office as governor.[30]

Roosevelt clearly wanted the UNO to be supported in its work by a whole range of international functional organizations that would contribute to better international relations and thus help maintain the peace. He was interested in what came to be the Food and Agriculture Organization.[31] He wanted the United States to join the International Labor Organization and to assist in the development of other forms of international economic cooperation.[32] Having become president after years of economic turmoil in the world and during the Great Depression, Roosevelt was especially interested in the establishment of mechanisms to prevent the kind of economic warfare and dislocation that had preceded the war and, in his opinion, had contributed to its coming. It was in this context that he asked his personal friend and secretary of the treasury, Henry Morgenthau, to take the lead in the international negotiations that led up to the Bretton Woods Conference that established the World Bank and the International Monetary Fund.[33] Afraid that the dollar would replace sterling as the key international currency, Churchill initially had doubts and delayed action but eventually came around.[34] The Soviets were doubtful about the whole project from the start and refused to join the new institutions in spite of every American effort to bring them around. Clearly Stalin wanted no part of what he appears to have thought was some capitalist plot—but his rejection of a project close to Roosevelt's heart also reduced the chances of any American consideration of assisting the Soviets with their massive task of postwar reconstruction.[35]

nother field in which Roosevelt had quite definite views about post-war developments was in the field of international air travel. He favored competition not only within each country but also on international routes. He foresaw a dramatic change in the way in which the great distances in the Pacific would effectively shrink with future technological developments in aviation. Furthermore, he was especially concerned about the rights of American airlines to fly international routes in essentially open competition with other countries' airlines. On this subject, there would be enormous difficulties with the British, who feared that American advances in air transport during the war would put their own airlines at a severe disadvantage in future competition. These issues were ironed out after very heated debates at the Civil Aviation Conference in Chicago in November–December 1944, with both Roosevelt and Churchill actively involved in instructing their delegations. The compromise reached satisfied neither side entirely, but the basis for the postwar world travel system was laid at that time—and the British would discover that they could operate quite successfully within it.[36]

If Roosevelt was as interested as the evidence shows in the creation of a new international organization to help maintain peace as well as a host of subsidiary international agencies to supplement and support the United Nations Organization, how did he see the future place of the United States in the world? There was never any doubt in his mind on two subjects. He was confident that the strength of the country would grow as it built up its military power during hostilities and that the possession of atomic weapons, whose development he personally pushed forcefully at all times, would provide an adequate margin of strength once the country demobilized its armed forces, as he correctly assumed that it would.[37] He was equally certain that the country should under no circumstances annex any territory. He had signed the legislation under which the Philippines were to become independent, and while the war initiated by Japan caused postponement of independence for the islands, Roosevelt never had the slightest doubt that independence would come as soon as possible. It was the other side of this fundamental anticolonialism attitude of Roosevelt that would lead to his greatest difference with Churchill. If the United States was not to acquire any colonies, what about the colonies of others?

Throughout the war Roosevelt held to the view that colonialism was, and most certainly should be, a fading institution. Whatever the problems and faults of the American occupation of the Philippines, he saw the process toward its independence not only as the correct path but also as something of a model for the colonial empires of others. Since the British Empire was by far the largest, and India constituted its most populous element, he had no doubt that India should move toward independence and hoped that the crisis there in the spring of 1942, as the Japanese army approached India's border, would help the process along. He quickly discovered that Churchill was essentially unyielding on this question. The president felt unable to take dramatic steps against an important ally in the military crisis, but he left no doubt about his views, sent a mission to India that shared his views, and generally let it be known that he favored independence for the subcontinent.[38]

During the war, he repeatedly expressed his preference for the end of colonialism, whether British, French, Dutch, or other.[39] He felt strongly that the French should not be allowed to return to French Indochina, made certain that the French army units being equipped and organized for participation in the Pacific War were under no circumstances to be utilized there, and only began to change his mind on the issue when the Japanese displaced French control by their coup in March 1945.[40] He tried unsuccessfully to persuade Churchill to return Hong Kong to Chinese sovereignty early, with the provision that it become a free port, a concession to which Chiang Kai-shek was prepared to agree.[41] His personal meetings with the sultan of Morocco and King Ibn Saud were designed to demonstrate his anticolonial sentiments—and were certainly so interpreted by the annoyed French and British governments.[42]

An aspect of the president's view of the future was especially obvious in regard to the fate of what must be seen as a residual legacy of the League of Nations: the mandate system established by the 1919 treaties and placed under a form of very tenuous League supervision. In direct opposition to Churchill's view that the mandates ought to be converted into colonies, Roosevelt wanted them converted into trusteeships on the road to independence under a far stricter and more intrusive supervision than had ever been assigned to the League.[43] It was in fact his belief that not only should the former mandates become trusteeships but also that the colonial possessions of all the European powers should become trusteeships, with the explicit or, if necessary, implied assumption that in ten or twenty years they too would attain independence.[44] In the final analysis, he could push through the trusteeship concept just for the former mandates, and American negotiators at San Francisco had to settle for a situation where other colonies would become trusteeships only by the voluntary agreement of the colonial masters—which was most unlikely. No one, however, had any doubts about the president's strongly held views on the subject.[45]

Ironically, the persons who found the president's attachment to the trusteeship principle most difficult to understand included many of his own political and military advisers. They believed, in many instances very strongly, that the United States should annex the former Japanese mandated islands in the Pacific that

> He repeatedly expressed his preference for the end of colonialism, whether British, French, Dutch, or other.

were being taken from the Japanese in bloody battles during the war and that they considered essential to the security of the United States in the postwar years. Repeatedly they argued that the United States should add them to its totally owned possessions in the Pacific, like Guam and Wake Island. In this position, they were invariably supported by the British, who no doubt saw this as a useful precedent for their own policy, since the islands in question had been Japanese mandates. Roosevelt would have none of this. He repeatedly and emphatically informed his military and political subordinates that the United States was not about to annex anything anywhere. The islands that had been mandates of Japan would become trusteeships of the United States, with the implicit, and occasionally explicit, assumption that they would move toward independence, though possibly with leased American bases of the sort that the United States had acquired in British territories like Bermuda under the destroyers-for-bases deal of 1940.[46] In the face of strong arguments to the contrary from inside the administration and from Great Britain, as president and commander-in-chief, Roosevelt would have his way on this issue, and Harry Truman, when president, would adhere to this clear policy preference of his predecessor.

Whatever the differences between Roosevelt and Churchill on the question of mandates and colonies, the president not only expected to continue working effectively with the British prime minister during the war but also looked forward to a future in which Great Britain would continue to play an important role. His initial assumption that Britain would be one of the "Four Policemen" and then a permanent member of the Security Council of the UNO reflects this perspective. He expected the continuation of some sort of partnership in the field of atomic weapons, though he did not anticipate a maintenance after hostilities had ended of the Combined Chiefs of Staff system. This structure, created right after the United States had been drawn into the war, had worked very effectively in the co-ordination of the military efforts of the two powers and would be needed in the immediate postwar transition period but hardly thereafter.[47] As the roles of the United States, the Soviet Union, and Great Britain shifted during the course of the war, Roosevelt was increasingly inclined to pressure Churchill on issues of military strategy where they differed—like the invasion of southern France. In Churchill's view, the president seemed to be more willing to accept the strategic views of Stalin, an inclination that the British prime minister very much resented. Nevertheless, Churchill and Roosevelt continued to respect one another highly and to work in basic harmony.

As already mentioned, Roosevelt expected that there would be a British zone of occupation in Germany, though he would have preferred both a different border with the Soviet zone and a geographical reversal of the expected American and British zones. Roosevelt's views on the future of Japan are examined below. Although the evidence is not absolutely clear, it is safe to assume that he expected that Great Britain would also have a zone of occupation there. While this had not been settled in detail by the time Roosevelt died, the assumption was that there would similarly be a British zone in Austria.

There were other British policies in addition to Britain's colonial stance on which the American president had his reservations. As an advocate of low tariffs in domestic American politics and as a strong believer in the need to open the channels of world trade symbolized by the reciprocal trade agreements pushed by Secretary of State Cordell Hull, Roosevelt very much wanted the British system of imperial preferences weakened, if not abolished. His strong interest in Article VII of the master Lend-Lease agreement with Britain, which dealt with this question, certainly needs to be understood in this context.[48] He was by no means reluctant to pressure the British government on an issue that had long been close to his heart.

Although Roosevelt, like Churchill, was a self-confident patrician, he had a very different outlook on the social issues of the past, present, and future. He thought of Britain as far too rigidly structured socially and expected, correctly as we now know, that there would be a shift to the left in Britain during and after the war. He therefore hoped to develop ties to the British Labour Party in a way not entirely unlike his development of ties to Churchill earlier in the war, when the latter was a member of the government of Neville Chamberlain.[49] It is not clear whether the Labour Party's anticolonial stand influenced him in this regard, but it would certainly not have diminished his hopes for that party's advance to power, which came three months after his death.

Roosevelt was one of the few American presidents—if not the only one—with a real interest in Canada and some familiarity with it. His ties to that country's wartime prime minister, William Lyon Mackenzie King, were unprecedented. If this relationship facilitated the growing independence of Canada in international affairs, the deployment of American troops to the Southwest Pacific at a time when Britain was unable to defend Australia and New Zealand from the advance of Japan led to an even more dramatic shift in the outlook of the governments and people in those portions of the British Commonwealth. Seeing themselves dependent on the United State in their time of greatest peril, the peoples and governments in Australia and New Zealand looked increasingly to the United States as the bulwark of their security in the future.[50] The role of Great Britain in the world of the future that Roosevelt anticipated—one in which Britain's colonies became independent states and its Commonwealth members asserted their full independence—was a subject that Roosevelt did not puzzle over any more than anyone else at the time.

What about the future relationship with the other major ally of the United States, the Soviet Union? Roosevelt had no doubt about the character of its government, having described it as a dictatorship as bad as any other on earth at one of the few public gatherings where his comments were met with booing.[51] He had kept the London government from recognizing the Soviet annexation of the Baltic States in 1940 and would do so again during the 1941–42 negotiations for an Anglo–Soviet alliance.[52] He knew that the Soviet Union was carrying the main burden of the fighting in the war and did what he could to help this critical ally. Just as he had Hull work on the subject of international trade and Lehman on relief, so he put his close assistant Harry Hopkins to work on getting supplies to the Soviet Union.[53] The concern that the Soviets might either collapse or make a separate peace with Germany haunted him as it did Churchill. How did Roosevelt see the future place of the Soviet Union? There was never any doubt in his mind that it would be one of the major powers and that its cooperation would be essential if new wars were to be avoided. His willingness to concede to the Soviets their preference on the question of the veto if that were necessary to obtain that country's participation in the United Nations has already been mentioned. On the questions of the future borders of the Soviet Union and its possible control of the countries of eastern and southeastern Europe, the president had a series of interrelated views that require review in detail.

As just stated, he opposed any recognition of the annexation of the Baltic States. He expected Finland to retain its independence, though he does not appear to have been especially interested in the details of any boundary settlement between it and its Soviet neighbor.[54] The Soviet border in Europe in which Roosevelt was, by contrast, very much interested was that with Poland. He was not willing to recognize the 1939 line that the Soviets had worked out with Nazi Germany, and he pressured the British not to accept it either. Since he favored the assignment of East Prussia to Poland as well as some additional gains for Poland

STALIN, ROOSEVELT, AND CHURCHILL AT THE TEHERAN CONFERENCE, 1943

at the expense of German Silesia, he thought that the Polish government in exile could agree to the cession of some lands in the East to the Soviet Union, but he argued strongly himself and through his ambassador in Moscow for the Poles to retain Lvov and also perhaps the nearby oil fields.[55] He adhered to this position right through the Yalta Conference but could not persuade Stalin to agree. Stalin did agree to a small shift of the 1939 line at the southern end of the border with Poland but insisted on retaining Lvov. Roosevelt's ideas about the Soviet borders with Romania and Czechoslovakia remain unknown, and the Soviet demand for the cession of Turkish territory did not come until after Roosevelt's death. In East Asia, the president believed that the southern half of Sakhalin as well as the Kurile Islands should be returned by Japan to the Soviet Union. Other aspects of his opinions on the future of East Asia will be reviewed in the context of his thoughts about China, Korea, and Japan.

Roosevelt had several other major concerns about the role of the Soviet Union in the future. He was sure that in the course of military operations the Red Army would overrun most of east and southeast Europe and that as a result, in addition to reannexing the Baltic States, the Soviets would in effect control the countries there in the postwar years. While he hoped that the Yalta Declaration on Liberated Countries might moderate Soviet policy in these states,[56] he was never very confident on this point. Repeatedly he pointed out to those he met that there was nothing except a third world war that could prevent Soviet domination of east and southeast Europe and that this was simply not a realistic prospect. There was only the hope that over a period of years the Soviet Union itself would evolve on more moderate lines and ease its hold on the countries taken under its control.[57] He objected to the allocation of percentages of British and Soviet control in southeast Europe of the sort Churchill and Stalin agreed to in October 1944, but he was not about to advocate war with the Soviet Union to prevent it from annexing or dominating what it intended to annex or dominate. Not a happy prospect, especially for the peoples directly affected. It was, after all, the insistence of Germany on starting a second world war that had ended the independence of these countries in the first place. Another world war was unlikely to increase their happiness—or that of anyone else.

The other two worries of the president concerned the implications of Soviet great-power strength for a Europe without a strong Germany, on the one hand, and an Asia without a strong Japan, on the other. It was the first of these concerns that led him to rethink policy toward France. This topic is reviewed in more detail below, but the president's reversal from envisaging a disarmed France to one that had substantial military forces was tied to his concern about a Soviet Union faced with an essentially defenseless western Europe.[58] Similarly, his constant advocacy of a major role for China in the postwar world was in part due to his belief that there needed to be a counterweight to the Soviet Union in East Asia after the defeat of Japan.[59]

While Roosevelt was worried about future domination of east and southeast Europe by the Soviet Union as well as its possible danger to other countries in Europe and Asia, he was always sensitive to the real security concerns of a Soviet Union that had suffered a terrible invasion. His insistence on completely disarming Germany, reducing its size, possibly dividing it into several states, and in any case keeping it under Allied occupation for some time was intended to assure the Soviets that there would not again be an invasion by Germany.[60] In accordance with his proclivity for raising all manner of ideas, he toyed with several rather extraordinary notions that he believed might make the Soviet Union less inclined toward aggression and more satisfied with its international position. On the one hand, he considered the idea of some sort of internationalization of the railway across Iran to the Persian Gulf to ensure the Soviets would have a route to the ocean; on the other, he toyed with the notion of providing the Soviet Union with free ports on the North Atlantic coast of Norway.[61] Nothing would come of either scheme, but they deserve mention as a reflection of the president's strong belief that a satisfied Soviet Union might be a more comfortable neighbor for the United States as well as for all others on the globe.

Roosevelt's vision of the future position of China has been mentioned repeatedly. He assumed that it would be a major power once it recovered from the ravages of war. The territory taken from China by Japan, especially Manchuria and Formosa (Taiwan), would certainly be restored to China, though there was an interest in the possibility of an American base on Formosa.[62] As mentioned in connection with his meeting with Chiang at Cairo, Roosevelt expected Chinese participation in the occupation of Japan. Since Korea was also to be released from Japanese control, Roosevelt anticipated that there would be a period of trusteeship for it on the road to independence and expected China and perhaps the Soviet Union to act as trustees.[63] His ending of extraterritorial rights in China and his failure to persuade Churchill to agree to an early return of Hong Kong have also been mentioned. As the position of the Chinese Nationalist regime of Chiang Kai-shek became obviously weaker during the war, a fact shown dramatically by the failure of Chinese re-sistance in the face of the Japanese 1944 Ichigo offensive, Roosevelt was increasingly interested in having the United States develop some sort of relationship with the Chinese Communists.[64] He thought that Chiang's position could be strengthened by getting the Soviet Union to focus its policy in China exclusively on the Nationalists, and his agreements with Stalin at Yalta on East Asia were designed with this end in mind.[65] It is too often forgotten that this policy succeeded temporarily but was sub-sequently aborted by the collapse of the Nationalist regime. Whatever happened in East Asia, Roosevelt until his death remained convinced of the long-term signifi-cance of China and expected that country to share in the occupation of Japan when victory over that country had been achieved.[66]

The other ally about whose future Roosevelt had definite ideas was France. His general opposition to colonialism and its implications for the French colonial empire have already been mentioned. While originally in favor of the future dis-armament of France, he changed his mind during the war, as also mentioned, and not only came to play a major role in pushing the rebuilding of the French army during the war but also expected that process to continue.[67] He was dubious, to put it mildly, about Charles de Gaulle, and the latter did everything possible and impossible to reinforce the president's negative opinion.[68] Although all his life Roosevelt was a Francophile who maintained his knowledge of the French lan-guage, he very much doubted that the country would regain its status as a great power "for at least 25 years."[69]

As a practical matter, Roosevelt expected the French to continue to control northwest Africa for years to come even if his meeting with the sultan of Morocco showed that this was hardly the arrangement he preferred.[70] As the situation in liberated France appeared to develop in a calm and democratic direction, without either the chaos or the de Gaulle dictatorship that had earlier worried him, Roose-velt slowly moved toward a more positive view of France's future role. In Novem-ber 1944 he agreed to France becoming a member of the European Advisory Council, which was supposed to play a role in developing plans for the continent's future.[71] That organization never played the major role anticipated for it, in part because of Roosevelt's own attitude toward it, but here was a symbol of a new fu-ture and a revived role for the country that had collapsed so dramatically in 1940.

Given the continued stability in France that winter, Roosevelt was persuaded by his advisers to reverse himself on other aspects of the future position of France.

At the Yalta Conference, he consented to a seat for France on the Allied Control Council for Germany and also agreed that France should receive occupation zones in Germany and Austria.[72] He did not live to see the French government utilize its position on the Control Council to counter American policy in postwar Germany, but it is unlikely that he would have been surprised. Since he had originally agreed to Henry Morgenthau's plan for the future of Germany, which included a substantial territorial cession to France in addition to the return of Alsace and Lorraine, it is safe to assume that he was willing to have the Saar area turned over to French control at the end of hostilities. Roosevelt's effort to meet with de Gaulle after the Yalta Conference, though rudely thwarted by the latter, certainly illustrates the president's slow turning from his negative view of the future of France to one more in keeping with his own earlier essentially pro-French attitude.

With these perspectives on the future of the United States' main allies, how did Roosevelt see the future of its wartime enemies other than Germany? His view of Italy has to be examined first. The prior discussion of Roosevelt's great emphasis on the concept of trusteeships for colonial territories heading for independence will serve to explain his views of the future of Italy's colonial empire. Abyssinia (Ethiopia) had already been liberated by British forces and would regain its recently lost independence. The rest of Italy's African colonies would become trusteeships on the road to independence as well. There appeared to be no need to discuss the return of Albania to independence and the transfer of the Dodecanese Islands to Greece. What about Italy itself?

Like Churchill, Roosevelt would almost certainly have been willing to exempt Italy from the demand for unconditional surrender had not the British cabinet insisted on its inclusion when the demand was to be announced publicly at the Casablanca Conference. After Italy did surrender, Roosevelt generally sided with the more liberal elements in the internal bickering among Italian politicians. As for the future of the monarchy, it was his view that the Italian people should decide that themselves.[73] It is certainly likely that domestic American politics influenced the president in his generally more generous attitude toward Italy than Germany, but then the Italians had arranged to dump Mussolini and the Fascist Party, unlike the Germans, whose enthusiasm for Hitler appeared to be undiminished. As the United States took steps to assist with the rehabilitation and reconstruction of the country, it was likely that Italy would turn more toward the United States than Britain in the future, but it is by no means clear that this was

PRESIDENT FRANKLIN D. ROOSEVELT AT THE WHITE HOUSE, 1936

Roosevelt's intention.[74] Roosevelt never could quite figure out why Italy had declared war on the United States, any more than he could understand why Hungary, Romania, and Bulgaria had, and he never was as concerned about the
possibility of its launching a war of aggression again in the future as he was about
the potential for Germany or Japan to start another such war.

There was never any doubt in Roosevelt's mind that Japan would have to surrender unconditionally and be occupied by Allied forces, with China as the fourth
power instead of France. The plans for the invasion of Japan were submitted to his
successor, not to him, but there is no reason whatever to believe that Roosevelt's

decisions in the final stages of the Pacific War would have been any different than those actually made by President Truman. The way to defeat Japan was essentially similar to the road to final victory over Germany.

There were, however, critical differences in Roosevelt's view of Japan and his view of Germany. He registered the dissimilarity between a German public that had turned to National Socialism and had become increasingly enthusiastic about it, on the one hand, and the series of coups, assassinations, and provoked incidents by which the militarists had shot their way to power in Japan, on the other. While Japan, therefore, was to lose all imperial acquisitions gained since its war with China in 1894–95, there is not the slightest indication that Roosevelt ever contemplated for Japan the sorts of territorial amputations imposed on Germany—at least to a considerable extent with his approval. Similarly, Roosevelt at no time considered dividing Japan into several separate states, a strategy very much part of his thinking about the future of Germany. The literature that attributes all manner of racist sentiments to American leadership in World War II has conveniently and consistently ignored the fundamentally positive view of Japan held by Roosevelt and his advisers as compared with their perspective on Germany.[75]

Since American strategy looked to the defeat of Germany before that of Japan, the president was less involved in the planning for Japan's future, but the basic outlines were laid down while Roosevelt was still alive. Surrender, as already mentioned, would be followed by four-power occupation, on the assumption that the Soviet Union would have joined in the war against Japan. The possibility of retaining the emperor, at least in the transition period, was also contemplated.[76] Although the evidence on this is not clear, Roosevelt may well have thought of this very controversial issue the way he did about the monarchy in Italy: let the people decide what they wanted once unconditional surrender had taken place. Although it is difficult to establish the extent to which the president was consulted about the planning for the occupation of Japan, that planning clearly reflected Roosevelt's broad differentiation between Germany and Japan: a raw deal for the former and a new deal for the latter.[77] The long-term implication clearly was an essentially territorially undiminished Japan back on the track toward a democratic structure at home, accompanied by a moderate foreign policy in a peaceful world and by good relations with the United States. The president's assumption appears to have been that such a development was likely to occur substantially sooner with an essentially intact Japan in East Asia than with a greatly reduced Germany in Europe.

Roosevelt did not have many specific views on other countries. As indicated, he expected East Prussia and extensive additional former German territory to be turned over to a revived Poland, and he tried hard but unsuccessfully to retain Lvov for Poland. He assumed that the other countries occupied by Germany during the war would regain their independence and that Austria would also be reestablished with its prewar borders. Because of the obvious internal problems of Yugoslavia, Roosevelt at one point thought that it might have to be divided into a Serb state and a Croat state and that a plebiscite might be necessary, but this was no certain project.[78] In any case, he assumed that there would be substantial American troops in Europe for only a few years after the end of hostilities, a prediction that only Stalin's authorization for North Korea to invade South Korea in 1950 would invalidate.

A very specific issue of a different sort on which Roosevelt had a firm opinion was the issue of punishing war criminals. In October 1941, before American participation in the war, the president had, like Churchill, denounced the German practice of shooting hostages.[79] In August 1942 he stated publicly that war criminals would have to stand trial in the very countries that they now occupied, and he made it clear later that year that he wanted public warnings of forthcoming trials to discourage war crimes but that there would be no mass shootings or reprisals.[80] Under these circumstances, it is easy to understand his approval of the inclusion of a statement on war crimes and trials to be issued by the meeting of Allied foreign ministers in Moscow in October 1943.[81] In spite of the views of some of his advisers, like Secretary of State Cordell Hull, and the clear preference of the British government for summary executions, Roosevelt always held to the concept that those accused of war crimes, however awful, ought to be tried by tribunals of some sort.[82] His successor held to this view and, in agreement with the French and Soviets, imposed it on the reluctant British.

One might say that Roosevelt's opinion on the future of Palestine was in a way another side to his insistence that war criminals be brought before tribunals. He had followed the process of German persecution of Jews on the basis of extensive reports and had been the only chief of state to withdraw the American ambassador from Berlin on the occasion of the pogrom of November 1938. Convinced by the Zionist leader Chaim Weitzmann at their meeting in February 1940 that, for Jews, there was no real alternative to Palestine as a place of their own, he came to accept the project called the Philby Plan. From Roosevelt's point of view, this project for an independent Arab confederation under Ibn Saud and the creation in it of a Jewish state in Palestine had a double advantage. It would provide a place for the survivors of the mass killing of Jews by the Germans and for any others who wanted to move to Palestine as a place to settle, and at the same time it was consistent with his belief that the mandates carved out of the Ottoman Empire should move toward independence.[83]

The project Roosevelt favored was impossible to implement because Ibn Saud rejected it when the president urged it on him at their meeting after the Yalta Conference. His basic conviction that there should be a "free and democratic Jewish commonwealth in Palestine," as he expressed it, remained unshaken.[84] In view of the large number of Jews then living in Arab countries, the president appears to have thought of a population exchange of some sort. As he explained his position to Undersecretary of State Edward Stettinius after his reelection to a fourth term in November 1944, "Palestine should be for the Jews, and no Arab should be in it."[85] Ironically, this was almost the very day on which Churchill abandoned his support for the creation of a Jewish state in a partitioned Palestine upon the assassination of Lord Moyne, but Roosevelt did not know that. He had wanted a provision for religious freedom included in the initial proclamation of the United Nations in December 1941.[86] In one way or another, the subject was always with him.

Another area in which the president had a continuing interest was Central and South America. He had taken a whole series of steps to try to alter the relationship of the United States with countries that frequently, and with good reason, viewed the United States with apprehension and distrust. There is no reason to doubt that what had come to be called the "Good Neighbor Policy" was to be continued. The fact that almost all the countries of the Western Hemisphere had joined the United States and Canada in the war could be seen as in part the fruit of that policy—and Roosevelt himself certainly thought so. There were, and were likely to continue to be, major troubles in the relationship of the United States with Argentina, but that was more of an exception than is often realized. The largest country, Brazil, had sent soldiers to fight alongside those of the United States in Italy, and the most immediate neighbor, Mexico, had contributed a small air force contingent to the war effort. All signs pointed to a continuation of a degree of hemispheric solidarity, occasionally troubled but nevertheless solid in any crisis brought about by an outside challenge.[87]

What about the domestic situation in the United States? Roosevelt had no doubt that the United States would dramatically reduce the huge military forces it had mobilized and would reconvert its industrial capacity to peacetime production. With the memory of the problems of World War I veterans, culminating in the bonus march on Washington and its dispersal by troops, there had been much discussion of what to do to assist the vastly greater number of veterans returning from service this time. The GI Bill of Rights, as it was called, with its provisions for college support, housing loans, and special unemployment benefits, had been enacted by the Congress and signed by the president on June 22, 1944. An increasing number of the prewar New Deal measures, like Social Security, a legal minimum wage, deposit insurance, and the Tennessee Valley Authority, had become an accepted part of the American scene, but there was not likely to be anything new added to the earlier measures. The election of November 1942 had brought massive Republican gains in both houses of Congress, and the president was fully aware of the fact that a coalition of Republicans and Southern Democrats controlled the Congress and would easily block any new initiatives in the field of social legislation. There is no reason to believe that in these circumstances he gave much consideration to the possibility of further social reform.

In the prewar years, the appearance of Washington itself had changed as the result of the construction of large buildings to accommodate the personnel of new and enlarged government agencies. During the war, there had been not only additions to these buildings but also the construction of the huge Pentagon on the other side of the Potomac River. Roosevelt, however, had no grandiose plans for future buildings in the nation's capital. For himself, he wanted only a plain stone placed in front of the new National Archives building. It is there today, at the corner of Pennsylvania Avenue and Seventh Street, but almost no passersby notice it. Instead, tourists visit the elaborate memorial to Roosevelt constructed decades later in the park along the Potomac, where children now pat the bronze statue of his dog next to the statue of the president.

Roosevelt was aware of the changes in the situation of African Americans in the country and had tried to push the armed services, especially the navy, to employ more African Americans in positions other than mess stewards, but without much success. In part under the influence of his wife, Eleanor, he had reversed the position of the last Democratic administration in Washington, that of Woodrow Wilson, on the general attitude of the federal government on matters of race from support for segregation to opposition to it. The practical implementation of such opposition, however, was only in its initial stages. Here was an enormous problem on which much of the effort would fall to Roosevelt's successors.[88]

During the war, Roosevelt, like other World War II leaders, had led a country in which women were increasingly drawn into industrial work to satisfy the demands of the war effort as more and more men entered military service.[89] To a substantial extent, though not as much as in the Soviet Union and Britain, women were also enrolled in the military itself. In one important way, however, Roosevelt's position on the role of women in society differed from that of the other leaders at the time, and this difference predated the time when the United States was drawn into the conflict. Having relied on her already when he was governor of New York, Roosevelt had appointed Frances Perkins to be secretary of labor in 1933 as the first woman ever to serve in the Cabinet. She remained in that position until shortly after Roosevelt's death in April 1945.[90] Here was a clearly visible symbol of a new perspective on an issue on which Roosevelt was ahead of many of his contemporaries.

By temperament and inclination, Roosevelt was an optimistic person. He was always certain that victory would come to the Allies. And if any one or more of the Allies dropped out voluntarily or were forced out because of Axis victories, then the United States would defeat Germany, Japan, and their associates by itself. He was determined that this time the United States would play an active and constructive role in the postwar world, and he had worked hard during the conflict to prepare the American people for that eventuality. Military strategy had been oriented toward achieving a complete victory and placing American forces in a strong position to influence the settlement to come. That settlement would prevent the recurrence of the catastrophes Roosevelt had seen twice in his lifetime so that the peoples of the earth could live in peace as they moved toward the better and freer life that he confidently expected.

Notes

1. Raymond G. O'Connor, *Diplomacy for Victory: FDR and Unconditional Surrender* (New York: Norton, 1971), though rather dated, remains the best work on the subject.

2. The text is in *A Decade of American Foreign Policy: Basic Documents, 1941–1949* (Washington, DC: USGPO, 1950), 2.

3. Albrecht Tyrell, *Grossbritannien und die Deutschlandplanung der Alliierten* (Frankfurt am Main: Metzner, 1987), 49.

4. Gerhard L. Weinberg, *A World at Arms: A Global History of World War II* (Cambridge: Cambridge University Press, 1994), 438–39. There is evidence that Churchill—and most likely Roosevelt—would have been willing to exempt Italy from the demand, but at the insistence of the British Cabinet, Italy was included.

5. *Dokumente zur Deutschlandpolitik*, ser. 1, vol. 4, 311–12.

6. Ibid., 499–500.

7. U.S. Department of State, *Foreign Relations of the United States* [hereafter *FRUS*, with appropriate year, topic]: *The Conferences at Malta and Yalta 1945* (Washington, DC: USGPO), 158. Roosevelt did, however, want the subject studied carefully, as he told Hull in March 1943 (Cordell Hull, *The Memoirs of Cordell Hull* [New York: Macmillan, 1948], 1248).

8. *Dokumente zur Deutschlandpolitik*, ser. 1, vol. 3, pt. 1, 201–2, vol. 3, pt. 2, 1137–38, 1195–96; Anita J. Prazmovska, *Britain and Poland, 1939–1943: The Betrayed Ally* (Cambridge: Cambridge University Press, 1995), 178.

9. Roosevelt's map is in Maurice Matloff, *United States Army in World War II: Strategic Planning for Coalition Warfare, 1943-1944* (Washington DC: USGPO, 1959), facing p. 341; cf. Earl F. Ziemke, *The U.S. Army in the Occupation of Germany 1944–1946* (Washington, DC: USGPO, 1975), 115–22; Tyrell, *Grossbritannien und die Deutschlandplanung der Alliierten*, 239–41, 481; John Q. Barrett, ed., *Robert H. Jackson, That Man: An Insider's Portrait of Franklin D. Roosevelt* (New York: Oxford University Press, 2003), 107; Daniel J. Nelson, *Wartime Origins of the Berlin Dilemma* (Tuscaloosa: University of Alabama Press, 1978): William M. Franklin, "Zonal Boundaries and Access to Berlin," *World Politics* 16 (1963): 1–31; John L. Harper, *American Visions of Europe: Franklin D. Roosevelt, George F. Kennan, and Dean G. Acheson* (Cambridge: Cambridge University Press, 1994), 91–93; *Dokumente zur Deutschlandpolitik*, ser. 1, vol. 4, 509–10.

10. Gretchen M. Skidmore, "The American Occupation of the Bremen Enclave, 1945–1947" (master's thesis, University of North Carolina at Chapel Hill, 1989).

11. The discussion of a French zone of occupation and seat on the Control Council for Germany can best be followed in *FRUS: The Conferences at Malta and Yalta 1945.*

12. John M. Blum, *From the Morgenthau Diaries: Years of War 1941-1945* (Boston: Houghton Mifflin, 1967), 352. For Roosevelt's earlier views, see Michaela Hoenicke, "Franklin D. Roosevelt's View of Germany: Formative Experiences for a Future President" (master's thesis, University of North Carolina at Chapel Hill, 1989).

13. The full text and map are printed in Henry Morgenthau Jr., *Germany Is Our Problem* (New York: Harper, 1945). The account in Tyrell (pp. 281–85) is helpful, but he ignores the differences between the maps there and on pp. 322–24. The main differences between Morgenthau's proposed map and the Oder-Neisse line were that Morgenthau left all of Pomerania, much of Brandenburg, and a part of Silesia to Germany while placing additional territory in the West under French administration. See also Bernd Greiner, *Die Morgenthau-Legende: Die Geschichte eines umstrittenen Plans* (Hamburg: Hamburger Edition, 1995).

14. Blum, *Morgenthau Diaries*, 348–49, 412–14; Tyrell, *Grossbritannien und die Deutschlandplanung der Alliierten*, 286–91.

15. Robert A. Divine, *Second Chance: The Triumph of Internationalism in America during World War II* (New York: Atheneum, 1971), 49; Robert C. Hilderbrand, *Dumbarton Oaks: The Origins of the United Nations and the Search for Postwar Security* (Chapel Hill: University of North Carolina Press, 1990), 6–7, 12–13; Harley A. Notter, *Postwar Foreign Policy Preparations 1939-1945* (Washington, DC: USGPO, 1949).

16. *FRUS 1941*, 1:364–67; Theodore A. Wilson, *The First Summit: Roosevelt and Churchill at Placentia Bay 1941*, rev. ed. (Lawrence: University Press of Kansas, 1991).

17. Hilderbrand, *Dumbarton Oaks*, 15–16.

18. Edgar Snow, "Fragments from F.D.R.," *Monthly Review* 8 (1957): 317. The remark was made on 24 February 1942.

19. Averell W. Harriman and Elie Abel, *Special Envoy to Churchill and Stalin, 1941–1946* (New York: Random House, 1975), 236; Hilderbrand, *Dumbarton Oaks*, 58–60; T. G. Fraser, "Roosevelt and the Making of America's East Asia Policy, 41–1945." in *Conflict and Amity in East Asia: Essays in Honor of Ian Nish*, ed. T. G. Fraser and Peter Lowe (London: Macmillan, 1992), 98–107; *FRUS 1942: China*, 185–87; Xiaoyuan Liu, *A Partnership for Disorder: China, the United States, and Their Policies for the Postwar Disposition of the Japanese Empire, 1941–1945* (Cambridge: Cambridge University Press, 1996), 21–23, 117. For Chinese records on this issue, see the references in Liu, p. 250.

20. Eggleston diary, entry for 14 November 1944, quoted in William Roger Louis, *Imperialism at Bay: The United States and the Decolonization of the British Empire, 1941–1945* (Oxford: Clarendon Press, 1977), 424. Roosevelt had made similar comments on 31 August 1943; see Liu, *A Partnership for Disorder*, 117.

21. Hilderbrand, *Dumbarton Oaks*, 60–61.

22. Divine, *Second Chance*, 48–49, 61–62; Ann Lane and Howard Temperley, eds., *The Rise and Fall of the Grand Alliance, 1941–1945* (Basingstoke: Macmillan, 1995), 4–5, 10–11. Note that originally Roosevelt had wanted a meeting of all four leaders—including Stalin and Chiang—in Khartoum in November 1942 rather than a meeting of just himself and Churchill at Casablanca in January 1943. See Geoffrey Ward, *Closest Companion* (Boston: Houghton Mifflin, 1995), 27 November 1942, 187; cf. ibid., 207.

23. Divine, *Second Chance*, 114; Ward, *Closest Companion*, 287; Harper, *American Visions of Europe*, 96–97, 107–8; *Dokumente zur Deutschlandpolitik*, ser. 1, vol. 4., 189–90.

24. Divine, *Second Chance*, 136–137; Tyrell, *Grossbritannien und die Deutschlandplanung der Alliierten*, 120–22.

25. Charles E. Bohlen, *Witness to History 1929–69* (New York: Norton, 1973), 144–45; Divine, *Second Chance*, 158.

26. Divine, *Second Chance*, 86, 93–97; Hilderbrand, *Dumbarton Oaks*, 27, 57–58, 64–65; See also Robert Dallek, "Allied Leadership in the Second World War: Roosevelt," *Survey* 21 (1975): 1–10. On Roosevelt's interest in the location of the UNO headquarters, see Hilderbrand, *Dumbarton Oaks*, 106.

27. Divine, *Second Chance*, 184–85, 206–08; Hilderbrand, *Dumbarton Oaks*, 34–37, 131, 180, 198–202. On Roosevelt's thinking about a permanent Council seat in the future for Brazil and for a Muslim state, see Hilderbrand, *Dumbarton Oaks*, 124–27.

28. Hilderbrand, *Dumbarton Oaks*, 226–28, 251–52.

29. Blum, *Morgenthau Diaries*, 16 February 1944, 373.

30. Divine, *Second Chance*, 117–18, 156–57; James Herbert George Jr., "United States Postwar Relief Planning: The First Phase, 1940–1943" (Ph.D. diss., University of Wisconsin, 1970).

31. Divine, *Second Chance*, 116–17.

32. Frances Perkins, *The Roosevelt I Knew* (New York: Viking, 1947), 339–41, 346; *FRUS 1944*, 2:14ff., 5:16–17.

33. Blum, *Morgenthau Diaries*, chap. 5; *FRUS 1944*, 2:107ff.; Alfred E. Eckes, *A Search for Solvency: Bretton Woods and the International Monetary System, 1941–1971* (Austin: University of Texas

Press, 1975), 56–57, 81–82, 110–17. For the conference itself in July 1944, see Eckes, *A Search for Solvency*, chap. 6.

34. *FRUS 1944*, 2:110–11.

35. Eckes, *A Search for Solvency*, 104–05, 113–14, 205–8. Earlier Roosevelt had been inclined to assist the Soviet Union with its reconstruction problems; see *FRUS 1944*, 4:1046–48.

36. Alan P. Dobson, "The Other Air Battle: The American Pursuit of Post-War Civil Aviation Rights," *The Historical Journal* 28 (1985): 429–39; Louis, *Imperialism at Bay*, 268; *FRUS: The Conferences at Cairo and Teheran*, 177–79; Beatrice Bishop Berle and Travis Beal Jacobs, eds., *Navigating the Rapids 1918–1971: From the Papers of Adolf A. Berle* (New York: Harcourt Brace Jovanovich, 1973), 483–84, 487, 496.

37. On Roosevelt and the development of the A–bomb, there is a forthcoming book by Lawrence Suid. For Roosevelt's critical decision of 9 October 41 on moving forward, see Robin Edwards, *The Big Three: Churchill, Roosevelt and Stalin in Peace and War* (New York: Norton, 1991), 397; Harper, *American Visions of Europe*, 108–12.

38. Richard J. Aldrich, *Intelligence and the War against Japan: Britain, America and the Politics of Secret Service* (Cambridge: Cambridge University Press, 2000), chap. 8; *FRUS 1942*, 1:599ff.; Warren F. Kimball, ed., *Churchill and Roosevelt: The Complete Correspondence* (Princeton, NJ: Princeton University Press, 1984), 1:400–04.

39. Wilson, *First Summit*, 108–10; Bohlen, *Witness to History*, 140; Aldrich, *Intelligence and the War against Japan*, 122ff.; Douglas Brinkley and David R. Facey-Crowther, eds., *The Atlantic Charter* (New York: St. Martin's, 1994), chap. 4; Louis, *Imperialism at Bay*, 9, 29–30, chap. 5; Ward, *Closest Companion*, 187, 197–200; John J. Sbrega, "Determination versus Drift: The Anglo–American Debate over the Trusteeship Issue, 1941–1945," *Pacific Historical Review* 55 (1986): 266, 275–77; Snow, "Fragments from FOR," 318–21. The article by Robert B. Looper, "Roosevelt and the British Empire," *Occidente* 12, no. 4 (1956): 348–63; no. 5 (1956): 424–36, is still very helpful.

40. Elliott Roosevelt, ed., *F.D.R.: His Personal Letters 1928–1945* (New York: Duell, Sloan and Pearce, 1950), 2:1489; Bohlen, *Witness to History*, 140; Aldrich, *Intelligence and the War against Japan*, 205ff.; Louis, *Imperialism at Bay*, 28, 41–42, 356–57, 436–38; *FRUS 1944*, 5:1206; John A. L. Sullivan, "The United States, the East Indies and World War II: The American Efforts to Modify the Colonial Status Quo" (Ph.D. diss., University of Massachusetts, 1969). On the subject of the Portuguese portions of Timor, Roosevelt held back in November 1944 because of the Allied need for bases in the Azores.

41. *FRUS: The Conferences at Cairo and Teheran*, 307–08, 887–88; Kit-cheng Chan, "The United States and the Question of Hong Kong, 1941–1945," *Journal of the Hong Kong Branch of the Royal Asiatic Society* (1979): 9–10, 16.

42. Berle and Jacobs, *Navigating the Rapids 1918–1971*, 476, 5 March 1945, quotes Roosevelt: "so far as he could see Churchill was running things on an 1890 set of ideas." See also Ward, *Closest Companion*, 207.

43. Louis, *Imperialism at Bay*, 95–96.

44. Ibid., chap. 22, 44–45, 147–58, 484–87; Lane and Temperley, *Rise and Fall of the Grand Alliance*, 10; *FRUS 1942*, 3:378–81; *FRUS 1942: China*, 185–87.

45. Sbrega, "Determination versus Drift," 275–77.

46. Hilderbrand, *Dumbarton Oaks*, 170–76; Louis, *Imperialism at Bay*, 262, 266–68, 211–12, 426; Jack Stokes Ballard, "Postwar American Plans for the Japanese Mandated Islands," *Rocky Mountain Social Science Journal* 3 (1966): 109–16.

47. Harper, *American Visions of Europe*, 108–12; *Dokumente zur Deutschlandpolitik*, ser. 1, vol. 4, 520.

48. Lane and Temperley, *Rise and Fall of the Grand Alliance*, 56–57; *FRUS 1942*, 1:525–37.

49. David Reynolds, "Roosevelt, the British Left, and the Appointment of John G. Winant as United States Ambassador to Britain in 1941," *International History Review* 4 (1982): 393–413. so. P. G. A. Orders, in his book, *Britain, Australia, New Zealand and the Challenge of the United States, 1939-1946: A Study in International History* (New York: Palgrave Macmillan, 2003), argues that there was no substantial turn of the two dominions toward the United States; but whatever the frictions and distrust, the governments tended to turn toward Washington in crises during and after the war.

51. The incident occurred in February 1940; see Robert E. Sherwood, *Roosevelt and Hopkins: An Intimate History* (New York: Harper, 1948), 138.

52. *Dokumente zur Deutschlandpolitik*, ser. 1, vol. 3, pt. 1, 201–2, 205–9.

53. Sherwood, *Roosevelt and Hopkins*, chap. 18. See also George C. Herring Jr., *Aid to Russia: Strategy, Diplomacy, the Origins of the Cold War* (New York: Columbia University Press, 1973), 11–137 passim.

54. R. Michael Berry, *American Foreign Policy and the Finnish Exception* (Helsinki: Societas Historica Finlandiae, 1987).

55. Harriman and Abel, *Special Envoy to Churchill and Stalin*, 369–70, 373, chaps. 14–15; Bohlen, *Witness to History*, 187, 189–92; Alexander Contrast, *The Back Room: My Life with Khrushchev and Stalin* (New York: Vantage Press, 1991), 139; *FRUS 1944*, 3:1282.

56. The text is in *FRUS: The Conferences at Malta and Yalta 1945*, 977–78.

57. Geoffrey Warner, "From Teheran to Yalta: Reflections on F.D.R.'s Foreign Policy," *International Affairs* 48 (1967): 530–36; Harper, *American Visions of Europe*, 88–89, 121; Snow, "Fragments from F.D.R.," 398–402; *FRUS 1944*, 2: 112–15, 117ff.; *Dokumente zur Deutschlandpolitik*, ser. 1, vol. 4, 509–10.

58. Lane and Temperley, *Rise and Fall of the Grand Alliance*, 12–14; Harper, *American Visions of Europe*, 90–91; *Dokumente zur Deutschlandpolitik*, ser. 1, vol. 4, 189–90.

59. Lane and Temperley, *Rise and Fall of the Grand Alliance*, 12–15.

60. *Dokumente zur Deutschlandpolitik*, ser. 1, vol. 2, 178.

61. Harriman and Abel, *Special Envoy to Churchill and Stalin*, 227–8; Olav Riste, "Free Ports in North Norway: A Contribution to the Study of F.D.R.'s Policy towards the USSR," *Journal of Contemporary History* 5, no. 4 (1970): 77–95.

62. Liu, *A Partnership for Disorder*, 74.

63. Ibid., 100–01, 240–45, 258–59; Snow, "Fragments from F.D.R.," 308; *FRUS 1942: China*, 185–87.

64. Liu, *A Partnership for Disorder*, 203, 208–9, 233; Snow, "Fragments from F.D.R.," 395–98; Fraser, "Roosevelt and the Making of America's East Asia Policy," 98–107.

65. Liu, *A Partnership for Disorder*, 243.

66. Ibid., 212–13.

67. Marcel Vigneras, *United States Army in World War II: Rearming the French* (Washington, DC: USGPO, 1957); *FRUS: The Conferences at Cairo and Teheran*, 5; *Dokumente zur Deutschlandpolitik*, ser. 1, vol. 3, pt. 1, 657–61.

68. Harper, *American Visions of Empire*, 113–16.

69. *FRUS: The Conferences at Cairo and Teheran*, 5.

70. *FRUS 1942*, 2:379–81.

71. *FRUS 1944*, 1:98. Roosevelt wanted the announcement of this decision to be made on November 11.

72. Bohlen, *Witness to History*, 184–85.

73. See Roosevelt's comment on the *Iowa* on 15 November 1943, *FRUS: The Conferences at Cairo and Teheran*, 6–97.

74. James E. Miller, "The Politics of Relief: The Roosevelt Administration and the Reconstruction of Italy, 1943–1944." *Prologue* 13 (1981): 193–208.

75. Rudolf V. A. Janssens, *"What Future for Japan ?" U.S. Wartime Planning for the Postwar Era, 1942-1945* (Amsterdam: Rodopi, 1995), 4–60. There is a generally unsatisfactory discussion of American planning and Roosevelt's role in it in Dale M. Hellegers, *We, the Japanese People: World War II and the Origins of the Japanese Constitution* (Stanford, CA: Stanford University Press, 2002), 1–7, 32–33; but see Roosevelt's approved but not issued statement to the Japanese of early December 1944 on pp. 86–87.

76. Fraser, "Roosevelt and the Making of America's East Asia Policy," 96–99.

77. There is a recent account in Hellegers, *We, the Japanese People*, chaps. 7–9.

78. See *Dokumente zur Deutschlandpolitik*, ser. 1, vol. 4, 197–98 (22 February 1943).

79. Tyrell, *Grossbritannien und die Deutschlandplanung der Alliierten*, 49. There is a useful but not entirely satisfactory account in Arieh J. Kochavi, *Prelude to Nuremberg: Allied War Crimes Policy and the Question of Punishment* (Chapel Hill: University of North Carolina Press, 1998).

80. *FRUS 1942*, 1:58–59; *Dokumente zur Deutschlandpolitik*, ser. 1, vol. 1, 483; vol. 4, 301–4.

81. *Dokumente zur Deutschlandpolitik*, ser. 1, vol. 4, 592–93.

82. Hull, *Memoirs of Cordell Hull*, 1289ff.; Barrett, *Robert H. Jackson*, 109–10.

83. Gerhard L. Weinberg, *World War II Leaders and Their Visions for the Future of Palestine* (Washington, DC: U.S. Holocaust Memorial Museum, 2002), 13–14.

84. *FRUS 1944*, 5:615–16.

85. Thomas Campbell and George C. Herring Jr., eds., *The Diaries of Edward R. Stettinius, Jr.* (New York: New Viewpoints, 1975), 10 November 1944, 170.

86. *FRUS 1942*, 1:13, 25.

87. For a general survey placing relations with Latin America in context, see Robert Dallek, *Franklin D. Roosevelt and American Foreign Policy, 1932-1945* (New York: Oxford University Press, 1979).

88. John Morton Blum, *V Was for Victory: Politics and American Culture during World War II* (New York: Harcourt Brace Jovanovich, 1976), is still very useful. For a helpful recent survey, see John W. Jeffries, *Wartime America: The World War II Home Front* (Chicago: Ivan R. Dee, 1996).

89. D'Ann Campbell, *Women at War with America: Private Lives in a Patriotic Era* (Cambridge, MA: Harvard University Press, 1984); idem, "Women in Combat: The World War II Experience in the United States, Great Britain, Germany, and the Soviet Union," *Journal of Military History* 57 (1993): 301–23.

90. The book by Frances Perkins cited in note 32 provides some insight.

U.S. infantrymen heading west through Kasserine Pass, February 26, 1943

CHAPTER FOUR

RICK ATKINSON

Prologue
from *An Army at Dawn*

RICK ATKINSON OPENS THE PROLOGUE TO the first volume of his Liberation Trilogy by stating the historian's task: "to authenticate: to warrant that history and memory give integrity to the story, to aver that all this really happened." With this volume on the American army's experience in North Africa, 1942–1943, he measures up to that task in vivifying detail: He knows what really happened at Carthage, and at such battlefields as El Guettar and Sidi Nsir, and even the names and hometowns of the boys killed there.

This first statement is a prescient microcosm of Atkinson's now completed 750,000-word "triptych" on World War II. All the elements of those three volumes are here: the telltale numbers (the 2,174-day war claimed an average of 27,600 lives per day), the grand strategic thinking, the trenchant words and stray thoughts of the leaders, the noise and stench of battle, the political debates and decisions, the common follies and venalities—and the individual leaps to greatness. It is an astonishing achievement: an 18-page précis of the whole war.

Twenty-seven acres of headstones fill the American military cemetery at Carthage, Tunisia. There are no obelisks, no tombs, no ostentatious monuments, just 2,841 bone-white marble markers, two feet high and arrayed in ranks as straight as gunshots. Only the chiseled names and dates of death suggest singularity. Four sets of brothers lie side by side. Some 240 stones are inscribed with thirteen of the saddest words in our language: "Here rests in honored glory a comrade in arms known but to God." A long limestone wall contains the names of another 3,724 men still missing, and a benediction: "Into Thy hands, O Lord."

This is an ancient place, built on the ruins of Roman Carthage and a stone's throw from the even older Punic city. It is incomparably serene. The scents of eucalyptus and of the briny Mediterranean barely two miles away carry on the morning air, and the African light is flat and shimmering, as if worked by a silversmith. Tunisian lovers stroll hand in hand across the kikuyu grass or sit on benches in the bowers, framed by orangeberry and scarlet hibiscus. Cypress and Russian olive trees ring the yard, with scattered acacia and Aleppo pine and Jerusalem thorn. A carillon plays hymns on the hour, and the chimes sometimes mingle with a muezzin's call to prayer from a nearby minaret. Another wall is inscribed with the battles where these boys died in 1942 and 1943—Casablanca, Algiers, Oran, Kasserine, El Guettar, Sidi Nsir, Bizerte—along with a line from Shelley's "Adonais": "He has outsoared the shadow of our night."

In the tradition of government-issue graves, the stones are devoid of epitaphs, parting endearments, even dates of birth. But visitors familiar with the American and British invasion of North Africa in November 1942, and the subsequent seven-month struggle to expel the Axis powers there, can make reasonable conjectures. We can surmise that Willett H. Wallace, a private first class in the 26th Infantry Regiment who died on November 9, 1942, was killed at St. Cloud, Algeria, during the three days of hard fighting against, improbably, the French. Ward H. Osmun and his brother Wilbur W., both privates from New Jersey in the 18th Infantry and both killed on Christmas Eve 1942, surely died in the brutal battle of Longstop Hill, where the initial Allied drive in Tunisia was stopped—for more than five months, as it turned out—within sight of Tunis. Ignatius Glovach, a private first class in the 701st Tank Destroyer Battalion who died on Valentine's Day, 1943, certainly was killed in the opening hours of the great German counteroffensive known as the battle of Kasserine Pass. And Jacob Feinstein, a sergeant from Maryland in the 135th Infantry who died on April 29, 1943, no doubt passed during the epic battle for Hill 609, where the American army came of age.

A visit to the Tunisian battlefields tells a bit more. For more than half a century, time and weather have purified the ground at El Guettar and Kasserine and Longstop. But the slit trenches remain, and rusty C ration cans, and shell fragments scattered like seed corn. The lay of the land also remains—the vulnerable low ground, the superior high ground: incessant reminders of how, in battle, topography is fate.

Yet even when the choreography of armies is understood, or the movement of this battalion or that rifle squad, we crave intimate detail, of individual men in individual foxholes. Where, precisely, was Private Anthony N. Marfione when he

> For more than half a century, time and weather have purified the ground at El Guettar and Kasserine and Longstop.

died on December 24, 1942? What were the last conscious thoughts of Lieutenant Hill P. Cooper before he left this earth on April 9, 1943? Was Sergeant Harry K. Midkiff alone when he crossed over on November 25, 1942, or did some good soul squeeze his hand and caress his forehead?

The dead resist such intimacy. The closer we try to approach, the farther they draw back, like rainbows or mirages. They *have* outsoared the shadow of our night, to reside in the wild uplands of the past. History can take us there, almost. Their diaries and letters, their official reports and unofficial chronicles—including documents that, until now, have been hidden from view since the war—reveal many moments of exquisite clarity over a distance of sixty years. Memory, too, has transcendent power, even as we swiftly move toward the day when not a single participant remains alive to tell his tale, and the epic of World War II forever slips into national mythology. The author's task is to authenticate: to warrant that history and memory give integrity to the story, to aver that all this really happened.

But the final few steps must be the reader's. For among mortal powers, only imagination can bring back the dead.

No twenty-first-century reader can understand the ultimate triumph of the Allied powers in World War II in 1945 without a grasp of the large drama that unfolded in North Africa in 1942 and 1943. The liberation of western Europe is a triptych, each panel informing the others: first, North Africa; then, Italy; and finally the invasion of Normandy and the subsequent campaigns across France, the Low Countries, and Germany.

From a distance of sixty years, we can see that North Africa was a pivot point in American history, the place where the United States began to act like a great power—militarily, diplomatically, strategically, tactically. Along with Stalingrad and Midway, North Africa is where the Axis enemy forever lost the initiative in World War II. It is where Great Britain slipped into the role of junior partner in the Anglo-American alliance, and where the United States first emerged as the dominant force it would remain into the next millennium.

None of it was inevitable—not the individual deaths, nor the ultimate Allied victory, nor eventual American hegemony. History, like particular fates, hung in the balance, waiting to be tipped.

Measured by the proportions of the later war—of Normandy or the Bulge—the first engagements in North Africa were tiny, skirmishes between platoons and companies involving at most a few hundred men. Within six months, the campaign metastasized to battles between army groups comprising hundreds of thousands of soldiers; that scale persisted for the duration. North Africa gave the European war its immense canvas and implied—through 70,000 Allied killed, wounded, and missing—the casualties to come.

No large operation in World War II surpassed the invasion of North Africa in complexity, daring, risk, or—as the official U.S. Army Air Forces history concludes—"the degree of strategic surprise achieved." Moreover, this was the first campaign undertaken by the Anglo-American alliance; North Africa defined the coalition and its strategic course, prescribing how and where the Allies would fight for the rest of the war.

North Africa established the patterns and motifs of the next two years, including the tension between coalition unity and disunity. Here were staged the first substantial tests of Allied landpower against Axis landpower, and the initial clashes between American troops and German troops. Like the first battles in virtually every American war, this campaign revealed a nation and an army unready to fight and unsure of their martial skills, yet willful and inventive enough finally to prevail.

North Africa is where the prodigious weight of American industrial might began to tell, where brute strength emerged as the most conspicuous feature of the Allied arsenal—although not, as some historians suggest, its only redeeming feature. Here the Americans in particular first recognized, viscerally, the importance of generalship and audacity, guile and celerity, initiative and tenacity.

North Africa is where the Allies agreed on unconditional surrender as the only circumstance under which the war could end.

It is where the controversial strategy of first contesting the Axis in a peripheral theater—the Mediterranean—was effected at the expense of an immediate assault on northwest Europe, with the campaigns in Sicily, Italy, and southern France following in train.

It is where Allied soldiers figured out, tactically, how to destroy Germans; where the fable of the Third Reich's invincibility dissolved; where, as one senior German general later acknowledged, many Axis soldiers lost confidence in their commanders and "were no longer willing to fight to the last man."

U.S. TANKS ADVANCE IN THE VALLEY OF EL GUETTAR, TUNISIA, 1943

It is where most of the West's great battle captains emerged, including men whose names would remain familiar generations later—Eisenhower, Patton, Bradley, Montgomery, Rommel—and others who deserve rescue from obscurity. It is where the truth of William Tecumseh Sherman's postulate on command was reaffirmed: "There is a soul to an army as well as to the individual man, and no general can accomplish the full work of his army unless he commands the soul of his men, as well as their bodies and legs." Here men capable of such leadership stepped forward, and those incapable fell by the wayside.

North Africa is where American soldiers became killing mad, where the hard truth about combat was first revealed to many. "It is a very, very horrible war, dirty and dishonest, not at all that glamour war that we read about in the hometown papers," one soldier wrote his mother in Ohio. "For myself and the other men here, we will show no mercy. We have seen too much for that." The correspondent Ernie Pyle noted a "new professional outlook, where killing is a craft." North Africa is where irony and skepticism, the twin lenses of modern consciousness, began refracting the experiences of countless ordinary soldiers. "The last war was a war to end war. This war's to start 'em up again," said a British Tommy, thus perfectly capturing the ironic spirit that flowered in North Africa.

Sixty years after the invasion of North Africa, a gauzy mythology has settled over World War II and its warriors. The veterans are lionized as "the Greatest Generation," an accolade none sought and many dismiss as twaddle. They are condemned to sentimental hagiography, in which all the brothers are valiant and all the sisters virtuous. The brave and the virtuous appear throughout the North African campaign, to be sure, but so do the cowardly, the venal, and the foolish. The ugliness common in later campaigns also appears in North Africa: the murder and rape of civilians; the killing of prisoners; the falsification of body counts.

It was a time of cunning and miscalculation, of sacrifice and self-indulgence, of ambiguity, love, malice, and mass murder. There were heroes, but it was not an age of heroes as clean and lifeless as alabaster; at Carthage, demigods and poltroons lie side by side.

The United States would send sixty-one combat divisions into Europe, nearly two million soldiers. These were the first. We can fairly surmise that not a single man interred at the Carthage cemetery sensed on September 1, 1939, that he would find an African grave. Yet it was with the invasion of Poland on that date that the road to North Africa began, and it is then and there that our story must begin.

September 1, 1939, was the first day of a war that would last for 2,174 days, and it brought the first dead in a war that would claim an average of 27,600 lives every day, or 1,150 an hour, or 19 a minute, or one death every three seconds. Within four weeks of the blitzkrieg attack on Poland by sixty German divisions, the lightning war had killed more than 100,000 Polish soldiers, and 25,000 civilians had perished in bombing attacks. Another 10,000 civilians—mostly middle-class professionals—had been rounded up and murdered, and twenty-two million Poles now belonged to the Third Reich. "Take a good look around Warsaw," Adolf Hitler told journalists during a visit to the shattered Polish capital. "That is how I can deal with any European city."

France and Great Britain had declared war against the German aggressors on September 3, but fighting subsided for six months while Hitler consolidated his winnings and plotted his next move. That came in early April 1940, when Wehrmacht troops seized Denmark and attacked Norway. A month later, 136 German divisions swept into the Netherlands, Belgium, Luxembourg, and France. Winston S. Churchill—a short, stout, lisping politician of indomitable will and oratorical genius, who on May 10 became both Britain's prime minister and defense minister—told President Franklin D. Roosevelt, "The small countries are simply smashed up, one by one, like matchwood." It was the first of 950 personal messages Churchill would send Roosevelt in the most fateful correspondence of the twentieth century.

France was not small, but it *was* smashed up. German tactical miscalculation allowed the British to evacuate 338,000 troops on 900 vessels from the northern port of Dunkirk, but on June 14 the German spearhead swept across the Place de la Concorde in Paris and unfurled an enormous swastika flag from the Arc de Triomphe. As the French tottered, Germany's partner in the Axis alliance, the Italian government of Benito Mussolini, also declared war on France and Britain. "First they were too cowardly to take part," Hitler said. "Now they are in a hurry so that they can share in the spoils."

After the French cabinet fled to Bordeaux in shocked disarray, a venerable figure emerged to lead the rump government. Marshal Philippe Pétain, the hero of Verdun in World War I and now a laconic, enigmatic eighty-four-year-old, had once asserted, "They call me only in catastrophes." Even Pétain had never seen a catastrophe like this one, and he sued for terms. Berlin obliged. Rather than risk having the French fight on from their colonies in North Africa, Hitler devised a clever armistice: the southern 40 percent of France—excluding Paris—would remain under the sovereignty of the Pétain government and unoccupied by German troops. From a new capital in the resort town of Vichy, France would also continue to administer her overseas empire, including the colonies of Morocco, Algeria, and Tunisia, which together covered a million square miles and included seventeen million people, mostly Arab or Berber. France could keep her substantial fleet and an army of 120,000 men in North Africa by pledging to fight all invaders, particularly the British. To enforce the agreement, Germany would keep one and a half million French prisoners of war as collateral.

Pétain so pledged. He was supported by most of France's senior military officers and civil servants, who swore oaths of fidelity to him. A few refused, including a forty-nine-year-old maverick brigadier general named Charles André Joseph Marie de Gaulle, who took refuge in London, denounced all deals with the devil, and declared, in the name of Free France: "Whatever happens, the flame of French resistance must not and shall not die." Hitler now controlled Europe from the North Cape to the Pyrenees, from the Atlantic Ocean to the River Bug. In September, Germany and Italy signed a tripartite pact with Japan, which had been prosecuting its own murderous campaign in Asia. The Axis assumed a global span. "The war is won," the führer told Mussolini. "The rest is only a question of time."

That seemed a fair boast. Britain battled on, alone. "We are fighting for life, and survive from day to day and hour to hour," Churchill told the House of Commons. But German plans to invade across the English Channel were postponed, repeatedly, after the Luftwaffe failed to subdue the Royal Air Force. Instead, the bombardment known as the Blitz continued through 1940 and beyond, slaughtering thousands and then tens of thousands of British civilians, even as RAF pilots shot down nearly 2,500 German planes in three months, killing 6,000 Luftwaffe crewmen and saving the nation.

Churchill also received help from Roosevelt, who nudged the United States away from neutrality even as he promised to keep Americans out of the war. Roosevelt's true sympathies were given voice by his closest aide, Harry Hopkins. "Whither thou goest, I will go," Hopkins told Churchill in January 1941, quoting the Book of Ruth. "And where thou lodgest, I will lodge; thy people shall be my people, and thy God my God." Hopkins added softly, "Even to the end." Roosevelt sent Churchill fifty U.S. Navy destroyers in exchange for the use of British bases in the Caribbean and western Atlantic, and by the spring of 1941 had pushed through Congress a vast program of Lend-Lease assistance under the charade of "renting" out the materiel. By war's end the United States had sent its allies 37,000 tanks, 800,000 trucks, nearly two million rifles, and 43,000 planes—so many that U.S. pilot training was curtailed because of aircraft shortages. In 1941, though, the British were "hanging on by our eyelids," as General Alan Brooke, chief of the Imperial General Staff, put it.

Hitler faced other annoying disappointments. Spain refused to join the Axis or abandon her neutrality to permit a German attack against the British fortress at Gibraltar, which controlled the mouth of the Mediterranean. Italian troops in-

---

The British were "hanging on by our eyelids," as General Alan Brooke, chief of the Imperial General Staff, put it.

---

vaded Greece without warning on October 28, 1940—"Führer, we are on the march!" Mussolini exclaimed—and immediately found themselves so overmatched that Wehrmacht divisions were needed to complete the conquest and rout an ill-conceived British expeditionary force sent to save the Balkans. Greece fell in April 1941, a week after Yugoslavia, where 17,000 people had been killed in a single day of Luftwaffe bombing.

Mussolini's legions had also been on the march in Africa, attacking from the Italian colony of Libya into Egypt, a former British protectorate still occupied by British troops. A British and Australian counteroffensive in December 1940 smashed an Italian army twice its size, eventually inflicting 150,000 casualties. With the Axis southern flank imperiled, Hitler again came to Mussolini's rescue, dispatching a new Afrika Korps to Libya under a charismatic tank officer who had previously commanded the führer's headquarters troops in Poland. General Erwin Rommel reached Tripoli in mid-February 1941 and launched a campaign that would surge back and forth across the North African littoral for the next two years, first against the British and then against the Americans.

Two monumental events in 1941 changed the calculus of the war. On June 22, nearly 200 German divisions invaded the Soviet Union in abrogation of the nonaggression pact that Hitler and Soviet leader Joseph Stalin had signed in 1939, which had allowed a division of spoils in eastern Europe. Within a day, German attacks had demolished one-quarter of the Soviet air force. Within four months, the Germans had occupied 600,000 square miles of Russian soil, captured three million Red Army troops, butchered countless Jews and other civilians, and closed to within sixty-five miles of Moscow. But four months after that, more than 200,000 Wehrmacht troops had been killed, 726,000 wounded, 400,000 captured, and another 113,000 had been incapacitated by frostbite.

The second event occurred on the other side of the world. On December 7, Japanese aircraft carriers launched 366 aircraft in a sneak attack on the U.S. Navy Pacific Fleet at Pearl Harbor, sinking or damaging eight battleships at their moorings, destroying or crippling eleven other warships, and killing 2,400 Americans. Simultaneous attacks were launched on Malaya, Hong Kong, and the Philippines. In solidarity with their Japanese ally, Hitler and Mussolini quickly declared war on the United States. It was perhaps the führer's gravest miscalculation and, as the British historian Martin Gilbert later wrote, "the single most decisive act of the Second World War." America would now certainly return to Europe as a belligerent, just as it had in 1917, during the Great War.

"I knew the United States was in the war, up to the neck and in to the death," Churchill later wrote. "I went to bed and slept the sleep of the saved and thankful."

Two years, three months, and seven days had passed since the invasion of Poland, and the United States had needed every minute of that grace period to prepare for war. Churchill's chief military representative in Washington, Field Marshal Sir John Dill, told London that, notwithstanding the long prelude, American forces "are more unready for war than it is possible to imagine."

In September 1939, the U.S. Army had ranked seventeenth in the world in size and combat power, just behind Romania. When those 136 German divisions conquered western Europe nine months later, the War Department reported that it could field just five divisions. Even the homeland was vulnerable: some coastal defense guns had not been test-fired in twenty years, and the army lacked enough antiaircraft guns to protect even a single American city. The building of the armed forces was likened to "the reconstruction of a dinosaur around an ulna and three vertebrae."

That task had started with the sixteen million men who registered for the

1778 1943

AMERICANS
will <u>always</u> fight for liberty

AMERICA'S TRADITION OF PATRIOTISM PROVED A POWERFUL INCENTIVE TO ENLIST.

draft in the fall of 1940, and who would expand Regular Army and National Guard divisions. By law, however, the draftees and newly federalized Guard units were restricted to twelve months of service—and only in the western hemisphere or U.S. territories. Physical standards remained fairly rigorous; soon enough, the day would come when new recruits claimed the army no longer examined eyes, just counted them. A conscript had to stand at least five feet tall and weigh 105 pounds; possess twelve or more of his natural thirty-two teeth; and be free of flat feet, venereal disease, and hernias. More than forty of every hundred men were rejected, a grim testament to the toll taken on the nation's health by the Great Depression. Under the rules of conscription, the army drafted no fathers, no felons, and no eighteen-year-olds; those standards, too, would fall away. Nearly two million men had been rejected for psychiatric reasons, although screening sessions sometimes went no further than questions such as "Do you like girls?" The rejection rate, one wit suggested, was high because "the Army doesn't want maladjusted soldiers, at least below the rank of major."

Jeremiads frequently derided the nation's martial potential. A Gallup poll of October 1940 found a prevailing view of American youth as "a flabby, pacifistic, yellow, cynical, discouraged, and leftist lot." A social scientist concluded that "to make a soldier out of the average free American citizen is not unlike domesticating a very wild species of animal," and many a drill sergeant agreed. Certainly no hate yet lodged in the bones of American troops, no urge to close with an enemy who before December 7, 1941, seemed abstract and far away. *Time* magazine reported on the eve of Pearl Harbor that soldiers were booing newsreel shots of Roosevelt and General George C. Marshall, the army chief of staff, while cheering outspoken isolationists.

Equipment and weaponry were pathetic. Soldiers trained with drainpipes for antitank guns, stovepipes for mortar tubes, and brooms for rifles. Money was short, and little guns were cheaper than big ones; no guns were cheapest of all. Only six medium tanks had been built in 1939. A sardonic ditty observed: "Tanks are tanks and tanks are dear / There will be no tanks again this year." That in part reflected an enduring loyalty to the horse. "The idea of huge armies rolling along roads at a fast pace is a dream," *Cavalry Journal* warned in 1940, even after the German blitzkrieg signaled the arrival of mechanized warfare. "Oil and tires cannot like forage be obtained locally." The army's cavalry chief assured Congress in 1941 that four well-spaced horsemen could charge half a mile across an open field to destroy an enemy machine-gun nest without sustaining a scratch. "The motor-mad advocates are obsessed with a mania for excluding the horse from war," he told the Horse and Mule Association of America, four days before Pearl Harbor. The last Regular Army cavalry regiment would slaughter its mounts to feed the starving garrison on Bataan in the Philippines, ending the cavalry era not with a bang but with a dinner bell.

To lead the eventual host of eight million men, the army had only 14,000 professional officers when mobilization began in 1940. The interwar officer corps was so thick with deadwood that one authority called it a fire hazard; swagger sticks, talisman of the Old Army, could serve for kindling. Secret War Department committees known as plucking boards began purging hundreds of officers who were too old, too tired, too inept. Not a single officer on duty in 1941 had commanded a unit as large as a division in World War I; the average age of majors was forty-eight. The National Guard was even more ossified, with nearly one-quarter of Guard first lieutenants over forty, and senior ranks dominated by political hacks of certifiable military incompetence. Moreover, Guard units in eighteen states were stained with scandal—embezzlement, forgery, kickbacks, and nepotism.

Yet slowly the giant stirred. Congress in 1940 had given the army $9 billion, more than all the money spent by the War Department since 1920. The fabled arsenal of democracy began to build steam, although nearly half of all military production in 1941 went to Lend-Lease recipients (including 15,000 amputation saws and 20,000 amputation knives to the Soviets). A remarkable cadre of promising professional officers began to emerge. The two-year, three-month, and seven-day preparation period was over. It was time to fight.

But where? American strategists since the early 1920s had considered Tokyo the most likely enemy, as the United States and Japan vied for dominance in the Pacific. But in 1938 a series of informal conversations with the British marked the start of an increasing Anglo-American intimacy, nurtured by a growing conviction in Washington that Germany was mortally dangerous and that the Atlantic sea lanes must always be controlled by friendly forces. Among potential adversaries, Germany had the largest industrial base and the greatest military capacity, and therefore posed the biggest threat. A U.S. strategy paper of November 1940 concluded that if Britain lost the war "the problem confronting us would be very great; and while we might not lose everywhere, we might, possibly, not win anywhere."

An evolving series of American war plans culminated in a strategic scheme called RAINBOW 5, which in the spring of 1941 posited joint action by the United States, Britain, and France in the event of America's entry into war, with the early dispatch of U.S. troops "to effect the decisive defeat of Germany, or Italy, or both." Forces in the Pacific would remain on the strategic defensive until European adversaries had been clubbed into submission. Even the debacle at Pearl Harbor failed to shake the conviction of Roosevelt and his military brain trust that "Germany first" was conceptually sound, and this remained the most critical strategic principle of the Second World War.

The smoke had hardly cleared from Pearl Harbor when Churchill arrived in Washington for extensive talks. The conference, code-named ARCADIA, failed to produce a specific Anglo-American plan of attack, but the prime minister and president reaffirmed the Germany-first decision. Moreover, on January 1, 1942, twenty-six countries calling themselves the "united nations" signed an agreement to forswear any separate peace without mutual concurrence and to make a common cause of "life, liberty, independence, and religious freedom, and to preserve the rights of man and justice."

The American idea of how to defeat the Third Reich was simple and obvious: drive straight for Berlin. "Through France passes our shortest route to the heart of Germany," declared Marshall, the army chief of staff. It was only 550 miles from the northwest coast of France to the German capital, over flat terrain with a sophisticated road and rail network that also sliced through the core of Germany's war industry. If Hitler was the objective, the American instinct was to "go for him bald-headed and as soon as possible, by the shortest and most direct route," a British general later noted. The Yanks, another British officer agreed, "wanted revenge, they wanted results, and they wanted to fight."

Direct, concentrated attack was an American strategic tradition often linked to Ulysses S. Grant in the Civil War. The surest route to victory was to obliterate the enemy's army and destroy his capacity to make war. As the world's greatest industrial power, with a military expanding to twelve million men, the United States could do that—particularly now that the nation belonged to a powerful alliance that included the British Empire, the Soviet Union, and China. The prevailing impatience to get on with it was voiced by a young American general from Kansas, whose diligence, organizational acumen, and incandescent grin had made him a rising star in the War Department. "We've got to go to Europe and fight," Dwight

David Eisenhower scribbled in a note to himself on January 22, 1942. "And we've got to quit wasting resources all over the world—and still worse—wasting time."

As the new chief of war plans for the army's General Staff, Eisenhower helped draft the blueprint that would convert these strategic impulses into action. A three-part American proposal evolved in the spring of 1942. Under a plan code-named BOLERO, the United States would ferry troops and materiel across the Atlantic for more than a year to staging bases in Britain. This massing of forces would be followed in April 1943 by ROUNDUP, an invasion across the English

---

Eisenhower helped draft the blueprint that would convert these strategic impulses into action.

Channel to the coast of France by forty-eight American and British divisions supported by 5,800 aircraft. The spearhead would then seize the Belgian port of Antwerp before wheeling toward the Rhine. If Germany abruptly weakened before that invasion, or if Soviet forces in the east appeared in danger of collapse and needed diversionary help, a smaller, "emergency" assault by five to ten divisions—code-named SLEDGEHAMMER—would be launched in the fall of 1942 to secure a beachhead in France, perhaps at Cherbourg or Calais, and tie up as many German soldiers as possible.

Churchill and his commanders concurred in principle with this strategy in April 1942, then immediately began backing away. The British already had been expelled from the Continent three times in this war—from Dunkirk, from Norway, and from Greece—and they were reluctant to risk a fourth drubbing with a hasty cross-Channel attack. "We shall be pushed out again," warned Alan Brooke. More than two dozen German divisions were now based in France, and the Germans could operate on interior lines to shift additional forces from the east and seal off any Allied beachhead.

SLEDGEHAMMER particularly discomfited the British, who would have to provide most of the troops for the operation while American units were still making their way across the Atlantic. Studies of Channel weather over the previous

ALLIED COMMANDERS CONFER IN NORTH AFRICA:
HAROLD ALEXANDER, DWIGHT EISENHOWER, GEORGE PATTON.

decade showed the frequency of autumn gales that could dismast an Allied expeditionary force as surely as the Spanish Armada had been wrecked in 1588. The Axis enemy would also have a 6-to-1 air advantage and could reinforce the point of attack three times faster than the Allies could; in all likelihood, the Wehrmacht defenders in France would need no reinforcement from the Russian front to bottle up or even massacre an Allied bridgehead that would be so weak some skeptics called the plan TACKHAMMER. Hitler had begun constructing formidable coastal fortifications from above the Arctic Circle to the Spanish border on the Bay of Biscay, and a few planners considered *Festung Europa*, Fortress Europe, impregnable: in their view, the Allies would have to land in Liberia—midway down the west coast of Africa—and fight their way up.

Churchill shared his military commanders' misgivings. "He recoiled in horror from any suggestion of a direct approach" in attacking Europe, one British general later recalled. A disastrous Allied defeat on the French coast, the prime minister warned, was "the only way in which we could possibly lose this war." If eager to accommodate his American saviors, he was also mindful of the million British dead in World War I. A French invasion, he believed, could cost another half million and, if it failed, accomplish nothing. "Bodies floating in the Channel haunted him," George Marshall later acknowledged. Marshall's own reference to SLEDGE-HAMMER as a "sacrifice play" to help the Russians hardly was comforting.

Whereas the dominant American strategic impulse was a direct campaign of mass and concentration, the British instinctively avoided large-scale land campaigns. For centuries, Britain had relied on superior naval power to protect the Home Islands and advance her global interests. She was accustomed to protracted wars in which she minimized her losses and her risks, outmaneuvering opponents and restricting combat to the periphery of the empire. The catastrophic stalemate in the trenches from 1914 to 1918 was an exception to the wisdom of the strategic rule. Churchill even hoped that, by encircling and squeezing Hitler's empire, Allied forces could foster rebellions by the subjugated peoples of Europe; with the Wehrmacht enervated by such revolts, an Anglo-American force could swiftly dispatch a depleted, exhausted Germany.

North Africa seemed a plausible place to start. British officers had first floated the possibility of joint Anglo-American action there in August 1941. Churchill raised the notion again during the ARCADIA conference in Washington at the end of the year, when the plan was assigned the code-name SUPER GYMNAST, and he continued to bring it up throughout the spring with the dogged enthusiasm of a missionary.

Punctuating each point with a stab of his trademark cigar, the prime minister ticked off the advantages to anyone within earshot: the occupation of Morocco, Algeria, and Tunisia could trap the Afrika Korps between the new Anglo-American force and the British Eighth Army already fighting Rommel in Egypt; Allied possession of North Africa would reopen the Mediterranean routes through the Suez Canal, shortening the current trip around the Cape of Good Hope by thousands of miles and saving a million tons of shipping; green American soldiers would get combat experience in conditions less harrowing than a frontal assault on France; the operation would require fewer landing craft and other battle resources than a cross-Channel attack; the Vichy government might be lured back into the Allied camp; and the operation could be mounted in 1942, in keeping with Roosevelt's wishes to help the Soviets as soon as possible and to expedite the entry of American soldiers into the war.

"This has all along been in harmony with your ideas," Churchill told the president. "In fact, it is your commanding idea. Here is the true second front of 1942."

The American military disagreed, adamantly and then bitterly. North Africa was a defeatist sideshow, a diversion, a peck at the periphery. Even before Pearl Harbor, a War Department memo warned that an attack in Africa would provide only an "indirect contribution to the defeat of the Nazis." That obdurate conviction hardened through the first six months of 1942. Another memo, in June 1942, concluded that the invasion of North Africa "probably will not result in removing one German soldier, tank, or plane from the Russian front."

To many American officers, the British proposal seemed designed to further London's imperial ambitions rather than win the war quickly. The Mediterranean for centuries had linked the United Kingdom with British interests in Egypt, the Persian Gulf, India, Australia, and the Far East. Old suspicions resurfaced in Washington that American blood was to be shed in defense of the British Empire, particularly after Japanese armies swept across Hong Kong, Singapore, and Burma to threaten India. U.S. Army officers recalled a bitter joke from 1917: that "AEF" stood not for "American Expeditionary Force" but for "After England Failed."

Following another visit by Churchill to Washington in mid-June 1942, the fraternal bickering intensified and the Anglo-Americans entered what turned out to be the most fractious weeks of their wartime marriage. On July 10, Marshall and the chief of naval operations, Admiral Ernest J. King, suggested to Roosevelt that if the British continued to insist on "scatterization" in North Africa, "the U.S. should turn to the Pacific for decisive action against Japan." The irascible King, who had once been accused by Roosevelt of shaving with a blowtorch, went so far as to predict that the British would never invade Europe "except behind a Scotch bagpipe band." Roosevelt likened this repudiation of Germany-first as "taking up your dishes and going away"; he asked Marshall and King to send detailed plans for "your Pacific Ocean alternative" that very afternoon—knowing that no such plans existed.

Roosevelt was so enigmatic and opaque that his own military chiefs often had to rely on the British for clues to his inner deliberations. But increasingly he seemed beguiled by Churchill's arguments rather than those of his own uniformed advisers. Although Roosevelt never had to enunciate his war principles—and they could surely have been scribbled on a matchbook cover—foremost among them was "the simple fact that the Russian armies are killing more Axis personnel and destroying more Axis materiel than all the other twenty-five United Nations put together," as he had observed in May. The War Department now estimated that the Red Army confronted 225 German divisions; six faced the British in Egypt. If Soviet resistance collapsed, Hitler would gain access to limitless oil reserves in the Caucasus and Middle East, and scores of Wehrmacht divisions now fighting in the east could be shifted to reinforce the west. The war could last a decade, War Department analysts believed, and the United States would have to field at least

Restive Americans wanted to know why the country had yet to counterpunch against the Axis.

200 divisions, even though it was now hard pressed to raise fewer than half that number. A gesture of Anglo-American good faith beyond Lend-Lease was vital to encouraging the Soviets. After promising Moscow in May that the United States "expected" to open a second front before the end of the year, Roosevelt in July told his lieutenants that "it is of the highest importance that U.S. ground troops be brought into action against the enemy in 1942."

Other factors also influenced the president's thinking. More than half a year after Pearl Harbor, restive Americans wanted to know why the country had yet to counterpunch against the Axis; November's congressional elections would provide a referendum on Roosevelt's war leadership, and polls indicated that he and his Democratic Party could take a drubbing. Demonstrators in London's Trafalgar Square and elsewhere were chanting "Second front, *now!*" in sympathy with the besieged Russians. By seizing Africa, the Allies would deny the Axis potential bases for attacking shipping lanes in the South Atlantic or even striking the Americas. The Pacific campaign, although hardly swinging in the Allies' favor, had stabilized, permitting the strategic defensive envisioned in the RAINBOW 5 plan; but unless another battlefront opened across the Atlantic, U.S. forces would drain into the Pacific. In May, the U.S. Navy in the Coral Sea had attacked a Japanese fleet escorting invasion troops bound for the Solomon Islands and New Guinea;

losses on the two sides had been nearly equal. A month later, four Japanese aircraft carriers were sunk at the battle of Midway, marking the first unambiguous American victory of the war. Operation WATCHTOWER, the first Allied counteroffensive against Japan, was about to unfold with the landing of 16,000 American troops on an island in the Solomons: Guadalcanal.

The campaign against Germany and Italy, on the other hand, was faltering. Wehrmacht troops had overrun the Don River in southern Russia and were approaching Stalingrad, on the Volga. Except for Britain and neutrals such as Spain, Sweden, and Switzerland, all Europe belonged to the Axis. In Egypt, the Afrika Korps was only sixty miles from Alexandria and the Nile valley, gateway to the Suez Canal and Middle East oil fields. In Cairo, refugees jammed the rail stations, and panicky British officers burned secret papers in their gardens. After a long siege, Rommel had captured 30,000 British Commonwealth troops in the Libyan port garrison of Tobruk. Hitler rewarded him with a field marshal's baton, to which prize Rommel replied, "I am going on to Suez."

By chance, the bad news from Tobruk reached Churchill on June 21, 1942, while he stood next to Roosevelt's desk in the Oval Office. Marshall's face was grimmer than usual as he strode in with a pink dispatch sheet. Churchill read the message and took a half step back, his ruddy face gone ashen. Roosevelt's response was a thrilling gesture of magnanimity to a friend in need. "What can we do to help?" the president asked.

In the short run, the Americans could, and did, strip 300 new Sherman tanks from the newly outfitted U.S. 1st Armored Division for shipment to British troops in Egypt. Marshall, Admiral King, and Harry Hopkins returned Churchill's visit by flying as a delegation to London for more strategic negotiations, but the talks bogged down even as the Americans conceded that an attack across the Channel that year was unlikely. In a limp gesture of mollification, the British took the three Yanks to see Oliver Cromwell's death mask and Queen Elizabeth's ring before they flew home.

Roosevelt had had enough. The time had come to end the protracted stalemate and get on with the war. After informing both Churchill and his own senior military advisers on July 25 that he intended to invade North Africa, he slammed the door on further discussion. At 8:30 p.m. on Thursday, July 30, he summoned his lieutenants to the White House and announced that, as he was commander-in-chief, his decision was final. North Africa was "now our principal objective."

There would be no SLEDGEHAMMER against France. The African offensive was to occur "at the earliest possible date," preferably within two months.

The president had made the most profound American strategic decision of the European war in direct contravention of his generals and admirals. He had cast his lot with the British rather than with his countrymen. He had repudiated an American military tradition of annihilation, choosing to encircle the enemy and hack at his limbs rather than thrust directly at his heart. And he had based his fiat on instinct and a political calculation that the time was ripe.

In choosing Operation TORCH, as the North Africa invasion was now called, Roosevelt made several miscalculations. Despite Marshall's warnings, he refused to believe that a diversion to North Africa in 1942 precluded a cross-Channel invasion in 1943. He failed to see that the Mediterranean strategy of encirclement precluded other strategies, or that more than a million American soldiers, and millions of tons of materiel, would be sucked into the Mediterranean in the next three years, utterly eviscerating the buildup in Britain. He continued to argue that "defeat of Germany means defeat of Japan, probably without firing a shot or losing a life."

Yet the president's decision was plausible, if not precisely wise. As Brooke had observed of the proposed cross-Channel attack: "The prospects of success are small and dependent on a mass of unknowns, whilst the chances of disaster are great." American planners considered the British argument for TORCH "persuasive rather than rational," but the American argument for SLEDGEHAMMER and ROUNDUP had been neither. Direct attack was premature; its adherents exemplified an amateurish quality in American strategic thinking that would ripen only as the war ripened.

The American military had been animated mostly by can-do zeal and a desire to win expeditiously; these traits eventually would help carry the day, but only when tempered with battle experience and strategic sensibility. One general later claimed that army logisticians kept insisting they could support ten Allied divisions in Cherbourg although they were not certain where the French port was, much less what the condition of the docks might be or whence those divisions would come. Moving a single armored division required forty-five troopships and cargo ships, plus warship escorts, and moving the fifty divisions needed to sustain an invasion required far more ships than the Allies now possessed. Similarly, the critical issue of landing craft had been blithely ignored. "Who is responsible for building landing craft?" Eisenhower had asked in a May 1942 memo. With

AFTER OPERATION TORCH: GRAVES OF SIX AMERICAN SOLDIERS
KILLED IN ACTION IN NORTH AFRICA

some planners estimating that an invasion of France required at least 7,000 landing craft, and others believing the number was really triple that, the hard truth was that by the fall of 1942 all the landing craft in Britain could carry only 20,000 men. Yet a U.S. War Department study had concluded that to draw significant numbers of German troops from the Russian front required at least 600,000 Allied soldiers in France. "One might think we were going across the Channel to play baccarat at Le Touquet, or to bathe at the Paris Plage!" Brooke fumed.

Roosevelt had saved his countrymen from their own ardor. His decision provoked dismay, even disgust, and would remain controversial for decades. "We failed to see," Marshall later said of his fellow generals, "that the leader in a democracy has to keep the people entertained." Eisenhower believed the cancellation of SLEDGEHAMMER might be remembered as the "blackest day in history"—a silly hyperbole, given the blackness of other days. The alienation many senior American officers felt from their British cousins could be seen in a War Department message of late August, proposing that "the Middle East should be held if possible, but its loss might prove to be a blessing in disguise" by giving the British their comeuppance and bringing them to their senses.

But the decision was made. The "thrashing around in the dark," as Eisenhower called it, was over; the dangerous impasse had been breached.

Much, much remained to be done. Problems ranging from the size and composition of the invasion force to the timing and location of the landings required solutions. In early August, TORCH planners moved into offices at Norfolk House on St. James's Square in London under the supervision of Eisenhower, who had recently been sent from Washington to Britain as commanding general of the European Theater of Operations. As a gesture of reconciliation, and in anticipation of the eventual American preponderance, the British proposed that the Allied expedition be commanded by an American. Churchill nominated Marshall, but Roosevelt was reluctant to give up his indispensable army chief. Eisenhower, already overseas, had demonstrated impressive diligence and energy, and on August 13 he was named commander-in-chief of TORCH.

As the days grew shorter and the summer of 1942 came to an end, few could feel buoyed by news from the front:

Wehrmacht troops had reached the Volga, and the first shots were exchanged in the battle for Stalingrad. German U-boats, traveling in predatory "wolfpacks," were sinking ships faster than Allied yards could build them; a supply convoy to northern Russia lost thirteen of forty vessels, despite an escort of seventy-seven ships. The Chinese war effort against the Japanese had disintegrated. The fighting over the Solomon Islands had made Guadalcanal a shambles. The fall of Suez seemed imminent. Four of the seven aircraft carriers in the American fleet when the United States entered the war had been sunk. And antipathy between British and American confederates threatened to weaken the alliance even before the fight against their common enemy was joined.

Only seers or purblind optimists could guess that these portents foreshadowed victory. The Allies were not yet winning, but they were about to begin winning. Night would end, the tide would turn, and on that turning tide an army would wash ashore in Africa, ready to right a world gone wrong.

PATTON'S 4TH ARMORED DIVISION UNITS NEAR LUTREBOIS, BELGIUM, ON NEW YEAR'S DAY 1945

CHAPTER FIVE

CARLO D'ESTE

"Patton's Finest Hour: The Battle of the Bulge"
adapted from *Patton, A Genius for War*

CARLO D'ESTE STANDS IN THE FRONT RANK OF American military historians, having devoted his career to enlarging our understanding of World War II. An army veteran, D'Este retired as a lieutenant colonel after serving in Germany and Vietnam, before turning to the study of twentieth-century warfare. His narrative accounts of major actions—*Decision in Normandy*; *Bitter Victory: The Battle for Sicily, 1943*; and *Fatal Decision: Anzio and the Battle for Rome*—are complemented by comprehensive biographies of that conflict's most influential warriors: Dwight D. Eisenhower, Winston Churchill, and George S. Patton.

In this selection, adapted from *Patton: A Genius for War*, D'Este brings cinematic immediacy to what he calls "the crowning achievement of Patton's career." In the 1944 Battle of the Bulge, when a massive German counterattack burst through the Allied lines, Patton was one of the few Allied commanders not surprised by the German offensive—and the only one prepared to move his divisions in immediately and cut the enemy's advance.

L ate autumn of 1944 was a bleak time for Lieutenant General George S. Patton and his Third Army. After their breakout from Normandy the previous summer, they had ground to a halt in the Lorraine region, near the border of Germany. Patton still hoped to make a triumphant thrust into the Third Reich, but hobbled by a sea of mud and logistical problems that left his tanks short of fuel, he was reduced to limited and costly infantry offensives. Lorraine would become Patton's bloodiest and least successful campaign.

Even as he fought in Lorraine and made plans for a mid-December attack through Germany's Saar region, Patton was obliged to pay close attention to ominous German movements to his north, in the Ardennes Forest. Since mid-November, the Third Army intelligence officer, Colonel Oscar Koch, had been reporting a buildup of panzer and infantry divisions, ammunition, and gasoline dumps west of the Rhine in the Saar, and from Aachen in western Germany to the southern extremity of the Ardennes, the area occupied by the U.S. First Army.

German rail movements were increasing both east and west of the Rhine. On December 9, Patton was informed that the Germans now had a two-to-one manpower advantage in the Ardennes. This sector was thinly guarded by Major General Troy Middleton's VIII Corps, then in defensive positions, resting and reequipping after two of its divisions had been bloodied in the Hürtgen Forest. As a precaution, Patton directed that planning begin at once to counter any potential threat in the Ardennes. "We'll be in a position to meet whatever happens," he told his staff.

Such foresight was the key to winning the deadliest, most desperate battle of the war in western Europe. As a young man, George S. Patton Jr. had believed he was destined to lead a great army in a pivotal battle. That moment arrived in the Ardennes in late December 1944 during what became known as the Battle of the Bulge, the crowning achievement of Patton's career.

The German attack that Patton prophesied arrived in the early morning hours of December 16, 1944. It was to be the Germans' only major counteroffensive of the war in northwest Europe. The invaders quickly overwhelmed American troops occupying the weakest link in the Allied front, the lightly defended VIII Corps sector in the eastern Ardennes. Adolf Hitler was gambling that he could yet take control of Germany's destiny by means of a surprise lightning thrust through the Ardennes. Against the advice of his generals, Hitler ordered his armies to drive to the vital Belgian port of Antwerp in the belief that it would compel the Allies to sue for peace. Three powerful armies attacked: General Hasso von Manteuffel's

Fifth Panzer Army, SS General Sepp Dietrich's Sixth Panzer Army, and General Erich Brandenberger's Seventh Army (largely infantry).

Patton's superior, Lieutenant General Omar Bradley, the 12th Army Group commander, had unknowingly played into German plans. Taking what he later termed "a calculated risk," he had lightly defended the eastern Ardennes with only two inexperienced infantry divisions and two battered veteran divisions then absorbing replacements.

American units were caught flat-footed and immediately became embroiled in desperate battles. The inexperienced 106th Division was quickly surrounded and nearly annihilated. Even in defeat, however, it bought time for others to organize hasty defenses against the German onslaught. Time was of the essence; it would be much harder to blunt the attack if the Germans secured crossings over the Meuse River. And to reach the Meuse, Manteuffel's Fifth Panzer Army first had to seize the towns of Saint Vith and Bastogne. Bastogne was particularly important, as all the main east-west roads converged there.

The suddenness of the German breakthrough left senior Allied commanders scrambling to determine Hitler's objective. Indeed, the German attack in the Ardennes revealed serious lapses in Allied intelligence. Most senior intelligence officers simply failed to draw the right conclusions despite ample evidence of an impending attack. Bradley initially misread the German counteroffensive, dismissing it as a spoiling attack to hamper Patton's offensive into the Saar.

Eisenhower disagreed, declaring, "That's no spoiling attack." Patton's was the only intelligence staff that correctly assessed German intentions. As early as November 25—a full three weeks before the Germans launched their attack—he had sensed that "First Army is making a terrible mistake in leaving the VIII Corps static, as it is highly probable that the Germans are building up east of them." Worried by the German activity, Patton relied heavily on daily intelligence reports and lost no opportunity to question anyone who might provide information on the enemy.

Major General Kenneth Strong, intelligence officer for Supreme Headquarters Allied Expeditionary Forces (SHAEF), recalled that Patton demonstrated "an extraordinary desire for information of all kinds. He invariably came to see me when he was at Supreme Headquarters and would quiz me on details about the enemy, usually to satisfy himself that the risks he intended to undertake were justified."

Colonel Koch questioned the implications of the German buildup long before other intelligence officers recognized the threat. Koch's reports were enhanced by information gathered by the 3rd Cavalry Group. Patton had attached elements of the 3rd Cavalry to each committed division and corps. Being close to the action, these units supplied critical information that was sent directly to Third Army headquarters rather than passing through slow-moving command channels. Although this was unorthodox, the results made Patton the best informed of the senior Allied commanders.

Not surprisingly, only Patton and Koch believed that the Germans could secretly mount a counteroffensive in the rugged Ardennes in the vile winter weather. As yet unaware of the events unfolding in the north, Koch issued a prophetic warning at the morning briefing on December 16. He reported that the enemy was in a state of radio silence that strongly suggested "the Germans are going to launch an attack, probably at Luxembourg."

The next morning, with the German action underway, Patton said that, "the thing in the north is the real McCoy." His operations officer, Colonel Halley Maddox, predicted the Germans would have to commit their entire reserves in the Ardennes. This, he said, is a perfect setup for the Third Army: "We can pinwheel the enemy before he gets very far. In a week we could expose the whole German rear and trap their main forces west of the Rhine." Patton thought Maddox was right, but merely observed, "My guess is that our offensive [in Lorraine] will be called off and we will have to go up there and save their hides."

Bradley summoned Patton to 12th Army Group headquarters in Luxembourg and reported that the German penetration was far deeper and more serious than even Patton had previously thought. Bradley asked what the Third Army could do. Patton replied that he would have two divisions on the move the next day and, if necessary, a third in twenty-four hours. With that, the proposed Saar offensive was canceled. Patton shrugged off his disappointment, saying, "What the hell, we'll still be killing Krauts." He grinned when Bradley assured him they would "hit this bastard hard." Later that evening Patton was ordered to report to Verdun the following morning to meet with Eisenhower and the other Allied commanders to work out a plan of action.

December 19, 1944, was a historic day for the Third Army. It began at 7 a.m. when Patton briefed his key staff officers and two of his corps commanders. An hour later he convened the full staff and told them the Third Army would likely be called upon to come to the relief of the First Army. How and where would be decided at Verdun. The only certainty, recalled Patton, was that "while we were all accustomed to rapid movement, we would now have to prove that we could operate even faster. We then made a rough plan of operation."

The plan assumed the Third Army would move from the Saar via one or more of three possible routes. When Patton learned his mission, he would telephone his chief of staff, Brigadier General Hobart R. "Hap" Gay, and announce the preestablished code for which route to take.

Eisenhower arrived at Verdun, according to one observer, "looking grave, almost ashen." The meeting took place in a dismal room of a French barracks warmed only by a potbelly stove. The mood was grim despite Eisenhower's fragile attempt at levity at the meeting's opening. "The present situation is to be regarded as one of opportunity for us and not of disaster," he said. "There will be only cheerful faces at this conference table." The smiles seemed forced. Half in jest, Patton immediately chimed in: "Hell, let's have the guts to let the sons of bitches go all the way to Paris. Then we'll really cut 'em up and chew 'em up."

Besides Eisenhower, Bradley, and Patton, attendees included the supreme Al-

101ST AIRBORNE TROOPS LEAVE BASTOGNE ON DECEMBER 31, 1944.

lied commander's deputy, Air Chief Marshal Sir Arthur Tedder; the SHAEF chief of staff, Lieutenant General Walter Bedell Smith; the commander of the 6th Army Group, Lieutenant General Jacob L. Devers; a handful of staff officers; and British Field Marshal Sir Bernard L. Montgomery's able chief of staff, Major General Francis de Guingand.

It was quickly agreed to stop offensive action in all Allied sectors and concentrate on blunting the German drive. Eisenhower drew a line at the Meuse, beyond which there would be no further retreat. Once the German attacks were

---

"When can you attack?" Eisenhower asked. "The morning of December 21, with three divisions," Patton replied instantly.

---

contained, the Allies would counterattack. Eisenhower said: "George, I want you to command this move…[and make] a strong counterattack with at least six divisions. When can you start?"

Patton replied, "As soon as you're through with me." He explained how he had left instructions with his staff and could put any one of three plans in motion with a phone call. "When can you attack?" Eisenhower asked. "The morning of December 21, with three divisions," Patton replied instantly. Forty-eight hours! Eisenhower was not amused, assuming that Patton had picked a very inopportune moment to boast.

"Don't be fatuous, George," he retorted, in obvious disbelief. "If you try to go that early, you won't have all three divisions ready and you'll go piecemeal. You will start on the 22nd and I want your initial blow to be a strong one! I'd even settle for the 23rd if it takes that long to get three full divisions."

Eisenhower was dead wrong; this was not Patton the boastful, but Patton the prepared. Though others came to Verdun with only vague ideas, Patton had devised plans tailored to meet any contingency. "This was the sublime moment of his career," wrote historian Martin Blumenson.

Accounts of the exchange vary, but certainly some at Verdun that day responded to Patton with skepticism. This was yet another of the general's smug predictions, and one that was quite out of place in this somber setting. According to notes from the meeting, "There was some laughter, especially from British officers, when Patton answered 'Forty-eight hours.'"

Patton's senior aide, Lieutenant Colonel Charles Codman, reported "a stir, a shuffling of feet, as those present straightened up in their chairs.... But through the room the current of excitement leaped like a flame."

According to author John Eisenhower (son of Dwight Eisenhower): "Witnesses to the occasion testify to the electric effect of this exchange." The prospect of relieving three divisions from the line, turning them north, and traveling along icy roads to prepare for a major counterattack in less than seventy-two hours was astonishing.

Only a commander with exceptional confidence in his subordinate commanders and in the professional skill of his fighting divisions could dare risk such a venture. It wasn't just that Patton never hesitated; he eagerly embraced the opportunity to turn a potential debacle into a triumph. While near panic spread elsewhere, he believed the German offensive offered a magnificent opportunity to strike a killing blow. It was as if destiny had groomed him, through more than thirty-five years in the army, for this single, defining moment. The fate of the war rested upon the right decisions being made in that dingy room.

Cigar in hand, Patton illustrated his intentions on the map, pointing to the obvious bulge of the German salient in the Saint Vith-Bastogne sector. Speaking directly to Bradley, he said, "Brad, the Kraut's stuck his head in a meat grinder." Turning his fist in a grinding motion, he continued, "And this time I've got hold of the handle." He then responded to the inevitable questions, delivering specific, well-rehearsed answers. Codman later said, "Within an hour everything had been thrashed out—the divisions to be employed, objectives, new Army boundaries, the amount of our own front to be taken over by Devers's 6th Army Group, and other matters and virtually all of them settled on General Patton's terms."

It was perhaps the most remarkable hour of Patton's military career. Bradley later acknowledged that this was a "greatly matured Patton"; the Third Army staff had pulled off "a brilliant effort." With considerable understatement, Patton wrote of the day: "When it is considered that [Third Army deputy chief of staff Paul] Harkins, Codman, and I left for Verdun at 0915 and that between 0800 and that hour we [held] a staff meeting, planned three possible lines of attack, and made

ON DECEMBER 30, 1944, LT. GEN. PATTON AWARDED THE DISTINGUISHED SERVICE
CROSS TO THE 101ST AIRBORNE'S BRIG. GEN. ANTHONY C. MCAULIFFE.

a simple code in which I could telephone....it is evident that war is not so difficult
as people think."

As they parted, Eisenhower, recently promoted to the five-star rank of General of the Army, remarked, "Funny thing, George, every time I get a new star I get attacked." Patton shot back affably, "And every time you get attacked, Ike, I pull you out."

Many years earlier Patton had said, "Ike, you will be the Lee of the next war, and I will be your Jackson." Whether or not Eisenhower qualified as Robert E. Lee, Patton was about to establish a definite resemblance to Thomas J. "Stonewall" Jackson. The Third Army was poised to pull off one of the most remarkable feats of any army in history.

After the meeting Patton began snapping out orders: "Telephone Gay. Give him the code number, tell him to get started....You know what to do." During the next three days Patton and his driver, Master Sergeant John L. Mims, constituted Lucky Forward, the Third Army's command post. With one pistol strapped to the outside of his parka, another tucked into his waistband, Patton sped from one division or corps to another. On December 20, he said, "I visited seven divisions and regrouped an army alone." Like a cattle drover, he pushed, pulled, and exhorted everyone to keep moving and to "drive like hell" toward Bastogne. At the end of perhaps the most dynamic day of his life, Mims remarked: "General, the government is wasting a lot of money hiring a whole General Staff. You and me has run the Third Army all day and done a better job than they do." Patton was pleased that he had earned his pay: "It was quite a day....Destiny sent for me in a hurry when things got tight. Perhaps God saved me for this effort."

After the Verdun conference, Eisenhower's staff advocated splitting the Ardennes front in two until the situation could be brought under control, with Montgomery to be given temporary operational command of all Allied forces (principally the U.S. First and Ninth Armies) in the northern half of the Bulge, and Bradley to command only the southern flank (Third Army). Bradley's contact with Lieutenant General Courtney Hodges, commander of the First Army, was tenuous and he was in no position to control the northern flank from his headquarters in Luxembourg City. Eisenhower saw this and telephoned with his decision, thus effectively reducing Bradley's role to that of an observer; the battle was really Patton's to mastermind and control.

His first order to his troops was Pattonesque: "Everyone in this army must understand that we are not fighting this battle in any half-cocked manner. It's either root hog—or die! Shoot the works. If those Hun bastards want war in the raw then that's the way we'll give it to them!"

When he met with the staffs of three of his four corps in Luxembourg the night before the attack began, Patton saw their doubts and tried to answer them: "I always seem to be the ray of sunshine, and, by God, I always am. We can and will win, God helping....I wish it were this time tomorrow night. When one attacks it is the enemy who has to worry. Give us the victory, Lord."

Until the Third Army could attack from the south, the strategy had been to hold ground for as long as possible, retreat, blow up bridges, and delay again. Middleton's battered VIII Corps was the last Allied force between Manteuffel's

Fifth Panzer Army and the Meuse. Although it was against his principles to give up anything, Patton saw an opportunity in letting the Germans become over-extended before he struck their vulnerable left flank.

As late as December 20, Patton contemplated ceding Bastogne to the Fifth Panzer Army. That afternoon he conferred with Middleton, greeting him with the admonition, "Troy, of all the goddam crazy things I ever heard of, leaving the 101st Airborne to be surrounded in Bastogne is the worst!"

A friend of long standing, Middleton rejoined: "George, just look at that map with all the roads converging on Bastogne. Bastogne is the hub of the wheel. If we let the Boche take it, they will be in the Meuse in a day." Patton understood the need to hold Bastogne, and the two friends worked out an axis of advance for launching an attack to reinforce the beleaguered crossroads town. As promised, on the morning of December 22, three divisions launched the first Allied coun-terstroke of the Ardennes campaign, operating across a twenty-mile front. The Third Army struggled against the weather and the Germans to reach Bastogne, where the 101st Airborne Division and elements of Patton's own 9th and 10th Ar-mored Divisions were now surrounded.

December 23 brought a day of fair weather, and the Allied air forces took ad-vantage, attacking the Germans and making more than two hundred supply drops into Bastogne, whose defenders repulsed a strong German attack. The 4th Ar-mored Division spearheaded the drive toward Bastogne but ran into increasing difficulty. "It is always hard to get an attack rolling," Patton observed, but he was pleased that "the men are in good spirits and full of confidence." Once again he was impressed by "how long it takes to really learn how to fight a war."

Bastogne remained surrounded, and when the Germans demanded its sur-render, the acting commander of the 101st Airborne, Brigadier General Anthony C. McAuliffe, replied, "Nuts!" Upon hearing the now famous response, Patton said: "Any man who is that eloquent deserves to be relieved [i.e., rescued]. We shall go right away."

On Christmas Eve, Patton judged that "the German General Staff is running this attack and has staked all on this offensive to regain the initiative. They are far behind schedule and, I believe beaten. If this is true, the whole army may surren-der." Patton was only partly right. Surrender was not an option for the Germans. Indeed, at that moment, both combatants were at serious risk. The Battle of the Bulge was far from over.

International News Service correspondent Larry Newman, who covered the Third Army, wrote:

> Patton was never disheartened. In the midst of the battle—perhaps the most desperate a U.S. Army has ever had to fight—Patton called a conference of all correspondents. As we filed into the room, the tenseness was depressing. But when Patton strode into the room, smiling, confident, the atmosphere changed within seconds. He asked: "What the hell is all the mourning about? This is the end of the beginning. We've been batting our brains out trying to get the Hun out in the open. Now he is out. And with the help of God we'll finish him off this time—and for good."

More than ever, Patton made it a point to be seen during the battle, always riding in an open jeep. The cold was so intense soldiers dressed in as many layers of clothing as they could manage, but Patton's only concession was a heavy winter parka or an overcoat. He spent little time in his headquarters and ventured out most of each day to see and be seen by his troops and to endure the same wretched conditions. He prowled the roads of the Ardennes, sitting ramrod straight, often with his arms folded, his face unsmiling. More than once his face froze. Word of his presence filtered through the GI grapevine with astonishing rapidity, as did his praise for his troops, which was invariably reported down the chain of command: "The Old Man says..." or "Georgie says..."

During his brilliantly orchestrated weeklong defense of Saint Vith, Brigadier General Bruce C. Clarke informed a sergeant manning a forward infantry outpost that he had heard Patton's Third Army was attacking from the south. Years later Clarke remembered: "The sergeant thought for a minute and said, 'That's good news. If Georgie's coming we have got it made.' I know of no other senior commander in Europe who could have brought forth such a response."

One cold, dark afternoon, Patton encountered a column of the 4th Armored moving toward Bastogne. Tanks and vehicles were sliding off the icy road. Someone recognized him and let out a shout that rolled down the column as soldiers in trucks and tanks cheered. After the war, a GI told Patton's wife, Beatrice: "Oh, yes, I knew him, though I only saw him once. We was stuck in the snow and he come by in a jeep. His face was awful red and he must have been about froze riding in that open jeep. He yelled to us to get out and push, and first I knew, there was General Patton pushing right alongside of me. Sure, I knew him; he never asked a man to do what he wouldn't do himself."

The soldiers who had to fight in the terrible winter weather did so with woefully inadequate uniforms and equipment. There were no shoepacks, parkas, or white camouflage uniforms, and no white paint for the Sherman tanks. When the army could not fill his needs, Patton commissioned a French factory to manufacture 10,000 white capes per week for the Third Army.

Patton was quick to give credit to his troops. When asked about the remarkable swing of the Third Army to the north, Patton grinned and replied:

Yes, we broke all records moving up here. It was all done by the three of us...me, my chauffeur, and my chief of staff. All I did was to tell my division commanders where they'd got to be tomorrow. Then I let the others do it....To tell you the truth, I didn't have anything much to do with it. All you need is confidence and good soldiers....If there is confidence at the top the soldiers all feel it. I know a lot of soft-headed armchair generals accuse me of killing off my men. They don't know their fat behinds from a tommy gun. I don't waste men. I believe in saving my men's lives. And, by God! I've done it...again and again. More often than not the best way to save men's lives is to risk them...to take chances, and make your men fight better.

Passionately, Patton continued:

Maybe the GI hates discipline, but only until he learns that that's what makes a winning soldier. I'll put our goddam, bitching, belching, belly-aching GIs up against any troops in the world. The Americans are sons of bitches of soldiers—thanks to their grandmothers! All you've got to do is to show them the value of discipline...give them the habit of obeying in a tight place. Yes. The American is a hell of a fine soldier.

Patton joyfully declared he did not give a damn what others thought of him: "You know what they can do. I've studied military history all my life. Georgie Patton knows more about military history than any person in the United States

Army today. With due conceit—and I've got no end of that—I can say that's true."

The dreadful weather notwithstanding, Patton ordered that every soldier in the Third Army have a hot turkey dinner on Christmas Day. To ensure this was done, he and Mims spent the holiday driving from one unit to another. Mims recalled, "He'd stop and talk to the troops; ask them did they get turkey, how was it, and all that." Patton had little faith in his mess sergeants, who, he complained, "couldn't qualify as goddam manure mixers. They take the best food Uncle Sam

The 4th Armored finally broke through to bolster the "Battered Bastards of Bastogne" on the afternoon of December 26.

can buy and bugger it all up." He also constantly checked for trench foot, and his troops inevitably heard the refrain, "Men we can get all kinds of equipment except we can't get more soldiers."

On Christmas Day, Patton had a brush with death. As he neared the headquarters of the 4th Armored Division, his jeep was strafed by an American plane, forcing him to take cover in a ditch.

By December 26 Patton was convinced that "the German has shot his wad." Prisoners were coming in who had not been fed in three to five days. "We should attack," he said, but SHAEF was holding three reserve divisions at headquarters in Reims. He felt that Eisenhower ought to be more aggressive. When he received a message that Eisenhower was "very anxious" that he put all efforts into securing Bastogne, Patton wrote in his diary, "What the hell does he think I've been doing for the last week?"

The 4th Armored finally broke through to bolster the "Battered Bastards of Bastogne" on the afternoon of December 26. Bastogne remained surrounded on three sides and would come under its most critical threat in the days ahead. But Patton elatedly proclaimed in a letter to his wife: "The relief of Bastogne is the most brilliant operation we have thus far performed and is in my opinion the outstanding achievement of this war. Now the enemy must dance to our tune, not we to his....This is my biggest battle."

In his diary, Patton also wrote, "I hope the troops get the credit for their great work."

Later, he told an assemblage of correspondents: "It's a helluva lot easier to sit on your rear end and wait than it is to fight into a place like this. Try to remember that when you write your books about this campaign."

On January 8, 1945, Patton was again out in an open jeep marked with only his three stars. As usual, the roads were clogged with columns of vehicles stretching for miles. It was six degrees below zero. One column was filled with trucks

To have fought and won in the horrific winter conditions of the Ardennes in December 1944 was unthinkable.

carrying infantry of the 90th Division forward to battle; on the other side of the narrow highway were ambulances bringing the wounded to the rear.

According to historian John Toland: "When the men recognized Patton, they leaned out of the trucks, cheering wildly. The general's face broke into a smile. He waved. But he could hardly hold back the tears. Tomorrow many of those now cheering would be dead—because of his orders." It was an incredible scene, wonderfully spontaneous. For Patton, it was "the most moving experience of my life, and the knowledge of what the ambulances contained made it still more poignant."

The Battle of the Bulge was far from over, and the bloodiest battles of the winter war in the Ardennes were yet to come. The Germans resisted furiously, and in the foul weather Allied attacks moved with the sluggishness of a bulldozer. Although it was clear by the end of December that Hitler's strategic aim of splitting the Allied front was going to fail, German morale remained high. As the fighting continued to rage around Bastogne, Patton observed in his diary on January 4: "We can still lose this war. The Germans are colder and hungrier than we are."

The pincers of the First Army and Third Army at last closed the famous Bulge on January 16, 1945, dooming some 15,000 of Hitler's best troops to capture. The battles that had raged for six weeks in the frozen hell of the Ardennes were among the bitterest and bloodiest of any fought in Western Europe or Italy. Casualties on both sides were staggering, but the German losses were irreplaceable. The stage was set for the climactic battles of the war and the Allied invasion of the German heartland in early 1945. In such terrible conditions Patton was awed: "How men live, much less fight, is a marvel to me."

On December 30, not long after the relief of Bastogne, the *Washington Post* ran an editorial titled "PATTON OF COURSE." The newspaper, which had gleefully savaged Patton earlier in the year, now declared: "It has become a sort of un-

35TH INFANTRY DIVISION MACHINE-GUNNERS DUG IN NEAR LUTREBOIS, BELGIUM

written rule in this war that when there is a fire to be put out, it is Patton who jumps into his boots, slides down the pole, and starts rolling."

Although pleased that he was no longer the target of media criticism, Patton wrote to a friend, "Fortunately for my sanity, and possibly for my self-esteem, I do not see all the bullshit which is written in the home town papers about me."

Patton gave the credit to the soldiers who fought the battle. He rated them magnificent; they both moved and astonished him. On January 29, 1945, he told the press: "We hit the sons of bitches on the flank and stopped them cold. Now that may sound like George Patton is a great genius. Actually he had damned little to do with it. All he did was to give orders."

Patton's reputation soared; the battle had seized the public's imagination like no other since the Normandy landings and the great Allied breakout in early August. To have fought and won in the horrific winter conditions of the Ardennes in December 1944 was unthinkable.

After being stymied and frustrated in Lorraine and the Saar, Patton found an opportunity in the Ardennes to display his genius for war and turn a precarious situation to his advantage. Patton's maneuvering of the Third Army to relieve Bastogne did not win the Battle of the Bulge. Indeed, as historians have pointed out, the relief of Bastogne was made in a sector occupied by inferior German formations, and the heaviest German attacks against Bastogne did not commence until December 26. They also note that credit must be given to the men of the First Army, who stubbornly held the northern shoulder against such overwhelming odds.

If the entire Ardennes campaign resembled Wagnerian melodrama to the Germans, it was for Patton a Western film. Like the cavalry of yore, the Third Army rode to the rescue in a dramatic cliffhanger, with Patton at its head, rallying his troops. No battle could have been more tailor-made for Patton's talents— or for his theatrics.

What the Battle of the Bulge demonstrated most of all was his tremendous vision—the ability to anticipate and react with impeccable foresight to an enemy move or countermove. His greatest role was not so much as a battlefield tactician but as an organizer, mover, and shaker. His "true genius lay in his ability to put the show on the road, to move men and machines," historian Gerald Astor has observed.

Bradley later offered the highest praise of Patton he would ever accord: "His generalship during this difficult maneuver was magnificent. One of the most brilliant performances by any commander on either side in World War II. It was ab-

LT. GEN. GEORGE S. PATTON, IN FRANCE, JULY 7, 1944

solutely his cup of tea—rapid, open warfare combined with noble purpose and difficult goals. He relished every minute of it."

If George S. Patton had never before or after commanded again, he had earned a place in history in the Ardennes. It was a short, brutish campaign that solidified his reputation for battlefield generalship and left no doubt of the quality of the army upon which he had put his imprint.

Perfection is not attainable in war, but this was Patton's and the Third Army's finest hour. No one else could have pulled off such a feat.

BRITISH ARMY MOTORCYCLE DISPATCH RIDERS TAKE COVER, SOMEWHERE IN FRANCE, 1914.

CHAPTER SIX

MAX HASTINGS

"The British Fight"
adapted from *Catastrophe 1914: Europe Goes to War*

MOST OF THE TWO DOZEN BOOKS WRITTEN by Sir Max Hastings are focused on World War II. But with his most recent book, *Catastrophe 1914: Europe Goes to War*, Hastings has reached back a full century to the first chaotic year of the Great War. That wild time when the great powers of Europe dragged each other into industrial warfare on a shocking new scale was different from the later years of trench-warfare stalemate. To his accounts of the early, obsolete tactics and maneuvers, and of the hasty attacks and headlong retreats, he brings his usual attention to detail, to individual experiences and to big-picture analyses.

In this chapter, "The British Fight," Hastings chronicles the arrival of the British Expeditionary Force in France and its initial engagement with the advancing German armies at Mons and Le Cateau in August. Those first fights were marked by top-to-bottom confusion and incompetence, needless casualties, failure to communicate with their French allies, and "the moral collapse" of BEF commander-in-chief, Sir John French.

On August 3, 1914, *The Times*'s military correspondent, the intelligent cad Colonel Charles à Court Repington, declared that the Franco-German frontier would become the focus of the war's first big military operations. He added fiercely: "If our troops fail at the rendezvous, history will assign our cowardice as the cause"—meaning the sluggishness of the British government in agreeing to deploy an army on the continent. On the 12th, Repington wrote somberly: "We should not be under any illusion that the approaching Massenschlacht will be anything less than the most frightfully destructive collision of modern history," adding on August 15: "It is at least possible that the war may last a long time."

It is an enduring British conceit that the First World War began in earnest only on the 23rd, when the "Old Contemptibles" of the British Expeditionary Force drubbed the kaiser's hosts at Mons, thus saving England by their exertions and Europe by their example. In truth, of course, the French army had been engaged in vastly more murderous strife for almost three weeks before the first of the king-emperor's soldiers fired a shot in anger; Serbia, Poland and East Prussia were already steeped in blood. In northern France during the first exchanges of the war the British contribution was entirely subordinate to that of the vastly larger Allied forces. Against 1,077 German infantry battalions, at the start of the campaign the French deployed 1,108, the Belgians 120 and the BEF...fifty-two. It is unlikely that the kaiser ever spoke of Britain's "contemptible little army," as popular myth asserts, but its absurdly inadequate size justified such an appellation.

Field Marshal Sir John French's initial force comprised sixteen regiments of cavalry, the aforesaid fifty-two battalions of infantry, sixteen brigades of field guns, five batteries of horse artillery, four heavy batteries, eight field companies of Royal Engineers, together with supporting contingents. Later in the war—from 1916, as France became increasingly exhausted—Britain assumed a major role on the western front. In August 1914, however, the BEF conducted only a long retirement interrupted by two holding actions. German miscalculation and bungling, together with French mass and courage, did much more than British pluck to deny the kaiser his victory parade down the Champs Elysee. But this does not diminish the fascination with which posterity views the BEF's first actions.

The Anglo-Saxon allies were warmly welcomed to the continent. After a march on August 13, Lieutenant Guy Harcourt Vernon wrote: "Last mile ½ battalion fell out to be seized by inhabitants & dosed with water and cider. Discipline appalling." A café in the Place Gambetta of Amiens adopted a custom that had spread through all Europe's warring camps: at 9 p.m. closing time, uniformed and civilian patrons alike rose and stood at attention while the band played in succession each Allied national anthem. But the old women who supervised the local public baths treated their foreign visitors—by no means mistakenly—as lambs destined for slaughter. Mopping their eyes as they distributed free tea, they said: "*Pauvres petits anglais, ils vont bientôt être tués.*"

The third week of August found the adjoining armies of Alexander von Kluck and Karl von Bulow, half a million men, plodding doggedly southward through Belgium toward the French frontier. Richard Harding Davis, an American novelist and journalist, described their triumphal entry into Brussels at 3:20 p.m. on August 20: "No longer was it regiments of men marching, but something uncanny, inhuman, a force of nature like a landslide, a tidal wave or lava sweeping down a mountain. It was not of this earth, but mysterious, ghostlike." Harding Davis marveled at the sense of power projected by thousands of men singing "Fatherland, My Fatherland" "like blows from a giant piledriver."

As for their commanders, Kluck was sixty-eight years old, of non-noble background, a leathery professional who had risen on merit. Bulow was also sixty-eight, a Prussian nobleman to whom Kluck was subordinate, though in the field the latter not infrequently ignored the fact. Germany's chief of staff, Helmuth von Moltke, considered Bulow the ablest of his generals, and had thus entrusted him with the most critical responsibilities, but both he and Kluck were old men, well past fitness to assume leading roles in the greatest military operations in history, as would soon become apparent. Within their two armies both men and beasts were already feeling their feet. In just one German cavalry division, seventy horses died of exhaustion in the first fortnight of the campaign, and few others could raise a trot. No system was adopted for regularly resting the troops, the better to husband their strength and succour blistered feet.

Toward them marched the columns of the British Expeditionary Force, advancing through gently undulating countryside, basking in welcomes as warm as those its soldiers had everywhere met since disembarking at the Channel ports. "These French people are certainly enthusiastic beyond British comprehension," wrote Lord Bernard Gordon-Lennox of the Grenadiers, "and it would do old England a world of good to see the unbounded patriotism and *bon camaraderie* displayed on all sides." Some men remarked on the profusion of mistletoe in the branches of roadside trees, though relatively few would live to kiss any woman beneath Christmas boughs. Recalled reservists made up half of the strength of every BEF unit: fresh from soft civilian life and wearing unbroken boots, they struggled to keep up.

Sir John French had agreed to halt his force at the Mons-Condé canal just inside Belgium, where it could protect General Charles Lanrezac's left. But then Fifth Army received a bloody nose at Charleroi, and yielded ground. The British and French thus became perilously misaligned, with the BEF still advancing blithely, even as Lanrezac's men were falling back. When the khaki columns reached Mons, some thirty-five miles south of Brussels, soldiers whose faces were already reddened by the summer sun stripped off their tunics and began to dig in—to no great effect, amid the clutter of a suburban and industrial region. Buildings limited fields of fire. Toward nightfall insects began to swarm out of the waterways, causing thousands of men to curse freely as they slapped bites. In the distance to the southeast, some heard the crump of guns on Fifth Army's front. Sir John French learned of the repulses that had been inflicted on his allies, but did not comprehend their scale—the facts that the French Army had lost a quarter of its mobilized strength.

The little field marshal remained buoyant about Allied prospects. He knew there were Germans in the offing, but displayed a bizarre insouciance about placing his troops in their path. The BEF's highly competent chief of intelligence, Colonel George MacDonogh, gave warnings based on air reconnaissance and messsages from Lanrezac's staff, that three German corps were bearing down upon him. Sir John dismissed this threat, proposing to continue an advance. When he personally interviewed a Royal Flying Corps pilot who had gazed down on Kluck's masses, French revealed obvious disbelief, and changed the subject to quiz the troubled young man paternalistically about his aeroplane.

The first British shots of the war were fired early on the morning of the 22nd. Cavalry from C Squadron of the 4th Royal Irish Dragoon Guards were deployed at the top of a gentle slope about three miles north of the Mons-Condé canal. They saw approaching from a dip ahead of them a German lancer patrol, including an officer smoking a cigar. Captain Charles Hornby led two troops cantering down the road, sparks flying from the cobbles, in pursuit of the enemy, who took flight. There was a melee a mile on, in which the British took five prisoners. Corporal Ted Thomas used his rifle: after years on the ranges, where one waited some seconds for a paper target to be marked, he was amazed by the promptness with which a German horseman dropped from his saddle—the first enemy to fall to a British bullet. Hornby returned exultant, reporting that his own victim had died like a gentleman, at the point of a sword. He gave his weapon to the regimental armorer to be sharpened, expressing idiot regret at the necessity to have the blood wiped off. Bridges's brigadier had promised a Distinguished Service Order to the first officer to kill a German with the new pattern cavalry sword, and Hornby duly received this decoration.

The two divisions of General Sir Horace Smith-Dorrien's II Corps spent the night of the 22nd bivouacked along the Mons-Condé canal with Sir Douglas Haig's I Corps deployed in a quarter-circle on their right. The BEF's positions were anything but ideal for meeting an attack: the sixteen-mile canal was neither wide nor deep enough to constitute a major obstacle—averaging barely twenty yards of breadth. On some stretches, the ground on the north bank sloped down to the water amid either woodland or clusters of buildings, both offering cover to an approaching enemy. The British were too few to man a continuous line—some battalions bore responsibility for two thousand yards—and thus they concentrated around the bridges, leaving wide gaps that an attacker could exploit, especially with the aid of the barges moored at intervals beside the towpath. As the light began to fade on the 22nd, Lieutenant Colonel Charles Hull of the Middlesex regiment, whose rigid notions of discipline inspired both respect and fear, rode around the battalion's positions. He exploded into anger when he heard a company commander urge his men to blaze away at a passing German aeroplane: the colonel said they would soon need every round of ammunition they carried.

Partly because they expected imminently to advance again, but chiefly because they had not yet been galvanized by the ruthless imperatives of war, the defenders failed to use their hours of grace before the Germans' arrival to prepare the canal's eighteen bridges for demolition. They merely erected some half-hearted barricades, and covered the approaches with machine guns. Engineers laid a few precautionary charges; a sapper at one bridge set off on a bicycle to find detonators, with which he was unprovided. Just before dawn on August 23, Sir John French conferred briefly with his subordinates at Smith-Dorrien's headquarters in the Château de Sars. He seemed in ebullient form, asserting against all evidence that only one or at most two German corps were at hand. Then he rattled away in his motor car to visit an infantry brigade at Valenciennes, playing no further part in the battle which now developed. Here was extraordinary behavior by a commander-in-chief responsible for Britain's only field army, starting its first continental campaign for a century, with the enemy known to be at hand. French seemed to lack any sense of the gravity of the moment. His subordinates, down to platoon level, were given no clear briefings whatsoever, save that they should expect to defend their positions for a day or so.

In the small hours, an order reached units in the line: "You will stand to arms at 4:30 a.m. today. Transport to be loaded up and horses saddled. Acknowledge." During the tense hour or two that followed, while they lay on their weapons awaiting the enemy, the Middlesex received a superbly inconsequential message from division, complaining that one of their officers had ridden away from a Belgian blacksmith's forge at Taisnières without paying for his horse's reshoeing. Some men used the pause to improve their positions under the friendly eyes of local people, clad in their Sunday best. Neither soldiers nor civilians displayed much sense of peril, which only devastation and death would provide. The first brushes with German patrols took place in a light drizzle, but soon afterward the sun broke through. Enemy artillery began to drop shells on Smith-Dorrien's units, rudely interrupting some men at breakfast.

These were soldiers of an army which, for half a century past, had known only colonial campaigns, most often against natives armed with spears, though the Boers had shown them what modern small arms could do. The average age of the BEF was twenty-five, and many younger soldiers had never shot to kill. But there were also present old sweats who had fought dervishes and Pathans: when a *Guards* sergeant major set about forming his battalion's baggage carts into a defensive circle outside a Belgian village, he dubbed it—echo of Kitchener's Sudan—a "zareba." The BEF was small, but its soldiers were the best-armed Britain had ever sent to war, with the superb .303 Lee-Enfield rifle and Vickers machine gun. Their most serious deficiencies were of numbers, heavy artillery and motor transport. In the autumn of 1914, French countryfolk grew accustomed to the sight of requisitioned lorries still bearing the names of their London store owners—Harrods, Maples, Whiteleys—and motor bicycles ridden by eager young civilian owners who had volunteered their services as dispatch riders. Vans belonging to the caterers J. Lyons soon bore wounded men from London stations to hospitals.

This was an army whose officers took it for granted that—with the exception of the Army Service Corps, pioneers and suchlike subspecies—they were gentlemen, members of the same club, many of whom knew each other. When Major Tom Bridges found himself unhorsed in the path of the enemy, he was rescued by a pass-

SIR JOHN FRENCH, COMMANDER-IN-CHIEF, BRITISH EXPEDITIONARY FORCE, 1914

ing staff officer in a Rolls-Royce, who proved to have been at school with him. But British commanders of 1914 regarded the horse, not the automobile, as their natural means of transport, whether at home or on a battlefield. After peacetime years in which promotion proceeded at a tortoise pace, more than a few captains aged thirty-six or -seven served at Mons, along with many majors in their forties. Their men were overwhelmingly drawn from the industrial underclass or rural peasantry. Charles

Edward Russell, a prominent American socialist and former newspaper editor who visited Britain during the summer of 1914, deplored manifestations of the class system in uniform. Watching recruits being drilled, he noted the disparity between the heights of officers and men—the former were an average five inches taller—and the poor appearance of the latter: "the dull eyes, the open mouths that seem ready to drool, the vacant expression, the stigmata of the slum—terrible spectacle."

Yet some, though by no means all, such victims of privation made resolute soldiers. It was rash to expect them to think much for themselves, but the same limitation afflicted most of their officers. Few would have been wearing khaki serge that day, had they been capable of scraping a meal ticket by any other means. "There was no hatred of Germany," wrote Tom Bridges, a veteran of the Boer War. "In the true mercenary spirit we would equally readily have fought the French." Beside the canal, they smashed the windows of homes and warehouses to create firing positions, some with a vestigial guilt about injuring property.

The first of Kluck's infantry began to push downhill toward the water, shielded along most of its unlovely length by drab houses, mine pitheads and industrial installations. Though the German army was a mighty instrument of war, at this critical moment it revealed its weaknesses, foremost among which was intelligence. In August, all the belligerents' commanders vied with each other in misjudgements of their opponents' strengths and intentions. Kluck's was the largest of the kaiser's seven armies in the west. Men of its leading regiments approached Mons aware that British soldiers were in the vicinity, but with no notion of their strength or deployments. Kluck himself was esteemed by his peers, but revealed no genius in this, his first battle of 1914.

Private Sid Godley was enjoying coffee and rolls brought to him by two Belgian children, with whom he made clumsy efforts at conversation, when their little party was interrupted by an incoming German shell. He recalled later: "I said to this little boy and girl: 'You'd better sling your 'ooks now, otherwise you may get hurt'. Well, they packed their basket up and left." Godley settled down behind his rifle. As the first Germans showed themselves, thousands of British soldiers opened fire, the rippling crackle of their musketry soon overborne by the crump of artillery. The Germans began crowding around the dangerous salient northeast of Mons, at Nimy where the bridges were defended by the Royal Fusiliers: legend has it that they were warned of the enemy's approach by the stationmaster's daughter. Colonel Hull, commanding the neighboring Middlesex, was a small arms enthusiast who had taken pains to ensure that his men could shoot straight, and that day they did him proud. Successive German rushes were checked by murderous rifle fire. Huddled gray-green corpses, surmounted by pickelhaube helmets, soon littered the north bank. But Kluck's men, in their turn, took up firing positions and were soon inflicting casualties on the ill-concealed British.

One of Hull's men, Private Jack, said later: "When the firing began, I was frightened by the noise. I'd never heard anything like it. Most of the shells were bursting well behind us, but there was also a strange whistling sound as the bullets came over. There were four of us in a rifle pit and our officer walked over to us and I remember thinking: 'Get down, you silly bugger.' Later on I heard the poor man was killed. Then the man next to me was hit. I was firing away and suddenly he gave a sort of grunt and lay still. I'd never seen a dead man before." Guy Harcourt-Vernon wrote: "Funny to notice how everyone ducks at the sound of a bullet. You know it is past you, but down goes your head every time." Soon, too many bullets and shells were passing for any man to have time to duck. Most concentrated upon ramming clip after five-round clip into their hot weapons, though too much has been made of the British rifleman's notional fifteen-rounds-a-minute capability. Any unit that sustained such a rate of fire would speedily have exhausted its ammunition.

Most of the Germans surging forward were as new to war as their foes. Some experienced brief euphoria, such as was later described by Walter Bloem, a captain in the 12th Brandenburg Grenadiers. As he advanced "a shout of triumph, a wild, unearthly singing surged within me, uplifting and inspiring me, filling all my senses. I had overcome fear; I had conquered my mortal bodily self." At first, Kluck's men advanced in masses, direct from their line of march, and suffered in consequence. A British NCO wrote: "They were in solid, square blocks, standing out sharply against the skyline, and you couldn't help hitting them....They crept nearer and nearer, and then our officers gave the word....They seemed to stagger like a drunk man hit suddenly between the eyes, after which they made a run for us, shouting some outlandish cry that we couldn't make out."

In the same spirit a Gordon Highlander recounted: "Poor devils of infantry! They advanced in companies of quite 150 men in files five deep, and our rifle has a flat trajectory up to 600 yards. Guess the result. We could steady our rifles on the trench and take deliberate aim. The first company were simply blasted away to Heaven by a volley at 700 yards, and in their insane formation every bullet was almost sure to find two billets. The other companies kept advancing very slowly, using their dead comrades as cover, but they had absolutely no chance." The war would become overwhelmingly a contest between rival machine guns and artillery, but for a brief season the rifle displayed its powers against bodies of men exposing themselves in plain view.

Lieutenant George Roupell of the East Surreys wrote: "The enemy came through the woods about 200 yards in front, they presented a magnificent target....The men were very excited as this was their first 'shot in anger'. Despite the short range a number of them were firing high but I found it hard to control the fire as there was so much noise. Eventually I drew my sword and walked along the line beating the men on the backside and, as I got their attention, telling them to fire low. So much for all our beautiful fire orders taught in peacetime!"

The British grossly overestimated the casualties their riflemen inflicted. Many Germans who dropped to the ground were merely taking cover. Kluck's units broke up into smaller groups and began to fight more subtly, supported by howitzers which inflicted steadily mounting casualties. Far from behaving like the mindless squareheads of British caricature, the Germans used fire and maneuver effectively. Smith-Dorrien's companies holding advanced positions beyond the canal fell back to the south bank. "God! How their artillery do fire!" exclaimed a Gordon: the shelling was a new and unwelcome experience for almost every member of the BEF. "The men were digging little holes for themselves to sit in," wrote Tom Wollocombe of the Middlesex, "and most of them were getting a bit jumpy, not being used to such living." By the standards of the French battles a few days earlier, far less those of Ypres two months later, British losses were slight. But for troops with no experience on the receiving end of modern firepower, that August day on the canal bank seemed terrible enough.

While Kluck led a much bigger army than Sir John French, the numbers of troops engaged on each side at Mons on August 23 was roughly equal. Much has been written about British heroism, less about equally notable German courage. While significant numbers of Kluck's men were shot down at the approaches to the water, scores worked forward to seek footholds on the south bank, some of which were secured within ninety minutes of the battle's beginning. Memorable among the Germans was Oskar Niemeyer, a Hamburger. East of the rail crossing at Nimy defended by Royal Fusiliers stood a pedestrian bridge, which could be swung across the canal by pedal power. The British had parked this along their own bank. Niemeyer dived into the water, swam across, and under fire pedaled the bridge almost to the north bank before being shot down, a feat that would have won him a Victoria Cross had he been wearing khaki that morning. The dead man's comrades were able to toss a rope to secure the bridge and pull it to their side; then they began to dash across.

Such actions at a dozen points during the course of the morning exposed some British units to enfilading fire, and indeed threatened them with isolation. Shortly before 1 p.m., the Middlesex received a foolishly belated message from division: "you will decide when bridges and boats within your zones should be destroyed." Tom Wollocombe wrote: "it was too late. The enemy were across or crossing." The defenders at Mons were much too thinly

GENERAL ALEXANDER VON KLUCK, COMMANDER OF THE LARGEST GERMAN ARMY IN FRANCE, 1914

spread to generate the intensity of violence necessary to halt Kluck's host. British artillery batteries, close behind the infantry suffered almost as severely from German fire as did riflemen. "Our faithful gunners stuck to their pieces magnificently," said Wollocombe. One of them, Sergeant William Edgington, wrote in his diary with notable understatement: "A very trying day—Germans seemed to be all around us."

Though the British mauled Kluck's leading regiments, as the day wore on their own casualties rose; meanwhile, the trickle of enemies crossing the canal swelled into strong streams. So accurate was enemy artillery fire in some sectors that Smith-Dorrien's soldiers, like those of every other nation that August, became morbidly convinced that spies must be spotting for Kluck's batteries. Eventually, unit by unit II Corps began to fall back, its men scrambling in small groups toward the rear, platoons taking turns to cover each other's withdrawal, some soldiers supporting wounded mates. The difficulty was to make retreat a disciplined maneuver, not a headlong flight. When Colonel Hull saw one of his platoons retiring under the orders of a sergeant, he told his adjutant to see who the NCO was. After a glance through his field glasses, Tom Wollocombe gave the name, causing Hull to say furiously: "if Sgt.—— had not had any order to retire, he would have him shot." In the event, the suspect proved to be on the battalion's "Missing" list that night, so escaped the threatened firing squad.

At 3 p.m. Captain Theodore Wright of the Royal Engineers began a brave but hopelessly tardy journey along the canal, to attempt demolition of five bridges along a three-mile front. Wright's party was under fire most of the way, and his driver was understandably alarmed by the experience of threading a path across a battlefield in a car containing eight crates of guncotton. Shot at from three sides, the engineer was eventually successful in destroying the crossing at Jemappes.

British artillery batteries, close behind the infantry suffered almost as severely from German fire as did riflemen.

While working on another at Mariette, he was grazed in the head by a shell fragment, then found himself without electricity to detonate his charges. He hastily ran a cable to the mains of a nearby house. Still getting no live current, he tried again and again to achieve a contact, while men of the Northumberland Fusiliers provided covering fire. Then exhaustion caused Wright to slip into the canal. An NCO fished his officer out, but by then it was 5 p.m., and the Germans were shooting at them from a range of thirty yards. The engineers abandoned their efforts and retired. For this gallant day's work, and others before he was later killed, Wright received a VC. It was all in vain; only one bridge on the British front was ever blown—the necessary orders had been given far too late.

By nightfall, the Germans held Mons. There is no reliable record of their losses, but Walter Bloem's battalion commander of the Brandenburgers lapsed into emotional lamentations: "You are the only company commander left...[it is] a mere wreck, my proud, beautiful battalion!" Their regiment had lost killed one battalion commander and his adjutant, three company and six platoon commanders; a further sixteen officers were wounded; other ranks had suffered in proportion. Bloem reflected miserably: "Our first battle is a heavy, an unheard-of heavy defeat and against the English, the English we laughed at."

Though this remark is often quoted in celebration of the BEF's achievement, it was a wild exaggeration, reflecting the writer's sensitivity to losses, common to all novice warriors. Bloem's battalion suffered much heavier casualties than any other German unit that day. The British had been unable to frustrate Kluck's advance, merely delaying it by a day before falling back. Another German regimental narrative recorded triumphantly that at nightfall "the spirit of victory was overwhelming, and was enjoyed to the full." The miracle of Mons was that enemy bungling allowed the BEF to withdraw almost intact, having lost an estimated 1,600 men, many of them taken prisoner. A former traveling salesman from Hamburg who spoke fluent English marshaled some of the latter good-humoredly: "Gentlemen, please, four by four!" Almost half of the losses fell on just two battalions—4th Middlesex with over four hundred and 2nd Royal Irish with more than three hundred; several units were obliged to abandon their precious machine guns. German total casualties were roughly similar, but with a much higher proportion of killed and wounded.

The British regarded their allies with contempt. Yet it was critical to the stand at Mons and subsequent escape of II Corps that a scratch force of French territorials covered Smith-Dorrien's left flank. And even as the little British action was being fought, Lanrezac's Fifth Army suffered far more heavily, at Charleroi. Farther south still, in the Ardennes on the 23rd and 24th the Fourth French and Fourth German armies lost 18,000 dead between them. By comparison with such engagements, British doings at Mons receded in significance—though not in the minds of Sir John French and his senior officers. At 3 o'clock on the afternoon of the 23rd, the C-in-C returned from his trip to Valenciennes, still prey to delusions that the Allies might soon renew their advance. By nightfall, however, he was forced to recognize reality, to accept Colonel McDonogh's assessment that his army faced an overwhelmingly powerful enemy. Kluck's men were crowding in upon II Corps's right—now south and west of Mons—and threatened to isolate it from I Corps; finally and most disturbing, Sir John knew Lanrezac had begun to withdraw Fifth Army from the Sambre valley. The BEF had started the day nine miles ahead of the French. Now, that gap was about to widen dangerously, inviting the Germans to fill it. Sir John acknowledged that his own command must pull back fast, to avert almost inevitable destruction.

The BEF bivouacked for the night some three miles south of Mons, its men expecting to fight on their new line next morning.

That evening Tom Wollocombe "even had time to think that a battle was a wonderfully exciting thing when it was in progress....our men, instead of being downcast, were much impressed with the superiority of their rifle fire and extended order maneuvering, over the enemy's fire and movements 'en masse.'" But at 1 a.m. on the 24th, GHQ issued new orders for a retreat, unassisted by guidance about how this was to be carried out. In the space of a few hours, the C-in-C himself had lapsed from jaunty confidence into gloom, even panic. Now, French talked at one moment of leading his force to take refuge in the old fortress of Mauberge; at another, of withdrawing northwestward to Amiens, severing all contact with his allies. A few days' experience of campaigning caused the British C-in-C to leap to the almost hysterical conclusion that French soldiers were not people with whom he could do business, not "proper chaps" with whom he wished to continue fighting a war. Such an attitude would merely have invited ridicule, did it not threaten grave consequences for the Allied cause.

Meanwhile in Paris that morning of the 24th, Joffre told Adolphe Messimy, the war minister, that for the time being the French army had no choice save to abandon the offensive, which had failed. The nation's strategy was discredited. The French army had almost spent itself in futile attacks; it could aspire only to a protracted defense. "Our object," the commander-in-chief told the politician, "must be to last out as long as possible, trying to wear out the enemy, and to resume the offensive when the time comes." Amid the news from the north, Joffre's vast illusions about German deployments and intentions were at last falling away. He grasped Moltke's purpose—to achieve a vast envelopment from the north.

Hitherto, the French C-in-C had paid only casual attention to the Allied left wing. Hereafter, it became the focus of all his fears—and then of his hopes. On the 25th, he issued his later-famous *Instruction Général No. 2*, declaring his intention to start transferring large forces northward, to create a new army on the left of the BEF. He was anxious to confront the peril on his flank with forces upon which he could rely to accept his orders—as the British would not. But Joffre's immensely complex redeployment could not be completed before September 2—an eternity away, in the circumstances of the moment. Much must necessarily happen, for good or ill, before that day came, some of it to the BEF.

At first light on the 24th, the Germans began once more to press II Corps. Many units that day experienced skirmishes before falling back to bivouacs a few miles farther southward. A notorious incident took place when the 9th Lancers and Dragoon Guards charged German guns at Audregnies across a mile of open ground, an extraordinary piece of folly even by the standards of British cavalry. They were led by Lieutenant Colonel David Campbell, a celebrated horseman who had once won a Grand National steeplechase riding his own horse The Soarer. An unexpected sunken road caused many fallers; German guns unhorsed more men, who sought cover behind corn stooks, from which they returned fire. Eventually the British withdrew, having suffered eighty human casualties—fewer than they deserved—and rather more equine ones. Fourteen-year-old German schoolboy Heinrich Himmler wrote exultantly in his diary: "Our troops advance to the west of the Meuse toward Maubeuge. An English cavalry brigade is there and is beaten, really beaten! Hurray!"

That day Major "Ma" Jeffreys of the Grenadiers—in Haig's corps—described "a long and trying march...in great heat and over very bad and dusty roads. The men very tired and rather puzzled as to what we are at." Major Bernard Gordon-Lennox deplored the secrecy which kept officers in ignorance of GHQ's plans and intentions: "most disheartening. No one knows what one is driving at, where anyone is, what we have got against us, or anything at all, and what is told us generally turns out to be entirely wrong." In truth, of course, this mystification derived not from GHQ's sense of discretion, but rather from its incompetence and indecision. Failure to brief subordinates about the purpose and context of their movements proved a chronic British command weakness throughout the campaign.

The same pattern was repeated on August 25. Beside the ruins of the old Roman forum in Bavay, southward paths divided. A single road could not possibly carry the entire BEF and a mob of civilian fugitives. It was decided to dispatch I Corps by the route which ran east of the great forest of Mormal, while II Corps took an almost parallel line on its west side. All day, a traffic jam persisted in Bavay, as French's jumbled formations struggled through. "I have never been so tired," wrote Captain Guy Blewitt of the Oxfordshire and Buckinghamshire Light Infantry, "as in the last 46 hours I had no sleep, covered 40 miles besides having the anxiety of a rearguard. At nearing Bavay it was evident that things were serious, the road being packed—cavalry with their horses, cavalrymen who had lost their horses, ambulance wagons, refugees, bicycles, perambulators, guns, infantry in fours, infantrymen who had lost their units and infantrymen whose units didn't know where they were required and were sleeping by the side of the road. The cobbles made one's feet sorer and we were very glad to be turned into a stubble field to bivouac; here fires were soon burning and we got some food to eat and straw to sleep on."

In those innocent first days the British lacked the ruthlessness necessary to clear their road of fleeing civilians and vehicles. Blewitt saw a very old Belgian, ob-

Are **YOU** in this?

PATRIOTIC POSTERS WERE AN IMPORTANT RECRUITING TOOL IN BRITAIN.

viously at his last gasp, being wheeled past on a cart. The Englishman winced at the irony when the old man summoned just enough strength to cry in a high, fluty voice "*Vive l'Angleterre!*" By contrast, some units which had been cheered when they advanced now found themselves booed as they retired: local people divined the price they would pay for Allied defeat when the Germans arrived.

Meanwhile some rearguards, pressed by the Germans, were obliged to fight. Guy Harcourt-Vernon wrote: "Germans come on in masses...our fire mows them down; can't see much except dark moving mass." The Royal Welch went into action for the first time near St. Python; after a brief exchange of fire and a sharp rainstorm, they were ordered to withdraw. Outside the little French town of Le Cateau that evening, pickets suddenly glimpsed enemy cavalry on their front. Some women said: "*Les allemands arrivent,*" as casually as they might have reported the advent of a tourist party. Orders were given to the lead platoon of the Welch such as the British Army had been issuing since time immemorial: "Right wheel; double march; halt, right turn; front rank kneeling, rear rank standing, three rounds rapid, fire." The German horsemen made off. On that same day of the 25th, 2nd Grenadiers marched almost fifteen miles, oppressed by the heat, troubled by blistered feet and impeded by refugees pushing barrows and handcarts. A British officer gazed with pity upon an old Frenchwoman torn between the urge to seek safety, and a deep peasant instinct against abandoning her farm. "But who will feed the pigs if I go?" she cried.

Brilliant late August sunshine, warming and lighting the French countryside, mocked the condition of the warring armies, milling amid a fog of misapprehensions and uncertainties. On the 25th the British II Corps suffered many frustrations: dense masses of refugee traffic enforced halts on its retreating columns; units fell behind amid local difficulties—the Royal Irish Rifles were delayed by a long train of artillery crossing the battalion's line of march. That evening their colonel, Wilkinson Bird, reported to his brigadier that the men were too exhausted both to march and fight through the night as rearguard. At 10 p.m. the battalion entered Le Cateau, some twenty-five miles south of Mons. Bird went to the post office and telephoned Corps HQ, who told him to keep marching to the village of Bertry, three miles west.

He emerged into the brilliantly lit town square, thronged with wagons, stragglers, soldiers eating and drinking in restaurants. One of his officers asked: "Are you going to halt, sir?" Bird answered tersely: "No—damned sight too dangerous."

He knew that once his men fell out, it would take hours to herd them onto the move again. The battalion trudged up the hill out of the town into rustic darkess—and became lost. At 2 a.m., they blundered into Reumont, a mile short of Bertry, where they found 3rd Division's headquarters. Bird asked for a meal for his men. A staff officer said: "You won't get it because we are retreating again at four, and yesterday it took five hours to get under weigh." The riflemen collapsed into sleep in a nearby cluster of farm buildings. Some officers procured a meal at a little café in nearby Maurois.

The previous evening, II Corps had issued Operation Order No. 6, which began "The Army will continue its retirement tomorrow." In the small hours of the 26th, however, Smith-Dorrien felt compelled to reconsider. Many of his units were in the same exhausted and hungry condition as the Irish Rifles, and some were still tramping through the darkness toward Le Cateau. He reckoned that if the corps tried to move on southward that day, its cohesion must collapse; lagging units would be overrun by Germans hard on their heels.

Generals' personalities sometimes lack color, but this could not be said of Sir Horace Smith-Dorrien. Twelfth in a family of sixteen children, as a young transport officer in Zululand he was one of the few survivors of the 1879 disaster at Isandlwana, following which he was nominated for a VC for his efforts to save other fugitives. Thereafter he gained extensive experience of colonial wars, and fought at Omdurman—he became a lifelong friend of Kitchener. He emerged from the Boer War with an enhanced reputation, and thereafter held a succession of commands. In July 1914, Smith-Dorrien was sent to address several thousand public school cadets at their summer camp, where he astonished an almost uniformly jingoistic audience by asserting that "war should be avoided at almost any cost; war would solve nothing; the whole of Europe and more besides would be reduced to ruin; the loss of life would be so large that whole populations would be decimated." At the time most of his cadet listeners recoiled from such heresy, but those fortunate enough to survive until 1918 came to look back with respect on Smith-Dorrien's frankness and independence of thought. Though generally calm and robust, he was prone to outbursts of extreme temper which caused subordinates to quail, and had indeed provoked his chief of staff to try to resign his post after Mons.

This, then, was the man in charge at Le Cateau on August 26. Early in the small hours, Smith-Dorrien consulted such senior officers as he could convene. Edmund Allenby, commanding the cavalry, reported that both his men and his horses were "pretty well played out." He said that unless II Corps began to withdraw before dawn, the enemy was so close that a battle at daybreak would be unavoidable. Hubert Hamilton, commanding 3rd Division, said his men could not possibly move before 9 a.m. The 5th Division was even more scattered, and 4th Division—which had detrained from the Channel ports only on the night of the 24th and still lacked most of its support units—was entangled in a rearguard action. Smith-Dorrien asked Allenby if he would accept his orders. Yes, said the cavalryman. "Very well, gentlemen, we will fight," said the corps commander in a manner that would read well to history.

All the officers present heaved a sigh of relief. Amid the chaos and confusion of purpose which had attended them for three days past, here was a clear decision, which they welcomed. So too, at first, did Sir John French when informed in a message delivered to GHQ by automobile that half his army was to conduct a second battle of the campaign without benefit of the commander-in-chief's guiding hand or assistance. French later very publicly recanted, castigating Smith-Dorrien in his memoirs. Given II Corps's situation, however, it is hard to see how its commander could have acted otherwise. He proposed to try to inflict "a stopping blow" on the Germans, to gain a breathing space in which to resume his retirement. He expected I Corps to support him and was given no hint by French that Haig was continuing to withdraw, leaving II Corps's right flank in the air.

At 7 a.m. Smith-Dorrien was summoned to take a call on the railway telephone network, which proved to be from Sir Henry Wilson. The army's subchief of staff said the C-in-C had now decided that II Corps should resume its retreat. Too late, said Smith-Dorrien; his troops were already in action, and could not disengage before dark. Wilson said: "Good luck to you; yours is the first cheerful voice I have heard for three days." But he appears also to have expressed gloom about II Corps's prospects. James Edmonds later that day met Smith-Dorrien, who complained how little he knew about what was going on, and about having been obliged to make so big a decision. Edmonds replied reassuringly: "You needn't bother your head about that, sir. You have done the right thing." The general said that GHQ appeared to differ: "that fellow Wilson told me on the telephone this morning that if I stood to fight there would be another Sedan"—referring to the disaster that befell the French there in 1870.

When Sir John French's chief of staff received Smith-Dorrien's message that he planned to make a stand at Le Cateau, Sir Archibald Murray was convinced that it was all up with the BEF. In a manner that might be deemed tiresomely theatrical had it not been authentic, he collapsed in a dead faint. A colleague implausibly named "Fido" Childs said, "Don't call a doctor: I have a pint of champagne." James Edmonds wrote sardonically: "And that they poured into Murray about 5 a.m.!... 'Curly' Birch, who was riding about the field in a towering rage looking for the cavalry brigades which Allenby had lost, told me that the instructions of GHQ were 'to save the cavalry and horse artillery.'" It was a time of near-madness at French's headquarters, which enjoyed no accession of sanity as the day unfolded.

Now and for days ahead, the C-in-C and his staff were prey to confusion, defeatism and indeed panic. Joffre witnessed this for himself when he arrived at St. Quentin later in the morning, to confer about his new campaign plan with the British and Lanrezac of Fifth Army, even as Smith-Dorrien's men were fighting for their lives a few miles northward. The generals met in a gloomy, overfurnished bourgeois mansion off the main street, where Sir John French had established himself. Lanrezac was in a vile temper, and had earlier that morning abused both Joffre and French to his own staff in a fashion that dismayed and even disgusted them. He professed agreement when Joffre said that it was essential for Fifth Army to keep counterattacking, to sustain pressure on the Germans.

The C-in-C was not to know that, in reality, Lanrezac had no intention of doing anything of the sort. On the 26th, while the British fought at Le Cateau, Fifth Army continued its drifting retreat; the only French forces which saw significant action that day were the scratch group of territorial divisions on Smith-Dorrien's left. Tom Wollocombe was one of the few British officers to acknowledge handsomely the contribution of their allies: "the French troops...took a lot of pressure off us." Meanwhile at St. Quentin, Joffre was shocked by the wild words of the British C-in-C, who railed at the fashion in which the BEF had been exposed to disaster since the moment it reached the front, for lack of French support. Their conference took place in a shuttered and thus darkened room where, according to the young British liaison officer Lieutenant Louis Spears, who was an eyewitness, "everyone spoke in an undertone as if there were a corpse in the next room." Protracted interpretation was necessary, since few of the British present spoke French, and neither Joffre nor his subordinates were fluent in English.

France's commander-in-chief began to explain his counteroffensive plan—General Instruction No. 2. He was dismayed to learn that the BEF's C-in-C knew nothing about this. Sir Archibald Murray, in a state of physical and mental collapse, had failed to show his chief the critical document. Joffre summarized his intention to create a new "mass of maneuver" with the French Fourth and Fifth Armies on the right of the BEF, then bring up fresh forces on its left. He urged upon his British allies the need to stand their ground and launch a counterattack, for which he promised French support.

Sir John was unmoved by any of this; he merely insisted that he intended to continue his own withdrawal. Spears wrote: "The sense of doom was as evident in that room as when a jury is about to return a verdict of guilty on a capital charge." When the meeting ended, Sir John French drove away southward, taking his headquarters with him, almost heedless of Smith-Dorrien's battle farther north. Spears again: "It was perhaps the worst day of all at GHQ. Nerves were bad, morale was low, and there was much confusion. The staff wanted heartening, and Sir John's departure had the contrary effect."

Joffre wrote in his memoirs: "I carried away with me a serious impression as to the fragility of our extreme left, and I anxiously asked myself if it could hold out long enough to enable me to effect the new grouping of our forces." The Allies' principal commander was confronted by the vast, looming German threat; by doubts about the nerve and competence of Lanrezac in the most heavily threat-

SUPPLYING THE BRITISH EXPEDITIONARY FORCE WAS A MAJOR CHALLENGE.

ened sector; and finally by a British C-in-C alienated from his allies and visibly un-manned by the crisis. One British corps was retreating on a different axis from that which GHQ had decreed, while the other had started a critical battle on its own initiative. The St. Quentin conference ended in indecision, its only outcome British acquiescence in Lanrezac's further retirement. Joffre departed without having made any attempt to impose his personality, to force Sir John French's hand. Both the Allied commanders-in-chief seemed bereft of that most vital of all battlefield qualities: grip.

In fairness to the BEF's commander, Joffre's assurances of Lanrezac's cooper-ation proved worthless. But this scarcely justified Sir John's growing deter-mination, in effect, to wash his hands of the campaign. To say that French's headquarters was not a happy place, his staff not a united team, would be an un-derstatement. Years later, Sir Archibald Murray wrote bitterly to an old comrade: "To me it was a period of sorrow and humiliation....As you know, the senior mem-bers of [GHQ] entirely ignored me as far as possible, continually thwarted me, even altered my instructions...Why did I stay with this War Office clique when I knew that I was not wanted? It was my mistake....I wanted to see Sir John through. I had been so many years with him, and knew better than anyone how his health, temper and temperament rendered him unfit, in my opinion, for the crisis we had to face." The only sentiment shared by French, Murray, Wilson and their corps commanders was lack of confidence in each other, an alarming state of affairs at the summit of an army in the field. The only band of brothers to which they might be likened was that of Cain and Abel.

Once early morning mist cleared on the 26th, a succession of RFC pi-lots landed back at their fields from scouting missions to report enemy forces clogging every approach road for miles in front of II Corps; "[the airmen's] maps were black with lines showing columns of German troops," in the words of a staff officer. Six infantry regiments were clos-ing fast upon Le Cateau, celebrated as the home of Matisse. "A sun-baked drowsy little place it seemed," in the words of a British officer, "on the eve of being flung into history to the accompaniment of the roar of great guns...unconscious of its fate, the little town looked as if nothing could ever rouse it." The action Smith-Dorrien fought on August 26 proved much bloodier than Mons—indeed, as costly to the British Army as was D-Day in June 1944, a world war later. It was utterly unlike almost everything that happened to its survivors in the ensuing four years. This was the last significant battle the British Army would ever fight in which a man standing on the rising ground a mile or so northwest of Le Cateau might have beheld most of the day's critical points within his own range of vision.

The little town lay in a valley where it was almost invisible to 60,000 troops who took up positions across ten miles of green and golden fields in the open, rolling countryside above. The corn had been cut, and stood in ordered stooks on the stubble, interspersed with fields of sugar beet and clover, together with occasional haystacks, reaching as far as the eye could see. One soldier thought the place resembled a familiar exercise ground—"Salisbury Plain without the trees." Smith-Dorrien deployed his exhausted corps as best he could, without benefit of much reconnaissance. Some units, especially those on the right nearest to Le Cateau, found themselves defending positions which were soon overlooked by the advancing Germans, who could bring up men in dead ground.

Some local townspeople came out to help the British entrench. Nearest to Le Cateau, the Yorkshires settled into shallow rifle pits dug by Royal Engineers, with the Suffolks on their right. The Norfolks struggled to cut down a lone tree on their position, which offered a conspicuous aiming point for enemy gunners. Signals detachments cantered across the appointed battlefield, laying telephone cable off whirling drums mounted on wagons. But wire was in desperately short supply, because so much had been used and lost at Mons. The most important means of military communication in August 1914 remained that of the past few thousand years: messengers mounted or afoot. On the field of Le Cateau, gallopers were a familiar sight, dashing from unit to unit, delivering orders at mortal hazard.

The battle unfolded piecemeal, broadly from right to left of the British line. German artillery began firing at 6 a.m., and soon afterward Kluck's men entered the town of Le Cateau, which was undefended, pushing British pickets back up the hill at its eastern edge. One of the attackers, Lieutenant Kuhlorn, recalled later: "I gave my platoon the orders 'On your feet! Forward! Go!' and we advanced in short rushes. When I looked around during a pause, I found I had about eight men and some NCOs with me. The remainder had stayed where they were." But a few yards at a time, he and his regiment gained ground. By 9 a.m., Kluck's guns were bringing down heavy fire on the Suffolks and Yorkshires and their supporting batteries, all in plain view, plunging them into an ordeal which continued for more than six hours thereafter. The Suffolks' colonel was among the first to fall;

before long one British battery had lost all its officers, and was firing only a single gun. By midmorning Smith-Dorrien's right was outflanked, so that for the rest of the day, the Germans were shooting at the Suffolks and Yorkshires from three sides, and had machine guns sited to enfilade the British positions.

Some of II Corps's units farther north were still marching to their appointed places in the line after the battle started. At 7 a.m. a panting cyclist orderly pedaled up to the farmhouse where Colonel Bird of the Irish Rifles had snatched an hour or two's sleep, with orders to march at once to Bertry. Bird was at first mystified about where to find his men. He roused Captain Dillon, the adjutant, fast asleep in an armchair. "I'm awfully sorry, sir," said Dillon, "I remember sitting down, then nothing more until you woke me." An hour later, with his companies trudging dozily behind him, Bird rode into Bertry, where outside Corps HQ he met Smith-Dorrien. "Will your men fight?" demanded the general. Yes, said Bird. Smith-Dorrien stared down the column. "Your men look very well....[They] just want a damned good fight and no more of this retreating." The Irish Rifles were dispatched two miles northwest to Caudry station, in the center of the British line.

A staff officer later reported that once the die was cast, Smith-Dorrien wanted no interference from his C-in-C. "[He] was most anxious that Sir John should not come—he spoke for quite a long time on this point, after which a few casual remarks about his left and right flanks both being in the air, but that he was confident of giving the German a good fight even if he was running the risk of being surrounded." Around 10 a.m., masses of German infantry began advancing across the stubble fields west of Le Cateau. Kluck believed that he was deploying against six BEF divisions, which were retreating southwestward. In consequence of this misapprehension, his formations stumbled upon the British in a succession of uncoordinated encounters that denied the Germans the chance to throw their full weight behind the punches.

Kluck's soldiers were quite as tired as their opponents, having marched thirty miles the previous day. Contrary to British claims of overwhelming numbers of attackers thrown against II Corps, only six regiments, together with three or four Jäger battalions of skirmishers and several thousand dismounted cavalry, came into action on August 26. This was a formidable force, backed by excellent artillery. But Le Cateau cannot credibly be portrayed as the David-and-Goliath clash of British myth: the respective forces were about equal. Just as at Mons, wherever enemy masses appeared within rifle range, they were mauled by the BEF's impressive musketry. "It is impossible to miss German infantry," wrote forty-three-year-old Major Bertie Trevor, a company commander in the Yorkshires. "They come on in heaps." But the defenders in their turn suffered from artillery fire, which caused especially severe losses in British batteries, deployed as conspicuously as were their forebears on the ridge of Mont St. Jean, at Waterloo in 1815. Indeed, the first Duke of Wellington would have seen much at Le Cateau that was familiar to him: enemy troops advancing in close-packed columns; drivers lashing lathered artillery teams forward to unlimber; gallopers bearing orders hither and thither.

A German officer wrote wonderingly: "I did not think it possible that flesh and blood could survive so great an onslaught....Our men attacked with the utmost determination, but again and again they were driven back by those incomparable soldiers. Regardless of loss, the English artillery came forward to protect their infantrymen and in full view of our own guns kept up a devastating fire." Another German participant, Lieutenant Schacht of a machine gun company, observed more sceptically: "We could see a [British] battery which, according to our doctrine, was located far too far forward, in amongst the line of infantrymen, to which we had already approached very closely. Right! Sights at 1,400 metres! Rapid fire. Slightly short. Higher! Soon we could observe the effect. There could not be greater activity around an upturned antheap. Everywhere men and horses were milling around, falling down and, in among all this brouhaha was constant tack-tack-tack."

When Smith-Dorrien ordered forward men of his slender reserve to reinforce the threatened right, few were able to cover the distance, across ground swept by German fire. Bertie Trevor of the Yorkshires later described the battle as "too terrible for words....We fired 350 rounds a man in my company, and did a good deal of execution. But we were in an absolute trap—it is a marvel that anyone there is alive & untouched. Until one has been for hours pelted at with lyddite & shrapnel, machine guns and rifles, one cannot understand war. Where the fun comes in, I don't know." A circling German aircraft, dropping spasmodic colored smoke bombs to mark targets for the artillery, contributed a contemporary touch to a 19th century battle. On the British side, by 10 a.m. one artillery battery had lost all its officers, and had only a single gun left in action. This was a day when the county battalions of the British Army—Yorkshires, Suffolks, Cornwalls, Argyll & Sutherland Highlanders, and East Surreys—conducted themselves with a stubborn steadiness and professionalism in which their higher commanders—with the notable exception of Smith-Dorrien—showed themselves deficient.

On the British left, the day began with a small disaster. The King's Own had marched all night. Dawn found them on the Ligny road, waiting in column of companies for a promised breakfast. Captain R. G. Beaumont spotted some horsemen on the horizon who looked neither British nor French, but he was sharply put down by his colonel for talking nonsense when he suggested they might be German. The enemy, snapped the CO, was at least three hours away. The welcome rattle of carts caused voices to call "Here come the cookers!" Men piled arms and took out their mess tins, even as the distant horsemen brought up wheeled vehicles of their own, and offloaded them in plain sight. These were German cavalry, and they were permitted unmolested to deploy machine guns. As almost a thousand British soldiers crowded around their breakfast, the Maxims opened fire.

The first bursts killed the King's colonel, and prompted a panic-stricken flight by three companies, who abandoned their piled rifles. Almost all those who tried to run were cut down—only men who embraced the earth escaped the slaughter. The unit's second in command eventually rallied enough survivors to retrieve their weapons and bring in most of the wounded. But in the space of a couple of minutes the King's had suffered four hundred casualties—a murderous demonstration of the price of exposure. The King's thereafter held their ground for a time, assisted by the fact that they faced only German cavalry and skirmishers. But as Marwitz's horsemen worked around behind their left flank, the British infantrymen were obliged to withdraw.

When Germans exposed themselves in their turn, they suffered as heavily as Smith-Dorrien's men: a battery unlimbered and opened fire in front of the Hampshires, whose rifles immediately obliged the gunners to retire. Both sides' field artillery labored under the handicap that crews needed to be able to see their targets—so-called direct fire. Forward observation officers linked by telephone to gun positions were not then available. It was a terrible business, recalling the British disaster at Colenso in the Boer War, to invite gunners and horse teams to deploy within sight and range of German rifle as well as artillery fire; yet this happened all day at Le Cateau, and again and again through that first campaign. British guns fired over open sights, as the term went, at ranges of 1,200 yards—no more than Wellington's artillery knew. The Germans were better equipped to deliver indirect fire from concealed positions through their heavier howitzers, but both sides were constrained by the limited ammunition supplies they carried. The barrages seemed brutal to those who endured them, especially without benefit of trenches, but were mere miniatures of those that followed in subsequent battles.

It is characteristic of even the fiercest actions, that not all the participants are engaged all the time. At Le Cateau, though some units were brutally punished, others had an astonishingly quiet morning in sectors initially untroubled by the Germans. Tom Wollocombe of the Middlesex recorded that around 11:30 he "got quite a good lunch" at the battalion mess in the rear. Once back in the forward positions, for some time "we sat there talking and joking, and began to get quite bored." Even when German shellfire began to fall around them, Wollocombe was chiefly fascinated by the spectacle of four black cows, grazing on unconcernedly. One eventually received a direct hit and was killed, but the other three chewed the cud until the end of the battle. A German participant was similarly bemused by seeing a herd of sheep, bleating furiously, cross the front amid the cannonade.

Lieutenant Roebbling, an infantryman, found that although he peered intently toward the British positions through his telescope, he could not identify an enemy to fire at: "At the same time things were whistling past or crashing into the ground. Then all of a sudden the man two to my right called out 'Adieu, Subenbach, I've had it!' Corporal Subenbach said: 'Don't say that, Busse! Keep your chin up!' A little later comes a groan: 'Oh, I've only got it in the shoulder and ear!'" Roebbling asked for the wounded man's rifle and some ammunition, but still could not see anything to shoot at. Shrapnel began to burst around them, and a bullet hit the sling of his weapon, tearing open the lieutenant's hand. One of his men applied a field dressing. The young officer sensed British fire slackening, as German shelling took its toll. But when Lieutenant Fricke leapt to his feet waving a sword and ordered his men forward, he was promptly shot down; their company commander, son of an officer of the Franco-Prussian war, met the same fate. Lieutenant von Davier raised a laugh to steady his men by wailing: "I have lost my monocle. Anybody who finds it should give it to me later!"

The Germans began to press the British center only around noon, and suffered considerably when they did so. Colonel Hull of the Middlesex made his men wait until the enemy closed to five hundred yards. Their rippling fire then took effect, but dismounted German cavalry meanwhile infiltrated Caudry. Soon after 1 p.m., shellfire began to fall around the Royal Irish Rifles. Colonel Bird saw soldiers of other units running toward the rear. All the transport horses of the Middlesex were killed, and houses were soon blazing. "There were a lot of men making their way back in a disgraceful manner, even NCOs," wrote signals officer Alexander Johnston. "It makes one sad and anxious for the future to see Englishmen behave like this, as the fire was not really heavy nor the losses great. Of course these were only the bad men or men whose officers had been hit and were therefore out of control, and one always found plenty of splendid fellows holding on gamely."

Wilkinson Bird was attempting to check fugitives from Caudry when he suddenly met his brigadier slumped in the saddle of his charger, being led rearward by two staff officers. "Hullo, sir," said the colonel. "I hope you are not hurt." The brigadier mumbled in response: "No, I am just going back for a while," and left the battlefield. The senior officer's retirement was excused by the fact that he had been concussed by a shell, but later in the war humble rankers would be shot for less. The Germans were temporarily evicted from the southern half of Caudry by a counterattack delivered by a scratch group of British troops led by the divisional commander's aide-de-camp.

Meanwhile on the right, II Corps's predicament was worsening. Smith-Dorrien had counted on Haig for support, and instead I Corps's formations were still retreating, scarcely pursued, while GHQ made no attempt to turn them back. Thus the German assault on the exposed flank at Le Cateau was unimpeded. Infantry and gun batteries faced a storm of artillery and machine gun fire from an enemy who could now observe almost every yard of the British positions. When there was a brief lull in German movements and firing, George Reynolds of the Yorkshires said wryly, "it was as if the referee had blown his whistle. We lay there and wondered what the second half would be like."

More of the same, was the answer. Soon after noon, it became plain that the British must pull back—some men were already trickling toward the rear. Several units withdrew intact, but others remained as German infantry worked around behind them, up the hill from Le Cateau. "About 2:30 the situation was as bad as it could be," wrote Bertie Trevor of the Yorkshires. "The ridge on our right...was shelled to pieces, and we were getting it from the Maxims half-right at about 900 yards, as well as volumes of shell—H[igh] E[xplosive] and shrapnel. Half the men were hit and the ammunition was running out....One battalion had held up its hands and I remember the German Guards coming up and taking them prisoners, and executing a parade march round them."

The most urgent problem became that of extricating the British artillery. Some batteries were firing from positions alongside the infantry; these needed to bring forward horse teams, limber up and retire, within range of every Ger-

AT THE WAR'S OUTSET, BRITISH SOLDIERS WERE WARMLY WELCOMED BY FRENCH ALLIES.

man within a mile. The men holding Smith-Dorrien's right witnessed a series of displays of extraordinary, absurdly old-fashioned gallantry as again and again gunners galloped forward to bring away their pieces amid a rain of shell and small-arms fire. Infantrymen sprang to their feet cheering at the spectacle of one battery's horses charging down a forward slope in plain view of the Germans. On the other side, Lieutenant Schacht and his fellow machine gunners watched in disbelief: "Half-right among the flashes appeared a dark mass. It was the [British] teams approaching at a mad gallop. We could not help but think: 'Are they mad?' No, with extraordinary bravery they were attempting to pull out their batteries at the last minute...In a hectic rhythm twelve machine guns poured bullets at the sacrificial victims. What a dreadful tangle it was up there...one [horse] remained standing among this wild hail of fire, started to graze; whinnied for water and shook its head tiredly."

Again and again bullets struck and shells exploded among horses and riders, who collapsed in thrashing, bloody heaps. Two guns were extracted from the carnage and taken to the rear, but the neighboring batteries had to be abandoned, breechblocks removed. The Suffolks, Argyll & Sutherlands, and Yorkshire Light Infantry covered the retirement of 5th Division at midafternoon, before those three units were progressively demolished where they stood. At 3 p.m., Major Trevor of the Yorkshires led back the survivors of his own company. Two men beside him were shot down as they crossed a cornfield, "however, we retired at a walk in true Aldershot fashion, and 3 times we turned and tried to answer the fire. Then it became a case of each man running to some trenches and so on under a murderous fire....We retired through the guns, the gunners lying dead all around."

Smith-Dorrien stood by the roadside watching his troops march by, units in unsurprising disorder, but most soldiers in remarkably good heart. "It was a wonderful sight," he wrote later, "men smoking their pipes, apparently quite unconcerned, and walking steadily down the road—no formation of any sort, and men of all units mixed up together. I likened it at the time to a crowd coming away from a race meeting." This was a wildly fanciful view; Smith-Dorrien's corps had been obliged to act out a nineteenth-century battle while exposed to the power of twentieth-century weapons, and no sane participant had enjoyed the experience. Moreover, it was mistaken to pretend that all his men played heroes' parts. Officers drew their revolvers to arrest the flight of some aspiring fugitives. In Caudry at midafternoon, Bird of the Irish Rifles—who had taken over command of his brigade—was asked to lead a new counterattack. When he gave the order to a major commanding the neighboring battalion, that officer looked the colonel in the eye and said: "I must warn you, sir, that the men will not again attack. They have been much shaken."

"Will they defend?"

"Yes—I think so."

Bird, desperate for information, checked an agitated staff officer who was galloping by: "Hi! Hi! Tell me what's happened!" The man cried: "5th Division smashed to pieces on our right. 4th Division being driven back on our left. Goodbye." This was an extravagant version of the fact that the British were under heavy pressure, but reflected the panic infecting some people who should have known better. Alexander Johnston was much dismayed when his brigadier ordered a withdrawal from Caudry: "I feel that we ought to have stuck onto the town somehow. The German infantry showed no inclination to assault." But shellfire had corroded the defenders' spirits. Wilkinson Bird told the major of his neighboring battalion that his men must act as rearguard. The officer responded: "I'll do my best, sir, but I must warn you that, after what they have been through, the men may not withstand a strong attack." Though some British gunners displayed great gallantry that afternoon, a battery commander whom Bird requested to support his brigade refused, saying that he could not expose his men to German small arms. This was inglorious, but prudent.

A mounted orderly at last gave Bird's brigade the signal to pull back. Almost at once, hundreds of soldiers rose from where they had lain in the stubble and began running southward toward a bridge beneath a railway line that lay behind the British front. One spectator thought the scene resembled "the start of a big cross-country champion foot race." Bird and his brigade's adjutants mounted, to ensure that their men could see them: "We sat and watched the panic. First came drivers wildly lashing their teams, which rushed past with guns and carriages covered with infantrymen clinging to them. Then, after an interval, a mob of men now walking because out of breath....Towards the end of the crowd came the officers walking singly or in pairs."

Hull, iron-willed CO of the Middlesex, was seen to be the last man of his division to retreat. Wilkinson Bird survived unscathed, but lost a leg in another action three weeks later. Lieutenant Siegener, a German infantryman, described how his men began to advance as they saw the British withdraw: "Our losses had been, and continued to be, great, but we wanted to get on. 200 metres to our front was a trench that was still occupied. But already white flags were being displayed over there. The men had their hands up and they surrendered. An officer came up and handed over his sword, but there was still fire coming down on us from further up. I pointed this out and threatened to shoot him immediately. The Briton waved toward the rear and the shooting stopped."

On the right, the Yorkshires conducted a sacrificial stand. By 4:30, they were cut off; a German bugler sounded the British cease-fire, seeking to avert further slaughter. The remains of the battalion fought on; one of its officers, forty-two-year-old Major Cal Yate, led nineteen survivors in a final bayonet charge, in which he fell badly wounded. It is always disputable whether such actions are heroic or merely foolishly futile; in this case, Yate was given a VC, awarded posthumously since he died as a prisoner in Germany, allegedly attempting to escape. Some Yorkshires were bayoneted when they were finally overrun, but most were spared by the Germans, who also treated the wounded humanely. When the few men who escaped to II Corps's main body reformed, Bertie Trevor assumed command of what was left.

In the British center, the Gordon Highlanders failed to receive the signal to withdraw, which was given around 5 p.m. by a galloper who thought 250 yards close enough to ride toward the embattled unit. Only one subaltern saw him wave; being heavily engaged at the time, he said nothing. Three platoons slipped away on their own initiative, and eventually regained the British lines. The rest continued firing from the Audencourt ridge until nightfall, along with some stragglers from the Royal Scots and Royal Irish. A young German lieutenant offered chocolate to a wounded Scottish officer, but demanded: "Why have you English come against us? It is no use. We shall be in Paris in three days." Enemy shells continued to fall upon some of II Corps's positions for hours after they were abandoned. "The British had withdrawn so skilfully that we had not noticed anything," wrote cavalryman Captain Freiherr von der Horst.

German gunner captain Fritz Schneider observed that August 26 was a glorious day in his regiment's history, "but the British also fought bravely. That must be recognized. Despite heavy and bloody losses, they held their positions....When, later that evening, we found ourselves on the road in Beauvois a group of forty to fifty prisoners were being led past. They were all tall, well-built men whose bearing and clothing made an outstanding impression. What a contrast to the short, pale and anxious Frenchmen in their grubby uniforms, whom we had captured two days previously in Tournai!" The most popular booty from the battlefield proved to be scores of discarded British greatcoats, whose quality the victors appreciated.

German failure to encircle and shatter Smith-Dorrien's command reflected poorly on Kluck's competence as well as upon the resistance his regiments faced. On August 26 II Corps held a position in which its most likely fate was destruction. Smith-Dorrien kept his nerve, and instead extricated his force in tolerable order. As at Mons, however, this was certainly no British triumph. His men had merely checked their pursuers for a few hours and escaped catastrophe, chiefly because their enemies were slow to concentrate superior strength against them. They had abandoned thirty-eight guns, and officially recorded the loss of 7,812 men at Le Cateau, a serious toll for a small army, though many stragglers reappeared in the days that followed. Around 5,000 seems a realistic British casualty total for the battle, of which perhaps five hundred were killed, 2,500 taken prisoner and the rest wounded.

As II Corps continued its withdrawal, staff officers stood by roadsides directing men toward their own units. Tom Wollocombe described the spectacle, and his own mingled emotions, as they retreated: "the road...was absolutely ghastly—dead and wounded horses and men strewn all the way, and limbers, guns, ambulances, wagons, carts and all kinds of things running away and bumping into one another without drivers. I marched on that evening feeling fit as a fiddle, although I was nearly done when I went into action. An enemy is a wonderful stimulant."

Smith-Dorrien's men had won a twelve-hour start on the Germans, who made no attempt at close pursuit. Analysis of German regimental casualty figures suggest that Kluck's losses at Le Cateau were about half those of Smith-Dorrien. The German First Army during the entire month of August admitted only 2,863 men killed or missing, with a further 7,869 wounded. Such losses had only marginal significance when his entire command numbered 217,384. Kluck was at least equally troubled by a sick list of 8,000, mostly men too footsore to march farther. The BEF fought staunchly in both its significant August battles, but its fire injured the enemy less grievously than optimists supposed then and romantics have imagined since.

German soldiers emerged from the two encounters with respect for British determination, but commanders saw nothing to make them flinch. Moltke expressed satisfaction at the outcome of Le Cateau: Kluck's formations continued to advance, while the BEF kept on retreating. The British constructed a heroic legend by focusing upon individual acts of courage, glossing over the stark "big picture." It was probably the case that Smith-Dorrien had no choice save to fight a battle. But he found himself in an unholy mess in the beet and stubble fields that day, from which he was fortunate to escape, with underacknowledged French assistance. The professionalism of Britain's soldiers narrowly sufficed to compensate for the follies and inadequacies of their senior officers. The most significant contribution of the two actions at Mons and Le Cateau was to check the momentum of Kluck's advance: every day the Germans failed to traverse more miles across France was a precious gain for Joffre's redeployment. Time was critical, and Moltke was running short of it.

The retreat of the French and British armies continued until September 6, 1914, when General Joseph Joffre launched the great French counteroffensive of the Marne, which drove back the Germans into eastern France, and destroyed the kaiser's hopes of swift victory in the First World War.

AN EXHAUSTED INFANTRYMAN IN THE FOG-SHROUDED A-SHAU VALLEY, 1969

CHAPTER SEVEN

TIM O'BRIEN

"How to Tell a True War Story"
from *The Things They Carried*

AN AWARD-WINNING NOVELIST WHOSE WORKS have received widespread critical acclaim, Tim O'Brien burst upon the literary scene in 1973 with *If I Die in a Combat Zone, Box Me Up and Ship Me Home,* an intense and visceral account of his tour as an infantryman in Vietnam. The book stunned and enthralled readers with its unflinching look at the nature of ground combat during America's seemingly endless war in Southeast Asia, and ensured O'Brien's place as a founding member of a gifted group of Vietnam veterans-turned-authors.

While his first book was a straightforward account of O'Brien's time in Vietnam, his 1990 novel *The Things They Carried* deliberately blends reality and fiction, belief and imagination, intellect and emotion. This excerpt—How to Tell a True War Story—is a characteristically sharp response to the challenge of how to convey in words on paper a true sense of the surreal blend of horror, boredom, humor, fear, love, brutality, and grace that was combat in Vietnam.

This is true. I had a buddy in Vietnam. His name was Bob Kiley, but everybody called him Rat.

A friend of his gets killed, so about a week later Rat sits down and writes a letter to the guy's sister. Rat tells her what a great brother she had, how together the guy was, a number one pal and comrade. A real soldier's soldier, Rat says. Then he tells a few stories to make the point, how her brother would always volunteer for stuff nobody else would volunteer for in a million years, dangerous stuff, like doing recon or going out on these really badass night patrols. Stainless steel balls, Rat tells her. The guy was a little crazy, for sure, but crazy in a good way, a real daredevil, because he liked the challenge of it, he liked testing himself, just man against gook. A great, great guy, Rat says.

Anyway, it's a terrific letter, very personal and touching. Rat almost bawls writing it. He gets all teary telling about the good times they had together, how her brother made the war seem almost fun, always raising hell and lighting up villes and bringing smoke to bear every which way. A great sense of humor, too. Like the time at this river when he went fishing with a whole damn crate of hand grenades. Probably the funniest thing in world history, Rat says, all that gore, about twenty zillion dead gook fish. Her brother, he had the right attitude. He knew how to have a good time. On Halloween, this real hot spooky night, the dude paints up his body all different colors and puts on this weird mask and hikes over to a ville and goes trick-or-treating almost stark naked, just boots and balls and an M-16. A tremendous human being, Rat says. Pretty nutso sometimes, but you could trust him with your life.

And then the letter gets very sad and serious. Rat pours his heart out. He says he loved the guy. He says the guy was his best friend in the world. They were like soul mates, he says, like twins or something, they had a whole lot in common. He tells the guy's sister he'll look her up when the war's over.

So what happens?

Rat mails the letter. He waits two months. The dumb cooze never writes back.

A TRUE WAR STORY IS NEVER MORAL. It does not instruct, nor encourage virtue, nor suggest models of proper human behavior, nor restrain men from doing the things men have always done. If a story seems moral, do not believe it. If at the end of a war story you feel uplifted, or if you feel that some small bit of rectitude has been salvaged from the larger waste, then you have been made the victim of a very old and terrible lie. There is no rectitude whatsoever. There is no virtue. As a first rule of thumb, therefore, you can tell a true war story by its absolute and uncompromising allegiance to obscenity and evil. Listen to Rat Kiley. Cooze, he says. He does not say bitch. He certainly does not say woman, or girl. He says cooze. Then he spits and stares. He's nineteen years old—it's too much for him—so he looks at you with those big sad gentle killer eyes and says cooze, because his friend is dead, and because it's so incredibly sad and true: she never wrote back.

You can tell a true war story if it embarrasses you. If you don't care for obscenity, you don't care for the truth; if you don't care for the truth, watch how you vote. Send guys to war, they come home talking dirty.

Listen to Rat: "Jesus Christ, man, I write this beautiful fuckin' letter, I slave over it, and what happens? The dumb cooze never writes back."

THE DEAD GUY'S NAME WAS CURT LEMON. What happened was, we crossed a muddy river and marched west into the mountains, and on the third day we took a break along a trail junction in deep jungle. Right away, Lemon and Rat Kiley started goofing. They didn't understand about the spookiness. They were kids; they just didn't know. A nature hike, they thought, not even a war, so they went off into the shade of some giant trees—quadruple canopy, no sunlight at all—and they were giggling and calling each other yellow mother and playing a silly game they'd invented. The game involved smoke grenades, which were harmless unless you did stupid things, and what they did was pull out the pin and stand a few feet apart and play catch under the shade of those huge trees. Whoever chickened out was a yellow mother. And if nobody chickened out, the grenade would make a light popping sound and they'd be covered with smoke and they'd laugh and dance around and then do it again.

It's all exactly true.

It happened, to *me*, nearly twenty years ago, and I still remember that trail junction and those giant trees and a soft dripping sound somewhere beyond the trees. I remember the smell of moss. Up in the canopy there were tiny white blossoms, but no sunlight at all, and I remember the shadows spreading out under the trees where Curt Lemon and Rat Kiley were playing catch with smoke grenades. Mitchell Sanders sat flipping his yo-yo. Norman Bowker and Kiowa and Dave Jensen were dozing, or half dozing, and all around us were those ragged green mountains.

Except for the laughter things were quiet.

At one point, I remember, Mitchell Sanders turned and looked at me, not quite nodding, as if to warn me about something, as if he already knew, then after a while he rolled up his yo-yo and moved away.

It's hard to tell you what happened next.

They were just goofing. There was a noise, I suppose, which must've been the detonator, so I glanced behind me and watched Lemon step from the shade into bright sunlight. His face was suddenly brown and shining. A handsome kid, really. Sharp gray eyes, lean and narrow-waisted, and when he died it was almost beautiful, the way the sunlight came around him and lifted him up and sucked him high into a tree full of moss and vines and white blossoms.

IN ANY WAR STORY, BUT ESPECIALLY a true one, it's difficult to separate what happened from what seemed to happen. What seems to happen becomes its own happening and has to be told that way. The angles of vision are skewed. When a booby trap explodes, you close your eyes and duck and float outside yourself. When a guy dies, like Curt Lemon, you look away and then look back for a moment and then look away again. The pictures get jumbled; you tend to miss a lot. And then afterward, when you go to tell about it, there is always that surreal seemingness, which makes the story seem untrue, but which in fact represents the hard and exact truth as it *seemed*.

IN MANY CASES A TRUE WAR STORY cannot be believed. If you believe it, be skeptical. It's a question of credibility. Often the crazy stuff is true and the normal stuff isn't, because the normal stuff is necessary to make you believe the truly incredible craziness.

In other cases you can't even tell a true war story. Sometimes it's just beyond telling.

I heard this one, for example, from Mitchell Sanders. It was near dusk and we were sitting at my foxhole along a wide muddy river north of Quang Ngai City. I remember how peaceful the twilight was. A deep pinkish red spilled out on the river, which moved without sound, and in the morning we would cross the river and march west into the mountains. The occasion was right for a good story.

"God's truth," Mitchell Sanders said. "A six-man patrol goes up into the mountains on a basic listening-post operation. The idea's to spend a week up there, just lie low and listen for enemy movement. They've got a radio along, so if they hear anything suspicious—anything—they're supposed to call in artillery or gunships, whatever it takes. Otherwise they keep strict field discipline. Absolute silence. They just listen."

Sanders glanced at me to make sure I had the scenario. He was playing with his yo-yo, dancing it with short, tight little strokes of the wrist.

His face was blank in the dusk.

"We're talking regulation, by-the-book LP. These six guys, they don't say boo for a solid week. They don't got tongues. *All* ears."

"Right," I said.

Understand me?"

"Invisible."

Sanders nodded.

"Affirm," he said. "Invisible. So what happens is, these guys get themselves deep in the bush, all camouflaged up, and they lie down and wait and that's all they do, nothing else, they lie there for seven straight days and just listen. And man, I tell you it's spooky. This is mountains. You don't *know* spooky till you been there. Jungle, sort of, except it's way up in the clouds and there's always this fog—like rain, except it's not raining—everything's all wet and swirly and tangled up and you can't see jack, you can't find your own pecker to piss with. Like you don't even have a body. Serious spooky. You just go with the vapors—the fog sort of takes you in....And the sounds, man. The sounds carry forever. You hear stuff nobody should ever hear."

Sanders was quiet for a second, just working the yo-yo, then he smiled at me.

"So after a couple days the guys start hearing this real soft, kind of wacked-out music. Weird echoes and stuff. Like a radio or something, but it's not a radio, it's this strange gook music that comes right out of the rocks. Faraway, sort of, but right up close, too. They try to ignore it. But it's a listening post, right? So they listen. And every night they keep hearing that crazyass gook concert. All kinds of chimes and xylophones. I mean, this is wilderness—no way, it can't be real—but there it is, like the mountains are tuned in to Radio fucking Hanoi. Naturally they get nervous. One guy sticks Juicy Fruit in his ears. Another guy almost flips. Thing is, though, they can't report music. They can't get on the horn and call back to base and say, 'Hey, listen, we need some firepower, we got to blow away this weirdo gook rock band.' They can't do that. It wouldn't go down. So they lie there in the fog and keep their mouths shut. And what makes it extra bad, see, is the poor dudes can't horse around like normal. Can't joke it away. Can't even talk to each other except maybe in whispers, all hush-hush, and that just revs up the willies. All they do is listen."

Again there was some silence as Mitchell Sanders looked out on the river. The dark was coming on hard now, and off to the west I could see the mountains rising in silhouette, all the mysteries and unknowns.

"This next part," Sanders said quietly, "you won't believe."

"Probably not," I said.

"You won't. And you know why?" He gave me a long, tired smile. "Because it happened. Because every word is absolutely dead-on true."

Sanders made a sound in his throat, like a sigh, as if to say he didn't care if I believed him or not. But he did care. He wanted me to feel the truth, to believe by the raw force of feeling. He seemed sad, in a way.

"These six guys," he said, "they're pretty fried out by now, and one night they start hearing voices. Like at a cocktail party. That's what it sounds like, this big swank gook cocktail party somewhere out there in the fog. Music and chitchat and stuff. It's crazy, I know, but they hear the champagne corks. They hear the actual martini glasses. Real hoity-toity, all very civilized, except this isn't civilization. This is Nam.

"Anyway, the guys try to be cool. They just lie there and groove, but after a while they start hearing—you won't believe this—they hear chamber music. They hear violins and cellos. They hear this terrific mama-san soprano. Then after a while they hear gook opera and a glee club and the Haiphong Boys Choir and a barbershop quartet and all kinds of weird chanting and Buddha-Buddha stuff. And the whole time, in the background, there's still that cocktail party going on. All these different voices. Not human voices, though. Because it's the mountains. Follow me? The rock—it's *talking*. And the fog, too, and the grass and the goddamn mongooses. Everything talks. The trees talk politics, the monkeys talk religion. The whole country. Vietnam. The place talks. It talks. Understand? Nam—it truly *talks*.

"The guys can't cope. They lose it. They get on the radio and report enemy movement—a whole army, they say—and they order up the firepower. They get arty and gunships. They call in airstrikes. And I'll tell you, they fuckin' crash that cocktail party. All night long, they just smoke those mountains. They make jungle juice. They blow away trees and glee clubs and whatever else there is to blow away. Scorch time. They walk napalm up and down the ridges. They bring in the Cobras and F-4s, they use Willie Peter and HE and incendiaries. It's all fire. They make those mountains burn.

"Around dawn things finally get quiet. Like you never even *heard* quiet before. One of those real thick, real misty days—just clouds and fog, they're off in this special zone—and the mountains are absolutely dead-flat silent. Like *Brigadoon*—pure vapor, you know? Everything's all sucked up inside the fog. Not a single sound, except they still *hear* it.

"So they pack up and start humping. They head down the mountain, back to base camp, and when they get there they don't say diddly. They don't talk. Not a word, like they're deaf and dumb. Later on this fat bird colonel comes up and asks what the hell happened out there. What'd they hear? Why all the ordnance? The

You can tell a true war story by the way it never seems to end. Not then, not ever.

man's ragged out, he gets down tight on their case. I mean, they spent six trillion dollars on firepower, and this fatass colonel wants answers, he wants to know what the fuckin' story is.

"But the guys don't say zip. They just look at him for a while, sort of funny like, sort of amazed, and the whole war is right there in that stare. It says everything you can't ever say. It says, man, you got *wax* in your ears. It says, poor bastard, you'll never know—wrong frequency—you don't *even* want to hear this. Then they salute the fucker and walk away, because certain stories you don't ever tell."

YOU CAN TELL A TRUE WAR STORY by the way it never seems to end. Not then, not ever. Not when Mitchell Sanders stood up and moved off into the dark.

It all happened.

Even now, at this instant, I remember that yo-yo. In a way, I suppose, you had to be there, you had to hear it, but I could tell how desperately Sanders wanted me to believe him, his frustration at not quite getting the details right, not quite pinning down the final and definitive truth.

And I remember sitting at my foxhole that night, watching the shadows of Quang Ngai, thinking about the coming day and how we would cross the river and march west into the mountains, all the ways I might die, all the things I did not understand.

Late in the night Mitchell Sanders touched my shoulder. "Just came to me," he whispered. "The moral, I mean. Nobody listens. Nobody hears nothin'. Like that fatass colonel. The politicians, all the civilian types. Your girlfriend. My girlfriend. Everybody's sweet little virgin girlfriend. What they need is to go out on LP. The vapors, man. Trees and rocks—you got to *listen* to your enemy."

AND THEN AGAIN, IN THE MORNING, Sanders came up to me. The platoon was preparing to move out, checking weapons, going through all the little rituals that preceded a day's march. Already the lead squad had crossed the river and was filing off toward the west.

"I got a confession to make," Sanders said. "Last night, man, I had to make up a few things."

"I know that."

"The glee club. There wasn't any glee club."

"Right."

"No opera."

"Forget it, I understand."

"Yeah, but listen, it's still true. Those six guys, they heard wicked sound out there. They heard sound you just plain won't believe."

Sanders pulled on his rucksack, closed his eyes for a moment, and let out a short, throat-clearing sigh. I knew what was coming.

"All right," I said, "what's the moral?"

"Forget it."

"No, go ahead."

For a long while he was quiet, looking away, and the silence kept stretching out until it was almost embarrassing. Then he shrugged and gave me a stare that lasted all day.

"Hear that quiet, man?" he said. "That quiet—just listen. There's your moral."

IN A TRUE WAR STORY, IF THERE'S a moral at all, it's like the thread that makes the cloth. You can't tease it out. You can't extract the meaning without unraveling the deeper meaning. And in the end, really, there's nothing much to say about a true war story, except maybe "Oh."

True war stories do not generalize. They do not indulge in abstraction or analysis.

For example: War is hell. As a moral declaration the old truism seems perfectly true, and yet because it abstracts, because it generalizes, I can't believe it with my stomach. Nothing turns inside.

It comes down to gut instinct. A true war story, if truly told, makes the stomach believe.

THIS ONE DOES IT FOR ME. I've told it before—many times, many versions—but here's what actually happened.

We crossed that river and marched west into the mountains. On the third day, Curt Lemon stepped on a booby-trapped 105 round. He was playing catch with Rat Kiley, laughing, and then he was dead. The trees were thick; it took nearly an hour to cut an LZ for the dustoff.

173RD AIRBORNE TROOPS REACT TO SNIPER FIRE ON A "HOT" LANDING ZONE, 1965.

Later, higher in the mountains, we came across a baby VC water buffalo. What it was doing there I don't know—no farms or paddies—but we chased it down and got a rope around it and led it along to a deserted village where we set up for the night. After supper Rat Kiley went over and stroked its nose.

He opened up a can of C rations, pork and beans, but the baby buffalo wasn't interested.

Rat shrugged.

He stepped back and shot it through the right front knee. The animal did not make a sound. It went down hard, then got up again, and Rat took careful aim and shot off an ear. He shot it in the hindquarters and in the little hump at its back. He

Rat Kiley was crying. He tried to say something, but then cradled his rifle and went off by himself.

shot it twice in the flanks. It wasn't to kill; it was to hurt. He put the rifle muzzle up against the mouth and shot the mouth away. Nobody said much. The whole platoon stood there watching, feeling all kinds of things, but there wasn't a great deal of pity for the baby water buffalo. Curt Lemon was dead. Rat Kiley had lost his best friend in the world. Later in the week he would write a long personal letter to the guy's sister, who would not write back, but for now it was a question of pain. He shot off the tail. He shot away chunks of meat below the ribs. All around us there was the smell of smoke and filth and deep greenery, and the evening was humid and very hot. Rat went to automatic. He shot randomly, almost casually, quick little spurts in the belly and butt. Then he reloaded, squatted down, and shot it in the left front knee. Again the animal fell hard and tried to get up, but this time it couldn't quite make it. It wobbled and went down sideways. Rat shot it in the nose. He bent forward and whispered something, as if talking to a pet, then he shot it in the throat. All the while the baby buffalo was silent, or almost silent, just a light bubbling sound where the nose had been. It lay very still. Nothing moved except the eyes, which were enormous, the pupils shiny black and dumb.

Rat Kiley was crying. He tried to say something, but then cradled his rifle and went off by himself.

The rest of us stood in a ragged circle around the baby buffalo. For a time no one spoke. We had witnessed something essential, something brand-new and profound, a piece of the world so startling there was not yet a name for it.

Somebody kicked the baby buffalo.

It was still alive, though just barely, just in the eyes.

"Amazing," Dave Jensen said. "My whole life, I never seen anything like it."

"Never?"

"Not hardly. Not once."

Kiowa and Mitchell Sanders picked up the baby buffalo. They hauled it across the open square, hoisted it up, and dumped it in the village well.

Afterward, we sat waiting for Rat to get himself together. "Amazing," Dave Jensen kept saying. "A new wrinkle. I never seen it before."

Mitchell Sanders took out his yo-yo. "Well, that's Nam," he said. "Garden of Evil. Over here, man, every sin's real fresh and original."

How do you generalize?

War is hell, but that's not the half of it, because war is also mystery and terror and adventure and courage and discovery and holiness and pity and despair and longing and love. War is nasty; war is fun. War is thrilling; war is drudgery. War makes you a man; war makes you dead.

The truths are contradictory. It can be argued, for instance, that war is grotesque. But in truth war is also beauty. For all its horror, you can't help but gape at the awful majesty of combat. You stare out at tracer rounds unwinding through the dark like brilliant red ribbons. You crouch in ambush as a cool, impassive moon rises over the nighttime paddies. You admire the fluid symmetries of troops on the move, the harmonies of sound and shape and proportion, the great sheets of metal-fire streaming down from a gunship, the illumination rounds, the white phosphorus, the purply orange glow of napalm, the rocket's red glare. It's not pretty, exactly. It's astonishing. It fills the eye. It commands you. You hate it, yes, but your eyes do not. Like a killer forest fire, like cancer under a microscope, any battle or bombing raid or artillery barrage has the aesthetic purity of absolute moral indifference—a powerful, implacable beauty—and a true war story will tell the truth about this, though the truth is ugly.

To generalize about war is like generalizing about peace. Almost everything is true. Almost nothing is true. At its core, perhaps, war is just another name for death, and yet any soldier will tell you, if he tells the truth, that proximity to death brings with it a corresponding proximity to life. After a firefight, there is always the immense pleasure of aliveness. The trees are alive. The grass, the soil—everything. All around you things are purely living, and you among them, and the aliveness makes you tremble. You feel an intense, out-of-the-skin awareness of your living self—your truest self, the human being you want to be and then become by the force of wanting it. In the midst of evil you want to be a good man. You want decency. You want justice and courtesy and human concord, things you never knew you wanted. There is a kind of largeness to it, a kind of godliness. Though it's odd, you're never more alive than when you're almost dead. You recognize what's valuable. Freshly, as if for the first time, you love what's best in yourself and in the world, all that might be lost. At the hour of dusk you sit at your foxhole and look out on a wide river turning pinkish red, and at the mountains beyond, and although in the morning you must cross the river and go into the mountains and do terrible things and maybe die, even so, you find yourself studying the fine colors on the river, you feel wonder and awe at the setting of the sun, and you are filled with a hard, aching love for how the world could be and always should be, but now is not.

Mitchell Sanders was right. For the common soldier, at least, war has the feel—the spiritual texture—of a great ghostly fog, thick and permanent. There is no clarity. Everything swirls. The old rules are no longer binding, the old truths no longer true. Right spills over into wrong. Order blends into chaos, love into hate, ugliness into beauty, law into anarchy, civility into savagery. The vapors suck you in. You can't tell where you are, or why you're there, and the only certainty is overwhelming ambiguity.

In war you lose your sense of the definite, hence your sense of truth itself, and therefore it's safe to say that in a true war story nothing is ever absolutely true.

OFTEN IN A TRUE WAR STORY THERE is not even a point, or else the point doesn't hit you until twenty years later, in your sleep, and you wake up and shake your wife and start telling the story to her, except when you get to the end you've forgotten the point again. And then for a long time you lie there watching the story happen in your head. You listen to your wife's breathing. The war's over. You close your eyes. You smile and think, Christ, what's the *point*?

U.S. SOLDIERS ENGAGE NORTH VIETNAMESE TROOPS NEAR DAU TIENG, 1967.

THIS ONE WAKES me up.

In the mountains that day, I watched Lemon turn sideways. He laughed and said something to Rat Kiley. Then he took a peculiar half step, moving from shade into bright sunlight, and the booby-trapped 105 round blew him into a tree. The parts were just hanging there, so Dave Jensen and I were ordered to shinny up and peel him off. I remember the white bone of an arm. I remember pieces of skin and something wet and yellow that must've been the intestines. The gore was horrible, and stays with me. But what wakes me up twenty years later is Dave Jensen singing "Lemon Tree" as we threw down the parts.

YOU CAN TELL A TRUE WAR STORY by the questions you ask. Somebody tells a story, let's say, and afterward you ask, "Is it true?" and if the answer matters, you've got your answer.

For example, we've all heard this one. Four guys go down a trail. A grenade sails out. One guy jumps on it and takes the blast and saves his three buddies.

Is it true?

The answer matters.

You'd feel cheated if it never happened. Without the grounding reality, it's just a trite bit of puffery, pure Hollywood, untrue in the way all such stories are untrue. Yet even if it did happen—and maybe it did, anything's possible—even then you know it can't be true, because a true war story does not depend upon that kind of truth. Absolute occurrence is irrelevant. A thing may happen and be a total lie; another thing may not happen and be truer than the truth. For example: Four guys go down a trail. A grenade sails out. One guy jumps on it and takes the blast, but it's a killer grenade and everybody dies anyway. Before they die, though, one of the dead guys says, "The fuck you do *that* for?" and the jumper says, "Story of my life, man," and the other guy starts to smile but he's dead. That's a true story that never happened.

TWENTY YEARS LATER, I CAN STILL SEE the sunlight on Lemon's face. I can see him turning, looking back at Rat Kiley, then he laughed and took that curious half step from shade into sunlight, his face suddenly brown and shining, and when his foot touched down, in that instant, he must've thought it was the sunlight that was killing him. It was not the sunlight. It was a rigged 105 round. But if I could ever get the story right, how the sun seemed to gather around him and pick him up and lift him high into a tree, if I could somehow re-create the fatal whiteness of that light, the quick glare, the obvious cause and effect, then you would believe the last thing Curt Lemon believed, which for him must've been the final truth.

NOW AND THEN, WHEN I TELL THIS STORY, someone will come up to me afterward and say she liked it. It's always a woman. Usually it's an older woman of kindly temperament and humane politics. She'll explain that as a rule she hates war stories; she can't understand why people want to wallow in all the blood and gore. But this one she liked. The poor baby buffalo, it made her sad. Sometimes, even, there are little tears. What I should do, she'll say, is put it all behind me. Find new stories to tell.

I won't say it but I'll think it.

I'll picture Rat Kiley's face, his grief, and I'll think, *You dumb cooze.*

Because she wasn't listening.

It *wasn't* a war story. It was a *love* story.

But you can't say that. All you can do is tell it one more time, patiently, adding and subtracting, making up a few things to get at the real truth. No Mitchell Sanders, you tell her. No Lemon, no Rat Kiley. No trail junction. No baby buffalo. No vines or moss or white blossoms. Beginning to end, you tell her, it's all made up. Every goddamn detail—the mountains and the river and especially that poor dumb baby buffalo. None of it happened. None of it. And even if it did happen, it didn't happen in the mountains, it happened in this little village on the Batangan Peninsula, and it was raining like crazy, and one night a guy named Stink Harris woke up screaming with a leech on his tongue. You can tell a true war story if you just keep on telling it.

And in the end, of course, a true war story is never about war. It's about sunlight. It's about the special way that dawn spreads out on a river when you know you must cross the river and march into the mountains and do things you are afraid to do. It's about love and memory. It's about sorrow. It's about sisters who never write back and people who never listen.

AUTHOR BIOGRAPHIES AND ADDITIONAL READING

From the recipients of the Pritzker Military Museum & Library Literature Award for Lifetime Achievement in Military Writing. For more information, visit www.pritzkermilitary.org.

James M. McPherson, 2007

James M. McPherson is a Pulitzer Prize–winning author of more than a dozen books. He taught U.S. history at Princeton University for 42 years. Works by McPherson include:

The Struggle for Equality: Abolitionists and the Negro in the Civil War and Reconstruction

The Negro's Civil War: How American Negroes Felt and Acted during the War for the Union

Marching toward Freedom: The Negro in the Civil War

The Abolitionist Legacy: From Reconstruction to the NAACP

Ordeal by Fire: The Civil War and Reconstruction

Lincoln and the Strategy of Unconditional Surrender
    (23rd Annual Fortenbaugh Memorial Lecture, Gettysburg College, 1984)

How Lincoln Won the War with Metaphors

Battle Cry of Freedom: The Civil War Era

Abraham Lincoln and the Second American Revolution

What They Fought For, 1861–1865

Drawn with the Sword: Reflections on the American Civil War

For Cause and Comrades: Why Men Fought in the Civil War

Is Blood Thicker Than Water? Crises of Nationalism in the Modern World

Crossroads of Freedom: Antietam

Hallowed Ground: A Walk at Gettysburg

This Mighty Scourge

Tried by War: Abraham Lincoln as Commander in Chief

Abraham Lincoln

War on the Waters: The Union and Confederate Navies, 1861–1865

Allan R. Millett, 2008

Allan R. Millett, a retired United States Marine Corps Reserve colonel, is the Ambrose Professor of History and director of the Eisenhower Center for American Studies at the University of New Orleans and the Raymond E. Mason Jr. Professor Emeritus of History at The Ohio State University. Works by Millett include:

The Politics of Intervention: The Military Occupation of Cuba, 1906–1909
*The General: Robert L. Bullard and Officership in the United States
    Army, 1881–1925*
Military Effectiveness (coeditor and contributor)
Semper Fidelis: The History of the United States Marine Corps
Calculations: Net Assessment and the Coming of World War II (coeditor)
*In Many a Strife: General Gerald C. Thomas and the United States
    Marine Corps, 1917–1956*
Military Innovation (coeditor and contributor)
A War to be Won: Fighting the Second World War (with Williamson Murray)
Their War for Korea
Commandants of the Marine Corps (coeditor and contributor)
The Korean War: The Essential Bibliography
The War for Korea, 1945–1950: A House Burning
The War for Korea, 1950–1951: They Came from the North
*For the Common Defense: A Military History of the United States
    from 1607 to 2012* (with Peter Maslowski and William B. Feis)

Gerhard L. Weinberg, 2009

Gerhard L. Weinberg, a veteran of World War II, is the prize-winning author of numerous books and articles on World War II. Since 1974, he has been the William Rand Kenan Jr. Professor of History at the University of North Carolina at Chapel Hill. Works by Weinberg include:

Guide to Captured German Documents
Soviet Partisans in World War II (with John Armstrong and others)
Germany and the Soviet Union, 1939–1941
Transformation of a Continent: Europe in the Twentieth Century (editor)
World in the Balance: Behind the Scenes of World War II
Germany, Hitler, and World War II: Essays in Modern German and World History
Hitler's Second Book: The Unpublished Sequel to Mein Kampf (editor)
Visions of Victory: The Hopes of Eight World War II Leaders
A World at Arms: A Global History of World War II
Hitler's Table Talk, 1941–1944: His Private Conversations (editor)
Hitler's Foreign Policy, 1933–1939: The Road to World War II

Rick Atkinson, 2010

Rick Atkinson is the best-selling author of six books. A winner of Pulitzer Prizes in history and journalism, he was a reporter and senior editor at the *Washington Post*. Works by Atkinson include:

The Long Gray Line
Crusade: The Untold Story of the Persian Gulf War
Where Valor Rests: Arlington National Cemetery
An Army at Dawn: The War in North Africa, 1942–1943
In the Company of Soldiers: A Chronicle of Combat
The Day of Battle: The War in Sicily and Italy, 1943–1944
The Guns at Last Light: The War in Western Europe, 1944–1945

Carlo D'Este, 2011

Carlo D'Este is a retired U.S. Army lieutenant colonel, military historian, and author of widely acclaimed biographies and books about World War II. Works by D'Este include:

Decision in Normandy
Bitter Victory: The Battle for Sicily, 1943
World War II in the Mediterranean, 1942–1945
Fatal Decision: Anzio and the Battle for Rome
Patton: A Genius for War
Eisenhower: A Soldier's Life
Warlord: A Life of Winston Churchill at War, 1874–1945

Max Hastings, 2012

Max Hastings is the author of more than 20 books. He has won numerous awards, and has served as a foreign correspondent and as the editor of Britain's *Evening Standard* and *Daily Telegraph*. Works by Hastings include:

America 1968: The Fire This Time
Ulster 1969: The Fight for Civil Rights in Northern Ireland
Montrose: The King's Champion
Bomber Command
The Battle of Britain (with Len Deighton)
Das Reich: Resistance and the March of the Second SS Panzer Division
    through France, June 1944
The Battle for the Falklands (with Simon Jenkins)

Overlord: D-Day and the Battle for Normandy
The Oxford Book of Military Anecdotes (editor)
Victory in Europe
The Korean War
Did You Really Shoot the Television?
Going to the Wars
Editor (memoirs)
Armageddon: The Battle for Germany, 1944–45
Warriors: Extraordinary Tales from the Battlefield
Nemesis: The Battle for Japan, 1944–45
Finest Years: Churchill as Warlord, 1940–45
Inferno: The World at War, 1939–1945
Catastrophe 1914: Europe Goes to War

Tim O'Brien, 2013
Tim O'Brien received the 1979 National Book Award in fiction for *Going After Cacciato*. He served in Vietnam from 1969 to 1970 as part of the Americal Division. Works by O'Brien include:

If I Die in a Combat Zone, Box Me Up and Ship Me Home
Northern Lights
Going After Cacciato
The Nuclear Age
The Things They Carried
In the Lake of the Woods
Tomcat in Love
July, July

ACKNOWLEDGMENTS

This book would not have been possible without the vision, guidance, and support of Colonel (IL) J. N. Pritzker, IL ARNG (Retired).

On War Anthology Team
Kenneth Clarke, Executive Editor
Roger L. Vance, Managing Editor
Michael W. Robbins, Editor
Wendy Palitz, Designer and Art Director
Lisa Marie Lanz, Associate Editor
Nancy Houghton, Associate Editor
Stephen Harding, Senior Editor
Elizabeth Howard, Copy Editor
Jennifer Berry, Photo Editor
Patty Kelly, Photo Editor
Guy Aceto, Photo Editor
John Glenn, Consulting Editor
Kat Latham, Photo Editor
Teri Embrey, Photo Editor
Guy Sellars, Financial Advisor
Michelle Nakfoor, Legal Advisor

Thank you to all who have served on the selection committees of the Pritzker Literature Award. Based on the recommendation of the Screening Committee, Colonel J. N. Pritzker, with the assistance of the Tawani Foundation Executive Council, selects the recipient of the award. The leadership and insight of the award committees is directly responsible for the high caliber of writers represented here.

Screening Committee
Rick Atkinson (2010–2013)
Maj. Gen. John L. Borling, USAF (Retired) (2007–2008)
Kenneth Clarke, Ex Officio (2011–2012)
Edward M. Coffman (2009)
Steve Coll (2008–2010)

Lew Collens (2007)

Carlo D'Este (2007–2010, 2012–2013)

Nathaniel C. Fick (2011–2012)

Brig. Gen. David L. Grange, USA (Retired) (2007–2008)

Max Hastings (2013)

Gary T. Johnson (2007–2011), Chair (2012–2013)

Lisa Marie Lanz, Ex Officio (2011–2012)

Karl Marlantes (2013)

James M. McPherson (2008–2012)

Donald L. Miller (2007–2011)

Allan R. Millett (2009–2012)

Williamson Murray (2007)

Joseph E. Persico (2008–2011)

Carol Reardon (2007–2008, 2010)

Thomas E. Ricks (2008–2010)

Mark A. Stoler (2007–2009)

Edward C. Tracy, Chair (2007–2011)

Gerhard L. Weinberg (2010–2012)

Tawani Foundation Executive Council

Colonel (IL) J. N. Pritzker, IL ARNG (Retired) (2007–2013)

Lew Collens (2008–2013)

Charley Dobrusin (2007–2013)

Jane Feerer (2007–2010)

Mary Parthe (2012–2013)

Colonel David R. Pelizzon, USA (Retired) (2007–2013)

Tal Pritzker (2013)

Guy Sellars (2012–2013)

Administration

Kareema Cruz (2009–2013)

Nancy Houghton (2007–2013)

Lisa Marie Lanz (2009–2013)

Kenneth Clarke (2011–2013)

Edward C. Tracy (2007–2011)

PRITZKER MILITARY
MUSEUM & LIBRARY

tawanifoundation
distinction through transformation

Tawani Foundation Board of Directors
Colonel (IL) J. N. Pritzker, IL ARNG (Retired)
Lew Collens
Charley Dobrusin
Mary Parthe
Colonel David R. Pelizzon, USA (Retired)
Tal Pritzker
Guy Sellars

Tawani Foundation Staff
Lisa Marie Lanz, Executive Director
Kareema Cruz, Executive Assistant

WEIDER
HISTORY
GROUP

Weider History Group
Eric Weider, President and CEO
Bruce Forman, Chief Operating Officer
Roger L. Vance, Group Managing Editor
Karen Johnson, Business Director
Rob Wilkins, Military Ambassador,
    Partnership Marketing Director
Michael W. Robbins, Editor
Wendy Palitz, Art Director
Stephen Harding, Senior Editor
Elizabeth Howard, Copy Editor
Jennifer Berry, Senior Photo Editor

"Too cold for flowers," birthday card from
Pfc Franz Altschuler, Battle of the Bulge, 1944

INDEX